PROBLEM – BASED MICROBIOLOGY

Swapan K. Nath, PhD, F(CCM)

Senior Lecturer and Course Director
Department of Microbiology
University of Texas Southwestern Medical Center
Dallas, Texas

Sanjay G. Revankar, MD

Associate Professor
Department of Internal Medicine
University of Texas Southwestern Medical Center
Dallas, Texas

SAUNDERS

ELSEVIER

SAUNDERS
ELSEVIER

1600 John F. Kennedy Blvd.
Ste 1800
Philadelphia, PA 19103-2899

PROBLEM-BASED MICROBIOLOGY   ISBN 978-0-7216-0630-9
Copyright © 2006, Elsevier Inc.   ISBN 0-7216-0630-X

Notice

Knowledge and best practice in this field are constantly changing. As new research and experience broaden our knowledge, changes in practice, treatment, and drug therapy may become necessary or appropriate. Readers are advised to check the most current information provided (i) on procedures featured or (ii) by the manufacturer of each product to be administered, to verify the recommended dose or formula, the method and duration of administration, and contraindications. It is the responsibility of the practitioner, relying on their own experience and knowledge of the patient, to make diagnoses, to determine dosages and the best treatment for each individual patient, and to take all appropriate safety precautions. To the fullest extent of the law, neither the Publisher nor the Editors assumes any liability for any injury and/or damage to persons or property arising out or related to any use of the material contained in this book.

The Publisher

Library of Congress Cataloging-in-Publication Data
Nath, Swapan Kumar
  Problem-based microbiology/Swapan K. Nath, Sanjay G. Revankar.
    p. cm.
  ISBN 0-7216-0630-X
    1. Medical microbiology–Case studies. I. Revankar, Sanjay G. II. Title.

QR46.N37 2006
616.9'041--dc22 2005042791

Acquisitions Editor: William R. Schmitt
Developmental Editor: Kevin Kochanski
Publishing Services Manager: Tina Rebane
Project Manager: Mary Anne Folcher
Design Direction: Gene Harris

Printed in China

Last digit is the print number: 9 8 7 6 5 4 3 2 1

To my wife, Margaret, and children, Elizabeth and Joshua,
whose enduring patience and loving support made this
project an enjoyable experience
To my father, Sushil Nath, for all his encouragement
And to the many medical students who inspired me
along the way

SKN

To my wife, Varsha, for putting up with
seemingly endless evenings and
weekends working
on this project

SGR

CONTRIBUTORS

Dominick Cavuoti, DO
Assistant Professor
Department of Pathology
University of Texas Southwestern Medical Center
Dallas, Texas
Diagnostic Methods

Michael Gale, Jr., PhD
Associate Professor
Department of Microbiology
University of Texas Southwestern Medical Center
Dallas, Texas
Pathogenic Viruses

Kevin S. McIver, PhD
Assistant Professor
Department of Microbiology
University of Texas Southwestern Medical Center
Dallas, Texas
Pathogenic Bacteria

Preface

Why write another book on medical microbiology? In the wake of an explosion of knowledge on this subject, medical students deserve help with filtering the knowledge that will be valuable to retain, usable in the study of medicine, and applicable to future clinical practice. *Problem-Based Microbiology* is an organ system-based study of medical microbiology using patient problems (cases). The unconventional format of the book places the core principles on microbes and diseases in the Appendix. It uses case studies to promote interactive learning and to build a foundation of concise knowledge for clinical practice. Each section is introduced by an overview of how infectious diseases affect a particular organ system. Then, it provides clinical case scenarios, differential diagnoses, a problem-solving approach, and succinct explanations of the infectious process. *Problem-Based Microbiology* includes 91 of the commonly encountered infectious diseases, bioterrorism agents, and emerging infectious diseases. The following features make this book unique:

- Each case study is presented as an unknown to promote active learning. Each line of text provides additional information, designed to trigger recall, to facilitate the application of knowledge gained to that point in the case, and to provoke appropriate questions for sequential learning issues and further investigation.
- Tables show differential diagnoses and rationales that lead the reader through the inclusion and exclusion of specific diseases, suggesting the association of likely causes with the presenting symptoms.
- The detailed clinical case study for each infectious disease covered includes treatment options and outcomes.
- The integration of basic and clinical sciences is designed to help develop a "big picture" of the disease.
- Over 350 full-color illustrations and images of clinical disease reinforce the text.

- The summary tables help sort through the important diseases caused by an infecting organism and aid in the retention of usable information.
- Highlighted words throughout the cases identify key clinical symptoms and signs. Important concepts in the microbiology, epidemiology, and pathogenesis sections are also highlighted, which facilitates rapid review for course and Board exams.
- Over 250 clinical vignette-based practice questions, carefully integrated with information given in the cases, can also be used to prepare for course and Board exams.

For specific ideas on the various ways *Problem-Based Microbiology* can be used, please see *How to Use This Book* which follows.

The end-of-book coverage of "just what the students need to know" core concepts of infectious processes due to viral, bacterial, fungal, and parasitic pathogens and the succinct coverage of diagnostic principles and key laboratory tests make this an all-inclusive book for use during a Microbiology course. The many study aids also make it a good review book for preparation for the USMLE step 1 and 2 exams.

Although medical students are anticipated to be the main benefactors of this book, we recognize that it can be useful to a variety of different readers, including advanced level undergraduate and graduate students, taking medical microbiology courses, students in allied health, nursing, and physician assistant programs, residents and clinical fellows, and instructors from community colleges, universities, allied-health professional schools, and medical schools.

It is our hope that all students will take away not only a better understanding and application of concepts of medical microbiology, but also enthusiasm for patient-based problem solving and for learning the practical issues on infectious diseases. It is also our hope that *Problem-Based Microbiology* will promote students' quest for new knowledge.

We invite the readers' suggestions for improvement of this work. The preferred way to submit suggestions or corrections is electronic mail. The enthusiastic readers are advised to include PBM as the subject in the email and send submissions to email: swapannath@ sbcglobal.net.

Swapan K. Nath, PhD, F(CCM)

Sanjay G. Revankar, MD

Acknowledgments

Any new project, such as *Problem-Based Microbiology*, cannot be completed without the help of many individuals. In this regard we are particularly thankful to Beverly Shackelford (University of Texas Southwestern Medical Center at Dallas), without whose dedication to this project and untiring efforts to perfection this book could not have been completed. Words cannot describe our deep appreciation and indebtedness for her help with maintaining the high quality, clarity, and consistency throughout the text.

We wish to salute our contributors. We are indebted to Drs. Dominick Cavuoti, Michael Gale, Jr., and Kevin S. McIver, who have given unstintingly of their time and expertise to help provide an authoritative text, particularly in the Appendices.

We are also extremely grateful to some of our colleagues at the University of Texas Southwestern Medical Center at Dallas for providing photographic gems from their personal collections. Our special heartfelt thanks go to Drs. Paul Southern and Dominick Cavuoti who have donated a large number of clinical and microbiologic photographic images. Our thanks also go Drs. Anthony DalNogare, James Luby, Borna Mehrad, Daniel Skiest, and Jeanne Sheffield for some of their collections.

Many colleagues at the University of Texas Southwestern Medical Center have helped us with enhancement of the text by providing suggestions for improvement in their areas of interest. We wish to express our appreciation especially to Drs. Anthony DalNogare, Victor Garcia, Doug Hardy, Ron Hall, James Luby, Robert Munford, Jerry Niederkorn, Michael Norgard, Daniel Skiest, and Paul Southern, whose comments helped keep this book accurate and current.

Many at Elsevier deserve recognition for their role in the production of this book. Our heartfelt thanks go to William Schmitt, Acquisitions Editor of Medical Textbooks, for entertaining our ideas of a problem-based text, for his unfailing courtesy and efficiency in seeing the book through to publication. Our appreciation also goes to Kevin Kochanski, for competently helping us in the demanding job of making the manuscript ready for production. We wish to acknowledge the professionalism of Mary Anne Folcher, Project Manager, who took this complicated new project and completed it with remarkable speed and flexibility. In her calm and composed voice, she managed to keep one of us (SN) "in line," while displaying a keen understanding of what we are trying to accomplish with our unconventional style of work. We also thank Sylvia Stafford, Copy Editor, and Paul Fry, Designer, for their excellent work on this project.

A new project like this inflicts a heavy toll on the families of the authors. We thank them for their perseverance and encouragement. We are blessed and strengthened by their unconditional support and love and for their sharing with us the belief that our efforts are worthwhile and useful.

Finally, we wish to acknowledge our medical students, past and present, who provided support for the framework of the text in the beginning of the inception of the ideas of this project and provided encouragement and stimulation for completion of *Problem-Based Microbiology*.

SKN
SGR

How to Use This Book

Problem-Based Microbiology was written for students with various study goals and learning styles and can be used both in a microbiology course and as a review resource for the USMLE Step 1 exam. It might also be used as a reference during the clinical years. There are three parts to the book: cases and appendices, including the Practice Questions. Each of these parts is described next. At the end is a special discussion with suggestions for three ways or tracks that this book might be used, depending on the study goals, preferred learning style, and background of the individual student.

FEATURES

Introduction

This is an overview of ways in which microbiologic diseases affect the specific organ system, providing a background that can be built upon using the problem-solving exercises in the cases that follow.

Cases

These are presented in an "unknown format" intended to immediately engage the reader in thinking about microbiology in clinical terms, avoiding rote memorization of facts. Each case includes an important piece of information about the case, patient, and organism under discussion.

CASE DESCRIPTION. To obtain the *greatest benefit* from the problem-based nature of this book, it is advisable that the reader focus separately on each part on the first page of the case, making sure that the information is synthesized and digested before going on to the next. The entirety of the case description has been kept to the first page of the case as much as possible. In those cases where it has by necessity flowed onto the next page, it is suggested that the reader purposely not reveal the Etiology until a

proper synthesis of the case description has been attained.

Scenario. Each case begins with a short clinical scenario.

Physical Examination, Laboratory Studies, Diagnostic Workup. These give further information about the specific Case patient and offer a framework for evaluating the situation.

Differential Diagnosis. A table lists in alphabetical order possible organisms, including those that are common as well as those that should not be missed. A brief rationale guides students' thinking about inclusion and exclusion of these organisms while establishing an association of likely causes with the presenting symptoms.

Course. The Case patient is followed through the initial evaluation.

ETIOLOGY. The responsible organism is revealed. As suggested, it is best that the Etiology not be revealed until all possible exploration has been done with the progressively disclosed information.

MICROBIOLOGIC PROPERTIES. Significant microbiologic features of the organism, as well as important related organisms, are discussed.

EPIDEMIOLOGY. A brief description is given of salient epidemiologic features.

PATHOGENESIS. The specific effects of the ogranism on cellular and organ systems are described.

TREATMENT. A brief discussion of current treatment modalities is presented.

OUTCOME. The Case patient is briefly revisited to show the effects of both disease and treatment.

PREVENTION. A discussion is presented of pertinent methods (if any) of prevention.

FURTHER READING. A short bibliography is given of current as well as historical articles.

APPENDICES

Core Microbiologic Principles

Reviews of core microbiologic principles with illustrations provide inclusive but succinct background information that can be studied separately from the case problems—affording rapid review in the preparation for the USMLE Step 1 exam.

    Bacteria
    Viruses
    Fungi
    Parasites

Diagnostic Methods

Review of Diagnostic Methods is an appendix section. The purpose of this review is to explain the usefulness of different diagnostic tests used in clinical microbiology. Students are advised to consult this section when studying diagnostic issues and preparing for the USMLE Step 1 exam.

Antimicrobial Drugs

High-yield information covers antimicrobial drugs in a chart format for pharmacology review.

USMLE-Style Practice Questions include color images and case question clusters and can be used for preparing for course examinations and for readiness assurance for the USMLE. The color images are selected to aid memory association and recall of information about the disease. Many question sets throughout this Appendix are in the format of "case clusters," wherein each question in the set addresses a somewhat different aspect of a case, looking at the clinical situation from a variety of perspectives. These question clusters are particularly appropriate for assessment of learning in problem-based learning (PBL) pathways and integrated curricula.

A Table of Normal Values inside the back cover can help in interpreting a laboratory value or vital sign.

Throughout the book, **bolded concepts and facts** emphasize key clinical features and information about microbiology, epidemiology, and pathogenesis and demonstrate distinctive features of a disease. The intention is to help the reader distinguish the most important details for preparation for course exams or for the USMLE Step 1 exam. **Multicolor figures and micrographs** demonstrate important clinical information. In many cases, radiographic images and laboratory and pathologic findings are shown. These resources provide important clues for defining etiology. They assist in retaining important clinical information and positive findings of key diagnostic tests.

Summaries of integrated information on diseases and pathogens are presented as concise tables throughout *Problem-Based Microbiology*.

TRACKS OF STUDY

Before plunging into the first case, readers are encouraged to evaluate their fund of knowledge and think about how they should best approach this book. Three "Tracks" of study are described here, designed to give readers an appropriate experience, based on level of understanding.

Self-motivated learners with a strong basic science background who enjoy the freedom of a nontraditional curriculum and who learn best through self-directed reading and problem solving may wish to approach the book using the **Track 1** philosophy. These individuals would read the Introduction to the organ system and then delve into the investigative study of a patient problem. The goal would be to focus on the clinical scenario and related information available on the first page of the case, using them to define the etiology before turning the second page of the case to verify the etiology. (The design of this book is to present on the first page of a case all of the information needed to determine etiology. In a few instance, however, the page must be turned to finish reading all of the key points.) Once the etiology has been determined, the more general aspects of disease associated with the selected organism can be studied: microbiologic properties, epidemiology, pathogenesis, treatment, and prevention of disease, as well as the outcome of the Case patient.

Track 2 can be used by small groups of students in a PBL setting, where student-centered, self-directed learning is emphasized. Here, a faculty facilitator might present the first page of a case over the course of three discussion sessions. For instance, the presenting symptoms and medical/social history could be presented and progressively disclosed in Session 1; physical examination, diagnostic hypotheses, and laboratory results in Session 2. This would leave the follow-up materials for student self study and group discussion in Session 3. During the first session, students might also discuss open-ended learning issues that are outside the scope of this book, such as family behavior, responses to therapy in the elderly and the very young, impact of certain diseases on the community, and epidemiology issues not already addressed. Students might further refine their learning in the context of the problem and the open-ended issues by focusing on biology, diagnostics, pathophysiology, and management in the second, and possibly a third, session. This PBL track is appropriate for students who are flexible in their learning goals and who want a strong clinical context for their learning.

In **Track 3,** students who learn well from a more traditional combination of presentations (e.g., lectures) and readings and are more comfortable in a teacher-directed environment, or who perhaps need some preliminary study in the core concepts of microbial pathogens, may wish to begin with the material presented in the appendices and then read the introductions to the various organ-system sections. Appendix G, Microbes/Diseases, is provided at the end of the book for those wishing to focus on the "parade" of microbes and diseases and the pathogenesis of specific organisms. Students using Track 3 may save the first-page patient scenario and other information for the end of their study, to be used for assessment of recall and application of processed information.

In all cases, using any of the Tracks, self evaluation using the practice quiz questions will help the reader recall important information and identify areas for further study.

Contents

SECTION I
Respiratory Tract Infections 1

Case 1: A 14-year-old girl presented with a
2-day history of fever, sore throat, and a red
left eye. 7

Case 2: A 21-year-old male college student
presented with fever, sore throat, severe
fatigue, and difficulty in swallowing. 11

Case 3: A 6-year-old girl presented with a
high fever, an itchy throat, and difficulty
swallowing. 15

Case 4: A 9-year-old girl presented with
low-grade fever, sore throat, and malaise for
2 days. 19

Case 5: A 7-year-old girl presented with
fever, headache, earache, and swelling and
tenderness at the parotid and submaxillary
areas. 23

Case 6: A 16-month-old boy presented with
runny nose, hoarseness, barking cough, and
low-grade fever. 25

Case 7: A 2-month-old girl with no history
of immunization turned blue after a series of
coughing spells. 29

Case 8: A 64-year old man with COPD
presented with a low-grade fever, productive
cough of yellow-green sputum, and
worsening shortness of breath. 31

Case 9: A 67-year-old man presented with
abrupt onset of shaking chills, high fever,
chest pain, and a productive cough. 35

Case 10: A 66-year-old homeless alcoholic
man presented with a cough, fever, night
sweats, and chest pain. 41

Case 11: A 21-year-old woman developed
fever, headache, and a gradually progressive
dry cough. 45

Case 12: A 41-year-old man was admitted with
a 3-day history of high fever and cough. 49

Case 13: A 71-year-old man developed
fever, chills, muscle aches, cough, and
prostration. 53

Casae 14: A 5-month-old girl was brought
to the clinic with a 2-day history of cough
and respiratory difficulty. 57

Case 15: A 32-year-old woman presented
with a cough, night sweats, and a 15-lb
weight loss. 61

Case 16: A 29-year-old woman developed
fever, chills, and bloody cough after
chemotherapy for leukemia. 65

Case 17: Five workers between the ages of
29 and 48 years presented with a 1-week
history of fever, chills, night sweats, cough,
and weight loss. 69

Case 18: A 38-year-old man presented with
a 2-week history of fever, productive cough,
right-sided pleuritic chest pain, a 12-lb weight
loss, and a painful lesion on his left arm. 73

Case 19: A 35-year-old man was admitted
with a 3-week history of fever, night sweats,
headache, joint pains, and a dry cough. 75

Case 20: A 44-year-old man presented with a 3-month history of intermittent fever, chills, and a cough productive of green sputum. 79

Case 21: A 20-year-old man presented with a 3-week history of moderate fever, cough, shortness of breath, weight loss, and anorexia. 83

Case 22: A 23-year-old man with AIDS was admitted for fever, nonproductive cough, and progressive shortness of breath. 87

Case 23: An 18-year-old woman with cystic fibrosis presented with a worsening chronic cough productive of greenish sputum. 91

Case 24: A 62-year-old man was admitted with fever, shortness of breath, productive cough, and chest pain. He also complained of thick, yellowish discharge in his eyes. 95

SECTION II
Urogenital Tract Infections 99

Case 25: A 24-year-old sexually active woman had a 1-day history of urgency and burning pain on urination. 103

Case 26: A 31-year-old woman presented with low-grade fever, malaise, and a rash. She had a history of painless ulcers on the vulva. 107

Case 27: A 26-year-old man was concerned about the painful, itchy sores on the shaft of the penis. 111

Case 28: A 23-year-old woman presented with a few small, raised lesions on the cervix but was otherwise asymptomatic. 115

Case 29: An 18-year-old sexually active man presented with a 48-hour history of painful urination and a yellowish penile discharge. 119

Case 30: A 27-year-old woman presented with lower abdominal pain, vaginal discharge, and dysuria for 1 week. 123

Case 31: A 26-year-old woman came to a clinic complaining of a profuse yellow, foamy vaginal discharge with foul odor. 127

Case 32: A 21-year-old woman complained of bothersome vulvar itching and thick, whitish vaginal discharge. 131

Case 33: A 27-year-old man presented with fever, headache, sore throat, malaise, and a rash. 135

SECTION III
Gastrointestinal Tract Infections and
Liver Diseases 141

Case 34: A 20-year-old man was brought for evaluation of severe abdominal cramping, diarrhea, and bloody bowel movements. 147

Case 35: Six individuals presented with a low-grade fever, abdominal cramps, vomiting, and diarrhea. 151

Case 36: A 41-year-old man was brought with a 3-day history of shaking chills, high fever, headache, abdominal pain, and generalized weakness. 155

Case 37: A 71-year-old man had an acute onset of fever, crampy abdominal pain, and diarrhea. 159

Case 38: A 75-year-old man had severe abdominal cramps and watery diarrhea that became bloody. 163

Case 39: A 31-year-old man presented with sudden severe, watery diarrhea after returning to the United States from Bangladesh. 167

Case 40: Twenty-four people had nausea, vomiting, and crampy abdominal pain within 3 hours after eating a meal at an office party. . . . 171

Case 41: A 3-month-old female infant was brought with a 5-day history of decreased activity, decreased oral intake, upper airway congestion, and general irritability. 175

Case 42: A 67-year-old man presented with frequent diarrhea after treatment for nosocomial pneumonia. 179

Case 43: A 9-month-old baby girl was brought with a 2-day history of vomiting, watery diarrhea, and fever. 183

Case 44: Within 48 hours of a college football game, 158 students presented with nausea, vomiting, and diarrhea. 187

Case 45: A 54-year-old woman presented with worsening abdominal pain and occasional heartburn. 191

Case 46: A 36-year-old man experienced intermittent diarrhea, with blood and mucus, and tenesmus after a trip to India. 195

Case 47: A 25-year-old man presented with nausea, flatulence, lack of appetite, and watery foul-smelling diarrhea. 199

Case 48: A 35-year-old man with AIDS presented with low-grade fever, nausea, anorexia, and several weeks of watery diarrhea. 203

Case 49: A 3-year-old girl experienced a 3-week history of nausea, poor appetite, abdominal pain, and a 2-day history of no bowel movements. 207

Case 50: A 42-year-old man experienced 3 weeks of worsening diarrhea, abdominal pain, and fevers, and 2 weeks of an itchy rash. 211

Case 51: A 49-year-old woman from Argentina experienced high fever and chills, jaundice, and upper abdominal pain for 3 days. 215

Case 52: A 41-year-old man from Kenya presented wth a 4-month history of worsening abdominal pain, diarrhea, nausea, and vomiting with blood. 219

Case 53: A 23-year-old man presented with a 5-day history of fever, jaundice, nausea, and vomiting. 223

Case 54: A 27-year-old woman and IV drug user complained of fevers, chills, headache, malaise, anorexia, and abdominal pain. 227

Case 55: A 48-year-old man was asymptomatic at physical exam, but routine chemistry panel showed elevated transaminases. 231

SECTION IV
Nervous System Infections 235

Case 56: A 20-year-old male college student was brought to an ER with a 12-hour history of high fever, chills, and severe headache. 241

Case 57: A 3 week-old baby boy was brought to the ER with a 24-hour history of fever, poor feeding, irritability, and a seizure. 245

Case 58: A 64-year-old woman with a history of rheumatoid arthritis presented to the emergency department with 5 days of fever, headache, and confusion. 249

Case 59: In a rural county in the Midwest, 29 people had a rapid onset of fever, headache, stiff neck, and photophobia. 253

Case 60: A 61-year-old homeless man had fever, malaise, worsening headache, nausea, vomiting, and diarrhea for the past 3 days in July. 257

Case 61: A 56-year-old man was brought to the ER for fevers, headache, left-sided weakness, and a seizure. 261

Case 62: A 45-year-old homosexual man was brought to the ER for fever, severe headache, nausea, vomiting, and mental status changes. 263

Case 63: A 30-year-old woman was brought to the ER for severe headache, nausea, vomiting, and seizures. 267

Case 64: A 28-year-old man was brought to the ER for severe headaches and two generalized seizures. 271

Case 65: A 31-year-old man was brought to the ER for fever, muscle pains, vomiting, and visual hallucinations. 275

Case 66: A 9-day-old female newborn had a 10-hour history of inability to nurse and difficulty in opening her jaw. 279

SECTION V
Skin, Wound, and Multisystem Infections . . . 283

Case 67: An 18-year-old male college student developed fever, chills, and pain while walking. He had boils on his left leg and fever. 287

Case 68: A 24-year-old man developed severe pain and swelling in his left thigh. He was unable to walk without assistance and developed a high fever. 293

Case 69: A 66-year-old man developed local edema and brownish discharge 2 days after surgery for colon carcinoma. 297

Case 70: A 27-year-old man had high, spiking fevers, severe diffuse pain over the lower abdomen, and loss of appetite. 301

Case 71: An 8-year-old white boy was brought to the dermatology clinic for raised lesions on his head. 305

Case 72: A 58-year-old man presented with a 3-week history of progressive, mildly painful skin lesions on his left arm. 309

Case 73: Six unvaccinated college students became sick with fever, cough, conjunctivitis, coryza, and a rash while in India. 313

Case 74: A 3-year-old unvaccinated boy developed a low-grade fever, swollen lymph nodes, and a rash. 317

Case 75: A 55-year-old woman experienced burning and pain over her left forearm. Several vesicles developed in a band-like distribution on her arm. 321

Case 76: A 62-year-old man with a history of a heart murmur developed low-grade fevers, night sweats, and fatigue. 325

Case 77: A 36-year-old man with AML was brought to the hospital for high fever, dry cough, and worsening shortness of breath for a week. 329

Case 78: A 63-year-old white man came to the ER for fever, headache, generalized myalgia, arthralgias, and a rash under the armpit. 333

Case 79: A 9-year-old boy, a new immigrant from West Africa, developed intense chills and daily high fever for 4 days. 337

Case 80: On a hot summer day, a 23-year-old man was brought to the ER for fever, severe headache, and spotted rash. 341

Case 81: A 35-year-old man developed high fever, myalgias, and severe headache after a boating trip. 345

Case 82: A 55-year-old woman developed high fever, chills, and pain in the axilla after having been bitten by her cat. 347

SECTION VI
Bioterrorism Agents and Emerging Infections 349

Case 83: A 63-year-old man developed nausea, vomiting, and confusion. 355

Case 84: A 24-year-old man complained of a chickenpox-like rash after he had fever, severe headache, and back pain. 359

Case 85: An 18-year-old man was taken to the ER for fever, weakness, pain in his left groin, and small rashes on his leg. 363

Case 86: A 38-year-old man had an 8-week history of fevers, headache, sweats, a 30-lb weight loss, and pain in his joints and lower back. 367

Case 87: A 51-year-old man developed fever, headache, muscle aches, and an ulcer a week after skinning and tanning a rabbit. 371

Case 88: In early summer, a 42-year-old man was admitted with a 1-week history of fever, muscle aches, and malaise. 375

Case 89: In August, a 69-year-old diabetic woman developed fever, headache, vomiting, weakness, and confusion. 379

Case 90: A 46-year-old man had a 6-month history of increasing forgetfulness, depression, and personality changes. 383

Case 91: A 44-year-old female nursing aide attending a patient with severe respiratory illness had complaints of fever, cough, myalgias, and mild shortness of breath. 387

Appendix A: Pathogenic Viruses: Concepts Michael Gale, Jr, PhD 390

Appendix B: Pathogenic Bacteria: Concepts Kevin McIver, PhD 401

Appendix C: Pathogenic Fungi and Parasites: Concepts 418

Appendix D: Diagnostic Methods: Concepts Dominick Cavuoti, DO 433

Appendix E: Antimicrobial Therapy 451

Appendix F: Practice Questions and Answers 462

Appendix G: Microbes/Diseases 514

Index 517

Respiratory Tract Infections

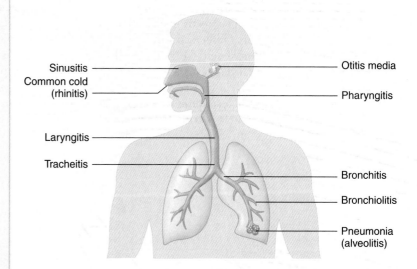

- Sinusitis
- Common cold (rhinitis)
- Laryngitis
- Tracheitis
- Otitis media
- Pharyngitis
- Bronchitis
- Bronchiolitis
- Pneumonia (alveolitis)

INTRODUCTION TO RESPIRATORY TRACT INFECTIONS

The respiratory tract is the most common site of infections (Fig. I-1) in humans. This is not surprising, because it is almost constantly exposed to a wide variety of potential pathogens from the environment during breathing and from hand contact with the nose and mouth. The organisms encountered include viruses, bacteria, and airborne fungi. As these are deposited in different regions of the respiratory tract, a variety of clinical syndromes is produced, including the common cold, pharyngitis, laryngitis, sinusitis, and otitis media in the upper tract, and tracheo-bronchitis, bronchiolitis (children), croup (children), and pneumonia in the lower tract. Viruses are by far the most common causes of upper respiratory infections (URIs). Whereas pneumonia may be caused by viruses, bacteria are more common causes, especially in adults. Certain fungal infections are also typically associated with pneumonia.

As one approaches the diagnostic work-up of a patient with a respiratory infection, the clinical syndrome and host factors should be considered to help narrow the differential diagnosis and assist with management of the infection. The age of the individual is also important in determining which clinical syndrome predominates for a particular group of pathogens. Other important factors include underlying diseases, geography, and travel history or other relevant history of exposures. The cases in the following section (and subsequent sections) will generally be presented in a classic clinical scenario, which may not be the most common one seen in practice but is a useful place to begin understanding various infectious syndromes. In the real world, however, the practitioner often makes clinical decisions based on presentations that are less than "classic."

The most common URI is the **common cold**. Rhinoviruses cause the vast majority of colds in adults, whereas parainfluenza and corona viruses are more common in children. Like many other respiratory infections, colds are more common in winter months. Contrary to popular opinion, this is more likely due to indoor crowding and contact than it is to colder weather. Most individuals will develop at least one episode of the common cold in a year, generally characterized by nasal discharge or congestion, mild sore throat, cough, and a lack of fever. Symptoms are self-limited, usually lasting a week or less. Most young children, especially those in day care, will have several episodes a year and may have mild fever. No specific therapy is available. Most URIs are transmitted **person to person**, often by **hand contact**, which underscores the need for regular hand washing as an important preventive measure for these infections.

Pharyngitis is also a very common URI. In young children (<3 years old) and adults, viruses are the usual

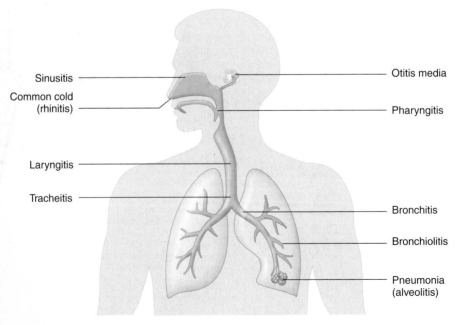

FIGURE I–1 Respiratory infections

cause (Table I-1). In older children (ages 5 to 15), *Streptococcus pyogenes* (Group A streptococcus) is the most common etiology, followed by viruses and *Mycoplasma pneumoniae*. *S. pyogenes* is the only common cause of pharyngitis that should be treated, because it can lead to several serious complications, including acute rheumatic fever. Unfortunately, it is not possible to reliably distinguish viral from bacterial etiologies on clinical grounds alone. As a result, a presentation with fever and pharyngitis usually prompts a microbiologic test to rule out Group A streptococcus infection.

Laryngitis is frequently an associated feature of infection with a variety of respiratory viruses. The characteristic symptom is hoarseness, usually associated with other URI manifestations. Laryngitis is self-limited and often resolves within a week. Rarely, bacterial causes may be found, such as *Moraxella catarrhalis*, or even *Mycobacterium tuberculosis*, which can be highly contagious. Noninfectious causes, such as acid reflux disease or voice abuse, should also be considered in patients without other associated symptoms.

Sinusitis is a typical manifestation of a viral URI, although it is often caused by a secondary bacterial infection. Patients usually present with fever, purulent nasal discharge, and sinus tenderness. Common bacterial causes are responsible, but often anaerobes will also be present. Approximately two thirds of cases of sinusitis recover without antibiotic therapy, and antibiotics are helpful in one half of the rest. Occasionally, fungi, such as *Aspergillus* or *Bipolaris*, may cause allergic fungal sinusitis. This is usually diagnosed in the setting of patients with chronic symptoms and often requires sinus cultures performed by an otolaryngologist to confirm.

TABLE I–1	Causes of Pharyngitis
Infectious Agents	**Risk (age) Group Affected**
Rhinoviruses	Young children (<3 years old) and adults
Coronaviruses	Young children (<3 years old) and adults
Adenoviruses	Young children (<3 years old) and adults
Enteroviruses	Older children (community outbreaks)
Influenza A, B, C	All age groups
Epstein-Barr virus	Older children and young adults
Streptococcus pyogenes	Older children
Mycoplasma pneumoniae	Older children and young adults
Corynebacterium diphtheriae	Unimmunized children

Otitis media may be a primary clinical manifestation of infection, but it is more commonly caused by a secondary bacterial infection following an initial viral URI. Otitis media is the most common reason for prescribing antibiotics in pediatrics. Young children, especially those in day care, may have several episodes in a year. However, by the time they reach 5 to 7 years of age, incidence drops off dramatically. Typical symptoms are fever, ear pain, and headache, although in very young children and infants, the only symptom may be irritability. Although acute otitis media usually resolves spontaneously, it is generally treated to prevent further suppurative complications. Chronic effusions of the middle ear are more difficult to manage and do not necessarily respond to antibiotics. The most common causes of otitis media are *Streptococcus pneumoniae* and nontypable *Haemophilus influenzae*.

The etiology of **tracheobronchitis** falls into two categories, depending on whether patients have or do not have chronic obstructive pulmonary disease (COPD). Tracheobronchitis in patients with normal lung function almost exclusively has a viral etiology. Adenoviruses, parainfluenza, and influenza are among the most common causes in adults, whereas respiratory syncytial virus occurs in children and infants. Normal individuals, especially older adults (whose post-vaccination-acquired immunity has presumably waned over time) may also develop bacterial infection, most frequently caused by *Bordetella pertussis*. Other, less common, nonviral causes include *Mycoplasma* and *Chlamydia*. Tracheobronchitis in patients with COPD typically has a bacterial origin. Because of their impaired mucociliary clearance, patients with COPD are susceptible to a variety of bacterial pathogens, namely nontypable *H. influenzae* and *M. catarrhalis*.

Croup (laryngotracheitis), a unique clinical syndrome seen in children younger than 3 years of age, is characterized by stridor and a cough described as a "seal's bark." It is unclear why croup is restricted to this particular age group, but significant factors may include primary viral infection, usually parainfluenza virus, a narrowed upper airway due to the pliant respiratory anatomy at this age, and various immunologic factors. Croup is usually self-limited, but it can be severe, leading to hospitalization.

Pneumonia is an inflammatory disease of the lungs with involvement of the lung parenchyma (alveolae). It is the most common infectious cause of death in the United States. Pneumonia should be distinguished from infections of the airways such as bronchitis and from URIs such as laryngitis and sinusitis. Pneumonia has many causes, but a few are responsible for the majority of infections seen clinically. Although sputum examination can be helpful, another useful approach for determining the etiology is to look at the patient's age, risk factors, and epidemiology. Pneumonia is conventionally divided into community-acquired pneumonia (CAP), which occurs in normal hosts who are not hospitalized, and nosocomial, or hospital-acquired, pneumonia, which occurs in sicker patients, who are often immunocompromised. The microbiology of CAP and nosocomial pneumonia differs considerably (Table I-2).

Traditionally, the clinical presentations of pneumonia have been subdivided into typical and atypical. However, there is often considerable overlap of the signs and symptoms of typical and atypical pneumonia. The clinical presentation of **typical pneumonia** is marked by the sudden onset of chills, fever, dyspnea, and a productive cough with purulent sputum (which may be blood tinged). The classic onset of a single rigor does not usually occur. Pleuritic chest pain may be present. The physical exam usually reveals fever, tachypnea, tachycardia, and rales on lung examination. The chest x-ray usually shows signs of consolidation. Typical pneumonia is usually caused by *S. pneumoniae*, *Klebsiella pneumoniae*, *H. influenzae*, *M. catarrhalis*, and *Staphylococcus aureus*. *S. pneumoniae*, the prototype of typical CAP, has been, and still remains, the most common cause of community-acquired pneumonia. *S. pneumoniae* and *S. aureus* are the most common causes of post-viral (influenza) pneumonia, particularly in the elderly.

The clinical presentation of **atypical pneumonia** is marked by gradual onset of cough (nonproductive) and dyspnea; extrapulmonary signs and symptoms may be more prominent, such as headache, sore throat, and diarrhea. The physical exam usually reveals fever, tachypnea, tachycardia, and various minimal findings. The chest x-ray may show only patchy or interstitial infiltrates with no signs of consolidation. *Mycoplasma*, *Chlamydia*, and *Legionella* are the major causes of atypical pneumonia. *Mycoplasma* pneumonia (often called "walking pneumonia") and *Chlamydia* pneumonia, more common in young adults, are often associated with pharyngitis and are usually milder than other forms of CAP. *Legionella* pneumonia may occur in healthy or immunocompromised patients and can be quite severe. When atypical pneumonia is suspected, additional history should be obtained to rule out the possibility of unusual causes (e.g., exposure to birds in psittacosis).

Viral causes of pneumonia are uncommon in adults but can be seen in infants and children, particularly

TABLE I-2 Common Causes of Pneumonia

Age Specificity and Other Risks	Infectious Agents
Neonate (birth-6 weeks)	Group B Streptococcus *Escherichia coli*
Children (6 weeks-18 years)	Viruses (e.g., respiratory syncytial virus) *Mycoplasma pneumoniae* *Chlamydia pneumoniae* *Streptococcus pneumoniae*
Adults (18-40 years)	*M. pneumoniae* *C. pneumoniae* *S. pneumoniae*
Adults (40-65 years)	*S. pneumoniae* *Haemophilus influenzae* Anaerobes Viruses
Elderly (≥65 years of age)	*S. pneumoniae* Viruses Anaerobes *H. influenzae* Gram-negative rods (e.g., *Enterobacteriaceae* and *Pseudomonas*)
Nosocomial (hospital acquired)	Gram-negative rods (e.g., *Klebsiella pneumoniae* and *Pseudomonas*; multiply drug-resistant bacteria) *Staphylococcus aureus* (drug resistant)
Immunocompromised	Gram-negative rods *S. pneumoniae* Fungi Filamentous bacteria (e.g., *Nocardia*) *Pneumocystis jiroveci* Viruses
Special Risks	
Gross aspiration	Anaerobes
Alcoholics	*S. pneumoniae*, *K. pneumoniae*, anaerobes
Intravenous drug use	*S. aureus*
Postviral (secondary infection)	*S. aureus*
Neutropenia	*Aspergillus* spp
Chronic steroids	*Nocardia*

respiratory syncytial virus (RSV) and influenza. Influenza is usually a protracted "flu-like" syndrome, and occasionally it causes fulminant pneumonia in all age groups. In adults, adenovirus and varicella zoster virus can cause acute respiratory distress syndrome (ARDS). Cytomegalovirus usually causes pneumonia in immunocompromised hosts (e.g., bone marrow transplant patients). Hanta virus, associated with contact with deer mice, is also a cause of a severe respiratory distress syndrome. These two diseases are discussed in a further part of the book.

Tuberculosis (TB) may manifest in either immunocompetent or immunocompromised (HIV) hosts. The diagnosis should be considered in patients with a possible exposure history (e.g., patients with a history of incarceration or homelessness, or foreign-born patients), predominantly upper-lobe disease with cavities, and in cases in which sputum culture and Gram stain are nondiagnostic and the patient is not improving on "antibacterial antibiotics." Patients with *Nocardia asteroides* can present with pneumonia with depressed cell-mediated immunity (generally owing to use of corticosteroids) and may manifest with fever, cough, and cavities or nodules in the lung (TB-like illness). Pneumonia due to *Pneumocystis jiroveci* is most commonly found in AIDS patients with CD4$^+$ cell counts less than 200 cells/μL. In normal hosts, fungi such as *Cryptococcus neoformans*, *Histoplasma capsulatum*, and *Coccidioides immitis* may manifest as a self-limited illness of unrecognized etiology. Infection of immunocompromised hosts with *Aspergillus fumigatus* usually manifests with progressive pneumonia with cavities, nodules, or interstitial infiltrates, sometimes severe enough to warrant hospital admission.

In summary, many infectious etiologies can be responsible for respiratory infections. However, the list for a differential diagnosis can usually be narrowed by considering factors including age, underlying diseases, geography, and travel history or other relevant exposures.

FURTHER READING

Bartlett JG, Mundy LM: Community-acquired pneumonia: Current concepts. N Engl J Med 333:1618, 1995.

Berman S: Otitis media in children: Current concepts. N Engl J Med 332:1560, 1995.

Bisno AL: Primary care: Acute pharyngitis. N Engl J Med 344:205, 2001.

File TM: Community-acquired pneumonia. Lancet 362:1991, 2003.

Jousimies-Somer HR, Savolainen S, Ylikoski JS: Bacteriological findings of acute maxillary sinusitis in young adults. J Clin Microbiol 26:1919, 1988.

Marik PE: Aspiration pneumonitis and aspiration pneumonia: Primary care. N Engl J Med 344:665, 2001.

Questions on the case (problem) topics discussed in this section can be found in Appendix F. Practice question numbers are listed in the following table for students' convenience.

Self-Assessment Subject Areas (book section)	Question Numbers
Upper and lower respiratory infections (Section I)	1-3, 44-47, 49-57, 59-61, 65-67, 71-74, 81-84, 96-99, 122-124, 126, 133-135, 155-157, 159-161, 167-168, 169, 170-172, 182-185, 192-194, 196-198, 205-208, 210, 234-246

A 14-year-old girl presented to the emergency department with a 2-day history of **fever, sore throat,** and a **red left eye, which felt like there was 'sand' in it.** She had been at a **summer camp** for the past 2 weeks where **several other children had a similar illness.** Activities at the camp included swimming in a local pool.

The patient had no significant past medical history and denied being sexually active. Family history was unremarkable.

PHYSICAL EXAMINATION

VS: T 38°C, P 94/min, R 12/min, BP 124/82 mmHg

PE: Examination revealed preauricular **lymph-adenopathy,** an **erythematous pharynx** (Fig. 1-1), and **conjunctivitis** of the left eye.

LABORATORY STUDIES

Blood

WBC: 10,000/μL
Differential: Normal
Serum chemistries: Normal

Imaging

No imaging studies were done.

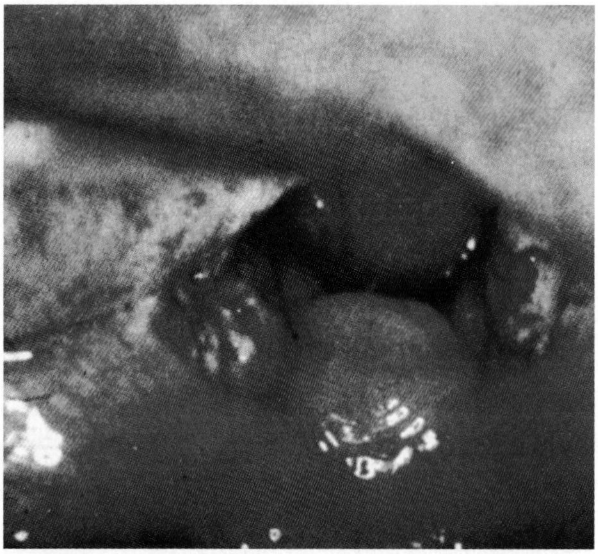

Diagnostic Work-Up

A clinical diagnosis of pharyngoconjunctival fever was considered. Table 1-1 lists the likely causes of illness (differential diagnosis). The investigational approach to delineating the etiology may include

- **Cultures** of bacteria and viruses (e.g., adenovirus or enterovirus) from the nasopharyngeal and conjunctival swabs
- **Monospot assay** for heterophile antibodies to rule out Epstein-Barr virus
- In failed diagnosis:
  - Antigen detection
  - Polymerase chain reaction assay
  - Serology

COURSE

A throat culture was taken, and the next day revealed no evidence of *Streptococcus pyogenes* or *Neisseria gonorrhoeae.* A respiratory virus was recovered from

FIGURE 1-1 Erythematous pharynx in this patient. *(Courtesy of Dr. John D. Nelson, Department of Medicine, University of Texas Southwestern Medical Center, Dallas, TX.)*

TABLE 1-1 Differential Diagnosis and Rationale for Inclusion (consideration)

Adenovirus
Coronavirus
Enterovirus
Epstein-Barr virus
Herpes simplex virus type 1
Influenza (in late fall through early spring)
Neisseria gonorrhoeae
Rhinovirus
Streptococcus pyogenes

Rationale: All of the above may cause pharyngitis. Influenza, a seasonal disease, is unlikely (also, no cough is present). Selected viral (e.g., adenovirus, enterovirus, herpes simplex) and bacterial agents (e.g., *N. gonorrhoeae*) cause conjunctivitis. Whereas rhinovirus, coronavirus, and Epstein-Barr virus cause upper respiratory symptoms, conjunctivitis is not usually seen. *S. pyogenes* and *N. gonorrhoeae* need to be ruled out as possible causes because they would require specific antibiotic therapy.

the cell cultures of the nasopharyngeal specimen and conjunctival swab and identified by fluorescence antibody staining of virus-infected cells using an antibody reagent.

ETIOLOGY

Adenovirus (pharyngoconjunctival fever)

MICROBIOLOGIC PROPERTIES

Adenovirus is a **nonenveloped, double-stranded, linear DNA virus** with an **icosahedral nucleocapsid**. It is the only virus with a fiber protruding from each of the 12 vertices of the capsid. The fibers consist of a slender shaft with a globular head (Fig. 1-2). The capsid components of adenovirus, the hexone and the fiber, display antigenic determinants, which together define the serotype specificity of adenovirus. A total of **49 serotypes** (in 6 subgenera, A through F) are found worldwide. The cultivable virus **replicates in the nucleus** of the host cell. Surrogate cells infected with the virus in cultures can be identified by fluorescent antibody staining and ultraviolet microscopy. The virus is unusually **stable to chemical or physical agents** and adverse pH conditions, allowing for prolonged survival outside of the body.

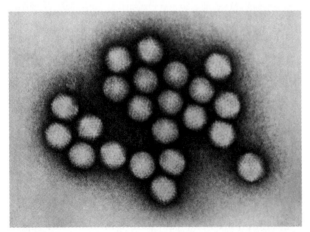

FIGURE 1-2 Electron micrograph of adenovirus. Note: naked capsids are seen. Fibers are shed and are not seen in this image. (*Courtesy of Centers for Disease Control and Prevention, Atlanta, GA.*)

EPIDEMIOLOGY

An extremely hardy virus, adenovirus is ubiquitous in human and animal populations; it survives for long periods outside a host and is endemic throughout the year. Although epidemiologic characteristics of the adenoviruses vary by antigenic type, **all are transmitted by direct contact, fecal-oral transmission,** and, occasionally, **waterborne transmission.** Aerosolized droplet exposure also causes pharyngitis and acute respiratory distress syndrome (ARDS). Spread also occurs via contact with ocular secretions during the acute illness, and indirectly through contaminated surfaces, instruments, or solutions. **Fecal-oral transmission** of enteric adenoviruses occurs among infants **in a day care setting.** This patient was thought to be a part of an outbreak of febrile disease with conjunctivitis, associated with waterborne transmission of an adenovirus type from an **inadequately chlorinated, turbid swimming pool** in the summer campsite. Turbid water contains organic molecules (e.g., humic and fulvic acids from plant decay) that react with chlorine, generating chloroform, which has no antiviral activity. Viruses may attach or embed in suspended particles in turbid water, and these virus-containing particles precipitate into the sediment on the bottom of the pool, where they may remain viable. The virus-containing particles may become resuspended when swimmers agitate the water. After swimming in the implicated pool, the onset of first symptoms of pharyngoconjunctival fever may take 5 to 12 days.

PATHOGENESIS

Pharyngoconjunctival fever is usually caused by adenovirus serotypes 3, 4, and 7. The virus has tropism for the mucosal epithelium of the upper respiratory tract, conjunctivae, and several other body sites, presenting with pathology in various organs (Table 1-2). Adenovirus gains entry into the target cells via attachment to the receptors on the cell surface using their fiber protein. The nucleoprotein complex enters the nucleus via endosomes. Adenovirus inhibits host cell protein synthesis and takes over the DNA synthesis machinery. Virus replicates, inducing a cytopathic effect on the infected cells via the lytic cycle of viral growth. The main immunologic defense against an established viral infection is cell-mediated immunity. Target-cell death and chemokine mediators released from the lysed cells contribute to fever, sore throat, coryza, and red eyes, generally lasting for 5 days.

TABLE 1–2 Other Clinical Features of Adenovirus Infections

Syndromes	Clinical Features
Acute respiratory distress syndrome (ARDS)	Predominantly caused by **serotypes 4 and 7** Occurs with greater frequency in spring and winter months; adenovirus accounts for 10% of all childhood lower respiratory tract infections. Lower respiratory tract infections, including tracheobronchitis, bronchiolitis, and pneumonia, may mimic respiratory syncytial viral infection in infants. Fever, rhinorrhea, cough, and sore throat usually lasting 3-5 days are typical symptoms of adenoviral ARDS.
Epidemic keratoconjunctivitis	Predominantly caused by **serotypes 8, 19, and 37** An insidious onset of unilateral red eye occurs, with spread to both eyes later on (highly contagious via hands and fomites). Patients have photophobia, tearing, and pain. Children may have fever and lymphadenopathy. Inflammatory pathology persists for weeks.
Gastroenteritis	Commonly associated with **serotypes 40 and 41** Fever and watery diarrhea occur in infants in the day care setting; **intussusception** occurs in rare complications (pathology is remarkable for **hyperplasia of Peyer patches**).
Acute hemorrhagic cystitis	Predominantly caused by **serotypes 11 and 21** Usually affects children aged 5-15 years, but it also may affect adults who are immunosuppressed (e.g., from renal or bone marrow transplantation or from AIDS). Dysuria, frequency, and grossly bloody urine are reported, predominantly in boys.

TREATMENT

There is **no specific antiviral therapy**. Fortunately, pharyngitis and most other infections are mild and require no therapy. Serious adenovirus illness (e.g., ARDS; see Table 1-2) is usually managed by treating symptoms and complications of the infection. Cidofovir, a nucleoside analogue, has been used with some success to treat severe disseminated infections.

OUTCOME

The patient was treated symptomatically. The other eye also developed conjunctivitis, although all symptoms resolved within a few days. The outbreak at the summer camp ended when the swimming pool was properly filtered and rechlorinated.

PREVENTION

Adenoviral pharyngitis, pharyngoconjunctival fever, and epidemic keratoconjunctivitis are preventable by strict asepsis and hand washing. Adequate chlorination of swimming pools may prevent waterborne outbreaks. A live vaccine (types 4 and 7) delivered in an enteric-coated capsule was used in military recruits to protect from ARDS. The cessation of vaccine production in 1995 led to a re-emergence of serious infections and outbreaks. Re-establishing adenovirus vaccine production may be necessary in the future.

FURTHER READING

Gray GC, Callahan JD, Hawksworth AW, et al: Respiratory diseases among U.S. military personnel: Countering emerging threats. Emerg Infect Dis 5:379, 1999.

Hierholzer JC: Adenoviruses in the immunocompromised host. Clin Microbiol Rev 5:262, 1992.

McNeill KM, Ridgely Benton F, Monteith SC, et al: Epidemic spread of adenovirus type 4-associated acute respiratory disease between U.S. Army installations. Emerg Infect Dis 6:415, 2000.

Ryan MA, Gray GC, Smith B: Large epidemic of respiratory illness due to adenovirus types 7 and 3 in healthy young adults. Clin Infect Dis 34:577, 2002.

A **21-year-old** male **college student** came to the emergency department of a local hospital complaining of **fever,** headache, and **sore throat** with severe fatigue and difficulty in swallowing of ten days' duration.

The patient had no significant past medical history. He was not sexually active, although he started dating a new girlfriend a week ago, and he admitted to **kissing** several girls in the past month. Family history was unremarkable.

PHYSICAL EXAMINATION

VS: T 39.3°C, P 104/min, R 14/min, BP 124/72 mmHg

PE: Pharyngeal erythema and bilaterally **enlarged tonsils with exudates** were noted; **posterior cervical lymph nodes** were enlarged and tender (lymphadenopathy); **spleen** was **enlarged** and a palpable **liver** was **tender (hepatosplenomegaly);** faintly jaundiced skin was also noted.

LABORATORY STUDIES

Blood

Hematocrit: 42%

WBC: 12,000/μL

Differential: 8% PMNs, 28% monos, 52% lymphs **(12% atypical cells)**

Platelets: 90,000/μL

Serum chemistries: AST 132 U/L, ALT 178 U/L, ALP 280 U/L, bilirubin 2.4 μmol/L

Imaging

Abdominal ultrasound revealed **hepatosplenomegaly.** No focal lesions were observed.

Diagnostic Work-Up

A clinical diagnosis of infectious mononucleosis (IM) was considered on the basis of the symptoms of fever, sore throat, lymphadenopathy lasting more than 1 week, and the age of the patient. Leukocytosis with lymphocytosis and monocytosis exceeding 50% (including ≥10% atypical cells) are typical signs of IM. Table 2-1 lists the likely causes of illness (differential diagnosis). The investigational approach to delineating the etiology may include

- **Throat cultures** to rule out *Neisseria gonorrhoeae* and *Streptococcus pyogenes*
- **HIV viral load** test

- **Monospot** test
- In a failed monospot test, Epstein-Barr (EBV)-specific **serology**, and detection of IgM for cytomegalovirus (CMV).

TABLE 2-1   Differential Diagnosis and Rationale for Inclusion (consideration)
Adenovirus
Cytomegalovirus (CMV)
Epstein-Barr virus (EBV)
Herpes simplex virus type 1 (HSV-1)
Human immunodeficiency virus (HIV)
Streptococcus pyogenes

Rationale: All of the above may cause pharyngitis. EBV, CMV, and HIV may cause fever, pharyngitis, cervical lymphadenopathy, atypical lymphocytosis, and hepatosplenomegaly. The remaining organisms are less likely to cause hepatosplenomegaly. *S. pyogenes* is also less likely to cause atypical lymphocytosis. Adenovirus and HSV-1 often cause pharyngitis, but they are not associated with hepatosplenomegaly. Viral hepatitis is not included because pharyngitis is not a typical feature.

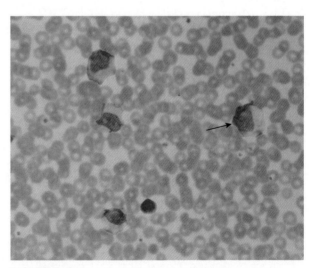

FIGURE 2-1   Blood film showing abnormal circulating cytotoxic T cells (atypical lymphocytes; *arrowhead*). Normal lymphocytes are also seen. (*Courtesy of Dr. Dominick Cavuoti, Department of Pathology, University of Texas Southwestern Medical Center, Dallas, TX.*)

COURSE

The patient was admitted to the hospital. No antibiotics were administered pending throat culture results. Throat culture results the next day were negative for *N. gonorrhoeae* and for *S. pyogenes* or any other β-hemolytic streptococci. A monospot test was positive, confirming the clinical diagnosis.

ETIOLOGY

Epstein-Barr virus (IM).

MICROBIOLOGIC PROPERTIES

EBV belongs to the herpesvirus family. The herpesviruses are **DNA viruses**, characterized by their **ability to persist indefinitely in a latent form after acute infection, with subsequent reactivation** during host immunosuppression. Herpesviruses are classified into three subfamilies based on the virus host range and other properties:

1. The α-**herpesviruses**, which include **HSV-1, HSV-2**, and **varicella** virus, grow rapidly in many types of cells including epithelial cells and neurons.
2. The β-**herpesviruses**, which include **CMV** and human herpesvirus types 6 and 7 (**HHV-6** and **7**), grow in specific cells types (human fibroblasts in the case of CMV).
3. The γ-**herpesviruses**, which include **EBV** and **Kaposi sarcoma-associated virus** (KSHV, also known as HHV-8), grow slowly in lymphoid cells.

EBV is an enveloped, linear, double-stranded-DNA virus like other herpesviruses but it is serologically distinct from other herpesviruses. The three viral antigens associated with EBV are viral capsid antigen (VCA), early antigen (EA), produced prior to viral DNA synthesis, and Epstein-Barr nuclear antigens (EBNAs), located in the nucleus bound to chromosomes. This **oncogenic virus** has **tropism for the B lymphocyte** and in certain geographic settings is associated with **Burkitt lymphoma** (in Africa) and **nasopharyngeal carcinoma** (in Asia).

EPIDEMIOLOGY

Transmission of EBV occurs by **salivary secretions** (close physical contact, kissing). Children infected at an early age usually present with subclinical disease, whereas adolescents or young adults (**15 to 20 years** of age) develop IM during acute infection. Typical **IM occurs primarily in developed countries**, where the major method of spread is kissing, whereas in non-industrialized countries, the majority of infections occur during childhood.

PATHOGENESIS

EBV enters through the oropharynx, and primary viral multiplication occurs in squamous epithelial cells. This phase is followed by secondary infection of B lymphocytes. A viral particle attaches via its major outer envelope glycoprotein (gp350/220) to the complement receptor CD21 on the B lymphocytes, which ultimately form the reservoir for the virus.

Normal uninfected B cells migrate to germinal centers where they proliferate and then undergo apoptosis. B lymphocytes infected with EBV, however, circumvent the signals for apoptosis, driving EBV-infected cells into the long-lived memory B-cell population. During latency, 1 to 50 memory B cells infected with EBV per million circulate back to the oropharynx, where they mature and undergo another round of viral amplification. The virus can be isolated from oral secretions of 20% to 30% of healthy latently infected individuals at any time. During latency, the virus minimizes gene expression, thus evading recognition and destruction by the immune system.

The clinical picture of IM is dependent on the host immune response. The timing and the strength of the immune response determine whether the acute infection will remain subclinical or will result in the clinical presentation of IM. The reasons for subclinical EBV infection at an early age may be due to the fact that young children can mount a vigorous cytotoxic T lymphocyte (CTL) response and control viremia, unlike adults, whose CTL reservoir is already jaded by having seen many other pathogens.

In acute infection, cytotoxic T cells aim to destroy EBV-infected B cells and thereby control viremia with a strong CTL response. Host immune response to the viral infection consists mainly of a subset of CD8+ T lymphocytes that have both suppressor and cytotoxic properties. These **CTLs are found in the peripheral blood** of a patient with EBV infection at a pronounced level and are called **atypical lymphocytes** (Fig. 2-1). The B-cell immune response to EBV infection consists of more specific IgM antibody to

VCA, which may persist for 1 to 2 months, and long-lasting IgG antibodies.

Diverse immunoglobulins resulting from B-cell stimulation include antibodies to the RBCs of several different animals. These **heterophile antibodies** serve as the basis of serologic diagnosis (**Monospot** test).

TREATMENT

Supportive care is usually all that is required for IM. Acyclovir is active in vitro against EBV and can reduce viral replication in tonsils, but it does not have significant clinical efficacy in uncomplicated IM. It is not used in acute EBV infection because the symptoms of IM are caused by an exuberant immune response and not by unimpeded viremia. Acyclovir cannot eliminate latent EBV infection. In some cases of reactivated EBV infection in immunocompromised patients (Table 2-2), the use of acyclovir may be beneficial, but the mainstay in these patients is to support or boost their immune system with EBV-specific cytotoxic lymphocytes. Steroids have been shown to be useful in severe cases of IM with airway obstruction, severe thrombocytopenia, or hemolytic anemia.

OUTCOME

The patient was discharged to home with the recommendation to avoid physical activity for 4 weeks (owing to the risk of splenic rupture). He recovered and returned to normal daily activities within 10 days.

PREVENTION

Prevention is difficult for this communicable disease. There is no vaccine. Initial viral load with EBV infections is not correlated with developing clinical symptoms of IM. The goal of vaccine development is, therefore, to prevent disease states.

TABLE 2–2 Other Clinical Features of EBV Infections	
Syndromes	**Clinical Features**
Burkitt lymphoma (BL)	EBV-encoded LMP-1 protein activates the TNF pathway and NF-κB, causing **immortalization of B cells.**
	BL is an endemic **B-cell lymphoma,** common in African children. *c-MYC* is a risk factor for BL, because it is known to regulate cell growth and apoptosis and is always translocated and deregulated in BL. The **oncogenic mechanism involves** transposition of **t(8;14)** *c-MYC* gene next to heavy-chain Ig gene.
	Histopathology of jaw tumor biopsy is characteristic with sheets of lymphocytes interspersed with macrophages ("**starry-sky**" appearance).
CNS lymphoma	**AIDS patients with very low CD4+ cell counts (< 50/μL) are at risk for primary CNS lymphoma,** presenting with one or more mass lesions in the brain following infection with EBV.
	The common signs and symptoms of primary CNS involvement include altered mental status (confusion, memory loss, lethargy), hemiparesis, dysphasia or sensory findings, seizures, cranial nerve findings, and headache.
	The lymphoma cells contain EBV DNA and the CSF EBV PCR is a useful diagnostic test for this disease.
Oral hairy leukoplakia (OHL)	OHL, found predominantly in patients with HIV, is clinically characterized by **elevated white lesions** on the lateral borders and dorsum of the tongue.
	Histopathology may reveal a profound increase in the thickness of the stratum spinosum of the epidermis (**acanthosis**) and perinuclear vacuolation of squamous epithelial cells (**koilocytic changes**). The acanthosis is believed to be associated with the combined action of the EBV-encoded proteins, and the koilocytic changes are due to intense replication of EBV.
Non-Hodgkin lymphoma	In approximately one half of AIDS patients with systemic non-Hodgkin lymphoma, the lymphoma is EBV associated. These patients often present with symptoms (weight loss, night sweats, and fevers) of B-cell lymphoma and often have extranodal disease.

FURTHER READING

Cohen JI: Epstein-Barr virus infection. N Engl J Med 343:481, 2000.

Cohen JI: Epstein-Barr virus and the immune system. Hide and seek. JAMA 278:510, 1997.

Okano M, Thiele GM, Davis JR, et al: Epstein-Barr virus and human diseases: Recent advances in diagnosis. Clin Microbiol Rev 1:300, 1988.

Pathmanathan R, Prasad U, Sadler R, et al: Clonal proliferations of cells infected with Epstein-Barr virus in preinvasive lesions related to nasopharyngeal carcinoma. N Engl J Med 333:693, 1995.

Thorley-Lawson DA, Gross A: Mechanisms of disease: Persistence of the Epstein-Barr virus and the origins of associated lymphomas. N Engl J Med 350:1328, 2004.

Triantos D, Porter SR, Scully C, Teo CG: Oral hairy leukoplakia: Clinicopathologic features, pathogenesis, diagnosis, and clinical significance. Clin Infect Dis 25:1392, 1997.

A **6-year-old** girl came home from school feeling miserable on a cold day in January. She had a **high fever** and complained of an **itchy throat**. She had **difficulty swallowing** any food, refused to eat, and cried almost all evening. The next day her grandpa took her to their family physician's clinic. It was noted that several children from her school had reported sore throats recently.

The patient had received all standard childhood immunizations at the appropriate times.

PHYSICAL EXAMINATION

VS: T 39.4°C, P 120/min, R 16/min, BP 110/60 mmHg

PE: Red throat (pharyngeal **erythema**) with petechiae (small red spots) on the soft palate and **patchy grayish-whitish tonsillar exudates** were seen (Fig. 3-1). **Enlarged and tender anterior cervical lymph nodes** were also noted. The patient **did not have any cough**.

LABORATORY STUDIES

Blood

WBC: 12,300/µL
Differential: 76% PMNs, 20% bands, 6% monos

Imaging

Chest x-ray was not done.

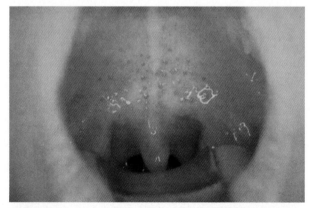

FIGURE 3-1 The redness and edema of the oropharynx, and petechiae (small red spots) on the soft palate in this patient. *(Courtesy of Dr. Paul Southern, Departments of Pathology and Medicine, University of Texas Southwestern Medical Center, Dallas, TX.)*

Diagnostic Work-Up

Table 3-1 lists the likely causes of illness (differential diagnosis). A monospot test can be useful to rule out Epstein-Barr virus (EBV) infection. Additional tests for delineating the etiology may include

- **Rapid antigen detection test** (RADT)
- **Throat culture**

COURSE

Throat swab was taken, and the RADT was slow positive. The patient was given oral penicillin V. In the laboratory, the throat culture on sheep blood agar yielded a significant bacterial isolate.

TABLE 3–1 Differential Diagnosis and Rationale for Inclusion (consideration)
Adenovirus
Coronavirus
Epstein-Barr virus (EBV)
Influenza (A, B, C)
Rhinovirus
Streptococcus (Groups C or G)
Streptococcus pyogenes (Group A streptococcus)

Rationale: All of the above may cause pharyngitis. The patient had pharyngitis. Because it affects therapeutic decision making, it is important to attempt clinical differentiation between viral and bacterial pharyngitis. The clinical presentations for viral and bacterial pharyngitis overlap broadly. Viruses cause more than 70% of cases of pharyngitis. Mild pharyngeal symptoms with rhinorrhea suggest a viral etiology. High fever, tonsillar exudates, anterior cervical lymphadenopathy, and the absence of cough are the best predictive clinical features for bacterial pharyngitis or EBV infection. EBV is less likely to be symptomatic in this age group.

ETIOLOGY

Streptococcus pyogenes (streptococcal pharyngitis)

MICROBIOLOGIC PROPERTIES

Streptococci are **Gram-positive cocci in chains**. *S. pyogenes* yield small colonies with **clear, sharp β-hemolysis on blood agar** culture (colonies are surrounded by a clear zone indicating complete lysis of red blood cells; Fig. 3-2). **Throat culture** is the standard for diagnosing streptococcal pharyngitis, and sheep blood agar incubated up to 48 hours is the culture medium of choice. The bacteria are **catalase negative** and **bacitracin sensitive** in a diagnostic disc susceptibility test. *S. pyogenes* is commonly known as **group A β-hemolytic streptococcus** (GABHS) for the presence of the cell surface **group A carbohydrate antigen**. RADT, a rapid (dipstick-based) test, which detects the presence of the cell-wall carbohydrate antigen in the throat swab, can be completed at the bedside in minutes. A positive RADT confirms the diagnosis, whereas a negative RADT requires confirmation with culture results.

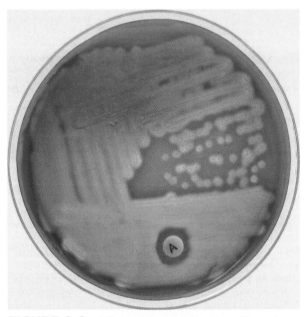

FIGURE 3-2 Culture of *Streptococcus pyogenes*. Note colonies are surrounded by a clear zone indicating complete lysis of red blood cells (β-hemolysis); *S. pyogenes* colonies are sensitive to **bacitracin** ("A" disc) in a diagnostic susceptibility test, shown here. *(Courtesy of Lisa Forrest, Department of Microbiology, University of Texas Southwestern Medical Center, Dallas, TX.)*

EPIDEMIOLOGY

Streptococcal pharyngitis ("strep throat") occurs worldwide, and in the temperate regions it occurs during the cooler months. Large **respiratory droplets** transmit *S. pyogenes* from **person to person.** Crowded households, schools, dormitories, and military barracks enhance the spread of infection. This patient may have acquired the illness by inhaling respiratory droplets from a sick classmate. Streptococcal pharyngitis can affect persons of all ages, with **a peak incidence in children aged 5 to 15 years.**

PATHOGENESIS

The acute infection of the oropharynx begins with the acquisition of large respiratory droplets. Several virulence factors (e.g., lipoteichoic acid, matrix-binding proteins that bind fibronectin, hyaluronate capsule, and M protein) mediate adherence to oral epithelial cells. The **extracellular pyogenic bacteria** directly invade the pharyngeal mucosa. PMNs are recruited to fight the invading organisms. The **M protein**, which is **a major virulence factor** and cell surface antigen (> **100 serotypes**), itself has an important **antiphago-cytic** effect by interfering with opsonization via the alternative complement pathway. Equipped with other antiphagocytic factors (such as hyaluronic acid capsule and a C5a peptidase), bacteria thwart destruction by phagocytosis (and evade killing by PMNs) in the course of early infection. Over-reacting local pyogenic, inflammatory, and cytokine responses cause symptoms and signs of "sore throat." Some strains of *S. pyogenes* also elaborate extracellular products such as streptolysin O, which is a potent pore-forming cytolysin, and streptolysin S, which is a hemolysin (also a cytolysin). These virulence factors play an important role in the invasion of bacteria in the adjacent tissue planes, leading to abscesses. Antibody to M protein facilitates host defenses against *S. pyogenes*. Long-term **type-specific humoral immunity** follows infection due to one M type, but **reinfection** with another M type **is common.**

TREATMENT

S. pyogenes is highly sensitive to penicillins and cephalosporins. No resistance to these antibiotics has been observed. A 10-day course of penicillin is required to eliminate throat carriage, but a shorter duration of other antibiotics may be just as effective. An alternative in penicillin-allergic patients is erythromycin. The main goal of treatment is to prevent postinfection sequelae (e.g., acute rheumatic fever [ARF]; see Table 3-2). Therapy for streptococcal infections initiated within 9 days of the onset of symptoms can prevent ARF. Antibiotic therapy affects only marginally the course of the streptococcal pharyngitis itself.

OUTCOME

The patient responded well to therapy using oral penicillin and recovered completely without any infection-associated complications.

PREVENTION

Hand washing is critical to preventing transmission of *S. pyogenes*. There is no vaccine against streptococcal diseases, as yet. The target for vaccine development is the surface M protein. The development of a vaccine has been hampered by the hypervariability of the M protein with limited cross protection. An octavalent group A streptococcal vaccine is in clinical trials.

TABLE 3–2   Other Clinical Features of the Upper Respiratory Infections due to *Streptococcus pyogenes*

Syndromes	Clinical Features
Abscess (pyogenic complication)	Can result from direct extension of the bacteria into adjacent structures and spread through the tissue planes (e.g., **peritonsillar abscess**, otitis media, sinusitis, mastoiditis, pneumonia).
Scarlet fever (toxigenic complication)	A characteristic systemic rash appears on the second day of pharyngitis; rash begins on the trunk and spreads outward, sparing the palms and soles. Other features include circumpolar pallor and red papillae on an infected patient's tongue (**strawberry tongue**).
	The **erythrogenic toxin** (the cause of the rash) is produced by certain strains of *Streptococcus pyogenes*, which have undergone lysogeny by a bacteriophage carrying the gene for this exotoxin. Cutaneous manifestations of scarlet fever result from delayed-type hypersensitivity to the toxin.
Acute rheumatic fever (ARF) and rheumatic heart disease (RHD) (immunologic complication)	ARF is a multisystem disease that affects many different organs. The major manifestations are **carditis, polyarthritis, chorea, erythema marginatum,** and subcutaneous nodules. Minor manifestations include fever, arthralgias, and an elevated ESR.
	Recurrent streptococcal infections may consequently cause **rheumatic heart disease** (late sequelae), which affects mitral and aortic heart valves. Histopathology may reveal Aschoff bodies (granuloma and giant cells) and Anitschkow cells (activated histiocytes).
	ARF and RHD are caused by **type II hypersensitivity reaction** due to **molecular mimicry** (cross reactivity). *S. pyogenes* may trigger an immune response against myocardial sarcolemma. M-protein antigens are immunologically similar to the cardiac antigens but differ sufficiently to induce an immune response when presented to T cells. As a result, the tolerance to autoantigens breaks down, and the pathogen-specific immune response that is generated cross reacts with cardiac tissue, causing **complement** (membrane attack complex)-**mediated damage (cytotoxicity) to heart valves.**
	A diagnosis of ARF is based on the supporting **evidence of a preceding streptococcal infection** (e.g., elevated antistreptolysin O antibodies).

FURTHER READING

Bisno AL: Acute pharyngitis. N Engl J Med 344:205, 2001.

Bisno AL: Group A streptococcal infections and acute rheumatic fever. N Engl J Med 325:783, 1991.

Bisno AL, Stevens DL: *Streptococcus pyogenes*. In: Mandell GL, Bennett JE, Dolin R (eds): Mandell, Douglas, and Bennett's Principles and Practice of Infectious Disease. Vol 2. 5th ed. New York, Churchill Livingstone, 2000, p 2103.

Cunningham MW: Pathogenesis of group A streptococcal infections. Clin Microbiol Rev 13:470, 2000.

Horn DL, Zabriskie JB, Austrian R, et al: Why have group A streptococci remained susceptible to penicillin? Report on a symposium. Clin Infect Dis 26:341, 1998.

A 9-year-old girl presented with low-grade **fever, sore throat,** and malaise for 2 days. These symptoms developed 10 days after arriving at a summer camp operated by a religious group. Recently, she was noted to have a dry cough and difficulty breathing. Her parents then brought her to the emergency department.

The family had **emigrated from Ukraine** one year before. The **child's immunization status** could not be determined. She had been otherwise healthy.

PHYSICAL EXAMINATION

VS: T 38.9°C, P 140/min, R 45/min, BP 92/50 mmHg

PE: The patient was in severe distress; **respiratory stridor** was present; **exudative pharyngitis** (Fig. 4-1) and **bilateral cervical adenopathy** ("bull neck") were noted. A **yellowish, leathery, thick membrane** extending to the uvula and soft palate was also seen (see Fig. 4-1).

LABORATORY STUDIES

Blood

Hematocrit: 42%

WBC: 18,400/μL

Differential: 92% PMNs

Platelet count: 320,000/μL

Serum chemistries: BUN 30 mg/dL, creatinine 1.0 mg/dL, O₂ saturation 84%

Imaging

Chest x-ray was normal.

Diagnostic Work-Up

Table 4-1 lists the likely causes of illness (differential diagnosis). Investigational approach for delineating the etiology may include:

- Routine bacterial and **selective cultures** of throat swabs
- Rapid tests (if necessary):
  - Enzyme immunoassay (for streptococcal antigen)
  - PCR
- Monospot test

TABLE 4–1 Differential Diagnosis and Rationale for Inclusion (consideration)
Adenovirus
Anaerobes (oral flora)
Corynebacterium diphtheriae
Epstein-Barr virus (EBV)
Haemophilus influenzae type b (Hib)
Influenza (A, B, C)
Mycoplasma pneumoniae
Parainfluenza virus
Streptococcus pyogenes

Rationale: The high WBC and the acute distress of the child with normal chest x-ray make a bacterial etiology more likely. Hib (for epiglottitis in questionable childhood immunization) and anaerobes (for Vincent angina) are reasonable. Diphtheria should be suspected in potentially unvaccinated patients with membranous nasopharyngitis or obstructive laryngotracheitis who emigrated or returned recently from areas where the disease is endemic or who were in close contact with persons who returned recently from such areas. The last three could cause pharyngitis, but generally do not lead to the toxic appearance of the patient as presented.

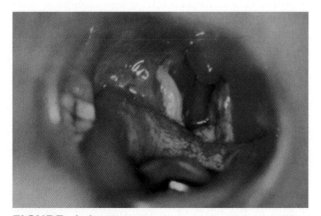

FIGURE 4-1 Exudative pharyngitis in this patient. Note marked swelling of the pharynx with a thick, adherent membrane. *(Courtesy of Dr. Paul Southern, Departments of Pathology and Medicine, University of Texas Southwestern Medical Center, Dallas, TX.)*

COURSE

A throat swab for a routine bacterial culture and one on special media was sent out. A streptococcal rapid antigen test and a Monospot test were performed. Because of her rapidly deteriorating condition, the patient was admitted to the hospital. She was placed on mechanical ventilation and empirical therapy. Culture of the throat swab on special isolation media containing tellurite grew dark colonies, which was significant.

ETIOLOGY

Corynebacterium diphtheriae (diphtheria)

MICROBIOLOGIC PROPERTIES

Corynebacteria are **small, club-shaped ("coryne") Gram-positive bacteria** with **metachromatic granules** (Fig. 4-2). They are noncapsulate, nonspore forming, nonmotile bacteria that grow aerobically on a differential medium containing potassium tellurite (colonies look grayish black). Most species are nonhemolytic on blood agar and are catalase positive.

Culture of pharyngeal exudates on selective tellurite media inhibits the growth of normal oral flora, facilitating the growth of *C. diphtheriae*, with characteristic gray-brown halo around the colonies. If diphtheria bacteria are isolated, they must be distinguished from normal diphtheroids, which commensally inhabit the pharynx and skin. In hosts colonized with diphtheroids (e.g., *Corynebacterium jeikium*), bacteria can be recovered from both skin (e.g., central

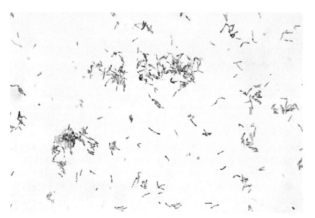

FIGURE 4-2 Gram stain of *Corynebacterium diphtheriae.* (*Courtesy of Centers for Disease Control and Prevention, Atlanta, GA.*)

line-associated infections in neutropenic patients) and mucosal surfaces. A modified plate test based on **immunoprecipitation on agar can** be performed to **detect the toxigenic strain** of *C. diphtheriae*, confirming the causality.

EPIDEMIOLOGY

Because of worldwide vaccination programs, diphtheria has been largely eradicated in the developed world. However, infections with *C. diphtheriae* still occur in countries with poor medical and socioeconomic conditions and in nonimmunized children or adults with waning immunity. *C. diphtheriae* is **transmitted via exposure to upper respiratory droplets and direct contact** with skin lesions.

PATHOGENESIS

C. diphtheriae colonizes the pharynx of healthy individuals. PMNs are recruited to fight the toxigenic *C. diphtheriae*. Toxin production occurs when *C. diphtheriae* is infected by a virus (phage) carrying the *tox* gene. The toxin, a potent 62-kd polypeptide exotoxin, is excreted locally and also goes into systemic circulation. The "B" fragment of the A-B toxin binds specifically to a cell-surface receptor (a membrane-bound form of the heparin-binding EGF-like growth factor). The **"A" fragment inhibits protein synthesis** intracellularly by ADP-ribosylation of elongation factor 2. Protein synthesis is blocked in the target cells locally in the posterior pharynx and in the end organs, resulting in cell death and tissue necrosis. Within the first few days of respiratory tract infection, a dense necrotic coagulum of organisms, epithelial cells, fibrin, leukocytes, and erythrocytes forms, advances, and becomes a **gray-brown adherent pseudomembrane.** Removal is difficult and reveals a bleeding, edematous submucosa.

TREATMENT

The use of **diphtheria antitoxin** (still made in horses) is critically important to neutralize the effect of circulating exotoxin. It is clinically efficacious if given within 4 days of the onset of illness. Antibiotic treatment (such as penicillin or erythromycin) is effective in reducing the severity of infection but has no effect on the outcome when effects of diphtheria toxin are present.

NOTE Untreated respiratory diphtheria may go on to develop complications, such as extension of the pseudomembrane into the larynx and trachea, causing airway obstruction and paralysis of the palate and hypopharynx as an early local effect of the toxin. Myocarditis with arrhythmias and circulatory collapse and recurrent laryngeal nerve palsy may develop as rare complications 2 to 10 weeks after the primary infection. These manifestations are due to circulating diphtheria toxin and represent cell and tissue death.

OUTCOME

The case patient received antitoxin and erythromycin for 2 weeks and progressively improved. She received the DTP vaccine after she recovered. Members of her family were also treated with erythromycin and received the DTP vaccine.

PREVENTION

The disease in the case patient could have been prevented by immunization. Active immunization with diphtheria toxoid (inactivated toxin) is administered to children 6 weeks to 7 years of age (DTP vaccine, including tetanus and pertussis toxins). Because the immunity wanes with time, adults should receive booster vaccinations every 10 years. Immunization does not prevent an asymptomatic carrier state. Clinical disease does not always produce protective immunity, so patients should also receive the vaccine after their illness has resolved. Antimicrobial prophylaxis for close contacts of patients with diphtheria is also recommended. Penicillin (IM) or erythromycin (PO) is recommended for all persons exposed to diphtheria.

FURTHER READING

Dittmann S, Wharton M, Vitek C, et al: Successful control of epidemic diphtheria in the states of the Former Union of Soviet Socialist Republics: Lessons learned. J Infect Dis 181:S10, 2000.

Farizo KM, Strebel PM, Chen RT, et al: Fatal respiratory disease due to *Corynebacterium diphtheriae*: Case report and review of guidelines for management, investigation, and control. Clin Infect Dis 16:59, 1993.

Golaz A, Hardy IR, Strebel P, et al: Epidemic diphtheria in the Newly Independent States of the Former Soviet Union: Implications for diphtheria control in the United States. J Infect Dis 181:S237, 2000.

A **7-year-old** girl presented with a 2-day history of fever, headache, earache, and **swelling** and **tenderness at the parotid** and **submaxillary areas.** She also found it **difficult to open her jaws** (to talk, eat, or swallow).

The child had been in good health. However, she received **only one dose of the usual childhood vaccinations** because the family had moved when she was four months old and she was lost to follow-up.

PHYSICAL EXAMINATION

VS: T 39.6°C

PE: Examination revealed enlargement of the parotid glands (**lymphedema of the neck and upward displacement of ears**); redness and swelling at the opening of the Wharton duct were also noted.

LABORATORY STUDIES

Blood

WBC: 4300/µL (**leukopenia**)

Differential: 71% lymphs (**lymphocytosis**), 9% monos

Serum chemistries: Serum amylase 300 U/L

Imaging

Chest x-ray: Not done (not indicated)

Diagnostic Work-Up

Table 5-1 lists the likely causes of illness (differential diagnosis). A swollen parotid gland, fever, and a negative history of immunization in a child has a limited differential diagnosis. If necessary, the diagnosis can be confirmed by virus culture or serology.

COURSE

The clinician made a diagnosis, prescribed analgesics for pain, and advised the mother to give a lot of fluid. The patient received the recommended supportive care at home.

TABLE 5–1 Differential Diagnosis and Rationale for Inclusion (consideration)
Coxsackievirus
Influenza virus
Mumps virus
Parainfluenza virus
Suppurative parotitis due to *Staphylococcus aureus* or other bacteria

Rationale: The lack of immunization and the clinical picture of the illness support a very narrow differential diagnosis. Viral etiologies may cause nonspecific constitutional symptoms, but parotitis is unusual, except in the case of mumps. Bacterial infection would cause a purulent infection that could extend to adjacent soft tissues.

ETIOLOGY

Mumps virus

MICROBIOLOGIC PROPERTIES

Mumps is an RNA virus and a member of the family **Paramyxoviridae**, which includes parainfluenza, measles (rubeola), and respiratory syncytial virus. The paramyxovirus genome is a **nonsegmented, negative sense, single-stranded RNA**. The helical ribonucleic-protein capsid has an envelope that contains two major surface glycoproteins: the hemagglutinin-neuraminidase and the fusion protein. Mumps virus can be cultured in a primary cell culture (e.g., rhesus monkey kidney cells). The appearance of **multinucleated giant cells** suggests a paramyxovirus. Hemadsorption-positive cultures can be confirmed by immunofluorescence staining.

EPIDEMIOLOGY

Few countries use mumps vaccine, so **mumps remains a common viral disease** in much of the world. Most infections in children younger than 2 years of age are subclinical. The greatest risk of vaccine-preventable infection has currently shifted toward older children, adolescents, and young adults. The virus is **acquired via exposure to respiratory secretions** or direct contact with the saliva of an infected person.

PATHOGENESIS

The mumps virus infects and **multiplies primarily in the epithelial cells of the oropharynx**. Secondary multiplication and viremia can extend into glands or nervous tissue. Postviremic **invasion occurs in the parotid and submaxillary glands**. The cell-mediated immune response and the cytokines cause symptoms of mumps. Gross pathology usually reveals enlargement of parotid glands. Affected glands show edema and necrosis **with perivascular mononuclear and lymphocytic infiltrates.**

TREATMENT

No specific antiviral treatment is available. **Management is supportive** with analgesics used for pain and swelling.

NOTE   Hematologic dissemination of the blood-borne virus may lead to complications such as **meningitis,** encephalitis, and pancreatitis (which were absent in the case patient). Other abnormalities in male patients include ductal obstruction and testicular interstitial edema (orchitis). Complications are treated based on symptoms and signs.

OUTCOME

The patient recovered gradually over the next 7 days without any complications.

PREVENTION

The illness in this patient could have been prevented by childhood immunization. Mumps vaccine, a component of *Mumps-Measles-Rubella* (MMR), consists of a **live, attenuated** viruses. MMR vaccine is generally given at 15 months of age and again at entrance into school. A single dose of MMR confers a 90% protective efficacy. About 10% of people receiving a single dose of the vaccine are not immunized against mumps virus.

FURTHER READING

Caplan CE: Mumps in the era of vaccines. Can Med Assoc J 160:865, 1999.
Chaiken BP, Williams NM, Preblud SR, et al: The effect of a school entry law on mumps activity in a school district. JAMA 257:2455, 1987.

Wharton M, Cochi SL, Williams WW: Measles, mumps, and rubella vaccines. Infect Dis Clin North Am 4:47, 1990.

On a cold day in December, a 16-month-old boy was brought to a clinic for **runny nose, hoarseness, barking cough,** and **a low-grade fever.** According to his mother, the child had developed the barking cough the night before. His breathing was forced and noisy, especially with inspiration.

He had been healthy and received **immunizations up to date.** There was no family history of note.

PHYSICAL EXAMINATION

VS: T 38.1°C, P138/min, R 28/min, BP 102/58 mmHg

PE: The patient was obviously distressed and had a runny nose, nasal congestion, sore throat, and cough. A hoarse cry, with intermittent stridor, and a **harsh, brassy, bark-like cough** were also noted. Suprasternal retractions were noted.

LABORATORY STUDIES

Blood

Not warranted for this presentation.

Imaging

Chest x-ray revealed **visible upper airway narrowing** (i.e., a steeple sign). (NOTE: X-rays of the neck were ordered to rule out epiglottitis, which is a more serious condition.)

Diagnostic Work-Up

Table 6-1 lists the likely causes of illness (differential diagnosis). The case presentation is, however, a clinical diagnosis. Further work-up for a specific clinical diagnosis is usually not pursued. If uncertain, the following tests can be performed:

- **Rapid antigen detection** for respiratory syncytial virus (RSV), or influenza, parainfluenza, and adenovirus
- **Viral cultures** from the nasopharyngeal swab

TABLE 6-1 Differential Diagnosis and Rationale for Inclusion (consideration)
Adenovirus
Corynebacterium diphtheriae
Haemophilus influenzae type b (Hib)
Influenza A and B
Mycoplasma pneumoniae
Parainfluenza viruses type 1 > type 3 > type 2 > type 4
Respiratory syncytial virus (RSV)
Superinfection with *S. aureus, S. pneumoniae, Moraxella catarrhalis* associated with bacterial tracheitis

Rationale: The above pathogens may cause tracheobronchitis. When this clinical presentation occurs, parainfluenza and RSV viruses should be considered. Influenza will cause the patient to be sicker than this patient was and usually have a higher fever. Adenovirus usually causes upper respiratory infection, but not with such a prominent cough. The main goal is to differentiate this presentation from bacterial tracheitis and acute epiglottitis, caused by Hib, where children are much sicker, present with drooling, and often have impending respiratory failure. *C. diphtheriae* and *M. pneumoniae* are not commonly seen in this age group.

COURSE

The pediatrician made a clinical diagnosis of croup. He administered a glucocorticoid using a nebulizer. He advised the infant's parents to take him home and call back in several hours if he did not respond to the treatment.

ETIOLOGY

Parainfluenza virus (PIV) or one of the other respiratory viruses, including RSV

NOTE The discussion that follows focuses on PIV.

MICROBIOLOGIC PROPERTIES

PIV is a paramyxovirus that has an enveloped virion (Fig. 6-1) with nonsegmented, negative sense, single-stranded RNA genome. The envelope contains **hemagglutinin** (H) and fusion protein (F). The **virus has four antigenic types** (1-4). It is a **cultivable** agent; cytopathic effects are not produced by most PIV strains. Infected cells produce **syncytia (cell-to-cell fusion)** and can be detected by **heme absorption** or direct fluorescence antibody staining.

EPIDEMIOLOGY

The disease occurs worldwide. Croup typically occurs in late fall and early winter in the temperate climates. PIV infection is acquired through **inhalation of infected respiratory droplets**. The case patient may have acquired the illness via exposure to respiratory droplets from an asymptomatic older sibling or an adult or by direct contact with respiratory secretions. The patient, being younger than 2 years of age, was susceptible to this infection. The incidence of croup decreases substantially after the sixth year.

PATHOGENESIS

The initial portal of entry of PIV is the nasopharynx. The virus initiates infection by attaching to the ciliated epithelial cells of these areas. It attaches to host cells via its H protein, which specifically combines with neuraminic acid receptors on the host cells. Subsequently, the F protein mediates virion entry into the cell via fusion of the envelope with the host cell membrane. The F protein is also responsible for cell-to-cell spread, manifested morphologically in cell cultures as syncytium. Viral replication occurs, and a local nonspecific host response (e.g., interferon) produces coryza.

The target for pathology primarily involves the larynx and may extend into the trachea and bronchi. Infection causes mainly **inflammatory changes in the superficial mucous membrane**, resulting in loss of cilia, cell damage, and edema, leading to significant airway compromise due to an already narrowed diameter of the airway, especially at the level of the cricoid cartilage. It is the narrowest part of a child's main airways and cannot expand outward. The narrowing worsens during inspiration and results in inspiratory stridor. In addition to luminal narrowing, **edema of the vocal cords and subglottic larynx leads to hoarseness, and to the characteristic bark-like cough**. The humoral response usually provides protective long-term immunity to subsequent infections with PIV.

TREATMENT

No specific antiviral therapy is available. Treatment may include humidification, epinephrine administered via a nebulizer, and glucocorticoids. Systemic steroids may provide significant benefit. Nursing the patient in a cool, humidified atmosphere is advised to prevent further distress.

NOTE Children with croup who have a bacterial super-infection (bacterial tracheitis) may reveal a necrotizing inflammatory reaction with mucosal ulcerations and microabscesses.

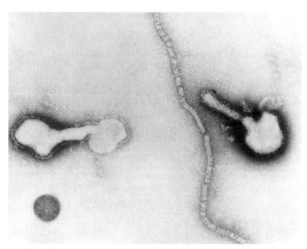

FIGURE 6-1 Electron micrograph of parainfluenza virus. Two intact particles and a free filamentous nucleocapsid are seen. *(Courtesy of Centers for Disease Control and Prevention, Atlanta, GA.)*

OUTCOME

Back at home, the patient's symptoms subsided, and the child was able to sleep. Over the next few days, all of the patient's symptoms gradually resolved.

PREVENTION

There is no vaccine to prevent this communicable illness. Good hand washing and cleanliness can help decrease transmission from an infected patient, particularly at day care centers or even in the home setting.

FURTHER READING

Henrickson KJ: Parainfluenza viruses. Clin Microbiol Rev 16:242, 2003.

Jones R, Santos JI, Overall JC: Bacterial tracheitis. JAMA 242:721, 1979.

Reed G, Jewett PH, Thompson J, et al: Epidemiology and clinical impact of parainfluenza virus infections in otherwise healthy infants and young children <5 years old. J Infect Dis 175:807, 1997.

A 2-month-old white girl suffered from **cough for more than 2 weeks.** The patient's mother became concerned when the child **turned blue after a series of coughing spells** that ended with vomiting. She brought her to a pediatrician.

The patient had **not yet received any vaccinations.** She had been healthy without any underlying medical problems.

PHYSICAL EXAMINATION

VS: Normal

PE: **Paroxysmal cough, whoop,** and **post-tussive vomiting** were noted during examination. Conjunctival hemorrhages and facial petechiae were also noted. There were no signs of lower respiratory illness.

LABORATORY STUDIES

Blood

WBC: 32,000/μL
Differential: 14% PMNs, 78% lymphs, 8% monos

Imaging

Chest x-ray **did not reveal any pulmonary infiltrates.**

Diagnostic Work-Up

Table 7-1 lists the likely causes of illness (differential diagnosis). Epidemiologic and clinical criteria of paroxysmal cough for 14 (CDC) or 21 (WHO) days are used for making the clinical diagnosis of pertussis because the presentation and duration of symptoms are so typical. However, laboratory tests may be useful in young infants and in atypical cases. Leukocytosis with absolute lymphocytosis occurs during the late catarrhal and paroxysmal phases and is a nonspecific finding. Investigational approach to delineating the etiology may include

- **Direct fluorescence antibody (DFA)** testing of nasopharyngeal specimens
- **Culture** of posterior **nasopharyngeal secretions**
- **PCR** testing of nasopharyngeal swabs or aspirates

COURSE

The patient's mother was told that the patient needed to be hospitalized for supportive treatment and

TABLE 7–1 Differential Diagnosis and Rationale for Inclusion (consideration)
Adenoviruses
Bacterial causes of bronchitis
Non type b *Haemophilus influenzae*
Mycoplasma pneumoniae
Chlamydophila pneumoniae
Bordetella parapertussis
Bordetella pertussis
Parainfluenza viruses
Respiratory syncytial virus (RSV)

Rationale: All of the above pathogens may cause tracheobronchitis. The classic presentation of paroxysmal cough and whooping and the prolonged duration of symptoms in an unimmunized patient suggest *Bordetella* species as the most likely agent. Other upper respiratory illnesses (viral or bacterial) generally lead to cough without the paroxysmal character.

observation. The patient was monitored for heart rate, respiratory rate, and oxygen saturation in relation to coughing paroxysms. Culture of a nasopharyngeal specimen was negative, but PCR confirmed the clinical diagnosis.

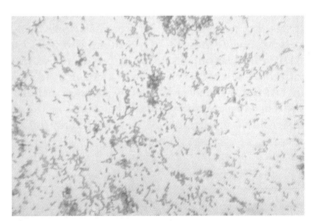

FIGURE 7-1 Gram stain if culture positive. *(Courtesy of Dr. Paul Southern, Departments of Pathology and Medicine, University of Texas Southwestern Medical Center, Dallas, TX.)*

ETIOLOGY

Bordetella pertussis (pertussis)

MICROBIOLOGIC PROPERTIES

B. pertussis are **small, nonmotile, aerobic Gram-negative rods** (Fig. 7-1). They are **nutritionally fastidious bacteria, requiring special growth medium for isolation**. *B. pertussis* produces multiple antigenic products, including pertussis toxin, filamentous hemagglutinin, agglutinogens, adenyl cyclase, and tracheal cytotoxin.

EPIDEMIOLOGY

Pertussis, although a vaccine-preventable disease, remains a worldwide problem. Adults and adolescents are an important reservoir for this pathogen and are often the source of infection for infants. Susceptibility of nonimmunized infants and young children is universal. **Transmission** most commonly **occurs by the respiratory route through contact with respiratory droplets.** This patient may have acquired the infection via exposure to respiratory droplets from an older sibling or an adult with coughing illness.

PATHOGENESIS

The inhaled bacteria attach to respiratory cilia, starting initially in the nasopharynx and ending up in the bronchi and bronchioles. A protein on the pili, called filamentous hemagglutinin, mediates attachment. These bacteria are noninvasive, but they produce toxins that paralyze the cilia. **Pertussis toxin irreversibly inactivates the G_i-protein complex via ADP ribosylation**, resulting in prolonged **stimulation of adenyl cyclase** and a consequent **rise in cAMP**. Elevated cAMP levels **increase cellular protein kinase activity**. Two other toxins that also inflict local damage are (1) tracheal cytotoxin (a peptidoglycan fragment that kills ciliated cells), and (2) hemolysin (which kills mucosal epithelial cells). A mucopurulosanguineous exudate is formed in the respiratory tract, compromising the small airways (especially those of infants) and predisposing the affected individual to paroxysmal coughs. Host response is mounted in the late acute phase of infection. This immunity does not last very long and may explain why adults can become symptomatic and why infants are at high risk (titer of maternal antibodies is low).

TREATMENT

Supportive care, which includes suctioning to remove mucus and the use of pressurized oxygen, is of primary importance. A macrolide (e.g., erythromycin) given before the onset of paroxysmal coughs may abort or eliminate pertussis.

NOTE Infants born prematurely and patients with underlying cardiac, pulmonary, neuromuscular, or neurologic disease are at high risk for complications of pertussis (e.g., pneumonia, seizures, encephalopathy, or death).

OUTCOME

The patient received supportive care for the management of coughing, paroxysms, and apnea. She gradually recovered during the course of the one-week hospital stay.

PREVENTION

The case patient had not yet been vaccinated. Active immunization against *B. pertussis* infection is usually effective for prevention of this illness. **Acellular pertussis vaccine** in combination with diphtheria and tetanus toxoids (DTaP) is available in the United States. The vaccine should be administered to all children aged 6 weeks to 6 years.

FURTHER READING

Bass JW, Stephenson SR: The return of pertussis. Pediatr Infect Dis J 6:141, 1987.

He Q, Viljanen MK, Arvilommi H: Whooping cough caused by *Bordetella pertussis* and *Bordetella parapertussis* in an immunized population. JAMA 280:635, 1998.

Wright SW, Edwards KM, Decker MD: Pertussis infection in adults with persistent cough. JAMA 273:1044, 1995.

A **64-year-old** man presented to a clinic with complaints of low-grade **fever, productive cough of yellow-green sputum,** and **worsening** of his chronic **shortness of breath** for several days. He had recovered from a mild cold just before the current symptoms began.

He had a **long history of chronic obstructive pulmonary disease (COPD)** and had been on home oxygen for the past 2 years. He had been taking his inhalers as directed.

PHYSICAL EXAMINATION

VS: T 38.1°C, P 108/min, R 28/min, BP 140/72 mmHg

PE: Thin male in moderate respiratory distress; lung exam revealed diffuse wheezes and rhonchi.

LABORATORY STUDIES

Blood

Hematocrit: 48%

WBC: 10,400/μL

Differential: Normal

Blood gases: pH 7.38, pCO_2 50 mmHg, pO_2 58 mmHg on room air

Serum chemistries: Normal

Imaging

Chest x-ray revealed hyperinflated lungs but no infiltrates.

TABLE 8–1 Differential Diagnosis and Rationale for Inclusion (consideration)
Adenovirus
Haemophilus influenzae
Influenza A or B virus
Moraxella catarrhalis
Parainfluenza virus
Streptococcus pneumoniae

Rationale: A clinical diagnosis of acute exacerbation of chronic bronchitis (AECB) should be considered. While many organisms can cause acute exacerbation of COPD, nontypable *H. influenzae* and *S. pneumoniae* make up a large majority of cases. *M. catarrhalis* is also seen and is indistinguishable from *H. influenzae*. *S. pneumoniae* is more likely to cause pneumonia. The respiratory viruses (e.g., influenza and adenovirus) are less likely to be productive of sputum.

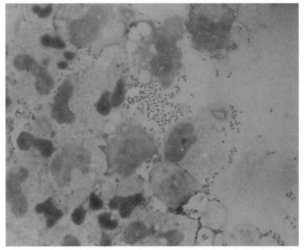

A

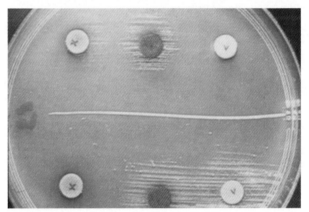

B

FIGURE 8-1 *A*, Gram stain of sputum. *B*, **Top:** Satellite colonies of the clinical isolate around XV (no growth around either X or V alone), showing dependence of these bacteria on both X and V factors. **Bottom:** Satellite colonies of the clinical isolate around XV and V, showing the dependence on X factor. (*A, Courtesy of Dr. Dominick Cavuoti, Department of Pathology, University of Texas Southwestern Medical Center, Dallas, TX; B, courtesy of Dr. Paul Southern, Departments of Pathology and Medicine, University of Texas Southwestern Medical Center, Dallas, TX.*)

Diagnostic Work-Up

Pneumonia can be reasonably excluded based on the absence of infiltrates on chest x-ray. Table 8-1 lists the likely causes of current illness (differential diagnosis). Investigational approach to delineating the etiology of the current episode may include

- **Gram stain** and **cultures of sputum**
- **Blood cultures** (Blood cultures are often negative in an acute exacerbation of chronic bronchitis).

COURSE

The patient was admitted, and sputum was obtained for Gram stain and cultures. Antibiotics were started with coverage against bacterial infections. Sputum culture yielded a significant isolate on chocolate agar (not on blood agar) that was biochemically identified and diagnostic.

ETIOLOGY

Haemophilus influenzae (acute exacerbation of chronic bronchitis ([AECB]))

MICROBIOLOGIC PROPERTIES

Haemophilus species are **small, Gram-negative coccobacillary rods** (Fig. 8-1*A*). The species implicated in human infections include *H. influenzae, H. parainfluenzae, H. haemolyticus,* and *H. parahaemolyticus.* These species are differentiated primarily by their requirements for X (heme) and V (nicotinamide adenine dinucleotide) factors. Species designated *para-* require V factor but not X factor for growth, whereas the others require either **X and V (*H. influenzae*)** or X only (see Fig. 8-1*B*). There are two different groups of *H. influenzae*, as determined by the presence or absence of a **polysaccharide capsule**. Encapsulated strains belong to group 1 and are subdivided into 6 subtypes (a, b, c, d, e, and f), based on the particular polysaccharide capsule. Unencapsulated strains belong to group 2 and are called nontypable *H. influenzae* (NTHi). All *H. influenzae* contain lipopolysaccharide (LPS) in the cell wall. ***H. influenzae* type b (Hib)** has a **polyribitol phosphate capsule** and **in the past was the major (invasive) pathogen** of this group (no association with the case; see Table 8-2 for a brief understanding of diseases caused by Hib). Group 2

(unencapsulated) strains are referred to as **nontypable *H. influenzae*** (NTHi) because they cannot be serotyped with antisera to the polysaccharide capsules. Gram-negative cell wall contains LPS with embedded endotoxin that is activated in body fluid when released in bacteremia and with other invasive infections.

TABLE 8–2   Other Clinical Features of *Haemophilus influenzae* Infections

Syndromes	Clinical Features
Otitis media/sinusitis	NTHi and other non-type b *Haemophilus influenzae* can easily **spread from the nasopharynx into the middle ear to cause otitis media or into the sinuses to cause sinusitis.**
Epiglottitis	Hib is virtually the exclusive cause of epiglottitis, and since the advent of an effective vaccine, this entity has been almost eliminated. The presentation is usually in children 2-7 years old, with sudden onset of fever, dyspnea, tachypnea, drooling, and inspiratory stridor. Respiratory failure often occurs without prompt therapy.
Meningitis	Before the availability of the Hib conjugate vaccine in the United States and other industrialized countries, more than one half of Hib patients presented as meningitis with fever, headache, and stiff neck. From its ecologic niche in the nasopharynx, Hib can penetrate through the mucosa into the blood stream, from which it can then penetrate the blood-brain barrier to reach the meninges. Once the organisms reach the cerebrospinal fluid, LPS cause intense inflammation, resulting in meningitis. It is fatal if untreated.
Other serious diseases	Hib can cause other systemic or invasive disease, including bacteremia (endotoxic shock), **pneumonia,** cellulitis, and septic arthritis, primarily affecting children younger than 2 years of age if they are unvaccinated. In some developing countries, NTHi are also associated with bacterial pneumonia in infants and elderly patients.

EPIDEMIOLOGY

Haemophilus species are components of the normal flora of the human upper respiratory tract. Group 2 (NTHi) strains cause most *H. influenzae* disease in adults, and group 1 strains cause most disease in children. Transmission of all types of *H. influenzae* (including NTHi) occurs via **direct contact with respiratory droplets** from a nasopharyngeal carrier. Adults with **COPD** are **at risk**; those who continue heavy smoking are at greater risk for AECB. Hib remains a major cause of lower respiratory tract and central nervous system infections in infants and children in developing countries where the conjugate vaccine is not widely used.

PATHOGENESIS

In the airways of adult patients with COPD are commonly found three bacterial species: *H. influenzae* (NTHi), *S. pneumonia*, and *Moraxella catarrhalis*. Increased numbers and new strains of these bacteria are associated with acute COPD exacerbations. Little is known about the virulence factors of NTHi or about immunity to disease caused by them. IgA1 protease has been associated with NTHi strains isolated from sputum and other sterile sites in patients with symptomatic disease. **IgA protease** allows these organisms to colonize the respiratory mucosa. Although NTHi have low pathogenic potential, they can spread from the nasopharynx into the bronchi to cause acute exacerbations of COPD, particularly chronic bronchitis, a condition associated with **excessive tracheobronchial mucus production** sufficient to cause cough with expectoration for a period of at least 3 months over a period of 2 years or more. Chronic cough and sputum production are associated with hypertrophy of the mucus-production glands in the mucosa of large airways.

TREATMENT

When COPD patients have an acute exacerbation, especially when they experience an increased volume of purulent sputum, antibiotic therapy is associated with improved outcomes. There are a number of effective antimicrobial agents for both Hib and NTHi infections, including the newer macrolides (clarithromycin and azithromycin) and cephalosporins. For invasive disease, particularly due to Hib (e.g., meningitis; see Table 8-2), parenteral administration of a third-generation cephalosporin (e.g., cefotaxime or ceftriaxone) is recommended.

OUTCOME

The patient was treated with cefotaxime and responded within 48 hours with improvement in cough and dyspnea. He was discharged home after 4 days to complete a 10-day course of oral cefuroxime.

PREVENTION

There is no vaccine to protect against NTHi disease, which remains very common in certain patient populations. However, there is a highly efficacious conjugate Hib vaccine for prevention of *H. influenzae* type b disease. **This vaccine has been responsible for a dramatic reduction in invasive disease due to this pathogen** (particularly meningitis in children), as well as for facilitating elimination of nasopharyngeal carriage of Hib in the developed countries. Mass immunization protocols are still needed to protect all children in developing countries against the life-threatening infections due to *H. influenzae* type b.

FURTHER READING

Adams WG, Deaver KA, Cochi SL, et al: Decline of childhood *Haemophilus influenzae* type b disease in the Hib vaccine era. JAMA 269:221, 1993.

Barnes PJ: Chronic obstructive pulmonary disease. N Engl J Med 343:269, 2000.

Everett ED, Rham AE, Jr, Adaniya R, et al: *Haemophilus influenzae* pneumonia in adults. JAMA 238:319, 1977.

Sethi S, Murphy TF: Bacterial infection in chronic obstructive pulmonary disease in 2000: A state-of-the-art review. Clin Microbiol Rev 14:336, 2001.

Vitovski S, Dunkin KT, Howard AJ, et al: Nontypable *Haemophilus influenzae* in carriage and disease: A difference in IgA1 protease activity levels. JAMA 287:1699, 2002.

A **67-year-old** white man was brought by his wife to the ED for the **abrupt onset of shaking chills, high fever,** and pain on the right side of his chest that began the prior evening. His wife reported that in the last 24 hours, he had experienced shortness of breath and a cough that was productive of **rust-colored sputum.**

The patient was **diabetic** and **smoked two packs of cigarettes** per day. He had been a chain smoker for the last 20 years. He had not sought medical care in the past and had **not received any vaccinations** in the last 20 years.

C A S E
9

PHYSICAL EXAMINATION

VS: T 39.6°C, P 130/min, R 32/min, BP 159/77 mmHg

PE: Examination revealed an ill-appearing, confused man in moderate respiratory distress. **Dullness to percussion over the right upper thorax,** associated with increased fremitus, was noted. Auscultation revealed bronchial breath sounds and crackles over this area.

LABORATORY STUDIES

Blood

Hemoglobin: 13 g/dL
WBC: 22,400/μL

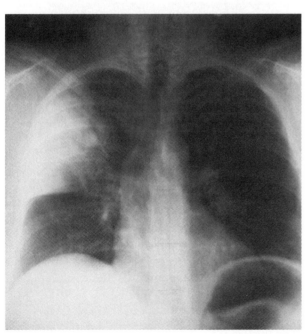

FIGURE 9-1 An admission chest film showing dense consolidation of the right upper lobe in this patient. (*Courtesy of Dr. Anthony DalNogare, Department of Medicine, University of Texas Southwestern Medical Center, Dallas, TX.*)

Differential: 65% PMNs, **24% bands**

Blood gases: pH 7.42, pO_2 58 mmHg, O_2 saturation 86% (on room air)

Imaging

A chest x-ray revealed **consolidation of the right upper lobe** (Fig. 9-1).

Diagnostic Work-Up

Chest x-ray is essential for confirming pneumonia and helps determine the location of pulmonary involvement (e.g., lobar, bronchial, interstitial), the nature of lung pathology (e.g., cavitary, necrotizing), whether a pleural effusion is present, and whether there is hilar adenopathy. Table 9-1 lists the likely causes of illness (differential diagnosis).

TABLE 9–1   Differential Diagnosis and Rationale for Inclusion (consideration)
Chlamydophila pneumoniae
Haemophilus influenzae
Klebsiella pneumoniae
Legionella pneumophila
Moraxella catarrhalis
Mycoplasma pneumoniae
Staphylococcus aureus
Streptococcus pneumoniae

Rationale: The clinical presentation of acute onset and severity of symptoms and the physical findings suggest typical pneumonia of bacterial origin. *H. influenzae* and *M. catarrhalis* are usually seen in patients with COPD, and *K. pneumoniae* is seen in alcoholics. *S. aureus* is a likely pathogen in postinfluenza pneumonia. *S. pneumoniae* often produces a sudden onset of symptoms and is the most common cause. The three agents *L. pneumophila*, *M. pneumoniae*, and *C. pneumoniae* are the least likely causes for the described presentation, but they should not be ruled out until the patient's pneumonia is determined to be atypical (see Case 11 for definition of atypical pneumonia).

Investigational approach to delineating the etiology may include

- **Gram stain of sputum**
- **Cultures of sputum** and **blood**
- In failed investigation, *Legionella* urinary antigen and **serology** for *Mycoplasma pneumoniae* and *Chlamydophila pneumoniae* (atypical pathogens)

COURSE

The patient was admitted to the hospital. Two sets of blood cultures and sputum were sent for microbiology investigation. He was started on empirical antibiotic therapy using erythromycin and cefotaxime. Gram smear of sputum revealed numerous PMNs and Gram-positive, lancet-shaped diplococci. Cultures of sputum and blood confirmed the microbiologic diagnosis.

> **NOTE** Empirical guidelines for treatment of community-acquired pneumonia are aimed at *Streptococcus pneumoniae, Haemophilus influenzae,* and atypical pathogens (*Chlamydia, Mycoplasma,* and *Legionella*).

ETIOLOGY

Streptococcus pneumoniae (pneumococcal pneumonia)

MICROBIOLOGIC PROPERTIES

S. pneumoniae (also known as pneumococci) are **Gram-positive, lancet-shaped diplococci** (Fig. 9-2*A*). In Gram stain examination of sputum, it is important to make sure sputum samples represent lower rather than upper respiratory secretions. Criteria for significant sputum Gram smear include less than 10 squamous epithelial cells per low-power field (10×), greater than 25 PMNs, and the presence of a single or predominant organism. Sputum meeting these criteria is cultured. Growth of *S. pneumoniae* on conventional sheep blood agar media yields α-**hemolytic colonies**. The organisms are **catalase negative** and **Optochin sensitive** (see Fig. 9-2*B*; **viridans streptococci** are α-hemolytic but **Optochin resistant**). They have a thick **capsular polysaccharide** (CPS) outer layer. Eighty-four different serotypes have been identified based on the antigenicity of CPS, with varying degrees of pathogenicity; **23 serotypes** cause 85% to 90% of pneumococcal infections in the United States. Antibody against one capsular type does not cross

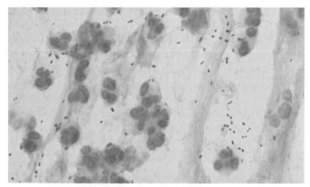

A

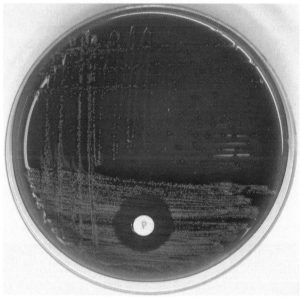

B

FIGURE 9-2 *A,* Gram stain of *Streptococcus pneumoniae* from sputum. *B,* Culture of *Streptococcus pneumoniae* with Optochin disc susceptibility (a diagnostic test for culture confirmation of a clinical isolate). Note the zone of inhibition around Optochin ("P") disc of the α-hemolytic colonies. (*A, Courtesy of Dr. Paul Southern, Departments of Pathology and Medicine, University of Texas Southwestern Medical School, Dallas, TX; B, Courtesy of Lisa Forrest, Department of Microbiology, University of Texas Southwestern Medical Center, Dallas, TX.*)

react to another type, signifying very little cross protection among the strains.

EPIDEMIOLOGY

Colonization is most common during the winter months, under crowded conditions such as jails, and after viral respiratory infections. *S. pneumoniae* is the most common cause of community-acquired pneumonia (CAP), afflicting all ages. Colonizing bacteria are transmitted (person to person) to susceptible individuals via aerosolized droplets. Conditions predisposing to pneumococcal pneumonia, and to a poor prognosis, are **AIDS, asplenia, influenza, sickle cell disease, multiple myeloma, alcoholism, smoking, diabetes, hypogammaglobulinemia,** and **nephrotic syndrome.**

PATHOGENESIS

The multifactorial pathogenesis follows a sequence of steps:

1. **upper airway colonization**
2. **aspiration** of a large bacterial inoculum into the lower airways
3. **failure of normal host defenses** to clear the aspirated bacteria
4. **bacterial proliferation**
5. **inflammatory response** (lung pathology)

If colonization is to take place, bacteria must be able to adhere to upper airway epithelial cells. However, capsular serotype-specific secretory IgA antibody, produced by airway epithelial cells, prevents bacterial adherence and colonization. Bacteria-generated **IgA protease** degrades IgA antibody and favors mucosal colonization. Pneumococcal adhesins bind to GlcNac β1,3-Gal disaccharide groups on epithelial cell glycolipids; they also bind to platelet activating factor (PAF) receptors on epithelial cells. The colonized bacteria aspirate down into the lung of patients, whose mucus and cilia functions are altered by predisposing factors (e.g., smoking).

The **polysaccharide capsule, a major virulence factor, has antiphagocytic properties**. The encapsulated bacteria aspirated into the lung initially encounter alveolar macrophages in the alveolar space. There, both antibody (IgG) and complement (especially C3) are important for effective opsonization and phagocytic killing. Normally, bacteria would be opsonized by C3 and cleared via the alveolar macrophage C3b receptor. In a nonimmune individual with an early lower respiratory tract infection, however, bacterial cell-wall glycopeptides stimulate the recruitment of leukocytes into the lung (the presence of teichoic acid in a cell-wall component enhances its inflammatory activity) and initiate the coagulation cascade and stimulate the production of PAF (thus offering the bacteria a convenient anchor).

Pneumococci proliferate in the lung. A hemolysin (also known as pneumolysin), released during bacterial lysis, is active in pore formation and is cytotoxic to virtually every cell in the lung. Pneumolysin-producing strains cause more severe infections than do strains that do not produce pneumolysin. Pathology of lobar pneumonia progresses in the following four stages:

1. **Congestion,** the earliest stage of lobar pneumonia, is characterized by extensive serous exudation, vascular engorgement, and rapid bacterial proliferation (Fig. 9-3*A*).
2. **Red hepatization** reflects the liverlike appearance of the consolidated lung: airspaces are filled with PMNs, vascular congestion occurs, and extravasation of RBCs causes a reddish discoloration on gross examination (see Fig. 9-3*B*).
3. **Gray hepatization** results from an accumulation of fibrin, associated with inflammatory WBCs and RBCs in various stages of disintegration; alveolar spaces are packed with inflammatory exudates (see Fig. 9-3*C*).
4. **Resolution** is characterized by resorption of the exudate.

Pneumococcal pneumonia is a classic airspace infection, with **intra-alveolar exudates** spreading rapidly within a lobe and through the pores of Kohn, until the **entire lobe is consolidated** (may involve the entire lung in severe pneumonia). The rapid interalveolar spread accounts for many of the clinical features, including chest pain and chest x-ray findings (lobar consolidation). The systemic inflammatory response, with the release of inflammatory cytokines such as TNF, causes high fever, chills, myalgias, and other systemic symptoms of pneumonia and complications.

TREATMENT

The treatment of choice for susceptible pneumococcal strains remains penicillin. In the United States (and in many other countries), some degree of resistance to penicillin is seen in more than 30% of clinical isolates. The prevalence of strains resistant to multiple classes of drugs is increasing, as well. Once an antibiotic susceptibility test of the clinical isolate is performed, therapy can be directed based on the susceptibility profile. For empirical therapy of community-acquired pneumonia (before the culture results are known), current guidelines suggest either a third-generation cephalosporin (cefotaxime or ceftriaxone) plus a

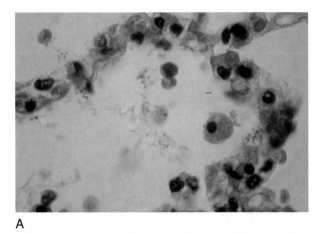

A

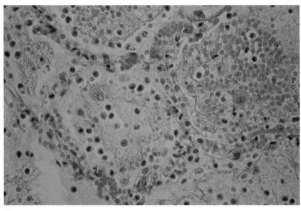

B

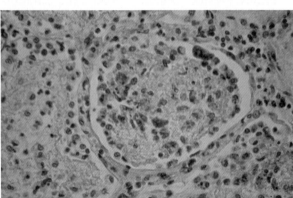

C

FIGURE 9-3 Lung pathology of pneumococcal pneumonia. *A*, Congestion. Note vascular engorgement (the capillaries are filled with RBCs), a few PMNs, and a small amount of intra-alveolar fluid. *B*, Red hepatization. Note exudate of RBCs, PMNs, and fibrin fills the alveolar spaces. *C*, Gray hepatization. Note the RBCs degrade and the alveolar exudate begins to organize. *(Courtesy of Dr. Anthony DalNogare, Department of Medicine, University of Texas Southwestern Medical Center, Dallas, TX.)*

macrolide or a newer quinolone (levofloxacin or gatifloxiacin). These regimens reasonably cover both resistant pneumococci and atypical organisms. For highly resistant strains of pneumococci, vancomycin can be used.

> **NOTE** Other diseases and complications caused by *S. pneumoniae* (e.g., meningitis; see Table 9-2) should be managed with appropriate choice of antibiotics and supportive care (when necessary).

OUTCOME

The patient improved dramatically over 48 hours. The clinical isolate was sensitive to penicillin, based on susceptibility testing. The patient's antibiotic regimen was switched to intravenous penicillin on the third day of hospitalization. After 7 days of intra-venous antibiotics, he was discharged home to complete an oral antibiotic regimen.

PREVENTION

Prevention of this disease in high-risk patients can be achieved by immunization. A polyvalent vaccine containing the capsular polysaccharides of the 23 most common pneumococcus types is administered. Administration of pneumococcal vaccine is indicated for people older than 60 years of age and other at-risk individuals listed earlier. Because of the high carriage rate of pneumococcus among attendees of child care centers, it has been recommended that children younger than 5 years of age who are attending child care centers be vaccinated with a heptavalent pneumo-coccal conjugate vaccine approved for pediatric use.

TABLE 9–2 Other Clinical Features of the *Streptococcus pneumoniae* Infections

Syndromes	Clinical Features
Acute otitis media	Very **common in young children.** Hearing impairment can result from recurrent otitis media.
Acute exacerbations of chronic bronchitis	Adults with chronic obstructive pulmonary disease may have chronic lower airway colonization with *S. pneumoniae*. An increased number of the colonizing bacteria may play a role in initiating an **acute exacerbation of chronic bronchitis.**
Empyema	Pneumonia can spread into the pleural space, resulting in an abscess (known as empyema), detected clinically by the presence of a pleural effusion on the chest radiograph. Requires drainage along with antibiotic therapy for cure.
Bacteremia	Defined clinically as a **positive blood culture;** bacteremia occurs in about 20% of cases of pneumococcal pneumonia. Loss of the normal intrapulmonary compartmentalization of the inflammatory response results in entry of pneumococci into the circulation. Subsequent **systemic inflammatory response is associated with cardiac failure and death.**
Endocarditis	Cardiac tissue is an unusual site of primary disseminated pneumococcal infection, with a predilection for the aortic valve.
Meningitis	*S. pneumoniae* causes meningitis in all age groups and is the **most common cause of meningitis in adults** (19-65 years of age). It is the second most common cause of meningitis in children 1 to 5 years of age. It is also the most serious pneumococcal infection (case fatality rate is approximately 21%) with or without lung pathology. **Neurologic sequelae** and/or learning disabilities (in children) can be seen as complications.

FURTHER READING

Artz AS, Ershler WB, Longo DL: Pneumococcal vaccination and revaccination of older adults. Clin Microbiol Rev 16:308, 2003.

Brown PD, Lerner SA: Community-acquired pneumonia. Lancet 352:1295, 1998.

Friedland IR, McCracken GH: Management of infections caused by antibiotic-resistant *Streptococcus pneumoniae*. N Engl J Med 331:377, 1994.

Giebink GS: The prevention of pneumococcal disease in children: Current concepts. N Engl J Med 345:1177, 2001.

Mandell LA, Bartlett JG, Craig WA, et al: Update of practice guidelines for management of community-acquired pneumonia. Clin Infect Dis 37:1405, 2003.

Tuomanen EI, Austrian R, Masure HR: Mechanisms of disease: Pathogenesis of pneumococcal infection. N Engl J Med 332:1280, 1995.

A **66-year-old** white man presented with a **cough, fever,** night sweats, and **chest pain.** He also noted a 12-lb weight loss over 3 weeks.

He was a **homeless** man, who admitted to drinking **2 quarts of vodka per day.** He vaguely remembered he was cough free a month ago, and his coughs had become progressively worse since then. In the last several days he had produced abundant, thick, tenacious, and **blood-tinged (currant jelly) sputum.**

PHYSICAL EXAMINATION

VS: T 39.7°C, P 110/min, R 34/min, BP 104/70 mmHg

PE: An ill-appearing man in soiled, torn clothes; he had rales and rhonchi at the right lung base and an enlarged liver with mild tenderness.

LABORATORY STUDIES

Blood

Hematocrit: 36%

Hemoglobin: 12 g/dL

WBC: 16,300/μL

Differential: 72% PMNs, 10% bands

Alcohol level: 185

Blood gases: pO_2 68 mmHg on room air

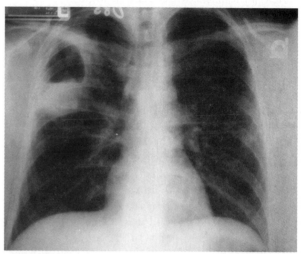

FIGURE 10-1 An admission chest film showing a cavitary right upper lobe infiltrate with an air-fluid level in this patient. *(Courtesy of Dr. Daniel Skiest, Department of Medicine, University of Texas Southwestern Medical Center, Dallas, TX.)*

Imaging

A chest x-ray revealed **right upper lobe infiltrate with cavitary lesion** (Fig. 10-1).

Diagnostic Work-Up

Table 10-1 lists the likely causes of illness (differential diagnosis). Investigational approach to delineating the etiology may include:

- **Blood cultures**
- **Sputum (Gram stain) examination** and **cultures**

COURSE

The patient was admitted to the hospital. Sputum was collected and sent for routine and acid-fast bacillus cultures, and two sets of blood were drawn for culture. Based on the initial sputum Gram stain that showed mixed microbes and many PMNs, he was initially treated with cefotaxime, erythromycin, and

TABLE 10–1 Differential Diagnosis and Rationale for Inclusion (consideration)
Haemophilus influenzae
Klebsiella pneumoniae
Legionella pneumophila
Mycobacterium tuberculosis
Staphylococcus aureus
Streptococcus pneumoniae

Rationale: A clinical diagnosis of pneumonia should be considered. The presence of cavitary lesions often implies an abscess with polymicrobial infection although TB is also an important consideration. *K. pneumoniae* is one of the more common causes of pneumonia in the homeless population (may be a part of the mixed anaerobic infection). Pneumococcal and staphylococcal pneumonias are commonly community acquired. The other causes listed above are less likely in this setting. Homelessness should generally prompt ruling out tuberculosis.

clindamycin. Within 24 hours he became hypotensive and was transferred to the ICU. Blood and sputum cultures yielded the diagnosis.

ETIOLOGY

Klebsiella pneumoniae (bacterial pneumonia)

MICROBIOLOGIC PROPERTIES

K. pneumoniae is a member of the family of *Enterobacteriaceae*. The bacteria are **short, plump, Gram-negative bacilli**. These are **lactose-fermenting** (Fig. 10-2), **urease-positive**, and **indole-negative** bacteria that are **nonmotile** and nonflagellated (no "H" antigens). A **prominent polysaccharide capsule** ("K" antigen; there are **77 "K" antigens**) is **a major virulence factor** and is **antiphagocytic**. The outer membrane contains **lipopolysaccharide** (LPS), whereas the lipid A portion is **endotoxic** and "O" (somatic) antigen is serotype specific (10 "O" antigens).

EPIDEMIOLOGY

One of the most frequently isolated pathogens in Gram-negative pneumonia, *Klebsiella* species are

FIGURE 10-2 Culture of *Klebsiella pneumoniae* on MacConkey agar medium. Note: the pink (lactose fermenting) colonies are highly mucoid, indicating encapsulated bacteria. (*Courtesy of Lisa Forrest, Department of Microbiology, University of Texas Southwestern Medical Center, Dallas, TX.*)

ubiquitous in nature. These bacteria live in water, soil, and, occasionally, food; they can form **part of the intestinal flora** of humans, with subsequent infections of the urinary tract, the respiratory tract, the biliary tract, and wounds. **Alcoholics** and people with a history of seizure disorders are at **increased risk** of infection. This man very likely acquired the infection via oropharyngeal colonization and aspiration. *K. pneumoniae* is a **common nosocomial pathogen**; hospitalized patients may become colonized and have a risk of nosocomial infection.

PATHOGENESIS

Oropharyngeal carriage of *K. pneumoniae* is associated with impaired host defenses in alcoholics. The bacteria gain access after the host aspirates colonizing oropharyngeal microbes into the lung. The bacteria adhere to target cells in the lower respiratory tract via the mediation of multiple adhesins, each with distinct receptor specificity. The adhesive properties are generally mediated by different types of pili (nonflagellar, filamentous projections, otherwise known as fimbriae) on the bacterial surface.

Host defense against bacterial invasion is dependent on phagocytosis by macrophages and PMNs and the bactericidal effect of serum—mediated, in large part, by complement proteins. The alternative complement pathway, which does not require the presence of immunoglobulins directed against bacterial antigens, contributes to host defense against the invading bacteria, which may overcome innate host immunity by several means. The bacteria possess a complex acidic **capsular polysaccharide** (CPS) as the **main determinant of virulence**. The CPS forms thick bundles of fibrillose structures covering the bacterial surface in massive layers that protect the bacterium from phagocytosis by PMNs.

The CPS inhibits the activation or uptake of complement components, especially C3b, and causes antigenic mimicry (presenting as autoantigen) by selective deposition of C3b onto LPS molecules. This results in inhibition of the formation of the membrane attack complex (C5b-C9), thereby preventing membrane damage and bacterial cell death.

Acute inflammatory infiltrates from bronchioles into adjacent alveoli that occur following *Klebsiella* infection of the lung usually **cause bronchopneumonia (patchy distribution of opacity on chest x-ray** involving one or more lobes). Lung abscess caused by the encapsulated bacteria in the polymicrobic infection results in necrotic destruction of alveolar spaces, **cavity formation**, and production of **blood-tinged sputum** due to endothelium damage.

TREATMENT

Extended-spectrum penicillins (e.g., **piperacillin** and ticarcillin), aminoglycosides, quinolones, and other antibiotics are useful for treatment of pneumonia and a variety of other infections due to *Klebsiella* species (Table 10-2). Newer generations of cephalosporins have also been widely used in combination with aminoglycosides. The acquisition of transferable plasmids possessing genes for **extended-spectrum β-lactamases** is increasing in *Klebsiella*. The plasmids confer resistance to third-generation cephalosporins and aztreonam and frequently contain linked resistance determinants for aminoglycosides. These resistant strains are generally found in hospital-acquired infections. *K. pneumoniae* pneumonia has a high mortality rate, even with adequate therapy. The prognosis is worse in patients with alcoholism and bacteremia.

OUTCOME

The patient received 10 days of antibiotics while in the hospital. He was discharged while on oral antibiotics, with follow-up to ensure that the abscess had completely resolved.

TABLE 10–2 Other Clinical Features of *Klebsiella pneumoniae* Infections

Syndromes	Clinical Features
Bacteremia	Bacteremia may be a major complication following lung infection, or it may be associated with another portal of entry (e.g., complicated UTI, line-associated infection).
	Hypotension due to envelope LPS **(endotoxin-mediated sepsis)** can complicate bacteremia and can result in shock; disseminated intravascular coagulopathy may also occur as a result of systemic inflammatory response.
Nosocomial infections	Klebsielleae are frequently involved in infections associated **with respiratory tract manipulations,** such as tracheostomy and mechanical ventilation.
	Urinary tract infections are frequently caused by hospital-acquired Klebsielleae that colonize indwelling urinary catheters. Highly resistant strains can also be seen.
	Often, hospital-acquired, drug-resistant strains also cause **wound infections.**
	Intra-abdominal infections postsurgery (e.g., following appendectomy) are polymicrobic (with *K. pneumoniae* as a major pathogen) and colonic (endogenous) in origin.

PREVENTION

No vaccines or hyperimmune sera are available. Risk avoidance is an important measure of prevention. Prevention of hospital-acquired infections is difficult; hand washing and barrier protection are the most effective ways to reduce transmission.

FURTHER READING

Bradford PA: Extended-spectrum β-lactamases in the 21st century: Characterization, epidemiology, and detection of this important resistance threat. Clin Microbiol Rev 14:933, 2001.

Donnenberg MS: Enterobacteriaceae. In Mandell GL, Bennett JE, Dolin E (eds): Mandell, Douglas, and Bennett's Principles and Practice of Infectious Diseases. Vol 2. 6th ed. New York, Churchill Livingstone, 2005, p 2567.

Podschun R, Ullmann U: *Klebsiella* spp. as nosocomial pathogens: Epidemiology, taxonomy, typing methods, and pathogenicity factors. Clin Microbiol Rev 11:589, 1998.

A **21-year-old** white woman developed **fever, headache,** and a gradually **progressive dry cough.** Over the next 2 days, her cough worsened, becoming productive of small amounts of **clear sputum.**

She was previously in good health. Her 19-year-old **brother had had similar symptoms** 2 weeks earlier.

PHYSICAL EXAMINATION

VS: T 39.3°C, P 105/min, R 28/min, BP 105/66 mmHg

PE: The patient appeared slightly pale. Mild **pharyngeal erythema** was noted with **minimal cervical adenopathy but no exudates.** Chest exam was completely normal.

LABORATORY STUDIES

Blood

Hematocrit: 32%
WBC: 8600/µL
Differential: Normal

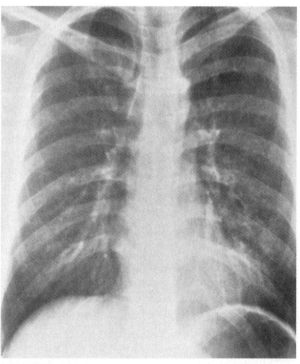

FIGURE 11-1 An admission chest film showing bilateral patchy infiltrates in this patient. *(Courtesy of Dr. Daniel Skiest, Department of Medicine, University of Texas Southwestern Medical Center, Dallas, TX.)*

Blood gases: Normal
Serum chemistries: Normal

Imaging

A chest x-ray revealed bilateral **patchy infiltrates** consistent with atypical pneumonia (Fig. 11-1).

Diagnostic Work-Up

Table 11-1 lists the likely causes of illness (differential diagnosis). Distinguishing bacterial pathogens of atypical pneumonia (e.g., *M. pneumoniae*) from other causes of acute respiratory infection is difficult because of a lack of reliable, widely available, rapid diagnostic tests. Investigational approach for delineating the etiology may include

- **Gram stain of respiratory specimen** to differentiate atypical presentation from typical lobar presentation (i.e., pneumococcal pneumonia)
- **Blood culture**

TABLE 11–1 Differential Diagnosis and Rationale for Inclusion (consideration)
Adenoviruses (ARDS)
Chlamydia psittaci
Chlamydophila pneumoniae
Coxiella burnetii (Q fever)
Influenza A and B
Legionella pneumophila
Mycoplasma pneumoniae
Streptococcus pneumoniae

Rationale: There are many etiologies for atypical pneumonia, and they are often difficult to differentiate clinically. *Mycoplasma* is unique in that clinical pulmonary findings are often absent. *Legionella* often causes gastrointestinal symptoms and a severe headache. *C. psittaci, Coxiella,* and *Legionella* may have a specific history of exposure (e.g., birds, domestic animals, or environmental). Truly purulent sputum, as is seen in *S. pneumoniae* (a major cause of typical pneumonia), is not consistent with atypical pneumonia.

● **Serologic tests**

  ◦ **cold agglutination**
  ◦ **four-fold rise in IgG antibody** titers between acute- and convalescent-phase serum specimens, ideally obtained 2 to 3 weeks apart may provide a retrospective diagnosis.

COURSE

A Gram stain of the woman's sputum revealed only sparse PMNs and no bacteria. Blood and sputum cultures were negative for routine bacterial pathogens. A retrospective serologic investigation based on the elevation of complement fixation antibody titers between acute- and convalescent-phase serum specimens confirmed the diagnosis.

ETIOLOGY

Mycoplasma pneumoniae (primary atypical, or "walking," pneumonia)

MICROBIOLOGIC PROPERTIES

Mycoplasmas are the smallest free-living, self-replicating organism (0.2 to 2 μm in diameter). The **wall-less bacteria** (no mucopolysaccharide cell wall) are bounded by a plasma membrane; they **do not react to Gram stain**. The three-layer **outer membrane contains cholesterol**. Isolation of this organism can be difficult and may take up to 6 weeks. Sixteen species of *Mycoplasma* have been recovered from humans (most are commensals).

> **NOTE** The **four most important characteristics of atypical pneumonia** are
>
> 1. nonproductive cough
> 2. variable chest x-ray (patchy, diffuse infiltrates)
> 3. no bacteria on smear
> 4. no response to β-lactam antibiotics

Gram stain and **cultures** (can be **grown with difficulty on Eaton's agar**) **are not useful for identifying *M. pneumoniae*** or any other atypical pathogen from sputum. Microorganism-specific IgG antibody response, although a retrospective diagnosis, is useful for confirmation of clinical diagnosis.

EPIDEMIOLOGY

M. pneumoniae is a common cause of acute upper and lower respiratory infection in children and young adults. *M. pneumoniae* accounts for **15% to 20% of community-acquired lower respiratory tract infection in adults**. Infections with *M. pneumoniae* occur sporadically throughout the year, and **person-to-person transmission** is the major mode of acquisition. Household infections are often the result of contact with siblings or children. In families, cases occur serially, with 2- to 3-week intervals between cases. The case patient might have acquired the infection **by inhalation of aerosol particles** or **contact with respiratory secretions** from her brother, who was recently sick with similar symptoms. **The general risk age group is 5 to 20 years (school age children to young adults). Outbreaks** are common among young adults, especially **in crowded military and institutional settings,** where the outbreaks can last several months.

PATHOGENESIS

Aerosolized particles less than 3 to 5 μm are able to bypass the upper airways and deposit in the lower respiratory tracts. The organisms have filamentous tips (flask-shaped appearance) that are complex, composed of a network of interactive proteins, designated adhesins, and adherence-accessory proteins. These proteins cooperate structurally and functionally to mobilize and concentrate adhesins at the tip and permit colonization of bacteria between cilia within the respiratory epithelium, probably through host sialoglycoconjugates and sulfated glycolipids. Bacterial adherence leads to **inhibition of ciliary movement** (known as ciliostasis), resulting in the **prolonged cough** seen in this disease. The organisms produce hydrogen peroxide, which is cytotoxic and is responsible for much of the initial cell disruption in the respiratory mucosa and, in uncommon cases, for damage to erythrocyte membranes (hemolytic anemia). *M. pneumoniae* stimulates T and B lymphocytes, inducing the formation of **IgM autoantibodies**, which react with a variety of host tissues, and antigen I on erythrocytes. **Cold agglutinin,** detected by agglutination of type O Rh-negative erythrocytes at 4°C, may be present in the acute serum of 30% to 60% of patients.

TREATMENT

Either **erythromycin** or **doxycycline** is the drug of choice. Cell-wall inhibitors (e.g., amoxicillin or cefotaxime) are **ineffective against *Mycoplasma*,** which are **wall-less bacteria.** Newer oral macrolides (e.g., clarithromycin or azithromycin) or fluoroquinolones (e.g., levofloxacin) are better tolerated than erythromycin and doxycycline and have comparable clinical efficacy against atypical pneumonia but are much more expensive. Therapy of confirmed primary atypical pneumonia is usually continued for 14 to 21 days, owing to relapses with shorter courses of treatment.

> **NOTE** In untreated patients with mycoplasma pneumonia, a range of complications (Table 11-2) may occur.

OUTCOME

The patient was given oral doxycycline. She started to improve rapidly and symptoms resolved within a week.

PREVENTION

There is no vaccine to prevent this person-to-person communicable disease.

TABLE 11–2 Other Clinical Features of *Mycoplasma pneumoniae* Infections

Syndromes	Clinical Features
Tracheobronchitis	Symptomatic (80%) disease is typically mild and is characterized by nonproductive cough, fever, malaise, and pharyngitis.
Extrapulmonary syndromes	Less common complications include adult respiratory distress syndrome, pericarditis, myocarditis, hemolytic anemia (associated with cold agglutinin IgM), and encephalitis. Neurologic complications (e.g., myelitis, encephalitis) may occur in up to 10% of cases. Some young male patients develop extensive rash, involving the mucous membranes and large areas of the body, known as "erythema multiforme" (**Stevens-Johnson syndrome**). Fatal cases are reported occasionally, primarily among the elderly and persons with sickle-cell disease.

FURTHER READING

Baseman JB, Tully JG: Mycoplasmas: Sophisticated, reemerging, and burdened by their notoriety. Emerg Infect Dis 3:21, 1997.

Clyde WA, Jr: Clinical overview of typical *Mycoplasma pneumoniae* infections. Clin Infect Dis 17:S32, 1993.

Feikin DR, Moroney JF, Talkington DF, et al: An outbreak of acute respiratory disease caused by *Mycoplasma pneumoniae* and adenovirus at a federal service-training academy: New implications from an old scenario. Clin Infect Dis 29:1545, 1999.

McIntosh K: Community-acquired pneumonia in children. N Engl J Med 346:429, 2002.

A **41-year-old** white man was admitted with a 3-day history of high fever and dry cough. His initial symptoms progressed to include headaches, muscle aches, and confusion.

Past history was unremarkable except for his being a chain smoker for the last 15 years. He had recently started working in a home-improvement center in the show room area with **whirlpools** and **spas.**

PHYSICAL EXAMINATION

VS: T 40.3°C, P 88/min, R 40/min, BP 110/60 mmHg

PE: Examination revealed a distressed patient with inspiratory rales. He had a cough that was productive of scanty, clear sputum.

LABORATORY STUDIES

Blood

Hemoglobin: 14 g/dL
Hematocrit: 44%

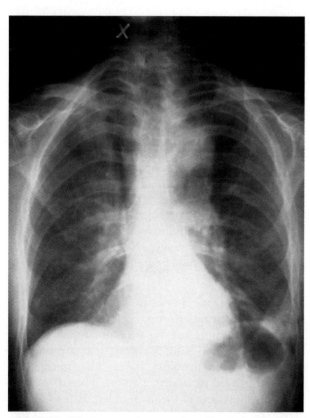

FIGURE 12-1 An admission chest film showing bilateral lower lobe interstitial infiltrates in this patient. *(Courtesy of Dr. Paul Southern, Departments of Pathology and Medicine, University of Texas Southwestern Medical Center, Dallas, TX.)*

WBC: 16,700/μL

Differential: Normal

Blood gases: pH 7.42, pO_2 70 mmHg, pCO_2 34 mmHg

Serum chemistries: BUN 26, creatinine 1.6, **sodium 126 mmol/L,** inorganic phosphorus 1.8 mg/dL

Imaging

A chest x-ray revealed bilateral lower lobe **patchy (interstitial) infiltrates** (Fig. 12-1).

Diagnostic Work-Up

Table 12-1 lists the likely causes of illness (differential diagnosis). Investigational approach to delineating the etiology may include

- Microscopic examination (Gram stain to rule out *Streptococcus pneumoniae*) and **direct fluorescence antibody** (DFA) staining of bronchoscopic (e.g., bronchoalveolar lavage [BAL]) specimen to rule out atypical pathogens

TABLE 12–1 Differential Diagnosis and Rationale for Inclusion (consideration)
Adenoviruses (acute respiratory distress)
Chlamydia psittaci
Chlamydophila pneumoniae
Coxiella burnetii (Q fever)
Influenza A and B
Legionella spp.
Mycoplasma pneumoniae
Streptococcus pneumoniae

Rationale: *Legionella* is associated with high fever, diarrhea, confusion, and headache, and *Mycoplasma* and *Chlamydophila* usually have a more indolent course. Viral pneumonias often present with typical symptoms of upper respiratory infection. *C. psittaci, Coxiella,* and *Legionella* usually have a specific history of exposure (e.g., birds, pigs, or environmental). Truly purulent sputum, as is seen in *S. pneumoniae* (a major cause of typical pneumonia) is uncommon with other listed pathogens.

- **Direct antigen in urine** to detect *Legionella pneumophila* serogroup 1
- **Cultures** of sputum for *S. pneumoniae*, *Legionella*, and so forth
- Failing all, **serology** and **PCR** (or DNA probes) for viruses, *Chlamydophila pneumoniae*, *Mycoplasma pneumoniae*, and so on.

COURSE

The patient was admitted to the hospital. Broncho-alveolar lavage specimen and blood were collected for microbiologic confirmation of the etiology. The Gram stain of BAL revealed a few PMNs but no bacteria. Culture of BAL, after 72 hours, yielded a significant isolate on a special agar medium. Identification of the isolate by DFA confirmed the clinical diagnosis.

ETIOLOGY

Legionella pneumophila (legionellosis)

MICROBIOLOGIC PROPERTIES

Legionellae are **motile, flagellated, pleomorphic rods**, which **stain faintly with Gram stain**, although the bacterial surface carries Gram-negative cell wall structures. Cytology tissue stains (e.g., Dieterle silver) provide improved microscopic visualization (Fig. 12-2). The definitive method for the diagnosis of legionellosis is culture of the organism. **Culture** of BAL or other respiratory specimens requires special media (the laboratory must be alerted that *Legionella* is a diagnostic consideration). The **nutritionally fastidious, aerobic** pathogen **grows slowly** in 3 to 7 days on **selective (buffered charcoal yeast extract) agar medium** supplemented with cysteine and iron. Fewer than 20 *Legionella* species (differentiated based on fermentative reactions) have been linked to disease in humans. *L. pneumophila* is the most pathogenic, accounting for the majority of the cases of legionellosis, followed by *Legionella micdadei*. Although more than 14 serogroups of *L. pneumophila* have been identified, LPS-specific serogroup 1 causes most cases of legionellosis.

EPIDEMIOLOGY

Legionellosis occurs worldwide. An estimated 8000 to 18,000 cases occur each year in the United States; 23% are nosocomial in origin. Most cases are sporadic; 10% to 20% can be linked to common-source outbreaks, usually originating in large buildings such as hotels and hospitals. Community-acquired disease most often occurs in the summertime. The organism survives for months in tap water and cooling towers in association with slimy growth of ameba and other protozoa. Primary sources of infection are

- **Environment**: showers, air-conditioning units, cooling towers, humidifiers, whirlpools, spas, etc. (Freshwater amebae appear to be the natural reservoir for the organisms.)
- **Aerosolizing equipment**: nebulizers, humidifiers, water faucets, etc.
- **Aspiration**: from contaminated water or via nasogastric tubes

Organisms living in water are **aerosolized** and **spread via airborne routes**. This patient may have contracted the infection via exposure to aerosolized bacteria from the work environment surrounding the whirlpools in the show room. The organism is **not transmitted person to person**. The risk groups for the illness are individuals older than 50 years of age, **smokers**, alcoholics, patients with COPD and malignancy, immunocompromised patients with organ transplants, and patients on corticosteroids.

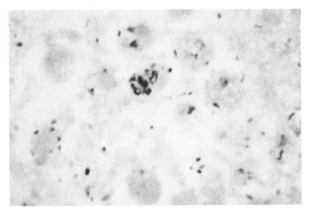

FIGURE 12-2 Micrograph of *Legionella pneumophila* from bronchoscopic biopsy. Note black rods of *L. pneumophila* stained with Dieterle silver stain. *(Courtesy of Dr. Paul Southern, Departments of Pathology and Medicine, University of Texas Southwestern Medical Center, Dallas, TX.)*

PATHOGENESIS

Legionella organisms are cleared from the upper respiratory tract by mucociliary action. Any process that compromises mucociliary clearance (e.g., smoking tobacco) increases risk of infection. **Flagellated organisms penetrate the mucus layer in the lower respiratory epithelium** and adhere to the target cell surface via specific pili, and an outer membrane protein binds C3, favoring opsonization. The **strictly intracellular** bacteria are internalized via phagocytosis, but specific gene (*dot* [delayed organelle trafficking]) products prevent phagosome-lysosome fusion in macrophages, whereas other bacterial virulence genes prevent acidification of the phagosome. The bacteria are able to survive in phagosomes—scavenging the iron they need from transferrin—and multiply, destroying the cell. They also release toxins and enzymes (phospholipases, metalloproteases) that damage host cells.

Cell-mediated immunity appears to be the primary host defense mechanism against *Legionella* infection. Activation of macrophages produces cytokines that regulate antimicrobial activity against *Legionella* organisms. With cytokines as chemotaxins, phagocytes and T cells continue to be attracted to the infected area. It is the **cytokines and other reactive mediators from PMNs and T cells that inflict the lung tissue damage**, and the complement and cellular elements of the host response lead to the **patchy, diffuse infiltrates on chest x-ray**.

TREATMENT

A macrolide (erythromycin, azithromycin, or **clarithromycin**) or a newer quinolone (**levofloxacin** or **gatifloxacin**) or **doxycycline** is clinically efficacious against this and other infections due to *L. pneumophila* (Table 12-2). These intracellular bacteria are not susceptible to β-lactam antibiotics. Penicillins, cephalosporins, and aminoglycosides cannot penetrate host cells, and therefore they are ineffective against *Legionella*.

T ABLE 12–2 Important Clinical Features of *Legionella* Infections

Syndromes	Clinical Features
Legionnaires' disease	Legionnaires' disease, the more severe form of legionellosis, commences 2-10 days after exposure, presenting as acute, severe fibrinopurulent pneumonia with alveolitis and bronchiolitis. In addition to inflicting the lungs, Legionellae may infect the lymph nodes, brain, kidney, liver, spleen, bone marrow, and myocardium. Hyponatremia and hypophosphatemia are commonly seen. Recovery is slow (weeks); case-fatality rate = 15%.
Pontiac fever	**Pontiac fever is an acute-onset, flu-like, non-pneumonic illness,** occurring within a few hours to two days of exposure to *L. pneumophila* (chest x-rays do not show pneumonic infiltrates). Often very high attack rate in exposed groups. Recovery after 2-5 days; case-fatality rate <1%

OUTCOME

The patient received erythromycin for 10 days. He gradually recovered from his illness after a stormy hospital course, although complete return to normal function took several weeks.

PREVENTION

Whirlpools, spas, and other similar risk environments for potential *Legionella* growth should be maintained slime free. Cooling towers should be cleaned and hyperchlorinated or flushed with hot water (70°C) for at least 2 hours during cleaning. Biocides should be used to limit the growth of slime-forming organisms. Tap water should not be used for respiratory therapy devices.

FURTHER READING

Fields BS, Benson RF, Besser RE: *Legionella* and legionnaires' disease: 25 years of investigation. Clin Microbiol Rev 15:506, 2002.

Luttichau HR, Vinther C, Uldum SA, et al: An outbreak of Pontiac fever among children following use of a whirlpool. Clin Infect Dis 26:1374, 1998.

Shuman HA, Purcell M, Segal G, et al: Intracellular multiplication of *Legionella pneumophila*: Human pathogen or accidental tourist? Curr Top Microbiol Immunol 225:99, 1998.

Stout JE, Yu VL: Legionellosis: Current concepts. N Engl J Med 337:682, 1997.

In **December,** a **71-year-old man from a nursing home** was brought to the hospital in **acute respiratory distress.** He had been in his usual state of health until 10 a.m. the previous day, when he suddenly developed **fever, chills,** muscle **aches, cough,** and **prostration.** Several other nursing home residents had developed a similar illness during the previous week.

His past medical history was unremarkable, and he had not seen a physician in the past year.

PHYSICAL EXAMINATION

VS: T 40°C, R 28/min, P 118/min, BP 140/90 mmHg

PE: An acutely ill, prostrated elderly man; lung exam was unremarkable. He had a frequent, weak cough.

LABORATORY STUDIES

Blood

Hematocrit: 44%

WBC: 7600/µL

Differential: Normal

Blood gases: pO$_2$ 74 mmHg

Serum chemistries: Normal

Imaging

Chest x-ray was normal.

Diagnostic Work-Up

Table 13-1 lists the likely causes of illness (differential diagnosis). Influenza was considered based on the clinical features and seasonality of the illness.

Investigational approach for delineating the etiology may include

- **Rapid antigen testing or direct immuno-fluorescence antibody (DFA)** of nasopharyngeal (NP) swab or nasal aspirate
- **Gram stain** and **cultures of sputum** to rule out bacterial pneumonia
- **Isolation of virus** (cell culture) of NP swab or nasal aspirate
- **Serology**

COURSE

A Gram stain of expectorated sputum was unremarkable. The patient was admitted to the hospital and given supplemental oxygen and intravenous hydration. He was also given antipyretics and was placed on cefotaxime, erythromycin, and amantadine because influenza A was predominant in the community at the time. Sputum specimens and a nasopharyngeal swab were obtained for bacterial and viral cultures. A DFA test of nasopharyngeal wash with monoclonal antibody directed against a seasonal virus yielded a positive result.

TABLE 13–1   Differential Diagnosis and Rationale for Inclusion (consideration)
Adenoviruses
Bacterial pneumonia
Chlamydophila pneumoniae
Influenza (types A, B, or C)
Mycoplasma pneumoniae
Coxiella burnetii (Q fever)

Rationale: In the appropriate season, typical clinical features are usually adequate to make a diagnosis of influenza. Other viral causes are generally not as acute or severe in onset. The presence of headache and myalgias is not as common with bacterial pneumonia. Atypical causes are generally associated with a more indolent presentation. Q fever is often associated with animal exposure.

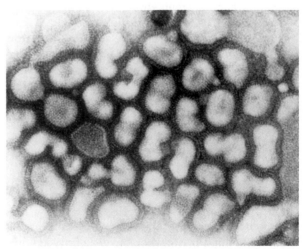

FIGURE 13-1 Electron micrograph of the pathogen. *(Courtesy of Centers for Disease Control and Prevention Atlanta, GA.)*

ETIOLOGY

Influenza

NOTE Influenza can be a clinical diagnosis and is recognized commonly by epidemiologic characteristics because it is a seasonal disease.

MICROBIOLOGIC PROPERTIES

This virus has a single-stranded **RNA genome with 8-segmented pieces** coding for 10 proteins; RNA has **negative polarity**. The virus has helical nucleocapsid symmetry, with an **outer membrane envelope** (Fig. 13-1) covered with two different types of spikes, **hemagglutinin** (H antigen; an attachment protein) and **neuraminidase** (N antigen; acts to sever the virus as it buds from plasma membrane or from mucus). The **influenza virus types A, B,** and **C** are based on antigenic characteristics of the nucleoprotein and matrix protein antigens. All have a similar complement-fixing antigen (ribonucleoprotein). Influenza A viruses are further subtyped on the basis of surface H and N antigens. **Current human types** are $A(H_1N_1)$ and $A(H_3N_2)$; multiple swine and avian types are also transmissible to humans.

Antigenic shift occurs due to **genetic reassortment** that results in a complete change in the configuration of a specific epitope on the surface of the influenza virion. This mechanism may lead to the **generation of new strains**. Many H and N antigens exist in the form of animal influenza viruses, particularly in aquatic avian species. Fifteen H and nine N antigens exist in animals at the present time. They can combine when different viruses from two different sources infect cells at the same time.

Antigenic drift occurs when **a point mutation results in a change in the configuration of a specific epitope** on the surface of influenza virion. Change in the antigenic composition of the H antigens from year to year may occur due to random mutations (**small changes**) affecting the cross reactivity of the H antigen in the human strains. A change in viral antigen usually occurs at a slow pace.

EPIDEMIOLOGY

Influenza viruses cause epidemic disease (influenza virus types A and B) and sporadic disease (type C) in humans. **Influenza viruses spread from person to person primarily through the coughing and sneez-ing** of infected persons. The incubation period ranges from 18 to 72 hours, with an average of 48 hours. This patient may have acquired the infection via exposure to aerosol produced by the coughing or sneezing of other nursing home residents, or by direct contact with infected respiratory secretions.

PATHOGENESIS

Influenza virus infection occurs after transfer of respiratory secretions from an infected individual to a person who is immunologically susceptible. Humoral immune response specific to the infecting strains develops after resolution of infection or active immunization. Mucosal immunity mediated by IgA in the respiratory tract is protective. If not neutralized by antibodies, the virus attaches to and penetrates respiratory epithelial cells in the trachea and bronchi. Each virus particle has approximately 500 H spikes, which bind to sialic acid receptors on host ciliated epithelial cells. Neuraminidase degrades the protective layer of mucus, allowing the virus to gain access to the cells of the upper and lower respiratory tract. Fusion of the viral envelope to the cell's plasma membrane initiates infection, which is limited primarily to the respiratory tract. Despite the systemic symptoms, viremia is absent. Virus replicates in **mucus-secreting, ciliated cells** and in other epithelial cells, resulting in cell dysfunction and degeneration (as viral replication and release of viral progeny occur). **Cytokines** liberated from damaged infiltrating leukocytes **cause systemic symptoms of influenza**.

TREATMENT

Amantadine or **rimantadine**, if given within 48 hours of onset of **influenza A (not B)** and given for 3 to 5 days, is clinically efficacious for primary influenza infection. Neuraminidase inhibitors **zanamivir** or **oseltamivir** have **activity against both influenza A and influenza B viruses** but are much more expensive. In general, therapy is more useful if given early in the course of the infection. Use of salicylates (aspirin) is not recommended in children (to avoid complications; see Table 13-2).

OUTCOME

The patient's symptoms initially improved and then progressively got worse following secondary bacterial

TABLE 13–2 Other Clinical Features of Influenza Infections

Syndromes	Clinical Features
Pneumonia	Among certain persons, influenza can exacerbate underlying medical conditions, leading to primary influenza viral pneumonia (uncommon) or secondary bacterial pneumonia (common).
	Influenza viral pneumonia is interstitial in location, with diffuse patchy inflammation localized to interstitial areas at alveolar walls.
	Infection of the lower respiratory tract causes severe desquamation of bronchial or alveolar epithelium to the level of the basement membrane. Necrosis of the superficial layers of the respiratory epithelium causes loss of primary host defense, predisposing hosts to secondary bacterial pneumonia, often due to *S. aureus*.
Extrapulmonary complications	Influenza has also been associated with transverse myelitis, Reye syndrome, myositis, myocarditis, and pericarditis.
	Reye syndrome is a rare, often fatal childhood hepatoencephalopathy associated with analgesic (salicylates) use and is predominantly associated with influenza B.

pneumonia due to *Staphylococcus aureus*. Rapid antigen testing was positive for influenza A. He was transferred to the ICU for more effective management of his serious infection. He was treated with vancomycin for secondary pneumonia, with gradual resolution of his symptoms.

PREVENTION

Annual immunization with current **killed virus vaccines** may provide 70% to 80% protection against influenza and is recommended, especially for those 65 years of age or older. It may reduce severity in 50% to 60% of elderly individuals. In the setting of an institutional outbreak, antiviral drugs can be given to prevent further cases in nonimmunized persons. Amantadine or rimantadine is appropriate if the circulating virus is influenza A; neuraminidase-inhibiting drugs (zanamivir or oseltamivir) can be considered if influenza B has been observed as well. These drugs can be given for 2 to 3 weeks (allowing time for vaccine immunity to develop).

FURTHER READING

Couch RB: Drug therapy: Prevention and treatment of influenza. N Engl J Med 343:1778, 2000.

Horimoto T, Kawaoka Y: Pandemic threat posed by avian influenza A viruses. Clin Microbiol Rev 14:129, 2001.

Reichert TA, Sugaya N, Fedson DS, et al: The Japanese experience with vaccinating school children against influenza. N Engl J Med 344:889, 2001.

Shaw MW, Arden NH, Maassab HF: New aspects of influenza viruses. Clin Microbiol Rev 5:74, 1992.

Stamboulian D, Bonvehi PE, Nacinovich FM, Cox N: Influenza. Infect Dis Clin North Am 14:141, 2000.

A **5-month-old** girl was brought to the pediatric clinic of a local general hospital **in February** with a 2-day history of **cough, respiratory difficulty** with nasal discharge, and low-grade fever. She had begun attending **a day care** center 4 weeks before. All of her immunizations were up to date, and no one else at home was ill.

PHYSICAL EXAMINATION

VS: T 38.1°C, P 135/min, R 60/min, BP 92/60 mmHg

PE: The patient was in respiratory distress. Rhinorrhea, expiratory and **inspiratory wheezes**, **hyper-inflation of chest**, and **atelectasis** were noted. Crackles were also noted bilaterally.

LABORATORY STUDIES

Blood

Hematocrit: Normal

WBC: Normal

Differential: Normal

Blood gases: pH 7.42, pO_2 50 mmHg, pCO_2 34 mmHg, oxygen saturation 72%

Serum chemistries: Normal

Imaging

A chest x-ray was remarkable for hyperinflation and peribronchiolar infiltrates.

Diagnostic Work-Up

Table 14-1 lists the likely causes of illness (differential diagnosis). Investigational approach for delineating the etiology may include:

- **Antigen detection** by enzyme immunoassay or direct fluorescence antibody staining
- In failing above, other tests
  - **Virus culture** of nasopharyngeal secretions
  - **Polymerase chain reaction**-based assay of nasopharyngeal secretions

COURSE

The patient was admitted to the hospital for the management of bronchiolitis (and possibly pneumonia).

TABLE 14–1 Differential Diagnosis and Rationale for Inclusion (consideration)

Adenoviruses

Human metapneumovirus

Influenza viruses (types A, B, or C)

Metapneumovirus

Mycoplasma pneumoniae

Parainfluenza viruses

Respiratory syncytial virus (RSV)

Rhinoviruses

Rationale: All the above viruses may cause indistinguishable illnesses in very young patients. Bronchiolitis should be considered based on the age and presentation. In certain times of the year, RSV, among the viral pathogens that cause respiratory illness, is simply the most common etiology, especially among infants less than 6 months old. *M. pneumoniae* only infrequently causes symptomatic respiratory illness in patients this young.

Concerned about the communicability of the suspected infectious agent, the pediatrician isolated her in a private bed on a medical teaching unit under contact isolation. The patient's nasopharyngeal secretion was tested by rapid antigen detection, and the test result confirmed the presumptive diagnosis of bronchiolitis.

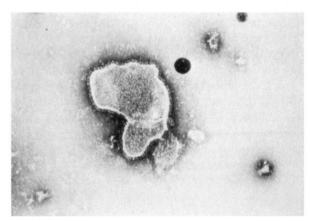

FIGURE 14-1 Electron micrograph of a respiratory viral agent. (*Courtesy of Centers for Disease Control and Prevention, Atlanta, GA.*)

ETIOLOGY

Respiratory syncytial virus (RSV) (bronchiolitis)

MICROBIOLOGIC PROPERTIES

RSV is a member of the Paramyxoviridae family and includes the genus *Pneumovirus*. It is an **enveloped virus** (Fig. 14-1), with a **negative-sense, single-stranded RNA** (nonsegmented). The genome codes for 10 virus-specific proteins, and is contained in a helical nucleocapsid surrounded by a lipid envelope. RSV has large envelope glycoproteins, which consist of a fusion protein (F) and a second glycoprotein (G). Unlike influenza virus, RSV does not have any hemagglutinin protein. RSV is so named because its replication in vitro leads to fusion of neighboring cells into a characteristic large, **multinucleated syncytium**. Its antigenic diversity is reflected by differences in its surface glycoproteins; two antigenic types are A and B. Cultivation and identification of RSV in the nasopharyngeal secretions of hospitalized patients are necessary for specific intervention and control of nosocomial spread in the unit. The virus is unstable in the environment (surviving only a few hours on environmental surfaces) and is readily inactivated with soap and water and disinfectants.

EPIDEMIOLOGY

In the United States, RSV infections occur as community outbreaks, often lasting 4 to 6 months from November to April with peak activity in **January** or **February**. The **most frequently affected age group is infants 2 to 6 months of age**. Infection occurs when infectious material contacts mucous membranes of the eyes, mouth, or nose. The case patient is believed to have contracted the infection by direct inoculation of contagious secretions from a symptomatic child in a day care center. **Nosocomial outbreaks** are not uncommon in the hospital setting, where breach in infection control (inadequate hand washing) and **contacts** with the health care providers facilitate transmission of the virus from an index case to predisposed infants (secondary cases).

PATHOGENESIS

After contagion and incubation for 2 to 8 days, RSV replicates in the nasopharyngeal epithelium, with spread to the bronchioles 1 to 3 days later. The disease may involve an immunopathogenic mechanism, with immune complexes, IgE antibody, and histamine playing a role in pathogenesis. Integral to immune response and pathogenesis are the F and G glycoproteins. **Bronchiolitis** is a result of **inflammation of the terminal bronchioles, necrosis, and sloughing of the epithelial cells lining the bronchioles.** The bronchioles of a young child have a small bore; when the lining cells are swollen by inflammation, the passage of air to and from the alveoli can be severely restricted, leading to **wheezing** and **hyperinflation.** Immunity to the initial infection is brief, and reinfections are common (although less severe), despite the presence of RSV-specific local and systemic antibodies and neutralizing antibody.

TREATMENT

For children with mild disease, no specific treatment is necessary other than the treatment of symptoms (e.g., acetaminophen to reduce fever). Children with severe disease may require oxygen therapy and sometimes mechanical ventilation. Ribavirin initially was reported to be an effective treatment; however, subsequent trials could not substantiate a benefit from this therapy. Some investigators have used a combination of immune globulin intravenous (IGIV) with high titers of neutralizing RSV antibody (RSV-IGIV) and ribavirin to treat patients with severe disease and compromised immune systems (e.g., bone marrow transplant patients).

NOTE Most children recover from illness in 8 to 15 days. Premature infants, children undergoing cancer chemotherapy, and infants with congenital heart disease or chronic lung disease are at risk for **serious pneumonia** and **encephalopathy** (case fatality rate 37%). Complicated respiratory disease also occurs among the elderly.

OUTCOME

The pediatrician started the patient on a treatment program consisting of ribavirin (in aerosol format), humidified oxygen, bronchodilator, and suctioning of secretions. Her symptoms gradually resolved over the next 7 days, and she returned home. She did not suffer from any complications.

PREVENTION

Vaccine is available on an experimental basis and is individualized (at the time of writing). RSV immune globulin (intravenous) and a humanized murine anti-RSV monoclonal antibody are available as prophylaxis for some high-risk infants and young children. **Contact isolation** procedures are recommended for prevention and control of nosocomial transmission of RSV. Hospital infection control measures include careful attention to hand washing between contacts with patients.

FURTHER READING

Anderson L, et al: Multicenter study of strains of respiratory syncytial virus. J Infect Dis 163:687, 1991.

Hall CB: Respiratory syncytial virus and parainfluenza virus: Medical progress. N Engl J Med 344:1917, 2001.

Falsey AR, Walsh EE: Respiratory syncytial virus infection in adults. Clin Microbiol Rev 13:371, 2000.

A 32-year-old Hispanic woman presented with a **cough** for several weeks and a 15-lb **weight loss.** She also had **night sweats** and subjective fevers and felt fatigued. Despite erythromycin treatment for suspected pneumonia given by her family physician, her fever and cough progressively got worse. She complained about **coughing blood-tinged sputum.**

She had **emigrated from Venezuela to the United States** three years before her illness, but she frequently returned to Venezuela to visit relatives.

PHYSICAL EXAMINATION

VS: T 38.6°C, P 96/min, R 18/min, BP 112/60 mmHg

PE: Examination was remarkable for bilateral rales and lymphadenopathy.

LABORATORY STUDIES

Blood

Hematocrit: 32%

WBC: 8500/μL

Differential: Normal

Blood gases: pO_2 78 mmHg

Serum chemistries: Normal

Imaging

Chest x-ray revealed **right upper lobe infiltrates** (Fig. 15-1).

Diagnostic Work-Up

Table 15-1 lists the likely causes of illness (differential diagnosis). Any patient with an exposure history (i.e., emigration from an endemic country), with apical infiltrates, and who does not respond to antibacterial agents for community-acquired pneumonia should, however, have appropriate studies to rule out tuberculosis. The investigational approach may include

- **Skin test**
- **Gram** and **acid-fast stain** of respiratory secretions
- **Cultures** of respiratory secretions

  - Routine aerobic and anaerobic cultures

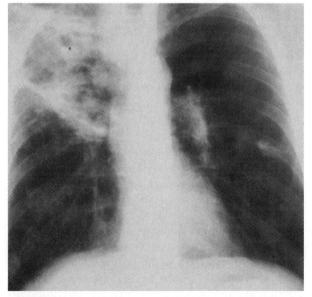

FIGURE 15-1 Admission chest film showing right upper lobe infiltrate in this patient. (*Courtesy of Dr. Anthony DalNogare, Department of Medicine, University of Texas Southwestern Medical Center, Dallas, TX.*)

TABLE 15–1 Differential Diagnosis and Rationale for Inclusion (consideration)

Actinomyces sp.
Anaerobes (aspiration pneumonia)
Endemic fungi (e.g., *Histoplasma capsulatum*)
Legionella penumophila
Mycobacterium tuberculosis
Mycoplasma pneumoneae
Nocardia sp.

Rationale: The chronic symptoms of fever, cough, night sweats, and weight loss should prompt the suspicion of tuberculosis (TB) in any patient particularly one with upper lobe disease. TB is also much more common in immigrants from developing countries than in those born in the United States. Chronic pneumonia due to endemic fungi (e.g., *H. capsulatum*) can be indistinguishable from TB and should also be considered, particularly in patients from the appropriate geographic region of the United States. *Nocardia* and *Actinomyces* may also cause chronic pneumonia. Aspiration pneumonia is usually seen in patients with poor dentition and altered mental status, such as dementia or chronic alcoholism. The remaining agents are less common causes of chronic pneumonia.

● **Broth- and agar-based mycobacterial cultures** and DNA probe-based identification

● In failed Gram and acid-fast stains, **fungal serology** and **fungal cultures**

COURSE

The patient was hospitalized for examination. The attending physician suspected that she might be suffering from pulmonary tuberculosis and administered a PPD (tuberculin) skin test. Forty-eight hours later, she showed a strong positive skin reaction with thickening of the skin and redness at the injection site. The physician referred her to an infectious disease consultant. Her sputum smear was strongly positive for "acid-fast bacilli." The patient was advised to also undergo HIV testing.

ETIOLOGY

Mycobacterium tuberculosis (post-primary tuberculosis)

MICROBIOLOGIC PROPERTIES

Mycobacteria are **acid-fast bacteria** (reactive to auramine O fluorescence and Kinyoun acid-fast stains; Fig. 15-2). The **cell wall contains 60% lipid** in the form of long-chain fatty acids called *mycolic acids*. *M. tuberculosis* is the major medically important mycobacterial agent (Table 15-2). *M. tuberculosis* **grows slowly** (doubling in 18 hours) on selective (e.g., **Lowenstein-Jensen**) agar media. Virulent strains grow in parallel and serpentine pattern due to the presence of **cord factor** (6,6′ trehalose-dimycolate), a virulence factor. The surface macromolecules of *M. tuberculosis* (e.g., purified protein derivative [PPD, glycolipids]), including species-specific mycosides, are highly antigenic. Phenolic glycolipids (e.g., lipoarabinomannan [**LAM**]) contribute to **enhanced virulence.** Mycobacteria are resistant to acid and alkali and are resistant to dehydration.

EPIDEMIOLOGY

People with pulmonary tuberculosis produce respiratory aerosol during coughing, sneezing, or singing. **Exposure to airborne organisms** from a symptomatic patient is the major mode of contagion. The risk

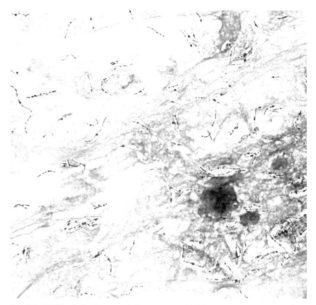

FIGURE 15-2 Acid-fast stain of *Mycobacterium tuberculosis* from sputum. *(Courtesy of Dr. Rita Gander, Department of Pathology, University of Texas Southwestern Medical Center, Dallas, TX.)*

is highest among children younger than 3 years of age, the lowest in later childhood, and high again among adolescents, young adults, the very old, and immunocompromised individuals. Risk is directly related to degree of exposure and is not related to genetic or other host factors.

PATHOGENESIS

The major determinants of the type and extent of disease are the patient's age and immune status, and the mycobacterial load. Fewer than 10 mycobacterial bacilli may initiate a pulmonary infection. Mycobacteria are **obligate aerobes** that cause disease only in highly oxygenated tissues, such as the upper lobe of the lung. The facultative intracellular bacteria seek refuge and hide in macrophages. Mycobacterial LAM is recognized by the macrophage mannose receptor. *M. tuberculosis* expresses a cell-wall C3 convertase activity, forming C3b on its surface that is recognized by the macrophage complement receptor CR4, triggering phagocytosis. *M. tuberculosis* evades phagocytic killing by inhibiting phagolysosome fusion; this is mediated by a tryptophan-aspartate-containing coat protein, which remains attached to the bacteria. Mycobacterial glycolipid antigens on MHC class II molecules do not produce a rapid immune response, adding an element of killing inhibition. The bacteria proliferate in the

TABLE 15–2 Classification of Major Mycobacteria and Association with Diseases	
Class/Species	**Disease Association**
Tuberculous (classic) mycobacteria	
M. tuberculosis	Causes tuberculosis (TB)
M. bovis	Causes TB of cattle and humans (rarely)
Nontuberculous (atypical) mycobacteria	
Mycobacterium kansasii	Causes pulmonary TB-like disease
Mycobacterium marinum	Found in water; causes "swimming pool granuloma" in some patients
Mycobacterium scrofulaceum	Causes cervical lymphadenitis in children
Mycobacterium xenopi	Found in hot water tanks; occasionally causes TB
Mycobacterium avium-intracellulare complex (MAC)	Both *Mycobacterium avium* and *Mycobacterium intracellulare* cause cervical lymphadenitis in children; cause disseminated disease in AIDS patients; may cause pulmonary disease in the elderly
Mycobacterium ulcerans	Primarily causes skin ulcers
Mycobacterium fortuitum-chelonae	Rapid growers; most are saprophytic and rarely cause disease
Leprosy bacteria	
Mycobacterium leprae	Uncultivable mycobacteria in laboratory media; in the United States, nine-banded armadillos are reservoir in Louisiana and Texas; causes tuberculoid and lepromatous leprosy (Hansen disease); pathology on skin and superficial nerves (areas of cooler temperatures)

phagosome, eventually bursting macrophages and spilling out to infect additional macrophages.

The initial interaction between *M. tuberculosis* and macrophages elicits both CD4+ and CD8+ T-cell responses. The main contribution of the CD4+ T cells comes from the T_H1 subset, which releases IFNγ to stimulate macrophage activation. Cord factor is toxic to PMNs, causing irreversible structural and functional damage to mitochondria. It also stimulates the formation of granulomas, a hallmark of tuberculosis. Granulomas are collections of macrophages that have taken on properties of epithelial cells (and are called epithelioid cells). Epithelioid cells may coalesce within the granuloma to form giant cells. When fully developed, a chronic granuloma encapsulated with fibrin (a **tubercle**) consists of a central area of large, multi-

nucleated giant cells containing tubercle bacilli, a midzone of epithelioid cells, and a peripheral zone of fibroblasts, lymphocytes, and monocytes (Fig. 15-3).

The infection may heal by fibrosis or calcification, and **TB granulomas may be visualized by chest x-ray** as **lobar (Ghon focus)** and **perihilar lymph node involvement**, reflecting **primary infection**. In a healthy adult exposed to relatively low numbers of bacteria, granulomatous inflammation stops the infection before appreciable damage to the lung occurs. Key T_H1 cytokines for controlling TB include IL-12, γ-interferon, and TNF. The key T_H1 effector molecule, required for efficient killing of *M. tuberculosis*, is nitric oxide, which is produced by the inducible nitric oxide synthase of macrophages. The number of viable bacteria falls, dissemination is curtailed, and the infection is now controlled. Such patients become skin-test positive. A **skin test** uses PPD from tubercle bacilli. Reddening and thickening of the skin (local erythema and induration; >15 mm) in 48 to 72 hours after intradermal injection of five tubercle units of PPD indicates positive reaction, although lesser reactions may be considered positive in certain high-risk groups (e.g., HIV patients with reactions >5 mm are considered positive).

The **majority (90%) of TB infections** are **latent**. Bacterial replication within granulomas is balanced by killing, so the bacterial burden remains at a constant low level and patients are asymptomatic. In certain infected individuals (as in the case patient) waning of

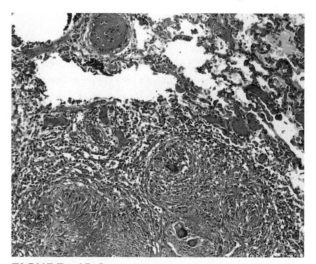

FIGURE 15-3 Caseating granulomatous inflammation containing a central area of large, multinucleated giant cells containing tubercle bacilli, a midzone of epithelioid cells, and a peripheral zone of fibroblasts, lymphocytes, and monocytes. *(Courtesy of Dr. Dominick Cavuoti, Department of Pathology, University of Texas Southwestern Medical Center, Dallas, TX.)*

cell-mediated immunity (from age, illness, or cumulative [overwhelming] exposure) leads to **reactivation (post-primary infection)** and increased bacterial growth. Some time after reactivation, the material within the granulomas becomes caseous (see Fig. 15-3). The central area of the enlarging granulomas undergoes necrosis, which may break into a bronchus, discharging *M. tuberculosis* into exhaled air. The infected patient coughs and infects another person. Phagocytes unsuccessfully trying to kill the bacteria also cause considerable damage to lung tissue.

TREATMENT

Therapeutic management involves (1) multiple antibiotics and (2) long-term therapy. In the United States, empirical therapy generally includes a four-drug regimen: **isoniazid (INH), rifampin (RIF), pyrazinamide (PZA), and ethambutol (EMB)**. After drug susceptibility results are available, a specific drug regimen is chosen. For susceptible isolates, therapy is usually completed with INH and RIF. The treatment of **multidrug-resistant tuberculosis** (MDR-TB, which implies resistance to INH and RIF at least) usually involves four or five drugs, including ciprofloxacin, amikacin, ethionamide, and cycloserine. HIV patients infected with nontuberculous mycobacteria receive a macrolide (azithromycin or clarithromycin), rifabutin, and other drugs based on susceptibility tests.

NOTE Older patients (≥ 50 years of age) are at greater risk of developing hepatic toxicity induced by INH. INH does induce microsomal enzymes and may have interactions with many drugs. Coadministration of these drugs may potientate the hepatotoxicity of INH. Without treatment, post-primary (reactivation) tuberculosis may progress to miliary tuberculosis (PPD-positive; **chest x-ray has the appearance of millet seeds scattered in all lung fields;** bacteriologic evidence [AFB smear positive]). The uncontrolled infection takes the **disseminated form in the elderly, HIV patients and other immunocompromised individuals.** In HIV patients, tuberculous meningitis may arise through blood-borne spread to CNS and typically involves the basilar meninges and, thus, is often complicated by obstructive hydrocephalus and cranial nerve palsies.

In some high-risk patients, untreated infection disseminates to invade skeletal tissues, usually involving the **midthoracic vertebral bodies, causing osteomyelitis (Pott disease).**

OUTCOME

Investigative cultures showed positive growth of *M. tuberculosis* in 3 weeks based on BACTEC culture and GenProbe-based identification of the isolate. The patient was placed on INH, RIF, PZA, and EMB for 2 months with a good clinical response. Later therapy was completed with isoniazid and rifampin for a total of 6 months. A mild, asymptomatic elevation in liver enzymes during therapy was noted.

PREVENTION

In exposed individuals younger than 35 years who are PPD positive, INH therapy is warranted. Bacillus of Calmette and Guérin (BCG) vaccine may be useful in superendemic areas. In a recent trial, BCG vaccine efficacy was found to persist for 50 to 60 years, suggesting that a single dose of an effective BCG vaccine can have a long duration of protection. However, BCG vaccination has no role in the United States because of the low prevalence of disease. Skin test can be difficult to evaluate in patients who received BCG vaccine, but the general recommendation is to interpret the PPD as if the patient had not received the BCG in the past. In endemic areas, BCG vaccine may be used in children who are constantly exposed to tubercle bacilli (children cannot be placed on INH therapy). The vaccine protects against meningitis in high-risk communities.

FURTHER READING

Aronson NE, Santosham M, Comstock GW, et al: Long-term efficacy of BCG vaccine in American Indians and Alaska natives: A 60-year follow-up study. JAMA 291:2086, 2004.

Frieden TR, Fujiwara PI, Washko RM: Tuberculosis in New York city—turning the tide. N Engl J Med 333:229, 1995.

Havlir DV, Barnes PF: Tuberculosis in patients with human immunodeficiency virus infection: Current concepts. N Engl J Med 340:367, 1999.

Issar Smith I: *Mycobacterium tuberculosis* pathogenesis and molecular determinants of virulence. Clin Microbiol Rev 16:463, 2003.

Small PM, Fujiwara PI: Management of tuberculosis in the United States: Medical progress. N Engl J Med 345:189, 2001.

van Crevel R, Ottenhoff THM, van der Meer JWM: Innate immunity to *Mycobacterium tuberculosis*. Clin Microbiol Rev 15:294, 2002.

A 29-year-old woman suffered from a dry cough, shortness of breath, and pleuritic chest pain. She continued to have these symptoms for more than a week and subsequently developed fever, chills, and bloody cough.

Two weeks before this episode, the patient had been admitted for **treatment of leukemia** to a hospital that was **in the midst of a major reconstruction** project. While hospitalized, she **received cytotoxic chemotherapy** and developed severe **neutropenia,** which had persisted to the current presentation.

C A S E

16

PHYSICAL EXAMINATION

VS: T 38.9°C, P 114/min, R 36/min, BP 106/62 mmHg

PE: An ill-appearing female in moderate respiratory distress. Bilateral rales over both lungs were heard.

LABORATORY STUDIES

Blood

Hematocrit: 24%

WBC: 1100/μL **(leukopenia)**

Differential: 5% PMNs (absolute neutrophil count 55), 80% lymphs, 10% monos, 5% blasts

Platelets: 12,000/μL **(thrombocytopenia)**

Blood gases: pO$_2$ 62 mmHg

Serum chemistries: Normal

Imaging

A chest x-ray showed a **wedge-shaped lesion in the left lung and a right middle lobe infiltrate** (Fig. 16-1). A CT scan revealed small pulmonary nodules and a hazy rim **(halo sign)** with ground-glass attenuation.

Diagnostic Work-Up

Table 16-1 lists the likely causes of illness (differential diagnosis). Investigational approach may include

- **Direct microscopy**

  - **Acid-fast stain, Gram stain** of sputum
  - **Cytology staining** and microscopy of broncho-scopic biopsy or sputum using KOH/calcofluor, Giemsa, and silver stain to demonstrate fungal elements

- **Fungal cultures**
- **Serum antigen** testing for *Aspergillus* galactoman-nan, which (in the appropriate clinical setting) can be useful
- **Blood cultures** to rule out bacteremia

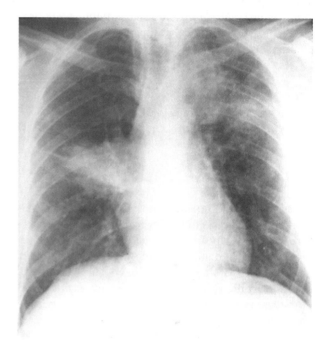

FIGURE 16-1 An admission chest film showing bilateral infiltrates in this patient. *(Courtesy of Dr. Paul Southern, Departments of Pathology and Medicine, University of Texas Southwestern Medical Center, Dallas, TX.)*

TABLE 16–1 Differential Diagnosis and Rationale for Inclusion (consideration)
Aspergillosis
Bacterial pneumonia (*Pseudomonas aeruginosa, Staphylococcus aureus*)
Legionellosis
Nocardiosis
Pneumocystis pneumonia (*Pneumocystis jiroveci*)
Zygomycosis (e.g., mucor)

Rationale: The clinical scenario described above is highly characteristic for invasive pulmonary aspergillosis in patients with neutropenia. While the other organisms listed can also cause severe pneumonia, the CT findings in this case suggest aspergillosis. Other filamentous fungi, or molds, such as mucor, can also cause similar presentations. Other causes of community-acquired pneumonia, as well as *Nocardia*, should always be considered in immunocompromised patients.

COURSE

Sputum and blood were collected. The patient was started on an empirical regimen of drugs to cover the major agents listed in the differential diagnosis. Gram stain and acid-fast stain of sputum did not reveal any significant pathogen. Invasive procedures, such as bronchoscopy, were not possible due to severe thrombocytopenia. A galactomannan antigen assay was positive. Blood cultures became negative in five days. Sputum cultures grew out a likely pathogen.

ETIOLOGY

Aspergillus fumigatus (invasive pulmonary aspergillosis)

MICROBIOLOGIC PROPERTIES

Aspergillus species are the most common mold that cause human disease. The classic morphologic appearance of the organism in stained tissue is that of **thin hyphae** of even diameter **(2 to 4 μm)** that **branch at a V-shaped (45°) angle** (Fig. 16-2). These molds are **not dimorphic**. Three major species are *A. fumigatus*, *A. flavus*, and *A. niger*. Of human diseases caused by the *Aspergillus* species, up to 90% is caused by *A. fumigatus*. In patients at high risk, the culture of *Aspergillus* species from respiratory samples, even in the absence of histology, is highly suggestive of infection and generally warrants institution of therapy. The mycologic culture of *A. fumigatus* is identified by its gross and microscopic appearance. In tissue, it is not possible to distinguish *Aspergillus* from other molds.

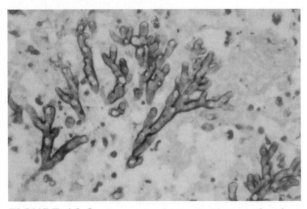

FIGURE 16-2 Silver stain of *Aspergillus fumigatus* from lung biopsy. Note acute angle branching of hyphae. *(Courtesy of Dr. Paul Southern, Departments of Pathology and Medicine, University of Texas Southwestern Medical Center, Dallas, TX.)*

EPIDEMIOLOGY

This fungus is ubiquitous in nature and is associated with decaying vegetation. **Inhalation of airborne conidia** (spores) is the major mode of acquisition; it is not transmitted from person to person. **Cytotoxic drugs** that lead to neutropenia greatly **increase susceptibility** to this infection, as is seen in the case patient. An increase in the environmental load of conidia (as may occur during building reconstruction and renovation projects) may lead to increased risk of disease. Other risk factors include corticosteroid use, organ transplantation, graft-versus-host disease, and advanced AIDS.

PATHOGENESIS

Infection occurs following the inhalation of conidia. *A. fumigatus* is an invasive pathogen. A number of **adhesins** (e.g, complement receptors, hydrophobins) are involved in the colonization of lower respiratory airways. The host immune response to the mold determines whether the organism is cleared or whether disease develops. Along with neutrophils (which are actively recruited during inflammation), the alveolar macrophages are the major cells involved in the phagocytosis of *A. fumigatus*. Lectin-like interactions are thought to be primarily responsible for the adherence and ingestion of conidia by alveolar macrophages. Hyphae are destroyed by neutrophils in normal hosts. Invasive disease develops in patients with phagocyte deficiencies (**chronic granulomatous disease** [defect in phagocytosis of neutrophils owing to lack of NADPH oxidase activity]) and primarily in patients with **profound neutropenia**. In the neutropenic patient, *A. fumigatus* invades the lungs, producing granulomas that are seen in chest x-rays. Fungal **hydrolases** (e.g., serine protease or phospholipase) and toxic molecules (e.g., **hemolysin**, secondary metabolites) of the angioinvasive pathogen are involved in endothelial damage, causing **hemoptysis** and other manifestations of invasive pulmonary aspergillosis.

TREATMENT

Amphotericin B has been the standard for therapy for many years, and various amphotericin B lipid preparations are available to help reduce nephrotoxicity. However, **voriconazole** (newer azole drug) is now considered the drug of choice for invasive pulmonary aspergillosis, based on a randomized study that showed improved survival over amphotericin B. Itraconazole also has activity against *Aspergillus* and may be considered for use in less immunocompromised patients. Another new anti-fungal drug, caspofungin, has activity against most *Aspergillus* spp. and is considered a second-line agent for invasive aspergillosis. The duration of treatment depends on improvement in host defenses and severity of disease.

> **NOTE** In uncontrolled invasive aspergillosis, **hematogenous dissemination** to various tissues (e.g., **brain and kidneys**) frequently occurs, especially in a profoundly neutropenic patient; it has a case-fatality rate of 50% to 90% even with appropriate therapy. Treatment of other diseases caused by *Aspergillus* (Table 16-2) is mostly symptomatic and may also include an azole drug.

OUTCOME

The patient was given granulocyte-colony stimulating factor (GCSF) to help improve her neutrophil counts, and she also received voriconazole. She gradually recovered after several weeks of therapy. She was started on chemotherapy again after her full recovery.

PREVENTION

No immunization is available. **Recovery from neutropenia** is necessary. It is achieved by reducing the levels of implicated drugs (e.g., cytotoxic chemotherapy) and administration of colony-stimulating factors such as GCSF. High-efficiency particulate air (HEPA) filtration has also been recommended for high-risk patients.

TABLE 16–2 Other Clinical Features of *Aspergillus* Infections

Syndromes	Clinical Features
Allergic disease	Aberrant host immune response to *Aspergillus* in an immunocompetent person can result in hypersensitivity diseases. **Allergic disease** (e.g., sinusitis, hypersensitivity pneumonitis) occurs following **repeated exposure** to *Aspergillus* conidia in patients without mycelial colonization or invasion.
Allergic bronchopulmonary aspergillosis (ABPA)	ABPA, an uncommon but distinct syndrome of **poorly controlled asthma,** is caused by a vigorous immune response to *Aspergillus* hyphae that have colonized the airways. In patients with **COPD** and **allergy to the mold,** ABPA may progress to bronchiectasis and permanent lung damage.
Aspergilloma	*Aspergillus* grows profusely in the **pulmonary cavities** of cavitary tuberculosis, producing a **"fungus ball,"** which can be **seen on chest x-ray.** Five to 10% of these patients develop massive, life-threatening hemoptysis.

FURTHER READING

Corey L, Boeckh M: Persistent fever in patients with neutropenia: Perspective. N Engl J Med 346:222, 2002.

Jean-Paul L: *Aspergillus fumigatus* and aspergillosis. Clin Microbiol Rev 12:310, 1999.

Powers CL: Diagnosis of infectious diseases: A cytopathologist's perspective. Clin Microbiol Rev 11:341, 1998.

Five workers between the ages of 29 and 48 years presented to an acute care clinic with a **1-week history of fever, chills, night sweats, cough,** headache, fatigue, myalgia, and **weight loss.**

Two weeks before the onset of symptoms, these five workers had begun partial **demolition of an abandoned building** in a small city near the **Ohio River.** At the time of demolition, a colony of pigeons was observed in the building, and **pigeon droppings** were found throughout the building. During the demolition, none of the workers wore personal protective equipment (i.e., respirators, eye protection, gloves, or protective clothing).

PHYSICAL EXAMINATION

Physical examination of one of the workers (Patient X) is presented here:

VS: T 38.3°C, P 96/min, R 24/min, BP 124/82 mmHg
PE: Crackles were noted on auscultation of the lung.

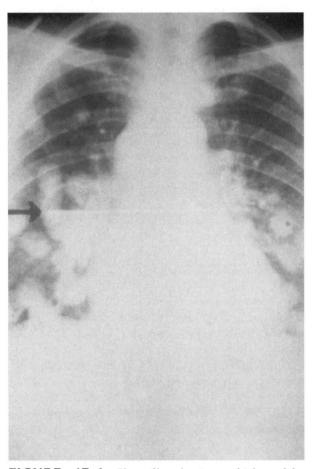

FIGURE 17-1 Chest film showing multiple nodular infiltrates in both lung fields and a cavitary lesion in the right middle lobe in this patient. Note popcorn calcification (*arrow*). (*Courtesy of Dr. Paul Southern, Departments of Pathology and Medicine, University of Texas Southwestern Medical Center, Dallas, TX.*)

LABORATORY STUDIES (PATIENT X)

Blood

Hematocrit: 36%
WBC: 12,200/μL
Differential: 70% PMNs, 15% lymphs, 15% monos
Platelets: 360,000/μL
Blood gases: pO_2 68 mmHg
Serum chemistries: Normal

Imaging

A chest x-ray from Patient X was remarkable for **enlarged hilar** and **mediastinal nodes** and multiple **nodular infiltrates** (Fig. 17-1).

Diagnostic Work-Up

Table 17-1 lists the likely causes of illness (differential diagnosis). Gram and acid-fast stains of sputum, blood

TABLE 17-1 Differential Diagnosis and Rationale for Inclusion (consideration)

Blastomycosis
Coccidioidomycosis
Cryptococcosis
Histoplasmosis
Legionellosis and other atypical pneumonia
Nocardiosis
Pneumococcal pneumonia
Tuberculosis

Rationale: This epidemiologic history is classic for histoplasmosis, often associated with pigeon droppings. Blastomycosis and cryptococcosis are less common, and coccidioidomycosis is not found in that region. The other agents in the differential diagnosis may be included until proven otherwise. The chest x-ray may be consistent with nocardiosis, legionellosis, or even primary tuberculosis, but the history is not. It would be an unusual presentation for pneumococcal pneumonia.

culture, and routine bacterial cultures of respiratory specimens may be necessary to rule out the agents included in the differential diagnosis. More specific investigational approach may include

- **Direct microscopy** of tissue (lung, spleen) biopsy sections or aspirate stained with Gomori methenamine silver stain
- **Antigen detection test** in serum or urine by immunodiffusion
- **Serology** (also by immunodiffusion) test
- **Mycologic cultures**

COURSE

All five workers (including Patient X) required treatment for acute respiratory illnesses, and three (including Patient X) were hospitalized. Lung biopsies were obtained from the three hospitalized patients; Giemsa-stained tissue from the lung biopsy of Patient X suggested the presence of an endemic fungal agent.

ETIOLOGY

Histoplasma capsulatum (histoplasmosis)

MICROBIOLOGIC PROPERTIES

Histoplasma capsulatum var. *capsulatum* is a dimorphic fungus. It **grows as a mold form in soil environment,** but **in the human host at 37°C, it converts to yeast form**. The **yeasts are seen in tissue as thin-walled, oval structures** (2 to 5 µm; Fig. 17-2), visualized by staining with Giemsa or Wright stains of smears of bone marrow, sputum, blood, or lung biopsy.

The mold form appears in cultures in the laboratory. Growth of the mold form on mycologic agar may take up to 4 weeks of incubation. **The aerial mycelial growth produces characteristic macroconidia (thick wall; finger-like projections)** and microconidia. The mold produces two glycoproteins, H and M, detectable by the agar immunodiffusion test. Fungal antigens can be detected in serum or urine. The test is very specific, and often results are available much sooner than culture results.

EPIDEMIOLOGY

H. capsulatum is **endemic** in parts of the central and eastern United States along **Ohio and Mississippi river valleys.** Microfoci are also found in Central and South America, Africa, India, and Southeast Asia. *H. capsulatum* grows in **soil contaminated with bat or bird droppings**. Environments around starling, blackbird, and pigeon roosts are commonly associated with fungal conidia. **Conidia become airborne when contaminated soil is disturbed (i.e., during demolition of old buildings); infection results from inhalation of conidia** from a common source. High-risk groups include immunologically naïve individuals going into endemic areas (e.g., construction or agricultural workers, spelunkers) and immunocompromised persons (e.g., persons with cancer, transplant recipients, persons with AIDS).

PATHOGENESIS

The respiratory tract is the portal of entry for *H. capsulatum* in most patients. After small microconidia have reached the alveoli, they bind to the CD2/CD18 family of integrins and are engulfed by both neutrophils and macrophages, where **microconidia transform into budding yeast** forms. The transition from the mycelial to the yeast phase is one of the most critical determinants for establishing infection. The **yeasts grow intracellularly in the inactivated alveolar macrophages**. With time, an intense granulomatous reaction occurs and produces calcified fibrinous granulomas with areas of caseous necrosis in lungs. The granulomas consist of a mixture of mononuclear phagocytes and lymphocytes, principally T cells. The yeasts migrate to local draining lymph

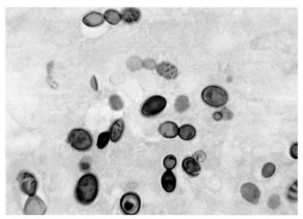

FIGURE 17-2 Yeasts of *Histoplasma capsulatum*. Note the typical budding yeast cells. Methenamine silver stain was used here.

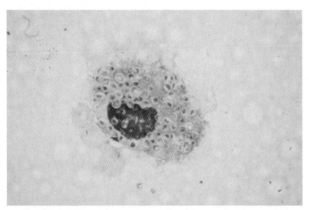

FIGURE 17-3 Histiocyte containing numerous yeast cells of *Histoplasma capsulatum*. Wright stain. *(Courtesy of Dr. Paul Southern, Departments of Pathology and Medicine, University of Texas Southwestern Medical Center, Dallas, TX.)*

nodes and, subsequently, to distant organs rich in mononuclear phagocytes, such as the reticuloendothelial system, including liver, spleen, and bone marrow (Fig. 17-3). Cytokine response leads to symptoms such as fatigue and weight loss.

TREATMENT

Most infections in individuals who are immunocompetent are self-limiting and do not require ther-

apy. Oral **itraconazole** is the drug of choice for the treatment of less severe manifestations in immunocompetent individuals.

> NOTE Some patients with preexisting lung diseases such as emphysema may fail to recover—even with therapy—developing a **chronic lung disease** that **resembles tuberculosis.** Permanent lung damage may occur. In immunocompromised patients (e.g., those with **AIDS**), **the initial pulmonary infection may disseminate,** producing extrapulmonary manifestations that affect mucosal surfaces, liver, spleen, adrenal gland, and meninges. For these patients with severe symptoms and disseminated histoplasmosis, intravenous **amphotericin B** is the drug of choice.

OUTCOME

All five patients received itraconazole for several weeks and recovered without complications, although symptoms took more than 2 weeks to resolve.

PREVENTION

There is no vaccine available, but risk can be lessened in highly contaminated environments by use of personal protective gear.

FURTHER READING

Bradsher RW: Histoplasmosis and blastomycosis. Clin Infect Dis 22:S102, 1996.

Wheat J: Histoplasmosis. Experience during outbreaks in Indianapolis and review of the literature. Medicine 76:339, 1997.

Wheat LJ, Chetchotisakd P, Williams B, et al: Factors associated with severe manifestations of histoplasmosis in AIDS. Clin Infect Dis 30:877, 2000.

Wheat LJ, Kohler RB, Tewari RP: Diagnosis of disseminated histoplasmosis by detection of *Histoplasma capsulatum* antigen in serum and urine specimens. N Engl J Med 314:83, 1986.

A **38-year-old man** was seen by his family physician for **two weeks of fever, productive cough, right-sided pleuritic chest pain, a 12-lb. weight loss,** and a **painful lesion on his left arm.**

The patient, a long time resident of **Tennessee,** had no significant past medical history. He had **worked outdoors as a landscaper** for the last 18 years. Family history was unremarkable. He had been **treated with oral cefuroxime** and erythromycin for suspected pneumonia 10 days prior to this visit, but his symptoms had **not improved.**

PHYSICAL EXAMINATION

VS: T 38.3°C, P 90/min, R 34/min, BP 126/76 mmHg

PE: Examination was remarkable for crackles in the left lung and a **verrucous skin lesion** (1 × 1 cm) **on his left arm that was tender and erythematous.**

LABORATORY STUDIES

Blood

Hematocrit: 38%
WBC: 19,000/µL
Differential: 79% PMNs
Platelets: 612,000/µL
Blood gases: Normal
Serum chemistries: Normal

Imaging

A **chest x-ray revealed multiple nodular lesions,** some of which were cavitating, in the left upper lobe. Left arm x-ray revealed no bony abnormalities.

Diagnostic Work-Up

Table 18-1 lists the likely causes of illness (differential diagnosis). Gram and acid-fast stains of sputum, blood culture, and routine bacterial cultures of respiratory specimens may be necessary to rule out the agents included in the differential diagnosis. More specific investigational approach may include

- **Direct microscopy** of sputum or bronchoscopic specimen and a tissue biopsy (lesions)
- **Mycologic cultures** of bronchoscopic specimen and a biopsy of lesion
- **DNA test** using chemiluminescent DNA probes

TABLE 18–1   Differential Diagnosis and Rationale for Inclusion (consideration)

Carcinoma
Endemic mycosis (e.g., histoplasmosis, blastomycosis, coccidioidomycosis)
Lung abscess
Nocardiosis
Sarcoidosis
Tuberculosis

Rationale: All of the endemic fungi may cause a pulmonary infection that is indistinguishable from tuberculosis based on clinical presentation. Upper-lobe disease is especially characteristic of tuberculosis. In appropriate geographic regions, endemic mycoses (histoplasmosis and blastomycosis) are the major agents included in the differential diagnosis. Lung abscess may also present in an indolent manner, as can nocardiosis. Sarcoidosis and malignancy should be considered when cultures are negative.

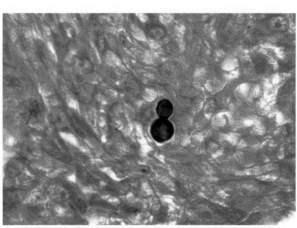

FIGURE 18-1 Yeast form of the pathogen undergoing broad-base budding. *(Courtesy of Centers for Disease Control and Prevention, Atlanta, GA.)*

COURSE

The patient was admitted to the hospital and placed on piperacillin/tazobactam and levofloxacin. Sputum cultures and Gram stains were negative for acid-fast bacilli and other classified organisms. An HIV sero-conversion test was nonreactive. Biopsy obtained by bronchoscopy revealed characteristic budding yeast. A culture of a biopsy from the left arm lesion on a nutritionally rich medium after 2 weeks of incubation at 37°C confirmed the diagnosis.

ETIOLOGY

Blastomyces dermatitidis (blastomycosis)

MICROBIOLOGIC PROPERTIES

B. dermatitidis is a **dimorphic fungus**. The fungus grows slowly at room temperature (25°C) on Sabouraud agar as a **white fluffy mold**. It produces a single terminal microconidium, which is round or oval. At 37°C on blood agar the fungus grows as brown, wrinkled colonies. The **large yeast forms (10 to 12 μm) with broad-based budding** in biopsies visualized by cytologic stains are virtually diagnostic (Fig. 18-1).

EPIDEMIOLOGY

Blastomycosis occurs sporadically in the United States, Canada, and Central and South America. The mycosis is **endemic in the southeastern region** of the United States **(states east of the Mississippi River)**, where the disease is also known as North American blastomycosis to differentiate it from South American blastomycosis, caused by *Paracoccidioides braziliensis*. Wooded areas along waterways with decaying vegetation in warm, moist soil have been found to harbor the fungus. Its precise ecology is still poorly understood. Infection results from inhalation of microconidia.

PATHOGENESIS

Inhalation of the microconidia and conversion to the yeast forms in the lungs leads to infection. The shift from the infective mold form to the pathogenic yeast form occurs at body temperature. Incubation time averages 4 to 6 weeks and varies widely. Yeasts increase in number in the lung parenchyma, where neutrophils are heavily recruited. The tissue response is a combination of **suppurative** and **granulomatous inflammation**. Lymphohematogenous spread to other organs commonly occurs. Skin involvement typically shows pseudoepitheliomatous hyperplasia with focal **microabscesses** in the papillary dermis. This often presents as **verrucous skin lesions** with pustular features.

TREATMENT

Itraconazole is the treatment of choice for adult immunocompetent patients with mild-to-moderate pulmonary or extrapulmonary disease, excluding CNS involvement. Amphotericin B is indicated in severely ill patients or in those with brain lesions.

> **NOTE** In some patients, pulmonary blastomycosis takes the form of chronic disseminated disease with spread to skin and bones. Skin lesions begin as erythematous papules that become verrucous, crusted, or ulcerated and spread slowly.

OUTCOME

The patient received itraconazole, and after several days, his fevers subsided. At discharge, he was given a maintenance course of itraconazole for a total duration of 6 months.

PREVENTION

There is no vaccine available. Prevention is difficult, as its environmental niche is not well known. Blastomycosis is not transmissible from person to person.

FURTHER READING

Meyer KC, McManus EJ, Maki DG: Overwhelming pulmonary blastomycosis associated with the adult respiratory distress syndrome. N Engl J Med 329:1231, 1993.

Yeo SF, Wong B: Current status of nonculture methods for diagnosis of invasive fungal infections. Clin Microbiol Rev 15:465, 2002.

A 35-year-old man was admitted to the hospital with a **3-week history of fever, night sweats,** headache, **joint pains, dry cough,** and severe fatigue. He had **lost 16 lb** in the past two weeks.

The patient had returned to his home in New York, one month after a brief visit with his mother in **Phoenix, Arizona.**

PHYSICAL EXAMINATION

VS: T 38.9°C, P 106/min, R 28/min, BP 130/58 mmHg

PE: The patient appeared ill and slightly pale. Rales were heard in the right lower lobe. **Erythema nodosum** lesions on his back were noted.

LABORATORY STUDIES

Blood

Hematocrit: 32%

ESR: 62 mm/h

WBC: 8,600/μL

Differential: 60% PMN, 18% lymphs, 6% monos, 16% eosinophils

Blood gases: pO_2 72 mmHg

Serum chemistries: Normal

Imaging

A chest x-ray revealed **infiltrates in both lung fields with a large cavity in the right upper lobe** (Fig. 19-1).

Diagnostic Work-Up

Table 19-1 lists the likely causes of illness (differential diagnosis). Gram and acid-fast stains of sputum, blood culture, and routine bacterial cultures of respiratory specimens may be necessary to rule out the agents included in the differential diagnosis. More specific investigational approach may include

- **Direct microscopy** to visualize fungal structures in silver-, Giemsa-, or Wright-stained biopsy or aspirate specimens
- **Serology:**
  - Immunodiffusion (ID) test
  - Complement fixation using paired sera
- **Fungal cultures** (State Laboratory or CDC)

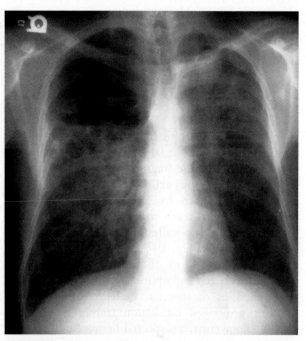

FIGURE 19-1 An admission chest film showing bilateral infiltrates with a large cavity in the right upper lobe in this patient. *(Courtesy of Dr. Daniel Skiest, Department of Medicine, University of Texas Southwestern Medical Center, Dallas, TX.)*

TABLE 19–1 Differential Diagnosis and Rationale for Inclusion (consideration)
Blastomycosis
Coccidioidomycosis
Histoplasmosis
Legionella, or other atypical pneumonia
Nocardiosis
Pneumococcal pneumonia
Tuberculosis (*Mycobacterium tuberculosis*)

Rationale: The most important issue when considering a differential diagnosis that includes endemic fungi is where the patient is living or has traveled to recently. Tuberculosis may manifest similarly as a chronic lower respiratory tract infection. Pneumococcus and *Legionella* are more common causes of acute pneumonia than fungi and *M. tuberculosis*. *Nocardia* can cause chronic symptoms but is less common in this low-risk patient.

COURSE

The patient was admitted to the hospital and placed on broad-spectrum antibiotics pending results of cultures and other diagnostic tests. Serology yielded the diagnosis in a reference laboratory.

ETIOLOGY

Coccidioides immitis (coccidioidomycosis)

MICROBIOLOGIC PROPERTIES

C. immitis is a **dimorphic** fungus. In soil, it grows as a mold with branching septate hyphae. When the soil is disturbed, the hyphae fragment and form extremely hardy structures called **arthroconidia**, which become airborne. If inhaled by animals or humans, the arthroconidia can reach the pulmonary alveoli and transform into **thick-walled, nonbudding spherules** (pathogenic form; Fig. 19-2), which form septa and produce hundreds to thousands of uninucleate **endospores**. Each endospore is capable of producing new spherules or mycelia. **Culture** of *C. immitis* requires prolonged incubation (rarely done when coccidioidomycosis is suspected because biohazard is a problem for the diagnostic laboratory) and yields white fluffy mold on most culture media. Therefore,

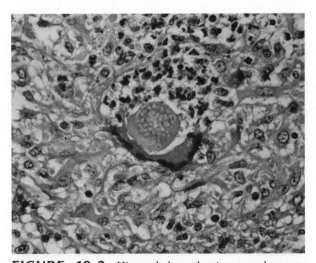

FIGURE 19-2 Histopathology showing granulomatous inflammation with a multinucleated giant cell and neutrophils surrounding an endospore-filled mature spherule of *Coccidioides immitis*. (*Courtesy of Dr. Dominick Cavuoti, Department of Pathology, University of Texas Southwestern Medical Center, Dallas, TX.*)

the test of choice when coccidioidomycosis is suspected is serology similar to the tests (e.g., agar immunodiffusion) undertaken for histoplasmosis or blastomycosis.

EPIDEMIOLOGY

C. immitis is endemic in the **southwestern United States, California (San Joaquin Valley;** local endemic disease is known as **"valley fever"),** and the western half of Texas. Other endemic areas are the regions of Mexico that border the southwestern desert of the United States. Arid soil in and around rodent burrows is a common reservoir for hyphae and arthroconidia. **Infection results from inhalation of arthroconidia** after disturbance of contaminated soil by humans or dust storms. Nonimmune individuals visiting endemic areas can develop this infection after overwhelming inhalation exposure.

PATHOGENESIS

A single *C. immitis* arthroconidium may be sufficient to produce a naturally acquired respiratory infection. The size of the **arthroconidium** (3 to 5 µm) allows it to be **deposited in the terminal bronchiole.** In this location, an arthrospore develops into a thick-walled spherule filled with endospores. As an arthroconidium transforms into a spherule, a pyogenic inflammation is mounted with an infiltrate of neutrophils. Cell-mediated immunity becomes a key factor in determining recovery from coccidioidomycosis and, eventually, a **chronic granulomatous inflammation** (Fig. 19-2) ensues similar to that of tuberculosis. Caseation without calcification may occur, encasing the granulomas and resulting in pulmonary lesions. Recovery is generally followed by lifelong immunity. Reactivation can occur in those who become immunosuppressed therapeutically or by HIV infection. Dissemination to skin causes characteristic **erythema nodosum** lesions (seen in the case patient).

TREATMENT

Oral **fluconazole** or **itraconazole** for 3 to 6 months in immunocompetent patients is efficacious to treat primary infection. Amphotericin B is required to treat severe pneumonia, particularly in immunocompromised patients.

NOTE Patients with depressed cellular immunity, such as pregnant women; those with lymphoma, HIV infection, or organ transplants; or those receiving high-dose corticosteroids also are more likely to develop dissemination. Skin, bones, joints, and meninges are the most frequent sites of dissemination. Meningitis may lead to permanent neurologic damage and is fatal if untreated. For patients with meningitis, fluconazole is preferred because of its excellent CSF penetration; therapy must be continued for life to prevent reactivation.

OUTCOME

The patient received intravenous amphotericin B for 10 days, and his symptoms subsided. He was discharged home with a maintenance course of oral fluconazole for 6 months to complete therapy.

PREVENTION

An effective vaccine is not yet available. Hospitalized patients do not require special infection control measures other than standard precautions. There is no person-to-person transmission.

FURTHER READING

Ampel NM, Mosley DG, England B, et al: Coccidioidomycosis in Arizona: Increase in incidence from 1990 to 1995. Clin Infect Dis 27:1528, 1998.

Arsura EL, Kilgore WB, Ratnayake SN: Erythema nodosum in pregnant patients with coccidioidomycosis. Clin Infect Dis 27:1201, 1998.

Rosenstein NE, Emery KW, Werner SB, et al: Risk factors for severe pulmonary and disseminated coccidioidomycosis: Kern County, California, 1995-1996. Clin Infect Dis 32:708, 2001.

Stevens DA: Coccidioidomycosis: Current concepts. N Engl J Med 332:1077, 1995.

A 44-year-old white man presented with a **3-month history** of intermittent **fever, chills,** and a cough **production of green sputum.** He also complained about weakness, weight loss, chest pain, and shortness of breath. He had been given several courses of **antibiotics without significant improvement** and had noted the presence of a **headache** for the past few weeks. He had a history of **COPD** and had been on chronic **steroids** for the past 6 months.

PHYSICAL EXAMINATION

VS: T 39°C, P 118/min, R 36/min, BP 148/90 mmHg

PE: The patient appeared ill. Decreased breath sounds in the right lung were heard. Slight weakness of the left upper extremity was also noted.

LABORATORY STUDIES

Blood

Hematocrit: 34%
WBC: 5800/µL
Differential: 63% PMNs, 5% bands, 32% lymphs

Blood gases: pO_2 62 mmHg
Serum chemistries: Normal

Imaging

A chest x-ray revealed **extensive nodular infiltrates in the right middle and upper lobes and cavitary disease** (Fig. 20-1).

Diagnostic Work-Up

Table 20-1 lists the likely causes of illness (differential diagnosis). Investigational approach for the etiology of chronic pneumonia may include

- **Gram** and **acid-fast stains** of respiratory or biopsy specimens
- **Cultures** of sputum or biopsy specimens (prolonged incubation of routine blood agar cultures and mycobacterial cultures are required)
- In failed tests, fungal cultures and serology

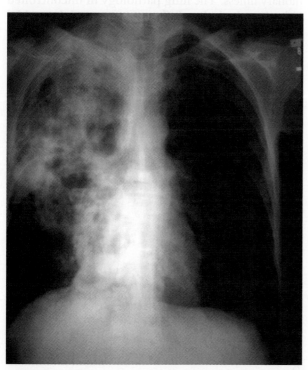

FIGURE 20-1 Admission chest film showing extensive nodular and cavitary disease in the right middle and upper lobes in this patient. *(Courtesy of Dr. Daniel Skiest, Department of Medicine, University of Texas Southwestern Medical Center, Dallas, TX.)*

TABLE 20–1   Differential Diagnosis and Rationale for Inclusion (consideration)
Actinomycosis
Aspergillosis
Endemic mycosis (e.g., histoplasmosis)
Mycoplasma pneumonia
Nocardiosis
Pneumococcal pneumonia
Tuberculosis

Rationale: The list includes chronic lower respiratory tract infections, but focus should be on the patient's immuno-compromised state, resulting from long-term steroid use. Tuberculosis should always be a consideration with upper lobe disease, and fungal causes and *Nocardia* are often associated with chronic symptoms. *Mycoplasma* does not typically cause such unilateral disease, and pneumococcal pneumonia is usually not a chronic infection.

COURSE

The patient was admitted to the hospital. Because of his headache, a head CT was performed, which revealed a lesion in the right parietal lobe. Sputum smear was positive for Gram-positive organisms that were also partially acid-fast positive.

ETIOLOGY

Nocardia asteroides (nocardiosis)

MICROBIOLOGIC PROPERTIES

The nocardiae are **filamentous (beaded)** bacteria belonging to the **aerobic** actinomycetes. They are weakly **Gram-positive** bacteria (Fig. 20-2*A*), and they are weakly **positive** on **acid-fast stain** (see Fig. 20-2*B*). The medically most important species are *N. asteroides* and *N. brasiliensis*. They **grow slowly** on the antibiotic-containing media usually used for fungi; clinicians must alert the clinical lab to the possibility of *Nocardia* so that specimens are cultured on media and held for an appropriate period (up to 4 weeks).

EPIDEMIOLOGY

Inhalation of contaminated dust from a soil environment is the major mode of acquisition. **Patients receiving cytotoxic or immunosuppressive drugs (particularly steroids) and patients with AIDS are at high risk** for nocardiosis. In the tropical countries, nocardiosis occurs as a result of skin inoculation in agricultural workers, presenting as a chronic, subcutaneous infection, characterized either by slow extension along lymphatics or by destruction of deeper tissues ("**Madura foot**").

PATHOGENESIS

Pleuropulmonary disease arises following inhalation of filamentous bacteria. The bacteria are phago-cytosed by neutrophils and alveolar macrophages but are not killed by inactivated phagocytic cells. The organisms prevent the phagosome-lysosome fusion and acidification of phagosome that are required for killing. The facultative intracellular bacteria grow within the inactivated phagocytic cells. Both activated macrophages and immunologically specific T lympho-cytes constitute the major mechanisms for host resist-ance to infection due to nocardiae. Granulomatous inflammation can contain the infectious process. Release of cytokines and other mediators, however, contributes to the symptoms and signs of the pul-monary illness. The **lung pathology in uncontrolled infection includes inflammatory endobronchial masses or diffuse pneumonitis, and abscess.** Chest x-rays often show nodules, which may go on to cavitate when large.

Unmanaged infection in a patient lacking T-cell immunity may extend into contiguous structures or disseminate to virtually any organ. **CNS involvement**

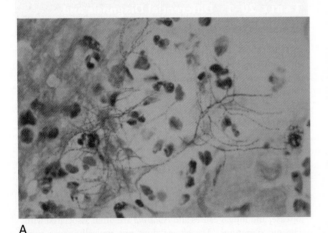

A

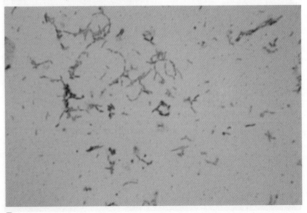

B

FIGURE 20-2 *A,* Gram stain of *Nocardia asteroides* from sputum. Note: long branching filaments that resemble the hyphae of fungi. *B,* Acid-fast stain of *Nocardia asteroides.* (*Courtesy of Dr. Paul Southern, Departments of Pathology and Medicine, University of Texas Southwestern Medical Center, Dallas, TX.*)

(as in this case) is common and usually takes the form of one or more abscesses (visualized by head CT or MRI). CNS disease symptoms are usually headache, lethargy, confusion, seizures, and sudden onset of neurologic deficit.

TREATMENT

Improvement of patient's immune status, if possible (e.g., tapering steroid use), is certainly useful, but prolonged antibiotic therapy is essential, usually for 6 to 12 months. **Sulfonamides** are first-line antimicrobial therapy for nocardiosis. Among the sulfonamides, trimethoprim/sulfamethoxazole is generally preferred because it is well tolerated. Imipenem and amikacin among others are alternatives.

OUTCOME

The patient was treated with high-dose trimethoprim/sulfamethoxazole. His symptoms improved over the course of a week, and antibiotics were continued for a total of 12 months, with complete resolution of the infection.

PREVENTION

There are no vaccines available. Improving immune status is necessary, and it can be achieved by reducing the levels of implicated drugs (e.g., steroids) when possible.

F U R T H E R   R E A D I N G

Beaman BL, Beaman L: Nocardia species: Host-parasite relationships. Clin Microbiol Rev 7:213, 1994.
Boiron P, Locci R, Goodfellow M: Nocardia, nocardiosis and mycetoma. Med Mycol 36 (Suppl 1):26, 1998.

Lerner PI: Nocardiosis. Clin Infect Dis 22:891, 1996.

A 20-year-old man presented to the emergency department with a **3-week history of moderate fever,** cough, shortness of breath, **weight loss,** and anorexia. He complained about a **draining lesion on his left chest** wall.

The patient had recently **emigrated from Pakistan** and had a history of **severe periodontal disease.**

PHYSICAL EXAMINATION

VS: T 38°C, P 82/min, R 24/min, BP 136/84 mmHg

PE: The patient was ill appearing and had a **productive cough with foul-smelling sputum**. Abnormal breath sounds were heard, and a **sinus tract** from the right chest wall **with drainage containing yellow granules** was noted.

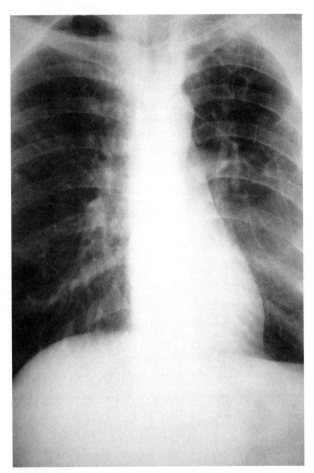

FIGURE 21-1 An admission chest film showing left lung infiltrates with upper lobe cavity in this patient. *(Courtesy of Dr. Borna Mehrad, Department of Medicine, University of Texas Southwestern Medical Center, Dallas, TX.)*

LABORATORY STUDIES

Blood

Hematocrit: 32%
WBC: 5200/μL
Differential: 78% PMNs, 18% lymphs
Serum chemistries: Normal

Imaging

Chest x-ray revealed left lung infiltrates with upper lobe cavity (Fig. 21-1).

Diagnostic Work-Up

Table 21-1 lists the likely causes of illness (differential diagnosis). Laboratory investigation may be initiated with microscopic **examination** of sulfur granules in pus (from sinus tract), crushed and examined by

- **Gram stain** for *Nocardia* or *Actinomyces*
- **Tissue stain** (e.g., H & E, Giemsa, silver)

The additional tests may include acid-fast stain to rule out TB by examining sputum specimens and aerobic and anaerobic cultures.

TABLE 21-1 Differential Diagnosis and Rationale for Inclusion (consideration)
Actinomycosis
Anaerobic lung abscess
Endemic mycosis (e.g., blastomycosis)
Nocardiosis
Tuberculosis (TB)

Rationale: Symptoms of chronic pneumonia can be difficult to differentiate between causative organisms. Tuberculosis is always a consideration with chronic pneumonia. Anaerobic lung abscess is associated with very foul-smelling sputum. Draining sinus tracts with yellow granules are practically diagnostic for actinomycosis. *Nocardia* is usually seen in immunocompromised patients. Blastomycosis can be associated with skin lesions, and other endemic fungi are also possible.

COURSE

The patient was admitted to the hospital. The next morning, he was taken to the operating room. Pus was drained from the chest lesion. Examination of the pus revealed sulfur granules that on microscopy also demonstrated Gram-positive filamentous bacteria. Culture of the pus on anaerobic blood agar yielded a significant anaerobic pathogen. All other tests yielded negative results.

ETIOLOGY

Actinomyces israelii (thoracic actinomycosis)

MICROBIOLOGIC PROPERTIES

Actinomyces are **filamentous, Gram-positive bacteria**. They form long, branching filaments that resemble the hyphae of fungi. They are nonspore forming, and, unlike *Nocardia*, they are **non-acid fast** and **anaerobic**. The most important species is *Actinomyces israelii*. In infected tissue, the organisms often form dense masses known as **sulfur granules.** Anaerobic blood agar culture of aspirated clinical specimens should be extended for a month to isolate the slow-growing *Actinomyces* species. **"Molar tooth"** colonies from sulfur granules from a draining sinus are diagnostic (Fig. 21-2).

EPIDEMIOLOGY

A. israelii organisms are found in the oral cavity (saliva, dental surfaces, and tonsillar crypts) and GI tract as the normal flora. Transmission is **endogenous**. Persons with **poor oral hygiene** are at risk for cervicofacial abscess. Women with **intrauterine contraceptive devices** are at risk for abdominal/pelvic actinomycosis.

PATHOGENESIS

Aspiration of oropharyngeal secretions containing *A. israelii* is the usual mechanism of thoracic actinomycosis. The organisms are unable to cause infection alone, requiring a synergistic presence of other commensals. The synergistic commensals include *Fusobacterium, Prevotella, Porphyromonas* (all anaerobes) and *Capnocytophaga* and *Actinobacillus* (facultative anaerobic oral bacteria). These companion bacteria appear to facilitate the anaerobic growth of the *Actinomyces* species. Once the **polymicrobic infection** is established, the host mounts an **inflammatory** (suppurative and granulomatous) response. A **chronic, suppurative abscess spreads mainly by direct extension to other tissue planes involving the chest wall,** ultimately forming a draining sinus that discharges **sulfur granules** (Fig. 21-3).

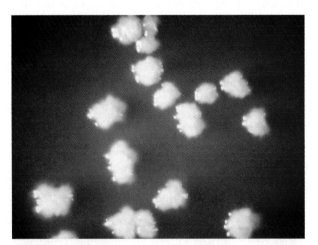

FIGURE 21-2 Plate culture of *Actinomyces israelii*. Note molar tooth colonies. *(Courtesy of Dr. Paul Southern, Departments of Pathology and Medicine, University of Texas Southwestern Medical Center, Dallas, TX.)*

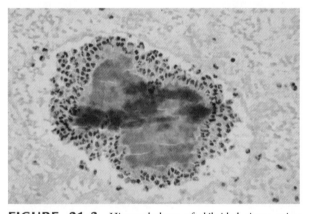

FIGURE 21-3 Histopathology of débrided tissue using H & E stain. Note inflammatory cells around the "sulfur granule" in the middle of the image. These granules actually represent clumped organisms of *Actinomyces israelii*. *(Courtesy of Dr. Paul Southern, Departments of Pathology and Medicine, University of Texas Southwestern Medical Center, Dallas, TX.)*

TREATMENT

Actinomyces are generally highly sensitive to appropriate antibiotics. Prolonged high-dose treatment with IV penicillin G is the mainstay of therapy. This is usually followed by oral amoxicillin for 6 to 12 months. Surgical débridement is often employed for extensive disease, but medical therapy alone has been found to be successful in most cases. Clindamycin or doxycycline can be used in penicillin-allergic patients. Antibiotics with broader spectrums may be used in cases in which other organisms are also isolated.

OUTCOME

The patient was started on high-dose IV penicillin for 4 weeks with substantial improvement in his symp-toms and chest x-ray. Treatment was then changed to oral amoxicillin to complete 12 months of therapy. No further surgery was necessary, and he recovered completely.

PREVENTION

Good oral hygiene can help prevent infection of the lung and cervicofacial region and other endogenous transmission.

FURTHER READING

Geers TA, Farver CF, Adal KA: Pulmonary actinomycosis. Clin Infect Dis 28:757, 1999.
Hsieh MJ, Liu HP, Chang JP: Thoracic actinomycosis. Chest 104:366, 1993.
Pulverer G, Schutt-Gerowitt H, Schaal KP: Human cervicofacial actinomycoses: Microbiological data for 1997 cases. Clin Infect Dis 37:490, 2003.

A 23-year-old man was admitted to the hospital for **fever, nonproductive cough, progressive shortness of breath,** and fatigue for 2 weeks.

He had been diagnosed as **HIV positive** 2 years before, at which time he presented with thrush. He had stopped taking all his HIV-related medications several months ago because of intolerance, and he **had progressed to AIDS.**

PHYSICAL EXAMINATION

VS: T 39.3°C, P 115/min, R 38/min, BP 124/80 mmHg

PE: Patient appeared ill and cachectic. **Thrush** was present, along with mild crackles bilaterally on lung exam.

LABORATORY STUDIES

Blood

Hematocrit: 29%

WBC: 3100/μL

Differential: 55% PMNs, 21% lymphs, 24% monos

CD4$^+$ T-cell count: **80/μL**

Platelets: 260,000/μL

Blood gases: pO$_2$ of 56 mmHg on room air

Serum chemistries: Normal

Imaging

A chest radiograph revealed **bilateral air-space consolidation** with **interstitial** and **alveolar markings** (Fig. 22-1).

Diagnostic Work-Up

Table 22-1 lists the likely causes of illness (differential diagnosis). Sputum samples should be obtained for Gram and acid-fast stains and routine cultures to rule out the likely fungi (including *P. jiroveci*), mycobacteria, and other bacterial causes. Additional tests for specific microbiologic diagnosis may include **methenamine silver stain** or direct fluorescence antibody (DFA) stain of **bronchoalveolar lavage** (BAL) specimens.

COURSE

The consulting doctor suspected interstitial pneumonia and performed bronchoscopy with the objective of obtaining BAL specimens. The patient was

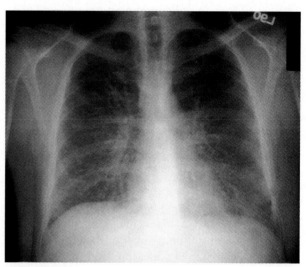

FIGURE 22-1 Admission chest x-ray showing bilateral airspace consolidation with interstitial and alveolar markings.

TABLE 22–1 Differential Diagnosis and Rationale for Inclusion (consideration)
Cryptococcus neoformans
Cytomegalovirus (CMV)
Histoplasma capsulatum
Mycoplasma or other atypical pneumonia
Nocardia spp.
Pneumocystis jiroveci
Respiratory viruses
Tuberculosis (TB)
Typical lower respiratory bacterial pathogens

Rationale: A diagnosis of pneumonia should be considered. In an AIDS patient (CD4$^+$ cell count <200/μL) who is not taking prophylaxis for opportunistic infections, the clinical presentation above is highly suggestive of *P. jiroveci* pneumonia. TB should always be considered in any HIV-positive patient with a respiratory syndrome, due to the greatly increased risk of TB in these patients. Other fungi (including *Histoplasma capsulatum*) and typical lower respiratory bacterial pathogens (including *Streptococcus pneumoniae*) are commonly seen. CMV and *Nocardia* are also likely to be considered, although CMV is usually seen with much lower CD counts (<50). The atypical respiratory bacterial pathogens (e.g., *M. pneumoniae*) and respiratory viruses (e.g., adenovirus) are less likely.

empirically treated. Microscopic examination of the BAL specimen was diagnostic based on DFA stain.

ETIOLOGY

Pneumocystis jiroveci (pneumocystis pneumonia)

MICROBIOLOGIC PROPERTIES

P. jiroveci, once thought to be a protozoon, is now regarded as a fungus, based on nucleic acid and biochemical analysis. Major developmental stages of the organism include the small (1 to 4 μm) trophic form and the **5 to 8 μm cyst**, which has **a thick cell wall** and contains up to **eight intracystic sporozoites** (Fig. 22-2). Ultrastructurally, *P. jiroveci* has a primitive organelle system, but little is known about its metabolism.

The standard method of diagnosis of pneumonia due to *P. jiroveci* is via cytologic examination of induced sputum specimens or bronchoalveolar lavage washings. Microscopic identification of *P. jiroveci* trophozoites and cysts is performed with stains that demonstrate either the nuclei of trophozoites and intracystic stages (such as Giemsa) or the cyst walls (such as the silver stain; see Fig. 22-2). In addition, direct immunofluorescence microscopy using monoclonal antibodies can identify the organisms with higher sensitivity than can conventional microscopy.

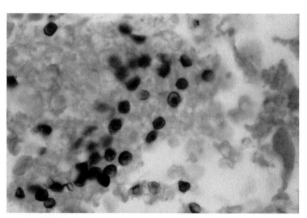

FIGURE 22-2 *Pneumocystis jiroveci* in this lung impression smear using Gomori methenamine silver stain. Note the intracystic bodies within the cysts. *(Courtesy of Dr. Paul Southern, Departments of Pathology and Medicine, University of Texas Southwestern Medical Center, Dallas, TX.)*

EPIDEMIOLOGY

P. jiroveci is ubiquitous in nature. Both an indirect (environmental) source and a direct (person-to-person) source have been proposed as a mode of transmission of *P. jiroveci* in humans. Primary acquisition early in life is likely. By the fifth year most children have antibodies to this organism.

PATHOGENESIS

P. jiroveci contains a major surface glycoprotein that is believed to play an important role in colonization. The organisms adhere avidly via fibronectin and glycoproteins to type I pneumocytes. In normal individuals, alveolar macrophages ingest and kill the organisms. Disease develops as a result of **reactivation of latent infection in individuals who are immunocompromised**. The cysts rupture after activation, and multiple trophozoites are released and fill the alveoli. Alveolar macrophages in patients with AIDS are ineffective in killing the organisms because HIV alters the mannose receptor-mediated binding and phagocytosis of *P. jiroveci*. The organisms do not invade the lung tissue but remain extracellular. As the organisms continue to propagate, basement membrane damage leads to alterations in alveolar capillary permeability. A series of complex events in the lung leads to increased phospholipase activity and a **deficiency of surfactant** (dipalmitoyl phosphatidylcholine [lecithin]) **secretion by type II cells** (as is seen in ARDS), explaining the **ventilation/perfusion mismatch** and manifestations of pneumocystis pneumonia. A **foamy exudate develops in the alveoli** (Fig. 22-3), and **interstitial pathology** occurs (radiology reveals a bilateral ground glass appearance).

TREATMENT

High-dose trimethoprim-sulfamethoxazole (**TMP-SMX**) is the drug of choice, given for 21 days. Alternative drugs include IV pentamidine; trimethoprim plus dapsone; or atovaquone. In addition, the use of steroids when the pO_2 goes below 70 mmHg has been shown to improve mortality.

OUTCOME

The treatment with TMP-SMX (begun empirically) was continued, and oral prednisone was added.

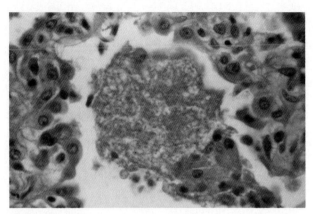

FIGURE 22-3 Histopathology of lung shows alveolar spaces containing exudates characteristic of infection with *Pneumocystis jiroveci*. Note eosinophilic foamy alveolar cast. H & E stain. *(Courtesy of Dr. Paul Southern, Departments of Pathology and Medicine, University of Texas Southwestern Medical Center, Dallas, TX.)*

Symptoms improved over the next 3 days and the patient was discharged 10 days later to complete an oral regimen of TMP-SMX and steroids at home. He was then placed on TMP-SMX once daily to prevent further episodes.

PREVENTION

Prophylaxis of patients with AIDS with oral **TMP-SMX**, dapsone, or aerosolized pentamidine has proved to be effective. This prevents endogenous reactivation in these patients. However, the most effective preventive method is reconstitution of the patient's immune system with antiretroviral therapy. **Once the CD4+ cell count recovers to greater than 200, the risk for developing pneumocystis pneumonia decreases** substantially, and prophylaxis may be safely discontinued.

FURTHER READING

Kovacs JA, Gill VJ, Masur M: New insights into transmission, diagnosis, and drug treatment of *Pneumocystis carinii* pneumonia. JAMA 286:2450, 2001.

Moe AA, Hardy WD: *Pneumocystis carinii* infection in the HIV-seropositive patient. Infect Dis Clin North Am 8:331, 1994.

Santamauro JT, Stover DE: *Pneumocystis carinii* pneumonia. Med Clin North Am 81:299, 1997.

Stringer JR: *Pneumocystis carinii*: What is it, exactly? Clin Microbiol Rev 9:489, 1996.

An 18-year-old woman presented with a **worsening of her chronic cough** for the past week. She had had a low-grade fever, as well as fatigue and shortness of breath. The cough was **productive of greenish sputum** that was thick and tenacious.

She was diagnosed with **cystic fibrosis** (CF) at age 4 and had had multiple hospital admissions for respiratory infections.

PHYSICAL EXAMINATION

VS: T 38°C, P 115/min, R 34/min, BP 104/56

PE: On examination, the patient was a pale, chronically ill-appearing young woman with increased respiratory effort and rapid breathing. Lung examination revealed **bilateral rales** and **wheezing**; heart exam demonstrated distant heart sounds.

LABORATORY STUDIES

Blood

Hematocrit: 34%

WBC: 18,400/μL

Differential: 64% PMNs, 14% bands, 20% lymphs

Blood gases: pO_2 58 mmHg

Serum chemistries: Normal

Imaging

A chest x-ray showed a small heart, hyperinflated lung fields, and **patchy bilateral infiltrates**.

Diagnostic Work-Up

Table 23-1 lists the likely causes of current illness (differential diagnosis). Investigational approach for a specific microbiologic diagnosis includes

- **Blood cultures** (may not yield the suspected agent)
- **Sputum examination and culture**
- **Antibiotic susceptibility of the isolate**

TABLE 23–1   Differential Diagnosis and Rationale for Inclusion (consideration)
Aspergillus fumigatus
Atypical mycobacteria
Burkholderia cepacia
Haemophilus influenzae
Pseudomonas aeruginosa
Staphylococcus aureus

Rationale: The patient has chronic pneumonia. Whereas CF patients can get common respiratory pathogens, the organisms listed above are most commonly associated with chronic infection, due to their ability to persist in respiratory secretions and in the abnormal lung environment. In particular, colonization and infection with *P. aeruginosa* and *B. cepacia* are very common. *S. aureus* and *H. influenzae* are also important pathogens. Unusual organisms such as *Aspergillus* and mycobacteria are less common and are often difficult to treat.

COURSE

The patient was readmitted to the hospital. By suctioning, a sputum specimen was obtained and sent for Gram stain and culture. The patient was started on empirical broad-spectrum antibiotics. A significant nonfermentative Gram-negative rod was isolated and tested for antibiotic susceptibility.

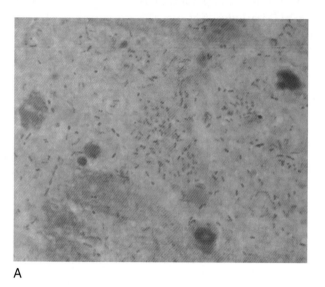
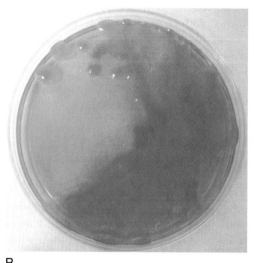

A   B

FIGURE 23-1 *A,* Gram stain of *Pseudomonas aeruginosa* from sputum of a CF patient. *B,* Culture of *Pseudomonas aeruginosa* on MacConkey agar from sputum of a CF patient. Note the mucoid (nonlactose-fermenting) colonies. *(Courtesy of Dr. Dominick Cavuoti, Department of Pathology, University of Texas Southwestern Medical Center, Dallas, TX.)*

ETIOLOGY

Pseudomonas aeruginosa (pneumonia)

MICROBIOLOGIC PROPERTIES

P. aeruginosa is a **Gram-negative rod** (Fig. 23-1*A*) in the family *Pseudomonadaceae*, which also includes *Burkholderia* and *Stenotrophomonas*. Pseudomonads are nonspore forming and actively **motile** by means of their single polar flagellum. These **strictly aerobic bacteria** are also **nonfermentative** and **oxidase** positive. The typical *Pseudomonas* bacteria in nature might be found in a biofilm, attached to some surface. The vast majority of strains are pigmented due to a **water-soluble pigment, pyocyanin** ("blue pus"). Colonies on routine blood agar plates have a **characteristic fruity odor**. *P. aeruginosa* isolates obtained from respiratory secretions of CF patients have a **mucoid** appearance (see Fig. 23-1*B*), which is attributed to its production of **alginate capsule**.

EPIDEMIOLOGY

Pseudomonads are ubiquitous pathogens, and they are flexible in their nutritional requirements. These hardy organisms are capable of growing in diverse environments and are **common inhabitants of soil and water** (tap water and ice). *P. aeruginosa* is one of the most important nosocomial (hospital-acquired) bacterial pathogens of humans. The bacteria are borne on the unwashed hands of hospital personnel; in the hospital setting, these ubiquitous organisms are commonly found in aqueous solutions, disinfectants, ointments, soaps, eye drops, dialysis fluids, and dialysis equipment. Equipment that requires a wet, body temperature environment, such as dialysis tubing and respiratory therapy equipment, is particularly susceptible to contamination. **Transmission** occurs **by ingestion of**, or contact with, **contaminated water** or ice; **aerosolization of contaminated liquids;** penetration by contaminated objects; and **ingestion of *Pseudomonas*-laden foods** (e.g., tomatoes).

PATHOGENESIS

CF patients, with a defective CFTR gene on chromosome 7, have an abnormal chloride channel with resultant secretion of abnormally thick mucus in the large airways. Early in life, the trachea becomes colonized with *P. aeruginosa*, mediated by **flagella** and **pili**. The receptor on tracheal epithelial cells for *Pseudomonas* pili is sialic acid (*N*-acetylneuraminic acid). Bacterial surface-bound **exoenzyme S** also serves as an adhesin for glycolipids on respiratory epithelial cells. At some time after colonization, bacteria move down to the bronchi and undergo a phenotypic shift, becoming mucoid owing to de novo alginate capsule. They lose their ability to move by shutting down the production of the polar flagellum, establishing a permanent and localized chronic infection. The alginate capsule, a repeating polymer of mannuronic and glucuronic acid, develops into a matrix of the *Pseudomonas* biofilm, which protects the colonizing bacteria from phagocytosis and other host defenses such as the ciliary action of the respiratory tract, antibodies, and complement. Host factors favoring persistence of *P. aeruginosa* in the lung are (1) the impaired ability of bronchial epithelial cells to clear *P. aeruginosa* and (2) increased mucin production by bronchial epithelial cells, stimulated by *P. aeruginosa* LPS.

A vigorous and chronic neutrophilic inflammatory response is mounted in the infected large airways by the bacterial cell-wall **LPS**. This **neutrophilic inflammatory response interferes with pulmonary function** and is, in fact, a major cause of morbidity in CF. Despite the intense inflammation within the tracheobronchial tree, bacteremia in these patients is very rare because of a high level of antibodies to various excretory antigens of *P. aeruginosa*. The three secreted products, **elastase, exotoxin A**, and phospholipases, also contribute to lung tissue damage. Elastase hydrolyzes elastin and collagen. **Exotoxin A causes ADP-ribosylation of EF-2**, resulting in inhibition of protein synthesis and ultimate cell death. Phospholipases hydrolyze phospholipids of eukaryotic membranes, resulting in host cell death.

TREATMENT

Empirical therapy usually includes **an extended–spectrum penicillin** (e.g., piperacillin), cephalosporin (e.g., ceftazidime), or a carbapenem (e.g., imipenem) and an antipseudomonal aminoglycoside (e.g., tobramycin). The proper empirical choice of antibiotics should be based on continuous surveillance of drug sensitivity. The resistance pattern among *Pseudomonas* strains varies from hospital to hospital and changes with time. Directed therapy of chronic pneumonia due to mucoid strains of *P. aeruginosa* or *Burkholderia cepacia* should be based on the susceptibility profile of the clinical isolates.

NOTE *P. aeruginosa* bacteria are resistant to many commonly used antibiotics, including first- and second-generation penicillins and cephalosporins, tetracyclines, chloramphenicol, and macrolides. However, newer quinolones with antipseudomonas activity have been useful in the management of some of the clinically diverse infections (Table 23-2) due to this opportunistic pathogen.

OUTCOME

The case patient was started on piperacillin and tobramycin for 2 weeks with gradual improvement in her symptoms. She was discharged home in good condition.

TABLE 23–2 Other Important Clinical Features of *Pseudomonas* Infections

Syndromes	Clinical Features
Normal hosts	**Otitis externa (swimmer's ear)** **Folliculitis (hot tubs,** swimming pools) **Puncture-wound osteomyelitis** (e.g., a nail puncture through a tennis shoe)
Abnormal hosts Local invasion	**Invasive otitis externa** (a slowly progressive, life-threatening infection if untreated). In patients with diabetes mellitus, *P. aeruginosa* colonizing the external auditory canal may invade along the cartilage-bony interface (theducts of Santorini) and infect the soft tissues below the temporal bone. **Urinary tract infections** (complicated). Whereas *P. aeruginosa* rarely causes UTIs in normal hosts, it is one of the more frequent etiologies of UTI in immunocompromised or **hospitalized patients,** often in association with the use of an indwelling bladder catheter (Foley catheter).
Invasion with systemic spread	**Bacteremia (sepsis).** Predisposing conditions include hematologic malignancies, immunodeficiency relating to AIDS, **neutropenia,** diabetes mellitus, severe burns, and **wound infections.** Most cases of *Pseudomonas* bacteremia (20% of nosocomial blood stream infections) are acquired in hospitals and nursing homes. **LPS-associated endotoxin causes a systemic inflammatory respiratory syndrome** (multiple-organ dysfunction), thrombocytopenia (**ecthyma gangrenosum**), **shock,** and death. **Pneumonia.** Respiratory infections caused by *P. aeruginosa* occur almost exclusively in individuals with a compromised lower respiratory tract or a compromised systemic defense mechanism, or in ICU patients supported by **respiratory therapy equipment (ventilator-associated pneumonia).**

PREVENTION

Strict prevention of acquisition of this opportunistic pathogen is not practical. There is no vaccine to prevent infection. CF patients have a high likelihood of developing colonization; they should be educated about the transmission of organisms from others. Proper maintenance of water reservoirs is important in reducing the likelihood of spread from environmental sources. In the hospital, contact and droplet precautions, with regular hand washing after patient contact, are important in preventing nosocomial infections. Careful attention to aseptic technique during procedures, and cleaning and disinfecting respiratory therapy equipment used in health care settings—as well as in the home—also help reduce risks.

FURTHER READING

Lyczak JB, Cannon CL, Pier GB: Lung infections associated with cystic fibrosis. Clin Microbiol Rev 15:194, 2002.

Morrison AF, Wenzel RP: Epidemiology of infections due to *Pseudomonas aeruginosa*. Rev Infect Dis 6 (Suppl):S267, 1984.

Pollack M: The virulence of *Pseudomonas aeruginosa*. Rev Infect Dis 6 (Suppl):S617, 1984.

Saiman L, Siegel J: Infection control in cystic fibrosis. Clin Microbiol Rev 17:57, 2004.

A 62-year-old white man was admitted to the hospital in the month of January with **fever, shortness of breath, productive cough,** and **chest pain.** He also complained of a **thick, yellowish discharge in his eyes** that prevented him from opening his eyes in the morning.

He was a **chain smoker** (three packs per day), did not drink alcohol, and had no chronic underlying diseases. The patient had **not received the current flu vaccine,** and 10 days prior to admission, the patient experienced **sudden onset of fever, chills, sore throat,** and **arthralgias.** His symptoms had gradually resolved over the following week, before the development of his current symptoms.

PHYSICAL EXAMINATION

VS: T 39.4°C, P 120/min, R 30/min, BP 140/80 mmHg.

PE: The ill-appearing man was in mild respiratory distress. On chest examination, **decreased breath sounds and rales** were heard at the left base. Unilateral **erythematous palpebral conjunctiva,** watery eye, and **purulent exudate** (Fig. 24-1) were also noted.

LABORATORY STUDIES

Blood

Hematocrit: 42%
WBC: 16,700 /μL
Differential: 55% PMNs, 16% bands, 22% lymphs
Blood gases: pO_2 64 mmHg on room air
Serum chemistries: Normal

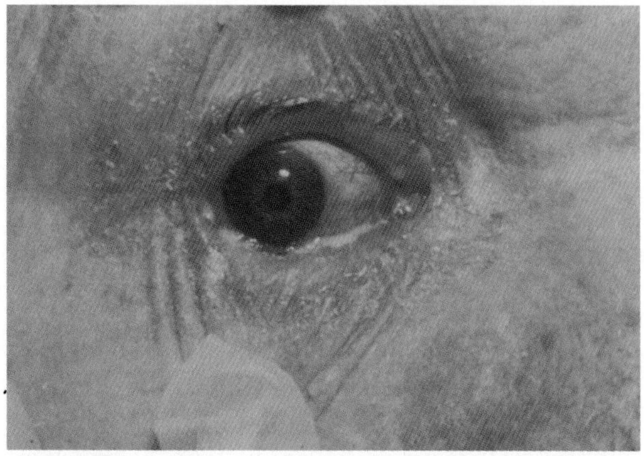

FIGURE 24-1 Erythematous palpebral conjunctiva, hyperemia, and purulent discharge in this patient. *(Courtesy of Dr. Paul Southern, Departments of Pathology and Medicine, University of Texas Southwestern Medical Center, Dallas, TX.)*

Imaging

A chest x-ray revealed an alveolar **infiltrate in the posterior segment of the left lower lobe.**

Diagnostic Work-Up

Table 24-1 lists the likely causes of illness (differential diagnosis). Viral serology (to retrospectively confirm influenza) may be undertaken, and additional tests to define the etiology of pneumonia and purulent conjunctivitis may include

- **Blood cultures**
- **Gram stain** and **cultures** of sputum and conjunctival swab
- In a negative Gram stain of conjunctival swab, direct antigen (ELISA) for *Chlamydia trachomatis*

TABLE 24–1 Differential Diagnosis and Rationale for Inclusion (consideration)
Atypical pneumonia due to intracellular pathogens (e.g., *Legionella* spp.)
Chlamydia trachomatis
Haemophilus influenzae
Moraxella catarrhalis
Staphylococcus aureus
Streptococcus pneumoniae

Rationale: A clinical diagnosis of post-influenza secondary bacterial pneumonia and concurrent bacterial conjunctivitis should be considered. Secondary bacterial pneumonia is often seen in older individuals seasonally. The common pathogens are *S. pneumoniae* and *S. aureus,* although others can be seen, such as *Haemophilus* or *Moraxella.* Microorganisms that cause atypical pneumonia would be less common in this setting. *C. trachomatis* may be considered as a major cause of conjunctivitis. Other pyogenic bacterial agents (including *S. aureus*) also cause acute conjunctivitis.

COURSE

Expectorated sputum was collected, and blood cultures were drawn. Sera for viral serology were sent for primary influenza that likely predisposed to the secondary pneumonia. A serologic study using a single serum revealed a significantly high IgG titer (1:512) specific for influenza virus A (H_3N_2). Cultures of peripheral blood, sputum, and conjunctival swab were positive for a significant isolate on blood agar media, revealing predominantly Gram-positive cocci in clusters. The patient subsequently developed left lung collapse (due to mucus plugging) and acute pulmonary edema.

ETIOLOGY

Staphylococcus aureus (secondary bacterial pneumonia and concurrent acute conjunctivitis)

MICROBIOLOGIC PROPERTIES

Species belonging to the genus *Staphylococcus* are **Gram-positive cocci** that occur individually, in pairs, and in irregular **grapelike clusters** (Fig. 24-2). They are nonmotile, nonspore forming, and **catalase positive**. The cell wall contains peptidoglycan and teichoic acid. These bacteria are resistant to temperatures as high as 50°C, to high salt concentrations, and to drying. The colonies are usually large (6 to 8 mm in diameter), smooth, and translucent, and colonies of

most strains are pigmented, ranging from cream-yellow to orange. Three medically important species are commonly isolated from the clinical specimens:

1. *Staphylococcus aureus* are **coagulase positive.**
2. *Staphylococcus epidermidis* are coagulase negative.
3. *Staphylococcus saprophyticus* are coagulase negative and novobiocin resistant.

The ability to clot plasma (**coagulase activity**) continues to be the most widely used and generally accepted criterion for the identification of *Staphylococcus aureus*. One such factor, bound coagulase, also known as clumping factor, reacts with fibrinogen to cause organisms to aggregate (a widely used key diagnostic test in the hospital laboratory). Another factor, extracellular coagulase, reacts with prothrombin to form thrombin, which can convert fibrinogen to fibrin.

EPIDEMIOLOGY

Human **nasal carriers** are the reservoir of *Staphylococcus aureus*. Approximately 20% of people have prolonged carriage, whereas 60% of persons have intermittent carriage, with the organism being present not only in the nasopharynx but also on the **surface of the skin**. A person can get the infection via endogenous or exogenous routes. *S. aureus* can **cause pneumonia by** either the **aspiration mode** (usually after the onset of influenza, seen in this case patient) or via the hematogenous mode (in patients with **illicit intravenous drug use** with concurrent heart valve infection).

PATHOGENESIS

In patients with influenza, **the virus destroys the ciliary defense** (and the mucociliary ladder, which normally keeps invading bacteria out of the lower respiratory tract), allowing **colonization of the opportunistic bacteria** in the lung. The colonization event requires that *S. aureus* adhere to host cells via the teichoic acid component of the cell wall and other staphylococcal factors. After the bacterium penetrates the respiratory mucosa, a chemotactic response occurs, involving host signals released via complement activation by staphylococcal factors, especially cell-wall components, and the host mounts an aggressive pyogenic inflammatory response. The **hallmark of staphylococcal infection is the abscess**, which consists of a fibrin wall surrounded by inflamed tissues enclosing a central core of pus containing organisms

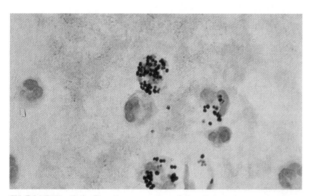

FIGURE 24-2 Gram stain of *Staphylococcus aureus* from sputum. Note Gram-positive cocci occurring individually, in pairs, and in irregular grapelike clusters. (*Courtesy of Dr. Paul Southern, Departments of Pathology and Medicine, University of Texas Southwestern Medical Center, Dallas, TX.*)

and leukocytes (in the lung alveoli), **consistent with the opacity of the chest x-ray** in the case patient.

The surface tissues of the eye are colonized by staphylococci (normal flora of nares) by spread from adjacent sites (runny nose and itchy throat) during influenza. Disruption of the epithelial layer covering the conjunctiva during rubbing and wiping can lead to infection. A profound pyogenic inflammation in response to extracellular staphylococci results after the infection is established, causing acute mucopurulent conjunctivitis (which occurred in this patient as an endogenous infection concurrent with pneumonia).

TREATMENT

A high dose of IV **antistaphylococcal penicillin** (e.g., **nafcillin**) is administered for a minimum of 2 weeks.

In the case of resistant isolates, which are becoming more frequent, **vancomycin** is used.

OUTCOME

The patient received 2 weeks of IV nafcillin and slowly improved over the course of the next several days.

PREVENTION

Administration of the current influenza vaccine before the flu season can prevent contagion of influenza and predisposition to secondary staphylococcal pneumonia.

FURTHER READING

Campbell W, Hendrix E, Schwalbe R: Head-injured patients who are nasal carriers of *Staphylococcus aureus* are at high risk for *Staphylococcus aureus* pneumonia. Crit Care Med 27:798, 1999.

Lowy FD: *Staphylococcus aureus* infections. N Engl J Med 339:520, 1998.

von Eiff C, Becker K, Machka K: Nasal carriage as a source of *Staphylococcus aureus* bacteremia. Study group. N Engl J Med 344:11, 2001.

SECTION II

Urogenital Tract Infections

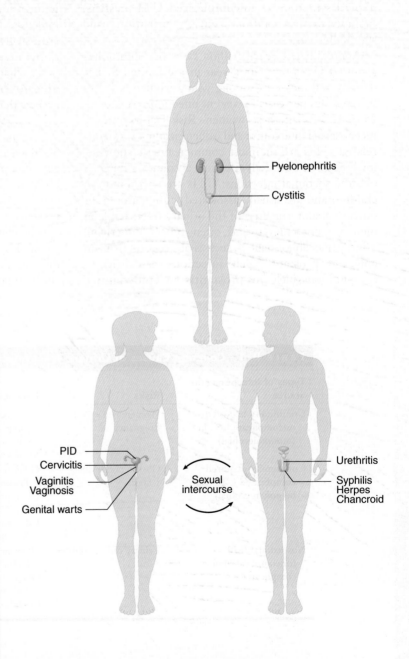

INTRODUCTION TO UROGENITAL INFECTIONS

The urogenital tract is a common source of infection in men and women. Although the pathogens involved are similar for both sexes, the clinical syndromes vary considerably depending on the anotomic location. Acute **infections of the urinary tract** (UTI) can be subdivided into two general anatomic categories: lower tract infection of the bladder (cystitis) and upper tract infection of the kidney (pyelonephritis). UTIs are very common in women because of the short distance organisms have to travel to reach the bladder. Typical symptoms of **uncomplicated UTI** (**cystitis**) include dysuria and/or frequent urination, and suprapubic pain.

Approximately seven million cases of community-acquired UTIs occur in a year in the United States. The overall incidence among young women is 0.5 to 0.7 episodes per year. Recurrent infections are seen in 25% to 30% of these women. Risk factors include recent sexual intercourse and catheterization. Catheter-related UTIs are the most common type of nosocomial (hospital-acquired) infection, accounting for 40% of all nosocomial infections. UTIs are uncommon in children and are usually associated with anatomic or functional abnormalities of the urinary tract. UTIs in men are uncommon, and often reflect underlying urinary tract abnormalities, such as obstruction from an enlarged prostate. Such infections can lead to complications, including **prostatitis** and **epidydimitis**.

Ascending infection may occur, leading to **complicated UTI** (**pyelonephritis**), which is characterized by fevers, flank pain, dysuria, nausea and vomiting, and malaise; 250,000 cases of pyelonephritis per year occur in the United States.

The most common causes of UTI are *Escherichia coli* (Gram-negative bacteria) and *Staphylococcus saprophyticus* (Gram-positive bacteria). Other microorganisms associated with uncomplicated and complicated UTI are listed in Table II-1.

Clinical signs and symptoms cannot be relied on for diagnosing UTI or localizing a site of infection. Diagnosis of UTI is based on the presence of bacteria in the urine and the host inflammatory response to the infection. Laboratory findings of uncomplicated UTI are usually pyuria and bacteriuria. **Pyuria** is defined as the presence of 10 or more neutrophils per high-power field of unspun, voided midstream urine. **Significant bacteriuria** is defined as the presence of greater than 10^5 colony-forming units (cfu) per mL of urine obtained from a clean-voided "midstream" specimen. It should be noted that colony counts might be much lower in women with cystitis or in patients with complicated UTIs. Laboratory findings of complicated UTI (pyelonephritis) include leukocytosis, pyuria, and bacteriuria (bacteremia may or may not be present).

The choice of antibiotics for treatment of UTIs is dictated by the likely organisms present in the urine, their antibiotic susceptibility profiles, how well the antibiotic penetrates renal tissue, and the levels it can

TABLE II-1 Common Causes of Urinary Tract Infections		
Types of Infections and Source	**Risks**	**Infectious Agents**
Community-acquired uncomplicated UTI (monomicrobial)	Women:men = 10:1 Pregnancy	Gram-negative enteric rods: *Escherichia coli* (major), *Proteus* spp, *Klebsiella* spp Gram-positive bacteria: *Enterococcus* spp, *Staphylococcus saprophyticus*
Community-acquired complicated UTI (can be polymicrobial)	Obstruction (urinary stasis)	Gram-negative enteric rods: *E. coli* (major), *Proteus* spp, *Klebsiella* spp Gram-positive bacteria: *Staphylococcus aureus* Uncommon pathogens (*Candida*, group B streptococci)
Nosocomial UTI (uncomplicated or complicated)	Urinary catheterization	Gram-negative enteric rods: *E. coli*, *Proteus*, *Klebsiella*, *Serratia*, *Pseudomonas* spp Gram-positive bacteria: *Staphylococcus epidermidis* *Candida* spp

achieve in the urine. Therapy for uncomplicated UTI is generally a 3-day regimen, whereas complicated UTI usually requires intravenous antibiotics in the hospital until the patient is able to take oral antibiotics, for a total of 2 weeks or longer.

Sexually transmitted diseases (STDs; Table II-2) are usually associated with relatively few clinical syndromes, such as urethritis, cervicitis, vaginitis (and vaginosis), and genital ulcers. STDs rank among the most common of all infectious diseases. In the United States, STDs are a major national health problem; an estimated 12 million new cases of STDs occur annually, giving this nation the highest STD rate in the industrialized world.

In men, **urethritis** and **penile ulcers** are typical manifestations of STDs. Urethritis is defined by painful urination or a penile discharge. The character of the discharge can suggest an etiology: purulent discharge is often due to *Neisseria gonorrhoeae*, whereas clear discharge is due to *Chlamydia trachomatis* infection. Epidydimitis can be a complication of this infection as well.

In women, STDs are not often associated with urethritis because organisms causing the STDs are transmitted through intercourse into the vagina and cervix. Cervical infection (**mucopurulent cervicitis**) is usually determined by significant tenderness on pelvic examination. Common causes are *Neisseria gonorrhoeae* and *Chlamydia trachomatis*. These are among the most prevalent of all STDs. In women, asymptomatic gonorrhea and chlamydia may result in pelvic inflammatory disease (PID), which is a major cause of infertility, ectopic pregnancy, and chronic pelvic pain. In addition, pregnant women infected with *C. trachomatis* can pass the infection to their infants during delivery, potentially resulting in **neonatal ophthalmia** and pneumonia.

Sexually transmitted herpes simplex virus (HSV) infections now cause most **genital ulcer** disease throughout the world. In men, penile ulcers can be painful, suggestive of **herpes simplex**, or painless, suggestive of **syphilis**. In genital ulcer disease, painful adenopathy can be seen with **lymphogranuloma venereum** caused by endemic strains of *C. trachomatis*. Painful ulcers are also caused by *Haemophilus ducreyi* (**chancroid**). Therapy is usually directed at a specific organism, although in males with urethritis or women with cervicitis, the coinfection rate of *N. gonorrhoeae* and *C. trachomatis* is so high that treatment for both organisms is recommended when either is detected. In women, external genital ulcers have similar significance as in men.

Vaginal discharge is a typical symptom of vaginitis and (noninflammatory) vaginosis. Symptoms of

TABLE II-2	**Common Causes of Sexually Transmitted Diseases**
Types of Infections	Infectious Agents
Inflammatory (Exudative) Infections	
Urethritis, cervicitis	*Neisseria gonorrhoeae* (gonorrhea), *Chlamydia trachomatis* (nongonococcal urethritis or cervicitis), *Mycoplasma hominis*, *Mycoplasma genitalium*
Vaginitis	*Trichomonas vaginalis* (trichomoniasis), *Candida albicans* (candidiasis), *Gardnerella* spp/*Mobiluncus* spp (bacterial vaginosis)
Genital Ulcers (Nonexudative Infections)	
Syphilis	*Treponema pallidum*
Herpes	HSV-2 >> HSV-1
Chancroid	*Haemophilus ducreyi*
Lymphogranuloma venereum	*Chlamydia trachomatis*
Genital warts (condylomata acuminata)	Human papillomavirus (HPV)
Sexually Transmitted Systemic Infections	
AIDS	HIV-1, HIV-2
Pelvic inflammatory disease	Polymicrobic; *Neisseria gonorrhoeae*, *Chlamydia trachomatis*, anaerobes
Cancer (Neoplasia)	
Cervical carcinoma	HPV
Kaposi sarcoma	Human herpesvirus type 8

vaginitis (caused by *Trichomonas hominis* or *Candida albicans*) and **vaginosis** (caused by *Gardnerella* and *Mobiluncus*) are nonspecific. The diagnosis of vaginitis or bacterial vaginosis is easily established by the finding of a change in vaginal pH and by direct microscopy.

Human immunodeficiency virus (HIV) infection has become the most deadly sexually transmitted disease in modern times. Unlike other STDs, there are no characteristic genital lesions, and it is usually not realized by the individual that infection has occurred. The major risk factors for HIV infection are promiscuous sexual activity and intravenous drug use. Heterosexual transmission of HIV occurs more commonly in Africa than in North America and Europe. Needle sticks and transfusions are also associated with transmission, although transfusion-related exposure is now very rare owing to screening of blood products. Primary HIV infection has an incubation period of 2 to 4 weeks and is often symptomatic, with a flu-like illness and a rash. This may not bring the patient to medical attention, and therefore many individuals are unaware that they are infected with HIV. There is a variable clinical latent period, followed in most patients by a decline in immune function due to progressive destruction of $CD4^+$ T cells. This leads to opportunistic infections with a wide variety of pathogens, often leading ultimately to the death of the patient. Therapy of HIV has improved dramatically in the past decade due to highly active antiretroviral therapy (HAART), and many patients can now expect to live with HIV infection as a chronic disease rather than a rapidly fatal one.

FURTHER READING

Gunn RA, Rolfs RT, Greenspan JR, et al: The changing paradigm of sexually transmitted disease control in the era of managed health care. JAMA 279:680, 1998.

Kahn JO, Walker BD: Acute human immunodeficiency virus type 1 infection: Current concepts. N Engl J Med 339:33, 1998.

McCormack WM: Pelvic inflammatory disease: Current concepts. N Engl J Med 330:115, 1994.

Sobel JD: Vaginitis: Current concepts. N Engl J Med 337:1896, 1997.

Trees DL, Morse SA: Chancroid and *Haemophilus ducreyi*: An update. Clin Microbiol Rev 8:357, 1995.

Questions on the case (problem) topics discussed in this section can be found in Appendix F. Practice question numbers are listed in the following table for students' convenience.

Self-Assessment Subject Areas (book section)	Question Numbers
Urinary tract infections and sexually transmitted diseases (Section II)	4-6, 21-23, 35-37, 44-48, 63, 78-80, 89, 94-95, 102-104, 108-110, 113, 116, 125

A **24-year-old sexually active woman** visited a family practice clinic because of **burning pain during urination, increased frequency,** and **urgency for 1 day.** She also complained about blood-stained debris at the end of urination. Symptoms had rapidly worsened in the past 6 hours.

PHYSICAL EXAMINATION

VS: Normal
PE: Mild suprapubic tenderness was noted.

LABORATORY STUDIES

Blood

Not done

Imaging

No imaging studies were done.

Diagnostic Work-Up

Table 25-1 lists the likely causes of this woman's illness. Investigational approach may include

TABLE 25–1   Differential Diagnosis and Rationale for Inclusion (consideration)

E. coli
Enterococcus spp
Other Gram-negative rods of enteric origin (e.g., *Pseudomonas, Klebsiella, Enterobacter, Serratia, Proteus* spp)
Staphylococcus saprophyticus
Uncommon pathogens (e.g., fungi, group B streptococcus)

Rationale: A clinical diagnosis of urinary tract infection (UTI; uncomplicated cystitis) should be considered. The most common cause of uncomplicated cystitis in women is *E. coli*, often in relation to recent sexual intercourse. Infections due to *S. saprophyticus* are common among sexually active adolescent girls and young adult women. The other organisms in the list are almost exclusively seen in patients with other risk factors, usually hospitalized patients with indwelling urinary catheters.

- Urinalysis
- Cultures of clean-voided urine to determine significant bacteriuria

COURSE

A diagnosis of UTI was made based on the clinical picture and the laboratory finding of bacteriuria. Routine culture of midstream urine on MacConkey agar yielded a significant isolate.

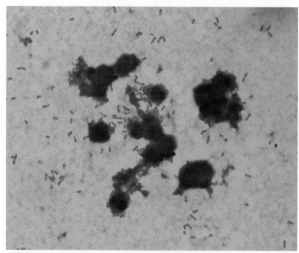

FIGURE 25-1   Gram stain of the pathogen from sediment of clean-voided urine. *(Courtesy of Dr. Dominick Cavuoti, Department of Pathology, University of Texas Southwestern Medical Center, Dallas, TX.)*

ETIOLOGY

Escherichia coli (uncomplicated UTI or cystitis)

MICROBIOLOGIC PROPERTIES

E. coli belongs to the family Enterobacteriaceae (see a partial list of species in Table 25-2). These bacteria are **Gram-negative rods** occurring singly or in pairs (Fig. 25-1). They are **facultatively anaerobic** bacteria; **all ferment glucose** (*E. coli* also **ferments lactose**; Fig. 25-2) and are **oxidase negative**. They

TABLE 25–2 Other Major Species in the Enterobacteriaceae Family and Their Clinical and Epidemiologic Features

General	Microbiologic/Clinical Features
Proteus (5 species)	Widespread in the environment; normal inhabitants of the large intestine
	Highly motile organisms, ranked as the fifth leading cause of urinary tract infections, causing complications and promoting the formation of struvite kidney stones by profuse urease activity
	Commonly associated with **nosocomial UTIs** (associated with **catheterization**), pneumonia, blood stream infections (BSIs), and surgical wound infections
Klebsiella	Discussed in detail earlier (Case 10)
Enterobacter (14 species)	Multiply antibiotic-resistant pathogen (known to harbor extended-spectrum β-lactamases [ESBLs], cephalosporinases, and carbapenemases)
	Commonly associated with nosocomial UTIs, pneumonia, BSIs, and surgical wound infections
Citrobacter (11 species)	Primarily inhabitants of the intestinal tract
	Multiply antibiotic-resistant pathogen (known to harbor ESBLs, cephalosporinases, and carbapenemases)
	Commonly associated with nosocomial UTIs, pneumonia, BSIs, and surgical wound infections
Serratia (9 species)	Many, but not all, strains of *Serratia marcescens* form pinkish-red colonies
	Cause typical opportunistic infections (e.g., UTIs), especially involving indwelling catheters of various types; commonly associated with nosocomial UTIs, pneumonia, BSIs, and surgical wound infections
	Multiply antibiotic-resistant pathogen (known to harbor ESBLs, cephalosporinases, and carbapenemases)

are either nonmotile or motile by peritrichous flagella. All species are fast growing and appear as **large, gray colonies on blood agar** (see Fig. 25-2). These species may have the following two to three antigens:

1. **Somatic O antigen** carrying lipopolysaccharide (LPS)
2. **H antigen** (flagellum; nonflagellated shigella do not have H antigen)
3. **K antigen** (a polysaccharide **capsule**; a major virulence factor)

Diagnosis of UTI is based on the presence of host inflammatory cells in the urine. Pyuria is defined as the presence of 10 or more WBCs/high-power field of unspun, voided midstream urine. **Leukocyte esterase** is a surrogate marker for the presence of WBCs using the urine dipstick test, which turns positive when 10 or more WBCs are present per high-power field. **Significant bacteriuria** is identified based on cultures of a clean-voided midstream urine specimen and the presence of greater than 10^5 colony-forming units per mL of urine (grown in laboratory media). In symptomatic patients, a smaller number of bacteria (10^2 to 10^4 cfu/mL) may suggest infection.

EPIDEMIOLOGY

UTI is much more common in women than in men because of differences in anatomic structure and changes during sexual maturation, pregnancy, and childbirth. *E. coli* is a major inhabitant of the large intestine; the common mode of transmission is, therefore, endogenous. Overwhelming inoculation

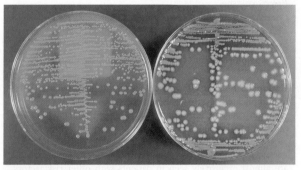

FIGURE 25-2 *Escherichia coli* cultures from clean-voided urine on (*left*) MacConkey agar (lactose-fermenting red colonies) and (*right*) sheep blood agar (large, moist gray colonies). Note the number of colonies on and around the streak (for semiquantitative urine culture), representing significant bacteriuria (>10^5 cfu per mL of urine). (*Courtesy of Dr. Dominick Cavuoti, Department of Pathology, University of Texas Southwestern Medical Center, Dallas, TX.*)

TABLE 25–3   Common Risk Factors for Cystitis in Both Sexes

Urethral stricture
↑ Residual volume
↓ Urinary flow
Stasis
↓ Urinary acidification/concentration
Instrumentation (catheterization)

from sexual intercourse and obstruction of urinary flow from any underlying cause is a major risk factor (Table 25-3) for the development of UTI. Benign prostatic hyperplasia in men older than 50 years of age, characterized by a nodular enlargement of the periurethral zones of the prostate gland, causes compression of the urethra with partial obstruction of the urinary flow and is the main contributor to the increase in UTI in men in that age group.

PATHOGENESIS

Uncomplicated cystitis (as seen in this case) occurs primarily in women who are sexually active and who become colonized by a uropathogenic strain of *E. coli*. **Sexual intercourse facilitates ascending passage of the bacteria, colonizing the urinary tract from the urethra to the bladder.** Only the aerobic and facultative anaerobic species, such as *E. coli*, colonize and infect the urinary tract. These uropathogenic *E. coli* are equipped with fimbriae, allowing them to adhere to the uroepithelium. Predominant among these are the **P-fimbriae** (so called because they share epitopes with the P blood group antigens) and **type-1 fimbriae** (prominent virulence factors of uropathogenic *E. coli*). Not only are they instrumental in colonizing the urinary tract, but they also stimulate local inflammatory responses, including the secretion of interleukin (IL)-6 and IL-8 and the induction of both apoptosis and epithelial cell desquamation. The chemotactic IL-8 ushers in PMNs, which are recruited to play a role in clearing bacteriuria. Local pyogenic inflammation contributes to the symptoms and signs of the uncomplicated UTI (cystitis). The uropathogens rarely invade bladder and urethral mucosa.

In untreated UTI, Gram-negative uropathogens (e.g., *E. coli*, *Proteus*, *Klebsiella*, and *Pseudomonas*) ascend **to the ureters and the kidneys**. The movement of bacteria against the urinary flow is facilitated by **flagella**. The **capsular polysaccharide** (K antigen;

one of the major virulence factors) associated with the cell surface of the particular bacteria resists phagocytosis and killing. Other virulence factors that contribute to kidney pathology are LPS endotoxin (associated with all Gram-negative bacteria), urease production (*Proteus*), hemolysin, and aerobactin (in selected uropathogens). **LPS** (endotoxin) increases the **pyogenic inflammatory response**. **Urease** promotes the formation of struvite kidney stones in the kidney by increasing the alkalinity of urine.

The **hemolysin** (with cytolytic activity) produced by some species may also be linked to kidney damage. After their arrival in the kidney, bacteria adapt to iron limitations by using **aerobactin**, a siderophore that increases iron uptake by bacteria. In severe pyogenic infection, IL-1 and tumor necrosis factor mount an inflammatory response, contributing to **high fever (>39°C), shaking chills,** and **localized flank or low back pain**. Elevated levels of C-reactive protein often accompany neutrophilic leukocytosis with left shift. If **white cell casts** are present in urine, the infection likely **represents pyelonephritis** (complicated UTI). Semiquantitative culture of a midstream urine specimen with more than 100,000 colony-forming units of a single organism and a positive blood culture usually confirm the diagnosis of pyelonephritis.

TREATMENT

The choice of antibiotics is dictated by the likely organisms present in the urine, their antibiotic susceptibility profiles, and how well the antibiotic penetrates renal tissue and the levels it can achieve in the urine. Usually, this means using an oral antibiotic, such as a β-lactam agent (amoxicillin and cephalosporin group), trimethoprim-sulfamethoxazole (Bactrim, Septra), or a flouroquinolone (ciprofloxacin, levofloxacin). In many areas, increasing resistance to amoxicillin and/or TMP/SMX makes these agents less useful as empirical therapy. Although the aminoglycosides (gentamicin and tobramycin) achieve high renal and urine levels and are potent against the common Gram-negative pathogens, their toxicity profile and need for parenteral administration limit their use to seriously ill patients. The duration of treatment is dictated by the severity of infection. Patients with prolonged symptoms (>1 week) and underlying risk factors for UTI, such as obstruction, reflux, immunocompromised state, or pregnancy, should be treated with longer courses of antibiotics (10 to 14 days) even if only lower tract infection is suspected. It is difficult to exclude upper tract infection in such patients. Uncomplicated UTI in women (as in this case patient) often responds to short course therapy of 3 days, and this is generally recommended for these patients.

NOTE *E. coli* causes a variety of other clinical syndromes (e.g., intra-abdominal infection; Table 25-4). Directed treatment of these syndromes is based on the culture and sensitivity of the clinical isolate from the appropriate specimens representing inflammatory pathology.

OUTCOME

The patient was treated with a 3-day course of trimethoprim-sulfamethoxazole (TMP/SMX), and her symptoms resolved. In the next 3 months, the patient had three more attacks, two with *E. coli* and one with *S. saprophyticus*. Urinary colony counts in these episodes ranged from 10^3 to 10^6/mL. Each attack responded either to amoxicillin or TMP/SMX given for 3 days.

PREVENTION

In sexually active women, a useful preventive measure is voiding after intercourse, which helps reduce bacterial numbers in the urethra. Use of a spermicide is also associated with higher rates of UTI in women because of changes in normal flora, so discontinuing this practice may be of benefit to women with recurrent UTI. Occasionally, patients with no obvious risk factors experience recurrent episodes of cystitis. If these episodes are very frequent, daily suppressive antibiotic therapy can be given. In hospitalized patients, limiting use of urinary catheters and removing them as soon as is reasonable will also help reduce the incidence of UTIs.

TABLE 25-4 Other Important Clinical Features of Infections due to *Escherichia coli*

Syndromes	Clinical Features
Acute bacterial meningitis	Pregnant women are at a higher **risk of colonization** with the **K1 antigen (encapsulated)** strain of *E. coli*. Commonly associated with neonatal sepsis. *E. coli* also accounts for up to 30% of cases of neonatal meningitis.
Pneumonia	*E. coli* pneumonia, although uncommon, may result from microaspiration of upper airway secretions that have been previously colonized with this organism in severely ill patients; hence, it is primarily a cause of nosocomial pneumonia.
Intra-abdominal infections	Intra-abdominal abscesses can occur secondary to spontaneous or traumatic GI tract perforation or after anastomotic disruption with spillage of colon contents and subsequent peritonitis. Pathogenesis is polymicrobial, and *E. coli* is one of the more common Gram-negative bacilli found together with anaerobes. Cholecystitis and cholangitis result from obstruction of the biliary system due to biliary stone or sludge, leading to stagnation and bacterial growth from the papilla or portal circulation.
Enteric infections	To be described under gastrointestinal infections (see Section III).
Other miscellaneous infections	Septic arthritis, endophthalmitis, sinusitis, osteomyelitis, endocarditis, or skin and soft tissue infections (in diabetics) are also caused by *E. coli*.

FURTHER READING

Daifuku R, Stamm WE: Bacterial adherence to bladder uroepithelial cells in catheter-associated urinary tract infection. N Engl J Med 314:1208, 1986.

Eisenstein BI, Jones GW: The spectrum of infections and pathogenic mechanisms of *Escherichia coli*. Adv Intern Med 33:231, 1988.

Fihn SD: Acute uncomplicated urinary tract infection in women. N Engl J Med 349:259, 2003.

Manges AR, Johnson JR, Foxman B, et al: Widespread distribution of urinary tract infections caused by a multidrug-resistant *Escherichia coli* clonal group. N Engl J Med 345:1007, 2001.

Ronald A: Sex and urinary tract infections. N Engl J Med 335:510, 1996.

Stamm WE, Hooton TM: Management of urinary tract infections in adults: Current concepts. N Engl J Med 329:1328, 1993.

A 31-year-old African American woman presented with low-grade **fever, malaise,** and a **rash.** She recalled having had **painless ulcers, which appeared on the vulva** one month before this new episode. She did not seek medical attention at that time, and the ulcers spontaneously resolved in 10 days.

The patient revealed that she had **had four sexual partners** in the month preceding the development of ulcerative lesions. She had not traveled outside the United States in recent months.

PHYSICAL EXAMINATION

VS: T 38.1°C, P 90/min, R 14/min, BP 124/72 mmHg

PE: Examination was remarkable for **inguinal lymphadenopathy,** a **generalized rash on palms** and **soles,** and **pustular cutaneous lesions** and **condylomata lata on her face** (Fig. 26-1).

LABORATORY STUDIES

Blood

ESR: Normal
Hemoglobin: Normal
WBC: Normal
Differential: Normal
Serum chemistries: Normal

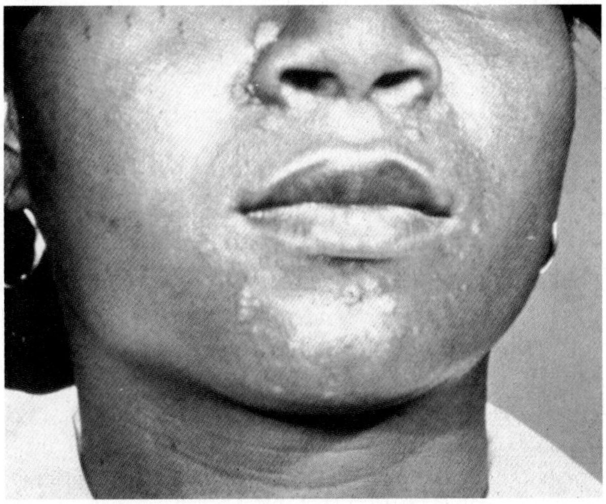

Imaging

Not necessary

Diagnostic Work-Up

Table 26-1 lists the likely causes of the woman's illness. Herpes can be ruled out by **Tzanck smear** (a cytologic examination to detect HSV-infected cells). Additional tests for confirmation of the clinical diagnosis may include

- **Serologic tests** (useful for the diagnosis of syphilis)
- **Culture** (to rule out chancroid)

COURSE

The patient was counseled for risks of serious sexually transmitted diseases (STDs) (i.e., HIV infection) and was advised to undergo an HIV seroconversion test. A rapid plasma reagin (RPR) serologic titer was 1:128 and was confirmed by a specific serologic test.

FIGURE 26-1 Pustular lesions and condylomata lata on the face of this patient. (*Courtesy of Dr. Paul Southern, Departments of Pathology and Medicine, University of Texas Southwestern Medical Center, Dallas, TX.*)

TABLE 26–1 Differential Diagnosis and Rationale for Inclusion (consideration)

Chancroid (*Haemophilus ducreyi*)
Donovanosis (*Calymmatobacterium granulomatosis*)
Genital herpes (herpes simplex virus [HSV])
Lymphogranuloma venereum (LGV; *Chlamydia trachomatis*)
Syphilis (*Treponema pallidum*)

Rationale: A clinical diagnosis of sexually transmitted disease (STD) should be considered based on the patient's medical and social history. Inguinal lymphadenopathy can be seen with any of the pathogens listed. Primary ulcers are seen with all of the organisms, with some being associated with painful ulcers (*H. ducreyi* and HSV) and others with painless ulcers (*T. pallidum*). A febrile episode with generalized rash postprimary genital lesion is a distinctive feature associated with one of the listed pathogens. LGV and donovanosis are extremely rare in the United States (exposure history in the endemic areas is required for consideration).

ETIOLOGY

Treponema pallidum (secondary syphilis)

MICROBIOLOGIC PROPERTIES

Treponemes are spirochetes, which are **thin-walled, flexible, spiral rods** and are invisible by light microscopy. Treponemes have a dual membrane system akin to, but dramatically different from, Gram-negative bacteria. They exhibit characteristic corkscrew motility due to axial filaments (**endoflagella**), which can be seen by darkfield microscopy (Fig. 26-2), used in clinical practice for visualization. Treponemes do not grow in bacteriologic media or in cell culture. Treponemal membrane lipids, particularly cardiolipin, induce **nonspecific antibodies** that cross-react with beef heart cardiolipin. This reactivity with lipoidal antigens is the basis of **nontreponemal screening** (rapid plasma regain [RPR] and Venereal Disease Research Laboratory [VDRL]) tests. These tests convert to reactivity early in disease, and titers wane with successful treatment. Patients with autoimmune diseases, such as SLE, may demonstrate false-positive reactions to nontreponemal tests. Fluorescence treponemal antibody-absorption (**FTA-Abs**) or micro-hemagglutinin-*T. pallidum* (**MHA-TP**), which **detects T. pallidum-specific antibodies,** is specific and confirmatory. These tests become positive later in disease and remain elevated, often for life.

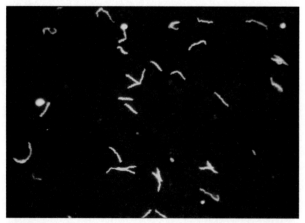

FIGURE 26-2 Darkfield microscopy of *Treponema pallidum* spirochetes. (*Courtesy of Dr. Paul Southern, Departments of Pathology and Medicine, University of Texas Southwestern Medical Center, Dallas, TX.*)

> **NOTE** *Treponema pertenue,* a related spirochete, causes yaws, which is endemic in tropical countries. It is not an STD. The serum from a patient with yaws is reactive in all treponemal or nontreponemal tests for syphilis.

EPIDEMIOLOGY

The incidence of syphilis has been increasing in the United States since 2000. *T. pallidum* is an obligate **human pathogen** and is transmitted either via **intimate contact** with infectious lesions (most common) or **transplacentally** from an infected mother to her fetus. Fetal (**congenital**) infection occurs with high frequency in untreated early infections of pregnant women and with lower frequency in latency. Kissing, transfusion, and inoculation may also transmit the infection. Health professionals have developed primary lesions on the hands following clinical examination of infectious lesions.

PATHOGENESIS

Treponemes penetrate intact mucous membranes or microscopic dermal abrasions. Constituent membrane **lipoproteins** are the principal **proinflammatory mediators of syphilis**. Overlapping acute and chronic inflammation accounts for the majority of symptoms for all stages of syphilis (Table 26-2).

A single lesion usually appears as an indurated painless ulcer (chancre) at the site of initial invasion (**primary syphilis**). Host humoral and cellular immune responses are insufficient to clear the organism. The treponemes **evade immune recognition and elimination by antibodies**, maintaining an outer membrane **rich in lipid** (major lipoprotein immunogens attached to the cytoplasmic membrane are not accessible to antibodies). The organisms enter the lymphatics and blood, and invasion is a critical virulence factor for *T. pallidum*, as demonstrated by its ability to traverse endothelial cell monolayers between cellular tight junctions. Pathologic changes occur with plasma cell–rich mononuclear infiltrates in the dermis, regardless of the stage of disease and the location of lesions. Lesions may appear as a maculopapular rash on skin or mucous membranes or may involve lymph nodes (**secondary syphilis**; seen in the case patient). Syphilis is one of the few infections that cause a rash on the palms and soles. The generalized mucocutaneous manifestations of secondary syphilis usually

TABLE 26–2 Important Clinical Features of *Treponema pallidum* Infections	
Syndromes	**Clinical Features**
Primary syphilis	Classically, an indurated, **painless, ulcer (chancre)** appears at the site of inoculation.
	Chancres heal within a few weeks.
	Patients are infectious.
Secondary syphilis	Disseminated disease appears within 6 months of primary infection.
	Constitutional symptoms (in more than 90% of patients) are fever, lymphadenopathy, **maculopapular rashes** (often on the palms and soles), and condylomata lata (eruptions of flat-topped papules, found wherever contiguous folds of skin produce heat and moisture, highly infectious)
	Patients are infectious.
Latent syphilis	One-third of untreated cases of secondary syphilis may become clinically **latent** for weeks to years.
	During this stage, there are no clinical manifestations of disease, yet the **infection may still be passed to newborns** by infected mothers; it may also progress, resulting in tertiary syphilis.
Tertiary syphilis	A delayed-type hypersensitivity to *T. pallidum* mediated by sensitized T lymphocytes and macrophages results in **gummatous ulcerations** and necrosis.
	Disabling lesions occur in the aorta (cardiovascular syphilis), in the skin and on mucosal surfaces, with chronic focal areas of inflammatory destruction (gummas), and in the CNS.
	Neurosyphilis causes degeneration of dorsal column and dorsal roots, resulting in locomotor ataxia (**tabes dorsalis**). Symptoms are shooting pain, Charcot joints, and **Argyll Robertson pupils**.
	A VDRL titer in the CSF is diagnostic of neurosyphilis, but not very sensitive.
	Patients with tertiary infection typically are not infectious.
Congenital syphilis	**Vertical transmission** usually occurs after 18 weeks of gestation and with women infected for 2 years or less (i.e., through early latency) and may cause **abortion or stillbirth**.
	Late manifestations in older children may include CNS abnormalities and blindness.

appear about 6 to 8 weeks after the healing of the chancre.

TREATMENT

Long-acting penicillin G (**benzathine penicillin G**) is effective for primary, secondary, and latent syphilis. Alternative therapy for penicillin-allergic patients is doxycycline (for adults only) or ceftriaxone (for congenital syphilis). Erythromycin can be used for penicillin-allergic pregnant women. However, the only effective therapy for neurosyphilis is high-dose intravenous penicillin G for 10 to 14 days. Repeated serologic testing is important to ensure adequate treatment. Titers of nonspecific antibodies drop, whereas titers of specific antibodies remain high for life. A four-fold VDRL or RPR serum titer increase indicates the need for retreatment. Syphilis in HIV-infected patients may be more difficult to treat.

OUTCOME

The patient was given benzathine penicillin G IM weekly for 3 weeks. Her symptoms gradually resolved. HIV testing was also performed, and results were negative.

PREVENTION

Early diagnosis and adequate treatment are effective in preventing transmission of the infection. Administration of antibiotic after suspected exposure is necessary. Serologic follow-up of infected individuals and their contacts is of value.

FURTHER READING

Augenbraun MH, Rolfs R: Treatment of syphilis, 1998: Nonpregnant adults. Clin Infect Dis 1:S21, 1999.

CDC: Primary and secondary syphilis, United States, 1998. MMWR 48:873, 1999.

Fiumara NJ: Treatment of primary and secondary syphilis. Serological response. JAMA 243:2500, 1980.

Larsen SA, Steiner BM, Rudolph AH: Laboratory diagnosis and interpretation of tests for syphilis. Clin Microbiol Rev 8:1, 1995.

Singh AE, Romanowski B: Syphilis: Review with emphasis on clinical, epidemiologic, and some biologic features. Clin Microbiol Rev 12:187, 1999.

A **26-year-old male** liberal arts college student visited a primary care clinic seeking medical attention. He was very concerned about the **painful, itchy sores** (blisters) that had developed on the **shaft of his penis.** He had low-grade fever, malaise, and a **mild headache.**

The patient had no significant medical history. He admitted to **unprotected sex with a new girlfriend** 3 days before arrival at the clinic.

PHYSICAL EXAMINATION

VS: T 38.2°C, P 90/min, R 14/min, BP 120/82 mmHg

PE: Examination revealed erythematous, **vesicular lesions on the penile shaft** (Fig. 27-1). There was also a clear discharge from the urethra and tender **inguinal lymphadenopathy.**

LABORATORY STUDIES

Blood

Hematocrit: 46%
WBC: 6800/μL
Differential: Normal

Imaging

Not necessary

Diagnostic Work-Up

Table 27-1 lists the likely causes of penile lesions (differential diagnosis). Rapid plasma reagin (RPR) serology can be performed to rule out syphilis. Investigational approach for delineating the etiology may include

- Tzanck smear
- Cultures for HSV-2 and *Haemophilus ducreyi*

COURSE

Viral cultures were taken from the base of the vesicles, which revealed cytopathic effects on the infected cells. The antigenic type of the virus isolate was subsequently delineated by direct fluorescence antibody (DFA) test.

FIGURE 27-1 Vesicular lesions of the penile shaft in this patient. *(Courtesy of Centers for Disease Control and Prevention, Atlanta, GA.)*

TABLE 27-1 Differential Diagnosis and Rationale for Inclusion (consideration)

Chancroid (*Haemophilus ducreyi*)
Donovanosis (*Calymmatobacterium granulomatosis*)
Genital herpes (herpes simplex virus)
Lymphogranuloma venereum (LGV; *Chlamydia trachomatis*)
Syphilis (*Treponema pallidum*)

Rationale: The differential diagnosis is very similar to that described in Case 26. Multiple etiologies should be considered in any patient with genital lesions. However, painful vesicular lesions are typical of sexually transmitted virus infection and are not associated with syphilis. Chancroid may produce painful ulcers, whereas the remaining infections generally produce painless ulcers.

ETIOLOGY

Herpes simplex virus (HSV) type 2 (genital herpes)

MICROBIOLOGIC PROPERTIES

Herpes simplex viruses HSV-1 and HSV-2 are two of the eight known viruses in the herpesvirus family (see Case 2 for classification). All herpesviruses are structurally similar, having large virions with an icosahedral nucleocapsid (Fig. 27-2A), a linear double-stranded DNA genome, and a lipoprotein envelope. **Virus-infected host cells reveal the presence of multinucleated giant cells on a Tzanck smear** (see Fig. 27-2B), although the findings are not specific for the type of herpesvirus. Identical findings are present in lesions caused by varicella-zoster virus. A Tzanck smear is prepared by scraping the floor of the herpetic vesicle. The development of cytopathic effects in cell cultures of HSV for the visualization of infected cells by DFA may take 2 to 4 days. HSV-2 is distinguished from HSV-1 by antigenicity, which is the basis of the laboratory differentiation. Monoclonal antibodies are labeled with fluorescein stain proteins specific for HSV-1 or HSV-2.

EPIDEMIOLOGY

Genital herpes is epidemic worldwide, with a high prevalence in industrialized countries, consistent with high seroprevalence of HSV infections in the western hemisphere. The prevalence of HSV-2 is greater than 20% among adults in the United States. **Transmission** of HSV-2 occurs predominantly by **sexual contact**. Unrecognized infections and subclinical viral shedding are believed to be major factors in transmission. Vertical transmission occurs during pregnancy from infected, seronegative women to newborns.

PATHOGENESIS

Intimate contact between a susceptible person (no antibodies) and an individual who is actively shedding the virus is required for HSV infection to occur. Contact must involve mucous membranes or open or abraded skin. Virus replicates in the skin or mucous membrane at the portal of entry. In men, lesions appear on the glans penis or prepuce, and in the anus and rectum of those engaging in anal sex. In women, the principal sites of primary disease are the cervix and

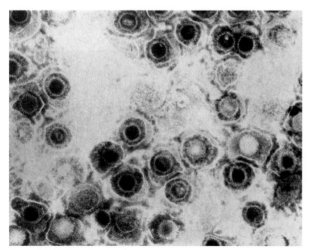

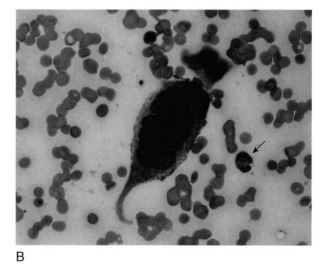

A   B

FIGURE 27-2 *A,* Electron micrograph of herpes simplex virus, an enveloped DNA virus. Note some nucleocapsids are empty, as shown by penetration of electron-dense stain. *B,* Tzanck smear of a specimen from a vesicle base. Note the giant, multinucleated squamous epithelial cell (keratinocyte). Compare the size of the giant cell to that of neutrophils (*arrow*), also seen in this smear. Diff-Quik (Romanowsky) stain was used here. (*A, Courtesy of Centers for Disease Control and Prevention, Atlanta, GA; B, Courtesy of Dr. Dominick Cavuoti, Department of Pathology, University of Texas Southwestern Medical Center, Dallas, TX.*)

the vulva. **Primary lesions** are due to the cytopathic (**cytolytic**) effect of the virus on **mucocutaneous epithelium**. Viruses travel from the mucocutaneous epithelium **to a small number of sensory neurons**, where multiplication occurs. Once quiescence has been achieved, the viral genome remains in a latent state unless reactivated. Cell-mediated immunity is important in controlling herpesviruses. Transient suppression of cellular immunity brought on by events such as physical or emotional stress, fever, UV light, and tissue damage causes the virus to be transported back down the axon to replicate again at or near the original point of entry into the body, often resulting in reactivation and episodic (recurrent) disease.

TREATMENT

Oral **acyclovir** (ACV) is effective in first-episode genital herpes and in recurrences. Maintenance therapy with oral ACV can prevent recurrences and reduce asymptomatic virus shedding. Topical ACV reduces viral shedding and hastens crusting in the first episode but not in recurrent herpes. Treatment has no effect on latency. Resistant viruses may cause significant complications, particularly in patients with AIDS. Famciclovir and valacyclovir, which are more recent drugs, can be taken less frequently, but they may not be effective against drug-resistant viruses. Foscarnet can be used in resistant cases.

NOTE A first-episode HSV-2 infection (60% asymptomatic) during pregnancy, without seroconversion, is associated with spontaneous abortion, prematurity, and congenital and neonatal herpes. Vaginal delivery **in pregnant women with active genital infections** carries a high risk of infection to the fetus or newborn. In women whose first episode of genital herpes occurred during pregnancy, treating for the last 4 weeks of pregnancy using acyclovir can reduce the risk of neonatal herpes. Like any other herpesvirus, HSV-2 establishes a lifelong latent infection that undergoes **episodic reactivation**, which has been associated with meningitis in adults with reduced cell-mediated immunity.

OUTCOME

The patient was given oral ACV for 10 days, with gradual resolution of the lesions.

PREVENTION

Using barrier contraception is an effective method of preventing transmission. In the pregnant patient, cesarean section may be performed to prevent transmission to the newborn. There is no vaccine.

FURTHER READING

Ashley RA, Wald A: Genital herpes: Review of the epidemic and potential use of type-specific serology. Clin Microbiol Rev 12:1, 1999.
Balfour HH, Jr: Antiviral drugs. N Engl J Med 340:1255, 1999.

Corey L, Spear PG: Infections with herpes simplex viruses. N Engl J Med 314:686, 1986.
Kimberlin DW: Neonatal herpes simplex infection. Clin Microbiol Rev 17:1, 2004.

A 23-year-old woman visited her family physician for her routine physical exam and Pap smear. She had a few **small, raised lesions on the cervix** but was otherwise asymptomatic.

She had been sexually active since she was 15, with many sexual partners.

PHYSICAL EXAMINATION

VS: Normal

PE: Labial venereal warts (condylomata acuminata) and a **friable, erythematous cervix** were noted during pelvic examination.

LABORATORY STUDIES

Blood

WBC: Normal

Differential: Normal

Serum chemistries: Normal

Imaging

Colposcopy confirmed the presence of **lesions on the cervix**.

Diagnostic Work-Up

Table 28-1 lists the likely causes of genital diseases. Papanicolaou (Pap) smear is necessary for any abnormal cell types, although it is not sufficient (50% false negative). Further investigation may include

- VDRL (Venereal Disease Research Laboratory) or RPR (rapid plasma reagin) serology for syphilis
- Tzanck smear for herpes
- DNA hybridization of biopsy (by colposcopy) for human papillomavirus (HPV)
- ELISA assay to detect HPV-specific IgG antibody

COURSE

The appearance was suggestive of venereal warts, and a Pap smear showed some atypical cells as well.

TABLE 28-1 Differential Diagnosis and Rationale for Inclusion (consideration)
Anogenital malignancy
Chlamydia
Gonorrhea
Human papillomavirus (HPV)
Genital herpes (HSV-2)
Syphilis

Rationale: Multiple etiologies are possible, generally sexually transmitted. Gonorrhea and chlamydia often produce significant discharge, and HSV-2 causes painful vesicles and shallow ulcers. Malignancy is always a concern with lesions as described, and should be ruled out. HPV often leads to verrucous lesions. Syphilis is often painless.

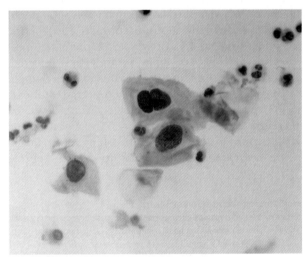

FIGURE 28-1 An abnormal Pap smear in this patient. Note the squamous cells show binucleation with enlarged hyperchromatic nuclei. *(Courtesy of Dr. Dominick Cavuoti, Department of Pathology, University of Texas Southwestern Medical Center, Dallas, TX.)*

ETIOLOGY

HPV

MICROBIOLOGIC PROPERTIES

HPVs are nonenveloped viruses of icosahedral symmetry with 72 capsomeres that surround a genome containing double-stranded circular DNA. HPV viruses do not grow in cell culture. There are more than 100 types of HPV, determined by molecular hybridization techniques using molecularly cloned HPV DNA of a known type as the standard. Each type affects certain parts of the body (for example, types 6 and 11 cause genital warts in some women; Table 28-2). **HPVs** may be classified on the basis of their tropism as either **genital** (mucosal) or **cutaneous. Genital HPVs** are subdivided into high-risk and low-risk types, according to their malignant potential and cell-transforming capacity in vitro. Cutaneous HPVs may be subdivided into the classic types associated with cutaneous viral **warts** (see Table 28-2).

EPIDEMIOLOGY

Humans are the only known reservoir for HPV. The types of HPV that infect the genital area are **spread primarily through genital contact.** Most HPV infections have no signs or symptoms; therefore, most infected persons are unaware they are infected. However, they can transmit the virus to a sex partner. Women with a history of sexually transmitted disease (STD) are at increased risk for acquiring HPV, and therefore may be at increased risk for cervical cancer. **Risk factors** for HPV infection leading to carcinoma of the cervix include **sexual activity before age 15,** **multiple sexual partners,** exposure to STD, mother or sister with cervical cancer, **smoking,** immunosuppression, HIV/AIDS, and chronic corticosteroid use (e.g., for asthma and lupus).

PATHOGENESIS

HPVs **infect the squamous epithelial cells** of the basal layer of the skin or mucous membrane. They enter basal cells that have been exposed through a disturbed epithelial barrier, as would occur during sexual intercourse. HPV infections have not been shown to be cytolytic; rather, viral particles are released with the degeneration of desquamating cells. Virus multiplication is confined to the nucleus. Consequently, **infected cells exhibit a high degree of nuclear atypia.** All types can cause mild Pap test abnormalities (koilocytosis, characterized as intracellular changes with **perinuclear clearing** [halo] and a **shrunken nucleus** [Fig. 28-1]) that do not have serious consequences. In benign or low-risk HPV lesions, such as those typically associated with HPV types 6 and 11, the HPV genome exists as a circular, episomal DNA separate from the host cell nucleus. The infection is characterized by **proliferation and thickening of the basal layer, leading to the appearance of a wart.**

TREATMENT

The primary goal of treating visible genital warts is the removal of symptomatic warts. In most patients, treatment can induce wart-free periods. Treatment may include physical therapy (e.g., cautery, cryotherapy with liquid nitrogen, or laser therapy) and chemical therapy (e.g.,10% to 25% solution of podophyllin

TABLE 28–2 Clinical Groupings of HPVs and Associated Diseases		
Clinical Grouping	**Anatomic Locations**	**Major HPV Types**
Cutaneous lesions in healthy host	Verrucae (common warts) on nongenital tissue	1, 2, 4, 26, 27, 29
	Plantar warts on nongenital skin	1, 2, 4, 63
Cutaneotropic lesions in an immunodeficient host	Epidermodysplasia verruciformis	3, 10, 27, 38
	Skin cancers in transplant recipients	48, 60
Mucosotropic lesions	Anogenital warts (condylomata acuminata)	6, 11
	Dysplasias and cancer	**16, 18,** 31, 33, 35

or 80% to 90% trichloroacetic acid). Determining whether treatment of genital warts will reduce transmission is difficult because no laboratory marker of infectivity has been established. No evidence indicates that the presence of genital warts or their treatment is associated with the development of cervical cancer.

NOTE The HPV types that cause genital warts are not the same as the HPV types that can cause cancer (see Table 28-2). **HPV 16 and 18** are considered the "bad viruses" that cause **carcinoma of the cervix.** In malignant lesions (dysplasia and carcinoma), the genomes of high-risk HPV types 16 and 18 typically are integrated into the host cell DNA. Integration of the viral genome into the host cell genome is considered a hallmark of malignant transformation. HPV proteins E6 and E7 of HPV types 16 and 18 inactivate the host's tumor suppressor proteins p53 and retinoblastoma (RB) and subvert the cell growth regulatory pathways. They modify the cellular environment in order to facilitate viral replication in a cell that is terminally differentiated and has exited the cell cycle. The result is an unregulated epithelial growth, which begins at the basal layer and extends outward (dysplasia) and may progress to invasive squamous cell carcinoma. A Pap test can detect precancerous and cancerous cells on the cervix (50% false negative). Staging of carcinoma in situ may be necessary. Management (which may include hysterectomy) is individualized based on staging results.

OUTCOME

Further investigation was encouraging. The HPV type was found to be not associated with malignancy, and the patient was reassured. The patient was treated with topical cryotherapy with good results.

PREVENTION

Barrier protection may reduce the risk of transmission of HPV and other STDs. Sexually active women should have a regular Pap test to screen for cervical cancer and other precancerous changes.

FURTHER READING

Burd EM: Human papillomavirus and cervical cancer. Clin Microbiol Rev 16:1, 2003.
Crum CP, Ikenberg H, Richart RM, Gissman L: Human papillomavirus type 16 and early cervical neoplasia. N Engl J Med 310:880, 1984.
Fife KH: New treatments for genital warts less than ideal. JAMA 279:2003, 1998.

Ho YF, Bierman R, Beardsley L, et al: Natural history of cervicovaginal papillomavirus infection in young women. N Engl J Med 338:423, 1998.
Sawaya GF, Brown AD, Washington AE, Garber AM: Current approaches to cervical-cancer screening. N Engl J Med 344:1603, 2001.

An 18-year-old male high school student presented with a 48-hour history of **painful urination** with a **yellowish penile discharge**.

He returned 2 days ago from Daytona Beach, where he had been **sexually active with several female partners** during Spring Break. He denied previous such episodes, and said he was generally in good health.

CASE
29

PHYSICAL EXAMINATION

VS: Normal

PE: Purulent urethral discharge was noted. No sign of genital ulcers was found. No rash or skin ulcers were seen, and the inguinal lymph nodes were not enlarged or tender.

LABORATORY STUDIES

Blood

Normal

Imaging

No imaging studies were done.

Diagnostic Work-Up

A clinical diagnosis of urethritis was suspected. Table 29-1 lists the likely causes of the illness (differential diagnosis). Investigational approach may include

- **Gram stain** and **culture** of urethral discharge
- **DNA probe**. A preferred testing method because of its sensitivity and rapid turnaround for busy sexually transmitted disease (STD) clinics

In failed tests above, a **direct antigen detection** test in urine can be pursued to demonstrate a chlamydia infection.

COURSE

The patient's urethral discharge was sent to the laboratory for Gram stain and culture, which yielded the diagnosis.

TABLE 29–1   Differential Diagnosis and Rationale for Inclusion (consideration)
Chlamydia trachomatis *Mycoplasma hominis* *Neisseria gonorrhoeae* *Ureaplasma urealyticum*

Rationale: Urethral discharge in men is related to a relatively few organisms. In sexually active individuals, purulent discharge is highly suggestive of gonorrhea. The other agents in the differential diagnosis, particularly *Chlamydia*, generally cause a clear discharge.

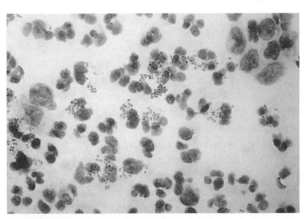

FIGURE 29-1   Gram stain of urethral exudate showing the pathogen associated with neutrophils.

ETIOLOGY

Neisseria gonorrhoeae (gonorrhea)

MICROBIOLOGIC PROPERTIES

All *Neisseria* species (pathogenic or nonpathogenic) are **Gram-negative diplococci.** A smear of urethral discharge can be examined under a microscope after Gram staining. The presence of **multiple pairs of bean-shaped, Gram-negative diplococci within a neutrophil** in a Gram smear of urethral discharge (Fig. 29-1) is **diagnostic of gonorrhea** (more sensitive in males versus females). The organisms have a typical Gram-negative bacterial cell wall with a peptidoglycan backbone. The genus *Neisseria* contains two pathogenic species, *N. gonorrhoeae* (also known as gonococcus) and *Neisseria meningitidis* (also known as meningococcus). *N. gonorrhoeae* **do not have a polysaccharide capsule but have pili, cell-wall lipo-oligosaccharide (LOS), and outer membrane proteins (OMP),** all of which **contribute to virulence.** The fastidious organisms **grow on selective media (e.g., Thayer Martin medium),** supplemented with antimicrobial agents that inhibit competing flora while allowing growth of *N. gonorrhoeae*. All *Neisseria* are **oxidase positive.** *N. gonorrhoeae* isolates from clinical specimens are confirmed by **sugar (glucose) fermentation,** immunoassay, or DNA probes.

EPIDEMIOLOGY

Gonorrhea is a very common infectious disease in the United Disease and worldwide. *N. gonorrhoeae* is a **strictly human pathogen,** with asymptomatic carriers being the largest reservoir. The infection is spread primarily by sexual contact. Any sexually active person can be infected. In the United States, the **highest reported rates** of infection are among **sexually active teenagers, young adults,** and African Americans. Individuals with multiple sexual partners (encounters) are at increased risk for this infection; other risks include inherited complement deficiencies, early age at onset of sexual activity, and lower socioeconomic status.

PATHOGENESIS

N. gonorrhoeae initiates infections by colonizing the mucosal epithelium, which serves as the site of entry.

N. gonorrhoeae binds to columnar epithelial cells mediated by bacterial **pili** and OMPs. Sophisticated genetic mechanisms enable the bacteria to control the presence or absence of these components, a phenomenon called *phase variation*. Host antibodies to pili and OMPs do not protect against gonococcal infection. **IgA protease,** an extracellular enzyme of the pathogenic *Neisseria*, hydrolyzes IgA$_1$ mucosal antibodies and inhibits the opsonization needed for phagocytic killing. **Gonococcal OMPs (e.g., protein I)** also protect against phagocytosis, and protein I interferes with neutrophil degranulation. Local invasion through or around epithelial cells allows *N. gonorrhoeae* to reach the subepithelial matrix, where it initiates an intense inflammatory reaction, ushering in PMNs. In men, pain, dysuria, and urethral discharge are brought on (as seen in the case patient). *N. gonorrhoeae* causes a variety of other clinical syndromes that are debilitating, particularly to women (Table 29-2).

TREATMENT

Penicillin-resistant *N. gonorrhoeae* (PPNG) originated in Southeast Asia and has become widespread in the United States and worldwide. Successful treatment of gonococcal infection requires the use of third-generation cephalosporins or quinolones. A single injection of **ceftriaxone** is the currently recommended drug of choice. Patients treated for gonorrhea should be simultaneously treated for chlamydia (e.g., with doxycycline or azithromycin) because coinfection rates are high. Presumptive treatment of sexual partners is also recommended.

OUTCOME

The patient was given intramuscular ceftriaxone and a 10-day course of doxycycline. His symptoms rapidly resolved.

PREVENTION

The surest way to avoid transmission of STDs is to abstain from sexual intercourse, or to be in a long-term mutually monogamous relationship with a partner who has been tested and is known to be uninfected. Latex condoms, when used consistently and correctly, can reduce the risk of transmission of gonorrhea. There is no vaccine.

TABLE 29–2 Other Important Clinical Features of *Neisseria gonorrhoeae* Infections

Syndromes	Clinical Features
Cervicitis	Relatively asymptomatic in women; when symptomatic, vaginal discharge, lower abdominal pain, and mucopurulent cervical discharge are seen.
Pharyngitis	Often asymptomatic; may occur in sexual abuse or in genital-oral sex.
Epididymitis	In men, gonorrhea can cause a painful condition of the testicles that can lead to infertility if left untreated.
	Without prompt treatment, gonorrhea can also affect the prostate and can lead to scarring inside the urethra, making urination difficult.
Pelvic inflammatory disease (PID; salpingitis)	In women, gonorrhea is a common cause of PID. Symptoms can be extremely severe and can include abdominal pain and fever.
	PID can lead to long lasting, chronic pelvic pain, with internal abscesses that may cause infertility or damage the fallopian tubes, increasing the risk of ectopic pregnancy.
Gonococcal conjunctivitis	Formerly a major cause of blindness in newborns until the institution of treating their eyes with silver nitrate, tetracycline, or erythromycin.
Disseminated gonococcal infection (DGI)	In untreated female patients, certain strains of *N. gonorrhoeae* may survive in the blood and settle in the joints (causing acute septic arthritis) or skin (causing rash).
	Sialylation of LOS allows host CMP-NANA to combine with LOS so the bacterium's surface mimics the host tissues. This protects the bacteria from complement-mediated killing. LOS endotoxin mediates a systemic inflammatory response, releasing cytokines and reactive amines.
	Absence of bactericidal antibody or terminal components of the complement pathway (C_5-C_8) also causes a host to be highly susceptible to recurrent infection and systemic disease.

FURTHER READING

Handsfield HH, McCormack WM, Hook EW: A comparison of single-dose cefixime with ceftriaxone as treatment for uncomplicated gonorrhea. N Engl J Med 325:1337, 1991.

Handsfield HH, Rice RJ, Roberts MC: Localized outbreak of penicillinase-producing *Neisseria gonorrhoeae*. Paradigm for introduction and spread of gonorrhea in a community. JAMA 261:2357, 1989.

Nassif X, So M: Interaction of pathogenic neisseriae with nonphagocytic cells. Clin Microbiol Rev 8:376, 1995.

Shirtliff ME, Mader JT: Acute septic arthritis. Clin Microbiol Rev 15:527, 2002.

A 27-year-old woman presented to the emergency department of a city medical center complaining of **lower abdominal pain, vaginal discharge,** and **dysuria for 1 week.** She also complained of **fevers and chills for the past 2 days.**

She had **had four sexual partners** in the previous year and used a contraceptive measure (condoms) only occasionally. She had never been treated for a sexually transmitted disease (STD), and she had not seen a physician in 2 years.

PHYSICAL EXAMINATION

VS: T 38.4°C, P 104, R 16, BP 112/68

PE: Pelvic examination revealed a **reddened cervical os** (Fig. 30-1). Lower abdominal tenderness, **adnexal tenderness,** and **cervical motion tenderness** were noted. The uterus, fallopian tubes, and ovaries were also tender but were not enlarged.

LABORATORY STUDIES

Blood

ESR: 31 mm/h
Hematocrit: Normal
WBC: 13,400/μL
Differential: 68% PMNs, 8% bands, 15% lymphs

Imaging

Not done

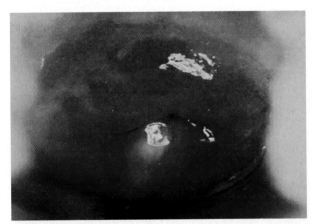

FIGURE 30-1 View through vaginal speculum showing reddened cervical os, through which mucopurulent secretion is exuding. (*Courtesy of Dr. Jeanne Sheffield, Department of Obstetrics and Gynecology, University of Texas Southwestern Medical Center, Dallas, TX.*)

Diagnostic Work-Up

A clinical diagnosis of PID was considered based on positive pelvic examination (adnexal tenderness and cervical motion tenderness). An endometrial biopsy or aspiration may reveal evidence of inflammation supporting the clinical diagnosis. Table 30-1 lists the likely causes of this PID. Investigational approach may include

- Microscopy

  - **Gram stain** of vaginal discharge for *Neisseria gonorrhoeae*
  - **Giemsa** or **direct fluorescent antibody** stain of cytobrush specimen of endocervix for *Chlamydia trachomatis* (urine antigen in male patients)

- Cultures for *N. gonorrhoeae* and *Chlamydia* and anaerobic cultures of endometrial aspiration to confirm polymicrobic infection

If all of the aforementioned fail, *N. gonorrhoeae* or *C. trachomatis* can be detected in endometrial biopsy

TABLE 30-1 Differential Diagnosis and Rationale for Inclusion (consideration)
Actinomyces israelii
Chlamydia trachomatis
Enteric Gram-negative rods (*Escherichia coli*)
Gardnerella vaginalis
Mycoplasma hominis
Neisseria gonorrhoeae
Polymicrobic infection
Streptococcus agalactiae

Rationale: A clinical diagnosis of pelvic inflammatory disease (PID) should be considered. There are many pathogens associated with PID, but the pathology is generally the same. *Gardnerella* and Group B streptococcus are common organisms in the female genital tract. *N. gonorrhoeae* and *C. trachomatis* should be considered in sexually active patients. Often, multiple organisms are present, particularly anaerobes along with facultative anaerobes. *A. israelii* is often associated with intrauterine contraceptive devices.

by (1) 16S rRNA probes, (2) polymerase chain reaction, or (3) ligase chain reaction.

COURSE

The patient was hospitalized. Gram stains of the secretions revealed numerous PMNs but no evidence of Gram-negative diplococci. The diagnosis of PID was made based on the positive culture results, and the patient was started on antibiotics. Cultures of endometrial aspiration under anaerobic conditions also grew three different anaerobic species.

ETIOLOGY

C. trachomatis (and anaerobes)

> **NOTE** The following sections deal with *C. trachomatis;* anaerobes are described in later chapters.

MICROBIOLOGIC PROPERTIES

C. trachomatis are **obligate intracellular bacteria**. These bacteria are energy dependent (cannot produce ATP on their own; dependent on host epithelial cells) and grow only in the monolayers of cell cultures. The bacterial cell contains Gram-negative cell layers. Genus-specific LPS is the common antigen among the species such as *C. trachomatis, C. psittaci,* and *Chlamydophila pneumoniae* (formerly known as *Chlamydia pneumoniae*). *C. trachomatis* has more than 15 serovars. *C. trachomatis* **serotypes D to K are considered the world's most common sexually transmitted bacterial pathogens**, and, following vertical transmission through an infected birth canal, they cause neonatal conjunctivitis and pneumonia. Respiratory infection with *C. pneumoniae* occurs worldwide around the year and the atypical pathogen is estimated to cause, on average, 10% of community-acquired pneumonia cases and 5% of bronchitis and sinusitis cases. In addition, avian strains of *C. psittaci* have long been known to cause psittacosis (atypical pneumonia) in humans.

EPIDEMIOLOGY

Chlamydia, one of the causative agents of PID, is the most frequently reported sexually transmitted bacterial infection in the United States. A high asymptomatic carrier rate (more than 25%) is predominantly associated with men. PID is caused by the same serovars as in exudative genital infections (D-K). Risk factors for PID include young age at first intercourse, multiple sexual partners, intrauterine device (IUD) insertion, and tobacco smoking.

PATHOGENESIS

Chlamydia organisms begin infection through minute abrasions (after sexual intercourse) on the mucosal surface as metabolically inert **elementary bodies** (EB), attaching to and stimulating uptake by the host cell. EBs target receptors on **columnar epithelial cells** (CECs) and enter into the CECs of the cervix. The internalized EBs remain within a host-derived vacuole, called a **cytoplasmic inclusion** (**seen on Giemsa** or fluorescence antibody-stained smear; Fig. 30-2), where they differentiate to larger, **metabolically active, reticulate bodies** (RB). The RBs **multiply by binary fission**, and after 8 to 12 rounds of multiplication, they reorganize to EBs asynchronously, and multiplication ceases. At 30 to 84 hours postinfection, many EB particles are released from the host cell to initiate another cycle of infection.

Primary genital lesions of chlamydial disease occur when CEC destruction during the acute disease process causes release of **proinflammatory cytokines**, which are chemotactic for neutrophils and mononuclear cells. First acute (neutrophilic leukocytes) and then chronic (macrophages and lymphocytes) inflammatory cells are attracted, and an ineffective humoral antibody response is mounted. The presence of

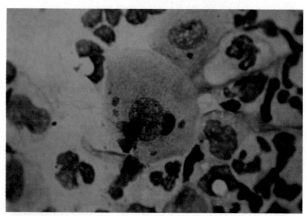

FIGURE 30-2 *Chlamydia trachomatis* cytoplasmic inclusions of reticulate bodies. *(Courtesy of Dr. Paul Southern, Departments of Pathology and Medicine, University of Texas Southwestern Medical Center, Dallas, TX.)*

γ-interferon may induce a persistent disease state, with the altered reticulate body growing, albeit more slowly. Tissue necrosis occurs, as well as subsequent infiltration of mononuclear cell aggregates surrounded by endothelial cells.

Ascending extension of chlamydiae from the endocervix to the endometrium and endosalpinx is dependent on many endogenous or exogenous factors (e.g., estrogen-dominated [thin] cervical mucus, attachment of the organisms to sperm that migrate upward into the tube, bacterial vaginosis, intrauterine contraceptive device). Arriving at the upper reproductive tract, *C. trachomatis* infect the CECs of the fallopian tube but produces little damage by direct effect (chlamydiae do not produce toxins). Systemic symptoms of PID are due to the **progressively chronic inflammatory disorder**, resulting in damage to the uterus, fallopian tubes, and adjacent pelvic structures.

| NOTE | *C. trachomatis* causes a variety of other clinical syndromes and complicated illnesses in specific risk groups (e.g., neonates; Table 30-2).

TREATMENT

Empirical therapy of PID must be comprehensive and cover all likely pathogens in the context of the clinical setting. A polymicrobial infection, such as occurred in the case patient, must be treated empirically with broad-spectrum antibiotics. A typical course of antibiotics would be cefoxitin (to cover organisms from lower bowel and anaerobes) plus doxycycline for chlamydia. **For confirmed (monomicrobic) chlamydia infection, doxycycline is the drug of choice** for adults. Azithromycin (in a single dose) is also effective for noncompliant patients.

TABLE 30–2   Other Important Clinical Features of *Chlamydia trachomatis* Infections

Syndromes	Clinical Features
Trachoma	Caused by chlamydial **serotypes A, B, and C.**
	Endemic in Middle East, India, and **North Africa; a leading cause of blindness** worldwide.
	Chronic infection of the conjunctiva leads to follicles of inflammatory cells; fibrosis of eyelid causes lashes to turn inward and abrade cornea.
Nongonococcal urethritis (NGU; in men)	Caused by chlamydial **serotypes D-K.**
	Marked symptoms (50% symptomatic) of the acute infection are discharges, urethral itching, and burning on urination.
Lymphogranuloma venereum (LGV)	Caused by four different **serotypes (L1, L2, L2a and L3)**, which are highly virulent and can disseminate systemically.
	In heterosexuals, the disease begins with a genital ulcer, often unnoticed, followed by **painful inguinal lymphadenopathy.**
	In homosexual males, the infection can cause granulomatous proctitis that closely resembles Crohn disease.
Neonatal conjunctivitis	**Caused by the serotypes D-K.**
	Newborns are infected **during the passage through the birth canal of infected mothers; mucopurulent conjunctivitis** ensues 2-25 days post birth.
	Ocular prophylaxis does not seem to be effective.
Neonatal pneumonia	Also results from passage through the *C. trachomatis*-infected birth canal; 11-20% of babies born to infected mothers will develop **interstitial pneumonia.**
Reiter syndrome	Reiter syndrome manifests with a **triad of urethritis, arthritis, and conjunctivitis** (and anterior uveitis) within 6 months of chlamydia infection.
	Pathogenesis is unknown but is presumed to be autoimmune because of the high prevalence of HLA-B27 positivity.

NOTE A delay in diagnosis and treatment of PID may result in long-term sequelae such as tubal infertility. Tubo-ovarian abscess is one of the major complications of acute PID, and it occurs in as many as 15% to 30% of women requiring hospitalization for treatment of PID.

OUTCOME

The patient was given cefoxitin and doxycycline for PID and was treated with IV therapy for 5 days. She was then given an oral regimen to complete a total of 14 days of therapy. Follow up showed no evidence of infection or infertility.

PREVENTION

The main preventable cause of PID is an untreated STD (caused by *C. trachomatis* or *N. gonorrhoeae*). Women can protect themselves from PID by taking action to prevent STDs or by getting early treatment if they do get an STD. The surest way to avoid transmission of STDs is to abstain from sexual intercourse, or to be in a long-term, mutually monogamous relationship with a partner who has been tested and is known to be uninfected.

FURTHER READING

Black CM: Current methods of laboratory diagnosis of *Chlamydia trachomatis* infections. Clin Microbiol Rev 10:160, 1997.

Centers for Disease Control and Prevention: 1998 Guidelines for treatment of sexually transmitted diseases. MMWR 47(RR-1):1, 1998.

Gaydos CA, Howell MR, Pare B, et al: *Chlamydia trachomatis* infections in female military recruits. N Engl J Med 339:739, 1998.

Hillis SD, Wasserheit JN: Screening for chlamydia—A key to the prevention of pelvic inflammatory disease. N Engl J Med 334:1399, 1996.

Scholes D, Stergachis A, Heidrich FE, et al: Prevention of pelvic inflammatory disease by screening for cervical chlamydial infection. N Engl J Med 334:1362, 1996.

A 26-year-old woman came to a city sexually transmitted disease (STD) clinic complaining of a **profuse yellow, foamy vaginal discharge** with **foul odor.** The discharge began 2 days before with vulvar irritation and itching. She said she was also having pain during intercourse.

The patient had been **sexually promiscuous with multiple partners.** The medical history was unremarkable.

PHYSICAL EXAMINATION

VS: Normal

PE: Pelvic examination revealed homogeneous vaginal discharge and vulvar erythema. A **diffuse macular erythematous lesion of the cervix** ("**strawberry cervix,**" Fig. 31-1) was noted. Lower abdominal tenderness was also noted.

LABORATORY STUDIES

Blood

Not done

Imaging

Not done

Diagnostic Work-Up

Table 31-1 lists the likely causes of the woman's illness (differential diagnosis). The vaginal pH is measured on Nitrazine paper, and a fishy odor released upon application of 10% potassium hydroxide to a vaginal swab sample ("KOH amine test") suggested trichomoniasis or bacterial vaginosis. The following additional studies may aid the specific microbiologic diagnosis:

- Wet mount examination of vaginal discharge
- **Gram stain** to reveal bacteria or yeasts

In **bacterial vaginosis** (caused by *Gardnerella vaginalis* and *Mobiluncus*), wet prep or Gram stain of discharge is usually performed. The presence of a vaginal epithelial cell with attached microorganisms ("**clue cell**" with a stippled appearance) is diagnostic.

COURSE

A wet mount of the vaginal discharge was performed, which revealed large, motile (flagellated) organisms in low-power fields. Clue cells were not observed.

TABLE 31–1 Differential Diagnosis and Rationale for Inclusion (consideration)

Bacterial vaginosis (*Gardnerella/Mobiluncus*)
Candidiasis (*Candida albicans*)
Trichomoniasis (*Trichomonas vaginalis*)

Rationale: Vaginal discharge can be caused by any of the above pathogens, and distinguishing them clinically is often difficult, if not impossible. Classically, the discharge from *Candida* is whitish, that of *Trichomonas* is yellow and frothy, and that of bacterial vaginosis is particularly foul smelling. Microbiologic examination is generally required for a definitive diagnosis.

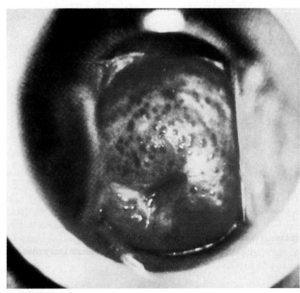

FIGURE 31-1 "Strawberry cervix" in this patient. *(Courtesy of Centers for Disease Control and Prevention, Atlanta, GA.)*

ETIOLOGY

Trichomonas vaginalis (trichomoniasis)

MICROBIOLOGIC PROPERTIES

T. vaginalis is a motile (with visible flagella), pear-shaped (10 μm by 7 μm) protozoan organism. Direct examination of a vaginal secretion usually demonstrates pus cells interspersed with the nucleated, **flagellated trichomonads**, which are slightly **larger than PMNs** and are easily **identified by their ameboid mobility** (Fig. 31-2). *T. vaginalis* favors the female lower genital tract and the male urethra and prostate, where it replicates by binary fission. These protozoa **do not have a cyst form**, and they do not survive well in the external environment. Direct contact is necessary for transmission.

EPIDEMIOLOGY

Trichomoniasis occurs worldwide. Its widespread prevalence in the United States is an important public health concern. *T. vaginalis* is transmitted among humans, its only known host, primarily by sexual intercourse. It is encountered in sexually active adolescents and adults, with a higher prevalence seen among persons with multiple sexual partners or those with venereal diseases.

PATHOGENESIS

In women, *T. vaginalis* is found in the vagina, cervix, urethra, bladder, and Bartholin and Skene glands. In men, the organism is found in the anterior urethra, external genitalia, prostate, epididymis, and semen. Women may be asymptomatic carriers, or they may experience a range of symptoms, including a mild-to-fulminant inflammatory disease. In the vaginal and urethral tissues, after an incubation period of 4 to 28 days, *T. vaginalis* causes **an inflammatory reaction (vaginitis)**; therefore, **numerous PMNs** usually are present. Vaginal pH is usually greater than 4.5. The protozoal pathogen may also cause direct damage to the epithelium, leading to microulcerations. Most men who harbor the organism are asymptomatic.

TREATMENT

The drugs of choice for treatment are metronidazole (in the United States) and tinidazole (not approved in the United States); therapy is usually highly successful. Treatment should also include all sexual partners of the infected person.

OUTCOME

Therapy was started with metronidazole, with rapid improvement in symptoms. The patient's partner was treated as well, although he was asymptomatic.

PREVENTION

The surest way to avoid transmission of STDs is to abstain from sexual contact, or to be in a long-term mutually monogamous relationship with a partner who has been tested and is known to be uninfected. Male condoms, when used consistently and correctly, can reduce the risk of transmission of trichomoniasis.

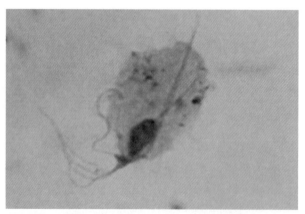

FIGURE 31-2 Giemsa stain of *Trichomonas vaginalis* protozoa (a single trophozoite is seen; there is no cyst form). Note the flagella that cause the organism to penetrate the mucus barrier at the urogenital mucosa. *(Courtesy of Dr. Paul Southern, Departments of Pathology and Medicine, University of Texas Southwestern Medical Center, Dallas, TX.)*

FURTHER READING

Catlin BW: *Gardnerella vaginalis*: Characteristics, clinical considerations, and controversies. Clin Microbiol Rev 5:213, 1992.

Centers for Disease Control and Prevention: Sexually transmitted diseases treatment guidelines 2002. MMWR 51(no. RR-6), 2002.

Krieger JN, Alderete JF: *Trichomonas vaginalis* and trichomoniasis. In Holmes K, Markh P, Sparling P, et al (eds): Sexually Transmitted Diseases, 3rd ed. New York, McGraw-Hill, 1999, p 587.

Petrin D, Delgaty K, Bhatt R, Garber G: Clinical and microbiological aspects of *Trichomonas vaginalis*. Clin Microbiol Rev 11:300, 1998.

A 21-year-old woman seen in a clinic complained of bothersome **vulvar itching** and **thick, whitish vaginal discharge** lasting for several days.

She had had no previous episodes. She denied being sexually active. She stated that she had recently completed a course of **antibiotics for a sinus infection.**

PHYSICAL EXAMINATION

VS: Normal

PE: **Vulvar erythema** was present, and a **thick whitish discharge** was also noted.

LABORATORY STUDIES

Blood

Normal

Imaging

Not done

Diagnostic Work-Up

Table 32-1 lists the likely causes of the woman's illness (differential diagnosis). Symptoms of vaginitis and vaginosis are nonspecific, and a diagnosis without laboratory investigation is unreliable. Investigative approach may include

- **Wet mount on saline or 10% KOH.** It can be used to demonstrate any large organisms (e.g., yeast or protozoa).
- **Cytology or Gram stain.** It can be used if the wet mount is uncertain, and also to rule out "clue cells" (i.e., vaginal epithelial cells with attached bacteria of *Gardnerella* and *Mobiluncus*, giving the cells a stippled appearance [diagnostic of bacterial vaginosis]).

In failed tests just mentioned, vaginal culture can be obtained.

COURSE

The KOH smear revealed a few yeast cells and pseudohyphae. The patient was sent home with advice on symptomatic care and use of an over-the-counter drug.

TABLE 32–1   Differential Diagnosis and Rationale for Inclusion (consideration)
Bacterial vaginosis (*Gardnerella/Mobiluncus*) Candidiasis (*Candida albicans*) Trichomoniasis (*Trichomonas vaginalis*) **Rationale:** Vaginal discharge can be caused by any of the above pathogens, and distinguishing them clinically is often difficult, if not impossible. Classically, the discharge from *Candida* is whitish, that of *Trichomonas* is yellow and frothy, and that of bacterial vaginosis is particularly foul smelling. Microbiologic examination is generally required for a definitive diagnosis.

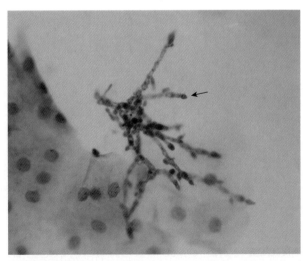

FIGURE 32-1 Vaginal smear identifying the overgrowth of microbe (Papanicolaou stain). Note pseudohyphae (*arrowhead*) in the smear. (*Courtesy of Dr. Dominick Cavuoti, Department of Pathology, University of Texas Southwestern Medical Center, Dallas, TX.*)

ETIOLOGY

Candida albicans (vulvovaginal candidiasis)

MICROBIOLOGIC PROPERTIES

Candida species are **yeasts. In secretions** (wet preparations) and **tissue** sections, however, they may appear as elongated structures **(pseudohyphae,** or even **true hyphae)** after Gram or Papanicolaou stain (Fig. 32-1). The organism grows on routine fungal media as well in blood culture bottles and on agar plates (Fig. 32-2). Yeasts **reproduce by budding** under most culture conditions. The most common species isolated from clinical specimens is *Candida albicans*, with *Candida glabrata, Candida tropicalis, Candida parapsilosis,* and *Candida krusei* found less commonly. *C. albicans* can be rapidly differentiated from other species by its production of **germ tubes** (short hyphal filaments) after 2 hours of cultivation of yeast cells at 37°C in human serum. *C. albicans* also characteristically produces **chlamydospores,** the reproductive, thick-walled structures of the fungus, larger than the standard spores produced by the molds.

> **NOTE** Diagnosis of nongenital deep infections requires isolation of the organism from culture of blood or other normally sterile sites. The histologic appearance of *Candida* in tissue is not sufficiently distinctive to differentiate it from other fungal infections.

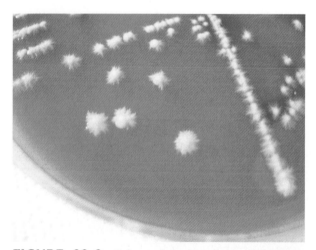

FIGURE 32-2 Culture of *Candida albicans.* Colonies are cream-colored, pasty, and smooth. Note radial furrows in these older colonies. *(Courtesy of Dr. Dominick Cavuoti, Department of Pathology, University of Texas Southwestern Medical Center, Dallas, TX.)*

EPIDEMIOLOGY

Candida species are the most common fungus causing human disease. *Candida* organisms are normal commensals of the entire gastrointestinal tract, the mouth, and the female genital tract, and most infections are endogenous in origin. Predisposing factors leading to vaginal candidiasis include feminine hygiene products, contraceptives, vaginal medications, **broad-spectrum antibiotics,** sexual intercourse, pregnancy, stress, and diabetes mellitus.

PATHOGENESIS

Vulvovaginal candidiasis is not considered a sexually transmitted disease in a classic sense, although an increase in the frequency of the disease is seen at the time most women begin regular sexual activity. A complex and intricate balance of microorganisms maintains women's normal vaginal flora. However, when an imbalance occurs, such as when normal flora changes because of **use of a broad-spectrum antibiotic** (as in the case patient), *C. albicans* **overgrows** on mucocutaneous surfaces. When that happens, symptoms of candidiasis appear, with a **white discharge that can be thick and/or curdy** (like cottage cheese).

> **NOTE** In infected uncircumcised males, intensely pruritic inflammation of the glans and prepuce with white cheesy exudate is usually present. *Candida* species causes a variety of other clinical syndromes and complicated illnesses in patients with risks (e.g., neutropenia with central venous line; Table 32-2).

TREATMENT

Vulvovaginal candidiasis responds to treatment with any of several available **topical antifungal agents,** such as **clotrimazole,** miconazole, or nystatin, or to systemic treatment with a single oral dose of fluconazole.

Deep tissue infections require treatment with systemic antifungal drugs. **Fluconazole** is effective for most cases of bloodstream infections, particularly those due to *C. albicans.* However, other species may be resistant. Intravenous **amphotericin B,** caspofungin, and voriconazole have a broad spectrum of activity; they are used **for severely ill patients** and for refractory infections. In addition, infections associated with foreign bodies (such as intravenous catheters)

TABLE 32–2 **Other Important Clinical Features of *Candida* Infections**

Syndromes	Clinical Features
Oropharyngeal candidiasis (OPC)	**Oral thrush** may be asymptomatic or it may manifest with oral discomfort and appear as white plaques adherent to the tongue and buccal mucosa, sometimes with angular cheilitis.
	Commonly seen in patients with AIDS (prior to HAART) or in patients using inhaled steroids.
Candidal esophagitis	An extension of OPC, usually occurring with dysphagia and chest pain.
	Esophageal disease occurs only in the immunocompromised and is the most common **esophageal disease in patients with AIDS**.
Cutaneous manifestations	The infection also affects moist and macerated skin, e.g., the diaper area in infants ("**diaper rash**"), perianal skin, and under large breasts of obese individuals.
Chronic mucocutaneous candidiasis (CMC)	Patients with T-cell dysfunction are predisposed to CMC, presenting with chronic skin and mucous membrane infection.
	These infections are often difficult to treat.
Systemic infections	*Candida* spp (e.g., *C. albicans, C. glabrata, C. tropicalis*) are the fourth most common cause of nosocomial blood stream infections. Risk factors involve the use of central venous lines and broad-spectrum antibiotics.
	Invasive disease occurs in critically ill **patients in the ICU, in patients with severe granulocytopenia,** and in hematopoietic stem cell and organ transplant recipients. *Candida* species **can contaminate the hub or skin site of a central venous line.**
	Complications include meningitis, endophthalmitis, endocarditis, osteomyelitis, and/or arthritis.
Urinary tract infections (UTI)	**Nosocomial UTI** occurs when *Candida* from the perineum enter the urinary tract **via an indwelling bladder catheter**.

uniformly require the removal of the foreign body in addition to systemic antifungal therapy.

OUTCOME

The patient was treated with a 3-day course of an over-the-counter topical azole, and symptoms resolved in a few days.

PREVENTION

Candida infections are generally acquired from endogenous, not environmental, sources, so prevention of deep-seated infection is accomplished by reducing risk factors such as intravenous lines, use of broad-spectrum antibiotics, and other risks for mucosal infection (e.g., uncontrolled diabetes mellitus and indiscriminate use of antibiotics).

FURTHER READING

Bodey GP: Candidiasis. Pathogenesis, Diagnosis, and Treatment. New York, Raven, 1993.

Corey L, Boeckh M: Persistent fever in patients with neutropenia. N Engl J Med 346:222, 2002.

Edwards JE, Jr, Bodey GP, Bowden RA: International Conference for the Development of a Consensus on the Management and Prevention of Severe Candidal Infections. Clin Infect Dis 25:43, 1997.

Fidel PL, Jr, Vazquez JA, Sobel JD: *Candida glabrata*: Review of epidemiology, pathogenesis, and clinical disease with comparison to *C. albicans*. Clin Microbiol Rev 12:80, 1999.

Sobel JD: Vaginitis: Current concepts. N Engl J Med 337:1896, 1997.

A 27-year-old man presented to an ambulatory care clinic with complaints of **fever, headache, sore throat,** and malaise for **over a week** and a **rash** for the past 2 days.

He admitted to having **unprotected sex with other men.** His last encounter was 3 weeks earlier. He denied prior transfusions or intravenous drug use.

C A S E

33

PHYSICAL EXAMINATION

VS: T 39.2°C, P 94, R 14, BP 136/82 mmHg

PE: Pharynx was erythematous; **cervical and axillary lymphadenopathy** was present. A **diffuse maculopapular rash** was observed on his abdomen.

LABORATORY STUDIES

Blood

Hematocrit: 42%

WBC: 5400/μL

Differential: 72% PMNs, 9% lymphs, 16% monos

Blood gases: Normal

Serum chemistries: ALT 102 U/L, ALP 185 U/L

Imaging

Not done

TABLE 33–1   Differential Diagnosis and Rationale for Inclusion (consideration)
Cytomegalovirus (CMV)
Human immunodeficiency virus (HIV) type 1
Infectious mononucleosis (EBV)
Primary herpes simplex (HSV-1)
Secondary syphilis (*Treponema pallidum*)

Rationale: Several primary viral infections have similar presentations. Fever, malaise, and lymphadenopathy are common symptoms. CMV is frequently associated with liver involvement. A maculopapular truncal rash can often be seen in primary HIV infection. Oral ulcers are commonly seen with primary HSV infection. Infectious mononucleosis may have associated splenomegaly and, in particular, a severe sore throat. Secondary syphilis is often characterized by a rash that also involves the palms and soles. As seen above, symptoms, signs, and epidemiology (e.g., host risk factors) are helpful in suggesting an infectious etiology, but laboratory investigation is needed for confirmation of the clinical diagnosis.

Diagnostic Work-Up

Table 33-1 lists the likely causes of the man's illness (differential diagnosis). A clinical diagnosis of acute human immunodeficiency virus (HIV)-1 infection was considered based on the risk and symptoms. Laboratory-based studies are necessary for confirmation of HIV-1 infection and may include

- **Serodiagnosis.** HIV seroconversion can be assessed by using two different types of antibody tests

  - Enzyme-linked immunosorbent assay **(ELISA)** of HIV-1 antibody (highly sensitive; screening test).
  - **Western blot** (highly specific; confirmatory test)

- **Plasma HIV RNA.** RT-PCR assay (sensitive and specific). May be used when acute infection is

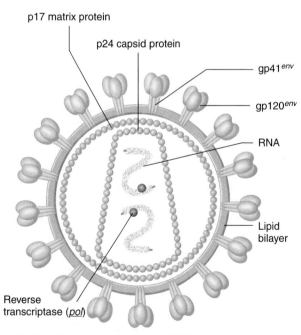

FIGURE 33-1 Architecture of HIV-1 virion. Structure of HIV-1, including the outer membrane glycoprotein (gp120) with transmembrane components (gp41) of the envelope, genomic RNA (diploid; 2 copies of identical genome), enzyme reverse transcriptase, p17 inner membrane (matrix), and p24 core protein (capsid).

135

suspected but the serology is negative. It is very sensitive and specific, but caution should be used in interpreting results in this setting, as false-positives can occur.

● **Cell culture** helps differentiate HIV-1 and HIV-2

COURSE

The patient consented to be tested for HIV antibodies, and the result was negative. However, HIV viral RNA level was very high, and a diagnosis of acute HIV infection was made.

ETIOLOGY

Human immunodeficiency virus type 1 (HIV-1)

MICROBIOLOGIC PROPERTIES

HIV belongs to a subgroup of retroviruses known as lentiviruses ("slow" viruses). The four recognized human retroviruses belong to two distinct groups: the human T lymphotropic viruses, HTLV-1 and HTLV-2, which are transforming retroviruses; and the human immunodeficiency viruses, **HIV-1 and HIV-2, which are cytopathic viruses**. The course of infection with retroviruses is characterized by a long interval between initial infection and the onset of clinical symptoms. The most common cause of HIV disease in the United States as well as worldwide is HIV-1. The virion structure of HIV-1 consists of a capsid and a surrounding matrix and envelope studded with virus-specific proteins (Fig. 33-1). Embedded in the viral envelope is a transmembrane protein (Env), consisting of a cap made of three molecules called glycoprotein (gp) 120, and a stem of three gp41 molecules that anchor the structure in the viral envelope. The envelope also incorporates a variety of host proteins, including major histocompatibility complex (MHC) class I and II antigens.

The bullet-shaped core (capsid) is made of a viral capsid protein, p24. The core surrounds two identical single-stranded RNA copies of the genome, each of which has a copy of the virus' nine genes, of which six are regulatory. The other three, genes *gag*, *pol*, and *env*, contain information needed to make structural proteins for new virus particles. The core of HIV also includes the HIV nucleocapsid protein and three enzymes that carry out later steps in the life cycle of the virus: reverse transcriptase (RT), integrase, and protease. The HIV inner membrane matrix protein (p17) lies between the viral core and the viral envelope.

EPIDEMIOLOGY

HIV is **transmitted by permucosal (sex), parenteral (intravenous drug use), and vertical (in utero to fetus) transmission.** HIV infection is currently a major sexually transmitted disease (STD) worldwide. The virus can enter the body through the lining of the vagina, vulva, penis, rectum, or mouth during sex. Having an STD such as syphilis, genital herpes, or chlamydia infection appears to make people more susceptible to acquiring HIV infection during sex with infected partners. This is at least partly due to the presence of genital ulcers. HIV frequently is spread among intravenous drug users by the sharing of needles or syringes contaminated with very small quantities of blood from someone infected with the virus. Women can transmit HIV to their babies during pregnancy or birth.

PATHOGENESIS

Infection typically begins when an HIV virion encounters a CD4+ T cell. Although **CD4+ T cells** appear to be the **main targets of HIV**, other immune system cells with CD4 molecules on their surfaces are infected as well. Among these are monocytes, macrophages, microglial cells, dendritic cells, and B lymphocytes, which are long-lived cells and apparently can harbor large quantities of the virus without being killed, thereby acting as reservoirs of HIV. One or more of the viral gp120 molecules binds tightly to CD4 molecule(s) on the cell surface. **The binding of gp120 to CD4** results in a conformational change in the gp120 molecule, allowing it to bind to a second molecule on the cell surface, **known as a coreceptor (CXCR4 or CCR5)**. CXCR4-binding viruses (X4) are T cell–tropic, and CCR5-binding viruses (R5) are T- and macrophage-tropic. Ninety-five percent of new infections are with R5 viruses (an apparent selective transmission of R5 viruses). A pH-independent conformational change in gp41 leads to fusion of cellular and viral membranes, leading to entry of the virus into the cell. The viral core is deposited into cytoplasm. In the cytoplasm of the cell, HIV reverse transcriptase converts ssRNA to proviral DNA, which moves to the cell's nucleus, where it integrates into the host's DNA (somewhat randomly in the host genome) by HIV integrase. Production of viral mRNA and viral genomic RNA occurs from integrated proviral double-stranded DNA.

Newly made HIV core proteins, enzymes, and genomic RNA gather just inside the cell's membrane, whereas the viral envelope proteins aggregate within the membrane. An immature viral particle forms and buds off from the cell (Fig. 33-2), acquiring an envelope that includes both cellular and HIV proteins from the cell membrane. The long chains of proteins and enzymes that make up the immature viral core are now cleaved into smaller pieces by a viral enzyme called protease. This step results in infectious viral particles.

A burst of **viremia** occurs in the early stages of acute infection. The immune system controls but **does not stop virus replication**. Loss of CD4$^+$ T cells (required for efficient cytotoxic T lymphocyte [CTL] response) early in infection is caused by direct lysis of cells by HIV, indirect killing of CD4$^+$ T cells, and lysis of infected CD4$^+$ T cells by immune response (CTL, antibody-dependent cellular cytotoxicity [ADCC]). CD8$^+$ cells are present in high numbers in HIV infection; they are involved in the initial control of viremia but decline with progressive CD4$^+$ T-cell loss. Depletion of CD8 numbers facilitates increased virus loads. Accelerated viral replication in conjunction with error-prone reverse transcriptase, creating many mutant viruses, allows HIV to escape control mechanisms. Cell-to-cell spread of HIV occurs through CD4-mediated fusion, with formation of syncytia between infected and uninfected cells. **Mononucleosis-like symptoms** in **acutely infected patients** occur 10 to 30 days after exposure and usually last less than 14 days.

Infection subsequently progresses to **clinical latency**, which is **not to say virologic latency**, because **active virus replication occurs throughout infection**. Latency ushers in **chronic infection**, and rates of disease progression vary widely. The time from initial infection to development of acquired immune deficiency syndrome (AIDS) may range from 2 years to 20 years (mean 10 years). Host factors that may affect disease progression are (1) coreceptor genotypes: CCR5Δ32 and CCR5 promoter mutations and (2) MHC genotypes: HLA class I alleles B*35 and Cw*04 are associated with rapid development of AIDS. Individuals with a homozygous defect in the gene that codes for CCR5, the cellular coreceptor for R5 strains of HIV-1, remain asymptomatic despite repeated exposure to HIV-1. During chronic HIV infection, the number of CD4$^+$ T cells progressively declines. When the CD4$^+$ T-cell count falls below 200/μL, susceptibility to the **opportunistic infections** and **cancers** (Table 33-2) that **typify AIDS** develops.

Virtually all of the immune defects in advanced HIV disease can ultimately be explained by the quantitative depletion of CD4$^+$ T cells. Any HIV-infected individual with a CD4$^+$ T-cell count of less than 200/μL has AIDS by definition, regardless of the presence of symptoms or opportunistic diseases. People with AIDS often suffer infections of the lungs, intestinal tract, brain, eyes, and other organs, as well as debilitating weight loss, diarrhea, neurologic conditions, and cancers such as Kaposi sarcoma and certain types of lymphomas. HIV-mediated destruction of the thymus and lymph nodes and related immunologic organs plays a major role in causing the immunosuppression seen in people with AIDS.

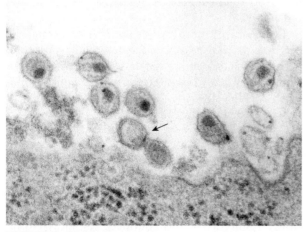

FIGURE 33-2 Electron micrograph of HIV. Note a few typical virions following budding from the surface of a CD4$^+$ T lymphocyte, together with a number of incomplete virions (*arrowhead*) in the process of budding from the cell membrane. (*Courtesy of Centers for Disease Control and Prevention, Atlanta, GA.*)

TREATMENT

Potent combinations of drugs are required to inhibit the emergence of resistant HIV during therapy. The goal of therapy is to stop virus replication, thereby improving the immune system and prolonging survival. Therapy has multiple targets:

1. The first group of drugs used to treat HIV infection, called **nucleoside reverse transcriptase inhibitors (RTIs)**, interrupts the virus in an early

TABLE 33–2	Common Opportunistic Infections and Cancers in AIDS
Agents	**Opportunistic Infections and Cancers**
Viral	
Herpes simplex virus 1 and 2 (1 > 2)	Mucocutaneous and disseminated infections
Epstein-Barr virus (EBV)	Hairy leukoplakia and lymphomas
Varicella-zoster virus	Pneumonitis and disseminated infection
Cytomegalovirus	Retinitis, pneumonitis, GI and CNS infections
Human herpesvirus type 8	Kaposi sarcoma and primary effusion lymphomas
JC virus	Progressive multifocal encephalopathy
Bacterial	
Mycobacterium tuberculosis	Miliary TB, meningitis, atypical pulmonary disease
Mycobacterium avium-intracellulare	Disseminated infections
Treponema pallidum	Tertiary syphilis (CNS and cardiac infections)
Fungal	
Candida albicans	Oral thrush and esophagitis
Pneumocystis jiroveci	Interstitial pneumonia
Cryptococcus neoformans	Meningoencephalitis
Histoplasma capsulatum	Pulmonary and disseminated infections
Protozoal	
Toxoplasma gondii	Encephalitis
Cryptosporidium parvum	Protracted diarrhea

stage of making copies of itself. Included in this class of drugs (called nucleoside analogs) are AZT (zidovudine), ddC (zalcitabine), ddI (dideoxyinosine), d4T (stavudine), 3TC (lamivudine), abacavir, tenofovir, and emtricitabine.

2. The second group of drugs is **non-nucleoside reverse transcriptase inhibitors** (NNRTIs), which act on the same enzyme as RTIs but at a different site. Nevirapine, delavirdine, and efavirenz are the currently available drugs in this class.

3. The third group of drugs, **protease inhibitors**, interrupt virus replication at a later step in its life cycle and include ritonavir, saquinavir, indinavir, amprenavir, nelfinavir, lopinavir, and atazanavir.

Because HIV quickly becomes resistant to monotherapy with any of these drugs, health care providers must use a combination to effectively suppress the virus. When at least three active drugs are used in combination, it is referred to as highly active antiretroviral therapy (HAART) and is the standard for therapy currently. This allows CD4 counts to increase, which decreases the risk of developing complications of AIDS.

The decision to begin therapy should be made after careful consideration and discussion with the patient, as side effects can be significant. Treatment early in the disease does not appear to have much benefit. Treatment of acute HIV infection is controversial. Guidelines are available that suggest when to begin therapy, based primarily on symptoms, CD4+ cell count, and HIV plasma load.

HAART has been credited as a major factor in significantly reducing the number of deaths from AIDS in the United States. While HAART is not a cure for AIDS, it reduces the amount of virus circulating in the blood to nearly undetectable levels and has greatly improved the health of many HIV-infected individuals.

Preventive measures are also important and include vaccines for *Streptococcus pneumoniae*, influenza, and hepatitis A and B, and prophylaxis against *Pneumocystis jiroveci* and *Mycobacterium avium-intracellulare* complex when CD4+ cells have declined below 200 and 50 cells/µL, respectively.

OUTCOME

After discussion with the patient, combination antiretroviral therapy was started, and his symptoms slowly resolved. He was continued on antiretroviral therapy and followed in clinic to further manage any disease symptoms and side effects of therapy.

PREVENTION

Because no vaccine for HIV is yet available, the only way to prevent infection by the virus is to avoid behaviors that put a person at risk of infection, such as sharing needles and having unprotected sex. Many people infected with HIV have no symptoms initially. Therefore, there is no way of knowing with certainty whether a sexual partner is infected unless he or she has repeatedly tested negative for the virus and has not engaged in any risky behavior for a period of time.

People should either abstain from having sex or use male latex condoms or female polyurethane condoms, which may offer partial protection, during oral, anal, or vaginal sex.

The risk of HIV transmission from a pregnant woman to her baby is significantly reduced if she takes antiretroviral therapy during pregnancy, labor, and delivery, and if her baby takes it for the first 6 weeks of life. Post-exposure prophylaxis with antiretroviral drugs is recommended in certain settings of potentially significant exposure to HIV.

FURTHER READING

Boden D, Hurley A, Zhang L, et al: HIV-1 drug resistance in newly infected individuals. JAMA 282:1135, 1999.

Hogg RS, Heath KV, Yip B, et al: Improved survival among HIV-infected individuals following initiation of antiretroviral therapy. JAMA 279:450, 1998.

Janssen RS, Satten GA, Stramer SL, et al: New testing strategy to detect early HIV-1 infection for use in incidence estimates and for clinical and prevention purposes. JAMA 280:42, 1998.

Kilby JM, Eron JJ: Mechanisms of disease: Novel therapies based on mechanisms of HIV-1 cell entry. N Engl J Med 348:2228, 2003.

Lawn SD, Butera ST, Folks TM: Contribution of immune activation to the pathogenesis and transmission of human immunodeficiency virus type 1 infection. Clin Microbiol Rev 14:753, 2001.

O'Brien WA, Hartigan PM, Martin D, et al: Changes in plasma HIV-1 RNA and CD4+ lymphocyte counts and the risk of progression to AIDS. N Engl J Med 334:426, 1996.

Gastrointestinal Tract Infections and Liver Diseases

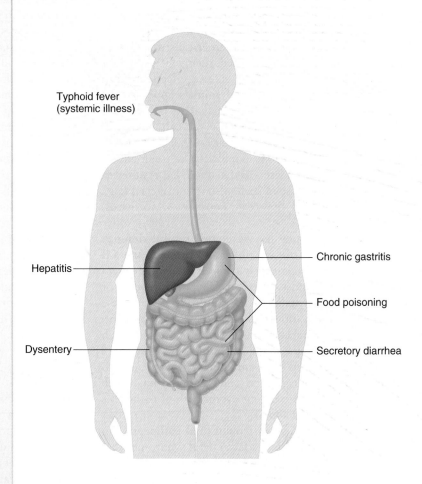

Typhoid fever
(systemic illness)

Hepatitis

Dysentery

Chronic gastritis

Food poisoning

Secretory diarrhea

INTRODUCTION TO GASTROINTESTINAL TRACT INFECTIONS

Gastrointestinal (GI) and hepatic infections are very common, probably second in frequency only to respiratory infections. A wide variety of microorganisms, with the notable lack of fungi, may cause these infections (Table III-1). The **route of infection** of most of these pathogens is fecal-oral, but important exceptions are *Strongyloides stercoralis* and *Schistosoma* species, which enter the body through the skin, and hepatitis viruses B and C, which are primarily blood-borne viruses. Enteric infections are transmitted by the fecal-oral route (person-to-person) or from an animal source. Animal-to-person transmission of enteric pathogens may, however, be followed secondarily by person-to-person spread. In the United States, person-to-person spread of enteric pathogens, which is an enormous problem for persons living in the developing countries (and for Americans traveling to these countries), has remained relatively stable for several decades. In contrast, enteric infections (e.g., *Salmonella*) derived from food animals appear to be increasing. Enteric infections may also be acquired directly from drinking water or recreational water (e.g., *Giardia lamblia* and *Cryptosporidium parvum*). In addition, the increasing popularity of seafood, in particular raw or undercooked seafood, has resulted in a substantial increase in seafood-borne bacterial (e.g., *Vibrio* species) and viral (e.g., hepatitis A, norovirus) illness.

The human body has developed several **defense mechanisms** to prevent infection through the GI tract. The most important is stomach acid, which effectively destroys most organisms. Hypochlorhydria or medications that reduce acid production can increase susceptibility to various pathogens (e.g., *Vibrio cholerae*). Other defense mechanisms are gut motility, which helps prevent adherence of microorganisms to mucosal surfaces; secretory IgA antibodies, which bind to recognized organisms; and endogenous flora, which are important in preventing overgrowth of pathogens such as *Clostridium difficile*.

Illnesses caused by gastrointestinal pathogens can generally be divided into four general syndromes that are based on the region of bowel that is involved and whether or not the organism is invasive. These clinical syndromes include **gastritis**, **food poisoning**, **infectious diarrhea (gastroenteritis)**, and **typhoid (enteric) fever** (see Table III-1).

Acute infection with *Helicobacter pylori*, acquired via the fecal-oral mode, leads to a severe **gastritis**, which is accompanied by a profound reduction in acid secretion. As a result of the immune response, the gastritis becomes less severe, and acid secretion returns in most patients. Most infected individuals remain asymptomatic their entire lives. However, individuals who are at risk (e.g., poor socioeconomic condition, stress) may develop **peptic ulcer disease**, atrophic gastritis, gastric adenocarcinoma, or gastric B-cell lymphoma if not treated to eliminate the organism.

Ingestion of preformed bacterial toxins in food (e.g., **food poisoning** by staphylococcal enterotoxins) can cause upper gastrointestinal symptoms of nausea and vomiting (and, less frequently, diarrhea). These are not true infections in the classic sense, because there is no pathogen causing disease inside the body, but rather they are the effects of specific toxins. This leads to rapid onset of symptoms from exposure (often 1 to 3 hours) and rapid resolution as the toxin is removed (usually 12 to 36 hours). Isolated (sporadic) cases of food poisoning are not usually diagnosed, owing to their self-limited nature; epidemiologic investigation is undertaken only in outbreak scenarios. Although adult botulism is considered a preformed toxin-associated illness, infant botulism results from toxigenesis in the gut after the spores of *Clostridium botulinum* are ingested in baby food (e.g., honey or syrup) and transform into the vegetative form for growth. Unlike other food poisoning, botulism can be life threatening without appropriate supportive care. It should be considered in any patient with descending paralysis and no risk factors for other infections.

Diarrhea is the **most common manifestation of gastrointestinal infection**, although a precise definition is difficult to agree on. Diarrhea is often divided into **secretory** (watery) **diarrhea**, **dysentery** (fever and bloody diarrhea with mucus), or **hemorrhagic colitis** (frank bloody diarrhea), which can be helpful in considering which pathogens are responsible (see Table III-1). Nausea, vomiting, and abdominal pain are common accompanying symptoms of gastrointestinal infection.

Secretory (watery) diarrhea generally is caused by viruses and is usually self-limited. The most common **viral cause** of diarrhea worldwide is rotavirus. It mainly affects the pediatric population and it is an important cause of infant mortality in developing countries. Norovirus (previously named Norwalk virus) is also common and is often associated with outbreaks in predominantly adult populations. Severe

TABLE III–1 Common Causes of Gastrointestinal and Hepatic Infections

Syndrome	Location of Infection/Key Clinical Features	Microbiologic Agents
Chronic gastritis (peptic and duodenal ulcer disease)	Stomach/abdominal pain (improves after meals); no diarrhea	*Helicobacter pylori*
Food poisoning (pre-formed toxin)	Upper GI/nausea and vomiting (less diarrhea) and rapid onset	*Staphylococcus aureus* *Bacillus cereus* *Clostridium perfringens* *Clostridium botulinum* (peripheral nervous system abnormalities)
Acute infectious diarrhea (gastroenteritis)		
Secretory diarrhea	Proximal small bowel (SB)/ watery diarrhea (no fecal leukocytes)	*Escherichia coli* Enterotoxic *E. coli* Enteropathogenic *E. coli* Enterohemorrhagic *E. coli* (frank bloody diarrhea) Enteroadherent *E. coli* *Vibrio cholerae* Rotavirus Norovirus *Giardia lamblia* *Cryptosporidium parvum*
Dysentery	Colon/fever, pain, bloody mucopurulent diarrhea (heavy to variable fecal leukocytes)	*Campylobacter* spp *Salmonella* spp *Shigella* spp *Yersinia enterocolitica* (also terminal ileitis and pseudoappendicitis syndrome) Enteroinvasive *E. coli* *Entamoeba histolytica* *Strongyloides stercoralis* *C. difficile* (antibiotic-associated colitis)
Hemorrhagic colitis	Colon/pain, diarrhea, grossly bloody stools, hemolytic uremic syndrome (few or absent fecal leukocytes)	Enterohemorrhagic *E. coli*
Typhoid (enteric) fever	Entry via SB, but systemic febrile illness (monocytic leukocytosis)	*Salmonella typhi* *Salmonella paratyphi* *Yersinia enterocolitica* (mimicking symptoms of appendicitis)
Chronic (infectious) diarrhea in AIDS and immunocompromised patients	SB/protracted watery diarrhea (no fecal leukocytes)	*Cryptosporidium parvum* Microsporidia *Cyclospora cayetanensis* *Cytomegalovirus* *Mycobacterium avium* complex
Hepatitis	Liver jaundice, fatigue, abdominal pain, loss of appetite, nausea, and fever (acute or chronic)	Enteric hepatitis: hepatitis A and E Serum hepatitis: hepatitis B, C, and D *Schistosoma mansoni* (hepatosplenomegaly and portal hypertension) *Echinococcus granulosus* (hydatid cyst disease of liver)

vomiting is a prominent feature in this infection. Other viruses that cause diarrhea include adenovirus and astroviruses, although these viruses are less common causes of diarrheal illness. Secretory diarrhea is also caused by organisms such as enterotoxic *Escherichia coli* (**common cause of traveler's diarrhea**) and *Vibrio cholerae*. These organisms colonize the proximal small bowel and stimulate fluid and electrolyte secretion (by pharmacologic imbalance), resulting in watery diarrhea that can be voluminous. **Protozoan parasites** are also important causes of secretory diarrhea. The most common protozoa are *Giardia*, *Cryptosporidium*, and *Cyclospora*. They may cause isolated cases or large outbreaks, especially in travelers. Day care centers and contaminated food and water are typical routes of exposure of these protozoa. The incidence of intestinal protozoal infections is increasing in the United States because of international travel, immigration from developing countries, more children in day care, and AIDS.

Dysentery, with fever, abdominal cramps and bloody, mucopurulent diarrhea, is produced by invasive organisms that target the colon. Enteric bacterial pathogens are the most common pathogens of dysentery. In the United States, the major enteric bacterial pathogens include *Campylobacter*, *Salmonella*, *Shigella*, *Yersinia*, and *Clostridium difficile*. The initial lesions caused by these inflammatory pathogens are confined to the epithelial layer; however, as the disease progresses, the lamina propria becomes involved extensively with an inflammatory response, and crypt abscesses are prominent. Frank dysentery is frequently preceded by a day or two of watery diarrhea. Most bacterial enteric infections are self-limited, whereas some (e.g., *Salmonella typhimurium*) may cause severe, life-threatening infections. *Campylobacter* is the most frequently identified agent of acute infectious diarrhea in the United States, and poultry is the major vehicle of transmission to humans, with approximately 90% of raw poultry being contaminated with *Campylobacter*. *Salmonella* are ubiquitous in the environment and tend to colonize the GI tracts of domestic animals, usually producing asymptomatic or subclinical infections. In recent years, major outbreaks of salmonellosis have been traced to milk, beef, poultry, and eggs. Humans are the only natural reservoir of *Shigella*, which is unique among enteric pathogens in the small infectious inoculum needed to transmit disease. Person-to-person contact facilitates transmission, especially in countries where sanitation standards are low.

Yersinia enterocolitica is a common cause of dysentery (more commonly known as **enterocolitis**) in certain European countries and Canada, but it is relatively uncommon in the United States. Most cases in the United States are associated with outbreaks involving contaminated milk. *Y. enterocolitica* organisms penetrate the surface epithelium of the distal small bowel and quickly make their way to deeper tissues, including the regional lymphatics. Clinical features include diarrhea and prominent right lower-quadrant abdominal pain. The stool generally contains few, if any, leukocytes. The clinical spectrum of these organisms includes mesenteric adenitis mimicking acute appendicitis and bacteremia with hematogenous spread.

C. difficile **colitis** is almost always related to earlier antibiotic use and should always be considered in hospitalized patients. Unlike most other enteric bacterial pathogens, *C. difficile* is Gram positive and produces toxins. Stool cultures are not useful for identifying *C. difficile* organisms; rather, the toxins, which cause the diarrhea, are detected in feces.

Intestinal parasitic diseases are highly prevalent in developing countries. *Entamoeba histolytica* is a common invasive protozoan that causes dysentery (**amebic colitis**) with complications (liver abscess). **Helminth infections** may be relatively asymptomatic in most individuals until specific organ systems become involved (e.g., hydatid cyst disease of liver). The parasites infecting the intestine include roundworms or nematodes (e.g., *Strongyloides stercoralis*) and flatworms or cestodes (e.g., *Echinococcus granulosus*). *S. stercoralis* may cause bloody dysentery, but *E. granulosus* causes extraintestinal infections (e.g., mass lesions of liver or lung).

As the most common aerobic organism in the gastrointestinal tract, *E. coli* plays an important role in normal intestinal physiology. Within this species, however, there are fully pathogenic strains that cause distinct syndromes of diarrheal disease. *E. coli* O157:H7 (also known as enterohemorrhagic *E. coli* [EHEC]) causes **acute hemorrhagic colitis** and typically manifests with severe abdominal cramping followed by grossly bloody stools. Fever occurs in a minority of patients, and fecal leukocytes are uncommon. The hemolytic uremic syndrome and thrombotic thrombocytopenic purpura are the most serious complications of EHEC, usually occurring in the very young or very old. In addition to many well-described outbreaks, *E. coli* O157:H7 appears to be a common cause of sporadic infectious diarrhea, especially in the northern states and Canada.

It is often difficult to determine the etiology of a patient's diarrheal syndrome on clinical grounds alone. Stool cultures can be helpful in defining the specific etiology of diarrheal illness. Most episodes are self-limiting, with mild-to-moderate symptoms, and require no specific evaluation or **treatment**. Microscopic examination of the stool for RBCs and leukocytes using a counterstain—such as methylene blue—is a useful technique for presumptively differentiating invasive bacterial pathogens from noninvasive pathogens, viruses, and protozoa. Decisions regarding stool cultures, treatment, and need for hospitalization can then be made based on the positive examination of feces for leukocytes. Most episodes of acute diarrhea with a confirmed etiology (in immunocompetent hosts) are due to enteric bacterial pathogens. Most clinical laboratories routinely culture stool for *Salmonella, Shigella, Campylobacter,* and *Yersinia.* Cultures for other organisms must be specifically requested as indicated by the clinical presentation and the presence or absence of fecal leukocytes. Stool should also be examined for ova and parasites (O & P) to diagnose amebic dysentery, giardiasis, and helminth infections, when appropriate. Interestingly, fecal leukocytes are usually absent in patients with amebic colitis, as the amebae engulf the white cells (and also RBCs) for food. If colitis (bloody diarrhea, tenesmus, fecal WBCs) persists and stool cultures and O & P tests are negative, sigmoidoscopy or colonoscopy should be performed to evaluate other causes, including those that are noninfectious.

Typhoid (enteric) fever is a systemic febrile illness of prolonged (3 to 5 weeks) duration marked by fever, persistent bacteremia, and metastatic spread—leading to multiple-organ dysfunction. Most patients do not have diarrhea. *Salmonella typhi* is the main cause of typhoid fever. *S. typhi* cohabits exclusively with humans (in the gall bladder) and is transmitted via the ingestion of material contaminated with human feces from chronic carriers. Typhoid fever is uncommon in the United States (~500 cases/year) but is common in South Asia and other developing countries where sanitation and hygiene are poor.

Immunocompetent individuals do not usually develop **chronic diarrhea** due to infectious agents. However, **immunosuppressed** patients, particularly those with **AIDS,** frequently have prolonged symptoms from a variety of nonvirulent organisms, ranging from viruses such as cytomegalovirus (CMV) to parasites such as *Cryptosporidium* and microsporidia. As the CD4$^+$ T-lymphocyte count drops, AIDS patients may develop enteric infections with organisms that rarely cause disease in healthy individuals. *Cryptosporidium,* which causes self-limited diarrhea in healthy hosts, causes chronic voluminous diarrhea in AIDS patients. Microsporidia is emerging as one of the most common intestinal infections among AIDS patients. Cytomegalovirus may involve any region of the GI tract and is a common cause of chronic diarrhea in AIDS patients. The clinical manifestations of CMV bowel disease result from mucosal ischemic ulceration and include diarrhea, abdominal pain, weight loss, and bowel perforation. In AIDS patients with CD4$^+$ counts less than 50, *Mycobacterium avium* complex may involve any region of the GI tract and may lead to diarrhea, abdominal pain, and weight loss. Stool specimens should be obtained from patients with AIDS for bacterial cultures (*Salmonella, Shigella* and *Campylobacter*), O & P examination, modified acid-fast stain (for *Cryptosporidium, Cyclospora*), and *C. difficile* toxin (because nearly all AIDS patients are frequently on antibiotics). Inflamed or ulcerated mucosa should be biopsied for identification of viral inclusions (CMV) and protozoa. Upper endoscopy should be performed if diarrhea and weight loss persist and stool studies are not diagnostic. Endoscopy is used mainly to obtain aspirates and biopsies from the small bowel for the detection of protozoa (especially *Cryptosporidium* and *Giardia*) that can be missed on stool studies.

In summary, infections of the gastrointestinal tract are common and often have specific features that distinguish them, although stool examination and culture are required for a definitive diagnosis. Most are self-limited and easily treated once a diagnosis is established. Immunocompromised patients may have more severe and prolonged disease, often with otherwise opportunistic pathogens.

Hepatitis, an inflammation of the liver, can be caused by infections with various organisms, including bacteria (e.g., *Leptospira* species), viruses (e.g., hepatitis A, B, and C), or parasites (e.g., *Schistosoma mansoni*). Viruses are the most common infectious causes of hepatitis. The disease may manifest as acute hepatitis (hepatitis A, B, or E) or chronic hepatitis (hepatitis B or C). In hepatitis B or C infection, progressive liver damage, liver failure, or even liver cancer may result. Viral hepatitis is a febrile illness of prolonged duration marked by jaundice, fatigue, abdominal pain, loss of appetite, and nausea. Chronic hepatitis can be associated with a rash, due to immune complex-associated vasculitis, and with arthritis. Common risk factors include eating contaminated seafood or imported berries (hepatitis A), multiple sexual partners and unprotected intercourse (hepatitis B), intravenous drug use (hepatitis C), or blood transfusion received prior to 1990 (before hepatitis C blood test was available). Hepatic pathology is indistinguishable for all forms and causes of hepatitis. Clinical diagnosis on the basis of jaundice must be confirmed by liver function tests (e.g., levels of serum aminotransferases, bilirubin, and alkaline phosphatase) and hepatitis virus serologies (virus-specific antigen and antibody markers in serum). Management is mostly symptomatic, although specific treatments are available for chronic stages of hepatitis B and C.

Despite the large numbers of pathogens associated with gastrointestinal and liver diseases, relatively few clinical syndromes are predominant. In most cases, these infections tend to be self-limiting and are rarely life-threatening. Notable exceptions are cholera and typhoid fever. Chronic hepatitis C is the most common cause of liver failure and subsequent liver transplantation.

FURTHER READING

Bacaner N, Stauffer B, Boulware DR, et al: Travel medicine considerations for North American immigrants visiting friends and relatives. JAMA 291:2856, 2004.

Di Bisceglie AM: Hepatitis C. Lancet 351:351, 1998.

Donowitz M, Kokke FT, Saidi R: Evaluation of patients with chronic diarrhea: Current concepts. N Engl J Med 332:725, 1995.

Hedberg CW, Osterholm MT: Outbreaks of food-borne and waterborne viral gastroenteritis. Clin Microbiol Rev 6:199, 1993.

Koff RS: Hepatitis A. Lancet 341:1643, 1998.

Nataro JP, Kaper JB: Diarrheagenic *Escherichia coli*. Clin Microbiol Rev 11:142, 1998.

Thielman NM, Guerrant RL: Acute infectious diarrhea. N Engl J Med 350:38, 2004.

Questions on the case (problem) topics discussed in this section can be found in Appendix F. Practice question numbers are listed in the following table for students' convenience.

Self-Assessment Subject Areas (book section)	Question Numbers
Gastrointestinal infections and liver diseases (Section III)	7-10, 24-26, 90-93, 111-112, 117-121, 127, 128, 132, 136, 140, 141, 144, 146, 150-151, 153-154, 158, 162, 166, 173-175, 179, 180, 186-189, 191, 195

A 20-year-old white man was brought to the emergency department of a local hospital for evaluation of **severe abdominal cramping** and **diarrhea.** The patient had lower abdominal discomfort and reported having had eight loose bowel movements per day in the 3 days before his arrival at the hospital. He had recently noticed **bloody** bowel movements as well.

During history taking, the patient related that he had had **a dinner of mixed green salad** and **BBQ chicken** at a local restaurant approximately 30 hours before the onset of symptoms, and that his roommate had experienced similar but much milder symptoms, which had since cleared up.

PHYSICAL EXAMINATION

VS: T 38.1°C, P 102/min, R 16/min, BP 122/72 mmHg

PE: A bowel movement during evaluation at the ER revealed **gross blood in the stool** specimen.

LABORATORY STUDIES

Blood

Hematocrit: 44%

WBC: 12,600/µL

Differential: 72% PMNs, 18%lymphs

Serum chemistries: BUN 24 mg/dL, creatinine 1.1 mg/dL

Imaging

X-ray of the abdomen was normal.

Diagnostic Work-Up

Table 34-1 lists the likely causes of illness (differential diagnosis). A clinical diagnosis of enteritis (or dysentery) was considered based on the febrile illness and bloody diarrhea. For enteritis, examination of feces, before ordering any stool test, may produce useful information. Feces should be examined for (1) too much fluid (watery or secretory diarrhea); (2) leukocytes (inflammatory enteritis); or (3) frank or occult blood and mucus (dysentery syndrome). In most cases of acute diarrhea, screening for fecal leukocytes is recommended. Diarrhea investigation may include

- Enteric bacterial cultures
- Wet mount examination to rule out parasitic causes

TABLE 34–1 Differential Diagnosis and Rationale for Inclusion (consideration)

Bacterial enteritis due to:
  Salmonella spp
  Campylobacter
  Shigella sonnei
  Yersinia enterocolitica
Crohn disease
Enteric protozoal disease
Hemorrhagic colitis (*Escherichia coli* O157:H7)
Pseudomembranous colitis (*Clostridium difficile*)
Ulcerative colitis (UC)

Rationale: There are many causes of enteritis, and it is difficult to arrive at a specific diagnosis on clinical grounds alone. The presence of bloody stools does narrow the etiologies to those that are more invasive, such as those listed above. *E. coli* O157:H7 does not usually manifest with fever because it is not invasive. *C. difficile* is almost always associated with prior antibiotic use. However, noninfectious etiologies such as Crohn disease or UC must also be considered because they may manifest in an identical manner.

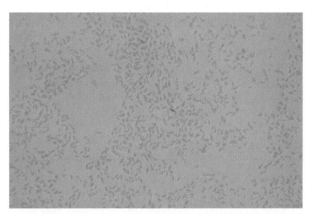

FIGURE 34-1 Gram stain of the enteric isolate. Note comma or seagull-shaped organisms. *(Courtesy of Dr. Paul Southern, Departments of Pathology and Medicine, University of Texas Southwestern Medical Center, Dallas, TX.)*

COURSE

The patient was admitted to the hospital and began receiving intravenous hydration. Positive fecal culture on special medium incubated under microaerophilic conditions at higher temperature yielded the diagnosis.

ETIOLOGY

Campylobacter jejuni (campylobacter enteritis)

MICROBIOLOGIC PROPERTIES

Bacteria in the genus *Campylobacter* are motile, **Gram-negative** curved rods (**comma** or **seagull-shaped organisms**; Fig. 34-1). Virtually all human *Campylobacter*-associated diarrheas are caused by *C. jejuni*. *Campylobacter coli* and *Campylobacter fetus* cause systemic infection such as bacteremia in neonates and young children. A major virulence factor for *C. fetus* is a proteinaceous capsule-like structure that renders the organism resistant to complement-mediated killing and opsonization in the blood stream. The major virulence of *C. jejuni* is its flagella.

For diarrhea investigation, stool is cultured on selective agar media to isolate one of the inflammatory diarrheal agents such as *Campylobacter*, *Salmonella*, and *Shigella* (and also, in some geographic areas, *Aeromonas*, *Escherichia coli* O157:H7, and *Yersinia*). **Growth of Campylobacter strains requires selective media, microaerophilic conditions** (5% oxygen, 5% to 10% CO_2), and **incubation at 42°C**. Isolated colonies are rapidly tested for catalase and oxidase activities. Differentiation of *Campylobacter* species is based on biochemical reactions (e.g., nitrate reduction test). Outer membrane LPS of *C. jejuni* is the major antigen; 90 different "O" antigens are recognized for serotyping (epidemiology).

EPIDEMIOLOGY

C. jejuni is the most common bacterial cause of diarrheal illness in the United States, as well as worldwide. In the developing countries, the diarrheal disease is extremely common among children younger than 2 years of age. In the United States, incidence is about 20 cases per 100,000 population, and most illnesses associated with *C. jejuni* infection are **sporadic**. Most cases are associated with **improper food handling and preparation of poultry**. Person-to-person spread of the organism is uncommon. All age groups are at risk for acquiring infection, and infants and young adults are particularly likely to be infected.

PATHOGENESIS

The infectious dose of *C. jejuni* is low; ingestion of only 500 organisms, easily present in one drop of raw chicken juice, can result in human illness. The incubation period after ingestion of incriminated food is 2 to 5 days, depending on the dose ingested. The microaerophilic organisms are adapted for survival in the gastrointestinal mucous layer, and they **colonize the intestinal mucosal layer, mediated by flagella** and putative adhesins. The organisms invade and/or translocate across the epithelial surface to the underlying tissue, where undefined virulence factors are released. A cytolethal distending toxin and an endotoxin likely contribute to tissue injury. **An acute, nonspecific neutrophilic and monocytic inflammatory reaction** causing tissue damage in the lamina propria and **jejunal epithelium** is seen, similar to that seen in **Crohn disease** and **ulcerative colitis**. Some strains of *C. jejuni* produce a heat-labile, cholera-like enterotoxin, which is important in the watery diarrhea observed in infections in children in developing countries and in adults as traveler's diarrhea.

> **NOTE** Complicated illness due to *C. jejuni* may manifest with bacteremia in the elderly and in patients with AIDS. **Reactive arthritis** has been associated with ***C. jejuni* enteritis in patients with HLA-B27. Guillain-Barré syndrome**—one of the most common causes of flaccid paralysis in the United States—manifests as symmetrical ascending muscle weakness and facial diplegia and may be associated with *C. jejuni*.

TREATMENT

Rehydration is the primary choice of management for *Campylobacter* enteritis, because this is generally a self-limited infection. Erythromycin may be administered if antimicrobial therapy is required; quinolones can also be used in the elderly or in complicated, bacteremic conditions. If antibiotics are indicated, they should be given early in the course for maximum benefit.

OUTCOME

The patient was rehydrated with intravenous saline. His symptoms improved greatly after 2 days in the hospital, and he was discharged home in good condition. There were no sequelae of his infection.

PREVENTION

Raw poultry and meat should be prepared on a separate countertop or cutting board from other food items. Poultry should be cooked to an internal temperature of 180°F or until the meat is no longer pink and juices run clear. Hands should also be washed thoroughly after preparing meat.

FURTHER READING

Blaser MJ, Reller LB: *Campylobacter* enteritis. N Engl J Med 305:1444, 1981.

Kosunen TU, Kauranen O, Martio J, et al: Reactive arthritis after *Campylobacter jejuni* enteritis in patients with HLA-B27. Lancet 1:1312, 1980.

Rees JH, Soudain SE, Gregson NA, Hughes RAC: *Campylobacter jejuni* infection and Guillain-Barré syndrome. N Engl J Med 333:1374, 1995.

Walker RI, Caldwell MB, Lee EC, et al: Pathophysiology of *Campylobacter* enteritis. Microbiol Rev 50:81, 1986.

Wassenaar TM: Toxin production by *Campylobacter* spp. Clin Microbiol Rev 10:466, 1997.

Six individuals from a single family presented over the course of 2 days with low-grade fever, abdominal cramps, vomiting, and diarrhea.

All six individuals had eaten Thanksgiving dinner together in a private home, and they had all eaten the turkey and stuffing approximately 24 hours before the onset of first symptoms.

PHYSICAL EXAMINATION

Physical examination of one (Patient X) of the six ill-appearing patients is presented here.

VS: T 38.5°C, P 98/min, R 18/min, BP 114/62 mmHg

PE: An ill-appearing patient with dry mucous membranes; abdominal exam revealed mild, diffuse tenderness.

LABORATORY STUDIES (PATIENT X)

Blood

Hematocrit: 42%

WBC: 8200/µL

Differential: Normal

Serum chemistries: BUN 21 mg/dL, creatinine 1.0 mg/dL

Imaging

Abdominal x-ray was normal.

Diagnostic Work-Up

Table 35-1 lists the likely causes of Patient X's illness and the outbreak (differential diagnosis). A clinical diagnosis of enteritis was considered. Microscopic examination of the feces demonstrating the presence of WBCs can support the diagnosis. Investigational approach may include

- **Stool cultures** on selective media
- **Blood culture** may be necessary for the febrile illness and sepsis syndrome.

In negative cultures, enteric viruses and protozoa, although highly unlikely, should be considered and sought.

TABLE 35-1   Differential Diagnosis and Rationale for Inclusion (consideration)

Bacterial enteritis due to:
  Campylobacter jejuni
  Salmonella spp
  Shigella spp
  Yersinia enterocolitica
Hemorrhagic colitis (*Escherichia coli* O157:H7)
Protozoal diarrhea
Viral gastroenteritis

Rationale: Enteritis is a broad category of illness, with multiple bacterial causes. Certain epidemiologic factors can suggest a particular etiology. Poultry exposure is often associated with *Campylobacter* or *Salmonella*. *E. coli* O157:H7 would be more likely to cause bloody diarrhea, as would *Shigella*. Enteric viruses and protozoa are the least likely to be considered in a presumptive common-source familial outbreak after eating a meal that includes poultry.

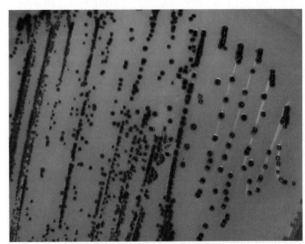

FIGURE 35-1 Cultural features of the pathogen. Note the growth on Hektoen agar showing green, nonlactose-fermenting colonies mostly with a black center due to H_2S production. (*Courtesy of Lisa Forrest, Department of Microbiology, University of Texas Southwestern Medical Center, Dallas, TX.*)

COURSE

Three of the six persons, including Patient X, were hospitalized because of dehydration and sepsis. Stool cultures were obtained, and leukocytes were present in all samples. Stool cultures from all three persons yielded the diagnosis.

ETIOLOGY

Salmonella typhimurium (salmonella enteritis)

MICROBIOLOGIC PROPERTIES

The bacteria in genus *Salmonella* are **Gram-negative rods**. Antigenic analysis of isolates (strains) based on cell wall (O) and flagellar (H) antigens has led to the identification of over 2300 serotypes (also known as species) of *Salmonella*. H antigens undergo **phase variation** via DNA rearrangements in certain serotypes, resulting in subspecies (strains). The salmonellae are **motile**, and **all serotypes except *Salmonella typhi*** (also known as typhoidal salmonella) **are noncapsulate**. Stool cultures on selective media that inhibit the growth of colonic commensals are the standard. All are facultative anaerobic organisms; 99% of *Salmonella* strains **do not ferment lactose** (Fig. 35-1). Like other *Enterobacteriaceae*, salmonellae produce acid on glucose fermentation, reduce nitrates, and do not carry cytochrome oxidase. Typing of *Salmonella* isolates from diarrheal stool, vomitus, or incriminated food using serologic and bacteriophage susceptibility methods is a useful epidemiologic tool for investigating outbreaks of food-borne disease.

EPIDEMIOLOGY

Nontyphoidal salmonellae are **second** only to *Campylobacter jejuni* as a cause of food-borne disease in the United States. Animals are the main reservoir (**poultry, eggs, dairy products**) for *S. typhimurium*, the most common species in the United States, and other nontyphoidal salmonellae. Pet reptiles can also be a source of salmonella infections. Infection is **acquired by ingestion of contaminated food** or water, by contact with infected animals, or by person-to-person transmission. Anyone consuming food contaminated with large numbers of salmonellae is at risk for enteritis, particularly **patients with reduced gastric acid**, children younger than 1 year of age, the elderly, and patients with AIDS.

PATHOGENESIS

Within 6 to 48 hours after ingestion of the organisms in food, gastric acid-sensitive nontyphoidal salmonellae (**requiring an infective dose of over 10^5 organisms**) penetrate the gastrointestinal mucus. The organisms adhere to a distal portion of small intestinal mucosa, often mediated by fimbriated adhesins. Salmonellae alter the normal structure of the brush border of intestinal cells within minutes following infection. A **localized invasion in the intestinal epithelial cells**, mediated by bacterial invasins, follows, resulting in an **overwhelming influx of neutrophils to the intestines.** The damage to the intestinal mucosa results in self-limited, but often bloody, diarrhea. In complicated enteritis, phagocytosis by macrophages results in the dissemination throughout the reticuloendothelial system and **invasion into the blood stream**, resulting in bacteremia (a not uncommon occurrence with *S. typhimurium*). **LPS and cytokines** mediate a **systemic inflammatory response syndrome** (sepsis).

TREATMENT

Antibiotics are not usually required to treat nontyphoidal *Salmonella* enteritis in an otherwise healthy individual because use of antibiotics may lead to protracted diarrhea and other symptoms of enteritis. However, for certain groups, including the immunocompromised, neonates, and individuals older than 50 years of age, antibiotic treatment (using a fluoroquinolone) should be considered.

OUTCOME

One of the elderly patients died from the infection and ensuing sepsis syndrome. The others, including Patient X, although requiring antibiotics and several days of hospitalization, fully recovered.

No leftover food was available for culture and outbreak investigation. A county health department investigator who interviewed the ill persons (including the cook) found that a 14-pound frozen turkey had been thawed for 6 hours in a sink filled with cold water. After thawing, the packet of giblets was removed, and the turkey was stored in a refrigerator overnight. However, the next day, despite the fact that parts of the turkey were noted to be frozen, the turkey was nevertheless filled with stuffing and then cooked for 4 hours in an oven set at 350°F. The turkey was removed from the oven when the exterior had browned. A meat thermometer was not used.

PREVENTION

A major source of outbreaks is improperly cooked eggs, particularly in bulk. Use of pasteurized eggs is recommended to avoid this. Adequate cooking of poultry products is also essential for prevention. However, one of the most effective preventive measures is simply good personal hygiene and regular hand washing, especially among food service workers.

FURTHER READING

CDC: Outbreak of *Salmonella* serotype Enteritidis infection associated with eating shell eggs—United States, 1999-2001. MMWR 51:1149, 2003.

Darwin HK, Miller VL: Molecular basis of the interaction of *Salmonella* with the intestinal mucosa. Clin Microbiol Rev 12:405, 1999.

Mølbak K, Baggesen DL, Aarestrup FM, et al: An outbreak of multidrug-resistant, quinolone-resistant *Salmonella enterica* serotype typhimurium DT104. N Engl J Med 341:1420, 1999.

Olsen SJ, Bishop R, Brenner FW, et al: The changing epidemiology of *Salmonella*: Trends in serotypes isolated from humans in the U.S., 1987-1997. J Infect Dis 183:756, 2001.

A 41-year-old white man was brought to the emergency department with a **3-day history of shaking chills, high fever,** headache, abdominal pain, and generalized weakness. Mild diarrhea had started 2 days earlier and had been improving when the fever began.

The patient had returned to the United States 10 days earlier, after a **3-week visit to India.** He recalls eating a variety of local foods, particularly from street vendors. He had not received any travel-related vaccines. Past medical history was unremarkable.

PHYSICAL EXAMINATION

VS: T 39.5°C, **P 82/min,** R 18/min, BP 94/58 mmHg

PE: On examination, the patient appeared ill and confused. **His abdomen was diffusely tender, and his liver** and spleen were enlarged, although there was no evident jaundice. **Erythematous maculopapular lesions** ("rose spots") were noted on his chest (Fig. 36-1).

LABORATORY STUDIES

Blood

Hematocrit: 36%

WBC: **3700/µL**

Differential: 30% PMNs, 28% lymphs, **38% monos**

Platelet: **76,000/µL**

Serum chemistries: ALT 344 U/L, AST 268 U/L

FIGURE 36-1 Maculopapular rash ("rose spots") on the chest of this patient. *(Courtesy of Dr. Paul Southern, Departments of Pathology and Medicine, University of Texas Southwestern Medical Center, Dallas, TX.)*

Imaging

Abdominal CT scan was remarkable for **enlarged liver and spleen** with no focal lesions.

Diagnostic Work-Up

Table 36-1 lists the likely causes of illness (differential diagnosis). A clinical diagnosis of sepsis was considered. Investigational approach may include

- **Cultures of blood** or bone marrow and stool (when diarrhea is present)
- Failing above, the additional tests are

  - Cultures of stool on selective media for *Salmonella* (if diarrhea is present)

TABLE 36–1 Differential Diagnosis and Rationale for Inclusion (consideration)
Amebic hepatic abscesses
Brucellosis
Dengue fever
Leishmaniasis
Malaria
Tuberculosis
Typhoid (enteric) fever
Salmonella paratyphi types A, B, and C
Salmonella typhi

Rationale: Travel-related infection should be considered. Dengue, malaria, and brucellosis may manifest as a nonspecific, sepsis-like illness; epidemiologic information is valuable in defining the illness and in further work-up. Leishmaniasis causes fever and skin lesions (often an ulcerated lesion) and is not generally acute in presentation. Tuberculosis is always a consideration, but there are usually some associated respiratory symptoms. Typhoid fever in returning travelers, particularly from India and Southeast Asia, presents with prolonged fever, abdominal pain, and maculopapular rash (a more specific finding). Nonspecific febrile symptoms are seen in some of the other diagnoses listed above.

- **Viral serology** for dengue fever
- Peripheral **blood smears** for malaria parasites

COURSE

The patient was admitted to the hospital for evaluation of increasing fever and hypotension. Blood cultures were drawn and smears for malaria were performed, in addition to serology for Dengue virus. Because of the severity of his illness, empirical antibiotics were started as soon as cultures were obtained. Blood cultures subsequently yielded a significant enteric bacterial pathogen.

ETIOLOGY

Salmonella typhi (typhoid fever)

> **NOTE** *Salmonella paratyphi* (consisting of types A, B, and C) are less common pathogens, causing less severe disease.

MICROBIOLOGIC PROPERTIES

All salmonellae **are Gram-negative rods**. All are facultative anaerobic organisms (99% of *Salmonella* strains **do not ferment lactose**). *S. typhi* is an **encapsulated** pathogen, carrying "K" (also known as "Vi" for virulence) antigen. The clinical isolates can be identified by biochemical reactions and by agglutination with O, H, and Vi antibodies.

EPIDEMIOLOGY

Typhoid fever is uncommon in the United States (approximately 400 cases/year, mostly among travelers). However, an estimated 21 million cases of typhoid fever and 200,000 associated deaths occur worldwide. *S. typhi* infects only humans, as do the paratyphoid strains. Carriers frequently have existing biliary tract abnormalities, including gallstones. The infection is **acquired by the fecal-oral route** (person-to-person transmission) or by ingestion of food and water contaminated either by a carrier or by an actively infected individual. **Risk is higher among international travelers** and highest among persons (most commonly young children) living in poverty in the developing world.

PATHOGENESIS

The incubation period for *S. typhi* may range from 3 days to 3 months, depending on the infecting dose. The typhoidal salmonellae bind to intestinal M cells by an unknown adhesin. Following invasion, the M cells die and deliver the invading salmonellae into the Peyer patch. The organisms then migrate through the intestinal mucosa of the **terminal ileum into the submucosal lymph nodes**. A distinctive feature critical to typhoid (enteric) fever is the ability of the organisms to **survive and multiply within macrophages**. An **infiltration of mononuclear cells into the colonic mucosa is a characteristic finding of enteric fever** (Fig. 36-2). The organisms (carrying **capsular Vi antigen**) **resist phagocytic killing** by inhibiting the oxidative burst and multiply within the mononuclear cells. Once through the mucosal barrier, the bacteria cause primary bacteremia (prodromic fever).

Facultative intracellular organisms are carried by the monocytes and delivered to the reticuloendothelial system, causing enlargement of liver and spleen. When secondary bacteremia has occurred, and a critical number of organisms have replicated in the

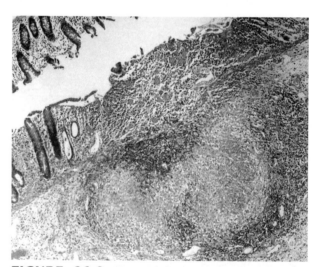

FIGURE 36-2 Histopathology of colonic mucosa with surface ulceration and submucosal lymphoid follicle with central necrosis and macrophages. H&E stain. Note a pink center due to necrosis in conjunction with an influx of macrophages that also have pink cytoplasm. *(Courtesy of Dr. Dominick Cavuoti, Department of Pathology, University of Texas Southwestern Medical Center, Dallas, TX.)*

blood stream, signs and symptoms of enteric fever, including abdominal pain, result, caused by secretion of cytokines and pyrogens by macrophages. The heart rate may be lower than expected based on the patient's fever. **LPS of the Gram-negative outer membrane elicits a systemic inflammatory response (septicemia).** Erythematous maculopapular lesions (**"rose spots"**), caused by **thrombocytopenia** and vascular capillary leakage, are also **characteristic of enteric fever** (see Fig. 36-1). The gallbladder is infected either from the blood or from the liver via the biliary tract. The **typhoidal salmonellae are particularly resistant to bile; they reenter the intestine, causing diarrhea** after days of febrile illness.

NOTE In nonfatal cases, humoral antibody and activated macrophages will eventually subdue the untreated infection over a period of about 3 weeks. A **chronic carrier state,** with excretion of organisms for more than 1 year, may occur in approximately 5% of infected persons. Some individuals with biliary obstruction or gallstones are particularly prone to becoming long-term (> 30 years) carriers.

TREATMENT

Chloramphenicol, ampicillin, or trimethoprim sulfamethoxazole (TMP/SMX) has been the standard therapy for typhoid fever for several decades. In recent years, multidrug-resistant strains have appeared, especially in India, Southeast Asia, and Africa. These isolates are generally susceptible to the fluoroquinolones (particularly **ciprofloxacin**), which are now the preferred choice of empirical antibiotics until susceptibility results are known.

OUTCOME

Therapy with ciprofloxacin was initiated and continued for a total of 10 days. The patient's fever gradually came down after 3 days of therapy, and he made an uneventful recovery.

PREVENTION

Vaccines for the international traveler to endemic areas are recommended to protect against typhoid fever. There is a live vaccine that is taken orally and consists of the attenuated mutant of *S. typhi*. There is also an IM vaccine based on the Vi capsular polysaccharide antigen (Typhim Vi). Good hygiene is a must for preventing infection. In addition, travelers to endemic areas should avoid eating food that has been prepared outside the home and, particularly, should not eat food prepared by street vendors.

FURTHER READING

Ackers M, Puhr N, Tauxe R, Mintz E: Laboratory-based surveillance of *Salmonella typhi* infections in the United States: Antimicrobial resistance on the rise. JAMA 283:2668, 2000.

Mermin JH, Villar R, Carpenter C, et al: A massive epidemic of multidrug-resistant typhoid fever in Tajikistan associated with consumption of municipal water. J Infect Dis 179:1416, 1999.

Parry CM, Hien TT, Dougan G, et al: Typhoid fever: Medical progress. N Engl J Med 347:1770, 2002.

A **71-year-old** male returned home after a 2-week stay in Mexico. The day after his return, he experienced an **acute onset of fever, crampy abdominal pain,** and **watery diarrhea.** By the next day, he had **tenesmus** and noticed **mucus** and a bloody tinge of the stool. The **stools became grossly bloody** and increased in number. Worried about his condition, his daughter took him to a hospital emergency department.

During his stay in Mexico the patient was in a rural area and had drunk water on several occasions from a well, but he had not come in contact with other sick persons.

PHYSICAL EXAMINATION

VS: T 38.8°C, P 118/min, R 16/min, BP 108/62 mmHg

PE: A sick-appearing, somnolent, elderly man with lower abdominal tenderness, mild dehydration, and hyperactive bowel sounds; **rectal exam** was very painful and showed **gross blood.**

LABORATORY STUDIES

Blood

Hematocrit: 42%
WBC: 5300/µL
Differential: 50% PMNs, 18% bands, 24% lymphs

TABLE 37–1   Differential Diagnosis and Rationale for Inclusion (consideration)

Amebic colitis
Clostridium difficile colitis
Crohn disease
Dysentery
  Salmonella spp
  Shigella spp
  Escherichia coli O157:H7
  Yersinia enterocolitica
Ulcerative colitis (idiopathic)

Rationale: Multiple enteric pathogens can cause the dysentery syndrome. Classically, a rapid, descending course of infection, with fever and abdominal pain progressing to mucoid diarrhea with bloody stools (colitis), is seen with shigellosis, but similar symptoms may also be seen with the other agents listed under dysentery. *E. coli* O157:H7 is usually not associated with fever. Amebic colitis has a gradual onset, manifesting with a 1- to 2-week history of abdominal pain, diarrhea, and tenesmus (fever is uncommon). *C. difficile* with earlier antibiotic exposure is also important to consider. Noninfectious causes should be ruled out by laboratory investigation, although it is uncommon for inflammatory bowel disease to occur at the age of the case patient.

Serum chemistries: BUN 26 mg/dL, creatinine 1.4 mg/dL

Imaging

Abdominal x-rays were unremarkable, and sigmoidoscopy showed **ulcers and an erythematous, friable mucosa.**

Diagnostic Work-Up

Table 37-1 lists the likely causes of illness (differential diagnosis). Leukocytes were present on fecal examination (Fig. 37-1). A diagnosis of dysentery was considered. Investigational approach may include

- **Culture** of the fecal specimen or rectal swabs on selective media
- In failed culture investigation
  - **Toxin testing** for *Clostridium difficile*
  - **Microscopic** (ova and parasite) **examination** for protozoal agent or stool antigens

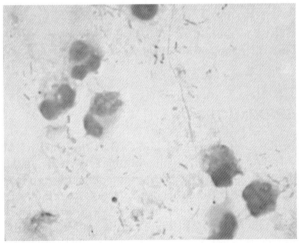

FIGURE 37-1 Leukocytes were seen on examination of fecal specimen from this patient. Gram stain. *(Courtesy of Dr. Paul Southern, Departments of Pathology and Medicine, University of Texas Southwestern Medical Center, Dallas, TX.)*

- **Colonic biopsy** and histopathology may be necessary to differentiate idiopathic ulcerative colitis and Crohn disease

COURSE

The patient was given IV rehydrating fluids. One day later, the diagnostic laboratory reported that a significant enteric bacterial species grew from the fecal culture.

ETIOLOGY

Shigella flexneri (bacillary dysentery)

MICROBIOLOGIC PROPERTIES

The bacteria in the genus *Shigella* are somewhat related to *E. coli*. All are **nonmotile, noncapsulate, and facultative anaerobes**. The *Shigella* genus is divided into the following four species: (1) *S. dysenteriae* (serogroup A; serotype 1 is highly pathogenic); (2) *S. flexneri* (serogroup B); (3) *Shigella boydii* (serogroup C); and (4) *S. sonnei* (serogroup D).

Selective media (e.g., Hektoen agar) are used for cultivation of any of the species from fecal specimens. With the exception of *S. sonnei*, which ferments lactose slowly, bacteria of other three species **do not ferment lactose** (Fig. 37-2). Species identification is achieved by using biochemical reactions and serologic confirmation (based on O and H antigens), using group-specific antibodies. *S. sonnei* is further subdivided into more than 30 individual serotypes on the basis of O antigens (because the strains lack flagella H antigen).

EPIDEMIOLOGY

In the United States, laboratory-confirmed cases of shigellosis occur predominantly due to *S. sonnei*. In the developing world, *S. flexneri* predominates (international travelers often acquire infection during travel). Epidemics of shigellosis due to *S. dysenteriae* type 1 (the most virulent of the four species) occur in the Indian subcontinent, Africa, and Central America. *Shigella* is **transmitted** mainly **by direct** or **indirect fecal-oral transmission** from a patient or carrier. In the United States, *S. sonnei* causes disease primarily

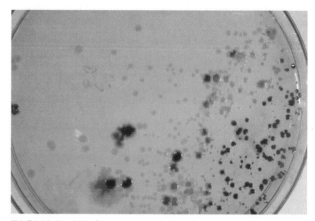

FIGURE 37-2   Culture of *Shigella* spp on Hektoen agar medium. Note the nonlactose-fermenting (pale) colonies of *Shigella* spp mixed with lactose-fermenting (*red*) colonies of *Escherichia coli*. *(Courtesy of Dr. Paul Southern, Departments of Pathology and Medicine, University of Texas Southwestern Medical Center, Dallas, TX.)*

among the **elderly** and **young children**, and it is a common cause of diarrhea outbreaks in **child care centers**, where personal hygiene is difficult to maintain.

PATHOGENESIS

Shigellae are unique among enteric pathogens in that **a very low infective dose** (of as few as 10 organisms) is needed to cause illness. The incubation period is 1 to 3 days. In the **colonic mucosa,** the invasive bacteria penetrate M cells and are taken up by macrophages in the lamina propria. The macrophages secrete inflammatory cytokines. Infected intestinal epithelial cells produce attractants for neutrophils, which migrate between the epithelial cells, resulting in a breakdown of the tight junctions and facilitating the **local spread** of the bacteria. Bacteria released from the dead macrophages also invade the intestinal epithelial cells via their basolateral membrane. Passage of the bacteria into adjacent epithelial cells occurs through finger-like projections from the surface of an infected cell to the surface of an uninfected cell. A host protein, known as cadherin L-CAM, is essential in the cell-to-cell spread of infection. When a sufficient number of **invaded cells die owing to intracellular multiplication of bacteria, the colonic mucosa sloughs off, causing an ulcer**, visible by sigmoidoscopy. Neutrophils that accumulate in large numbers in the mucosa are shed in the stool; blood is shed following endothelial damage, and mucus is secreted from ulcerated lesions caused by virulent strains of the three species (e.g., *S. flexneri*) but not by *S. sonnei*. The latter usually causes less severe and watery diarrhea, and most generally in very young children and the elderly.

NOTE In severe forms of shigellosis in highly endemic South Asia, **Shiga toxin** (Stx) contributes to the severity of disease caused by *S. dysenteriae* type 1. Stx is an inhibitor of protein synthesis, targeting the 23S ribosomal RNA. Stx kills intestinal epithelial cells and endothelial cells, resulting in blood loss. **Hemolytic-uremic syndrome** and **thrombotic thrombocytopenic purpura are the systemic complications** associated with **toxin-producing shigellae** (e.g., *S. dysenteriae* type 1) and also with *E. coli*.

TREATMENT

Fluid and electrolyte replacement using IV or oral rehydration solution is useful when diarrhea is watery. Antibiotic therapy may reduce the number of organisms and duration of carriage in symptomatic patients (thereby reducing person-to-person spread). Trimethoprim sulfamethoxazole (TMP/SMX) has been the treatment of choice for many years and is highly effective. However, areas in Southeast Asia,

Africa, and South America have a high prevalence of TMP/SMX-resistant strains. The quinolones are useful for these strains and are recommended for empirical therapy. Generally administration of antibiotics for 3 to 5 days is sufficient.

OUTCOME

The patient was given ciprofloxacin as soon as the culture results were known, and he recovered promptly.

PREVENTION

There is no vaccine currently, although an oral vaccine may hold promise for the future. The only means of preventing shigellosis is to disrupt the fecal-oral transmission. This involves hand washing and preventing fecal material from contaminating food and water supplies. Adequate chlorination is important in keeping water sources safe, both for drinking and in public water recreation areas.

FURTHER READING

CDC: Outbreak of gastroenteritis associated with an interactive water fountain at a beachside park—Florida, 1999. MMWR 49:565, 2000.
CDC: Outbreaks of *Shigella sonnei* infection associated with eating fresh parsley—United States and Canada, July-August 1998. MMWR 48:285, 1999.
Hyams KC, Bourgeois AL, Merrell BR, et al: Diarrheal disease during Operation Desert Shield. N Engl J Med 325:1423, 1991.
Lee LA, Shapiro CN, Hargrett-Bean N, Tauxe RV: Hyperendemic shigellosis in the United States: A review of surveillance data for 1967-1988. J Infect Dis 164:894, 1991.
Merson MH, Morris GK, Sack DA, et al: Travelers' diarrhea in Mexico. A prospective study of physicians and family members attending a congress. N Engl J Med 294:1299, 1976.
Sobel J, Cameron DN, Ismail J, et al: A prolonged outbreak of *Shigella sonnei* infections in traditionally observant Jewish communities in North America caused by a molecularly distinct bacterial subtype. J Infect Dis 177:1405, 1998.

A 75-year-old man experienced the acute onset of **severe abdominal cramps.** Later in the morning, watery diarrhea occurring every 15 to 30 minutes developed, initially with small amounts of visible blood. Diarrhea subsequently became **markedly bloody.** He was nauseated but not vomiting. Worried about his illness and his age, his son took him to a nearby hospital emergency department for evaluation.

Recent food intake history was remarkable for eating a **hamburger at a back yard BBQ** 2 days earlier. The patient recalled that the **meat inside was pink.** He said his teenaged grandson ate at the BBQ and had the same illness but with milder symptoms.

C A S E

38

PHYSICAL EXAMINATION

VS: T 37.4°C, P 108/min, R 16/min, BP 110/58 mmHg

PE: Clinical examination of the abdomen was unremarkable except for increased bowel sounds. Stool was **grossly bloody**.

LABORATORY STUDIES

Blood

Hematocrit: 42%

WBC: 6200/μL

Differential: Normal

Serum chemistries: BUN 18 mg/dL, creatinine 1.2 mg/dL

Imaging

Abdominal x-ray was normal.

Diagnostic Work-Up

Table 38-1 lists the likely causes of illness (differential diagnosis). A diagnosis of hemorrhagic colitis was considered. Investigational approach may include

- **Enteric bacterial stool cultures**
- In failed culture investigation
  - **Toxin testing** for *Clostridium difficile*
  - **Microscopic** (ova and parasite) **examination** for protozoal agent or stool antigens.

COURSE

The patient was hospitalized. Frank bloody diarrhea continued. Stool cultures were negative for *Salmonella*, *Shigella*, *Campylobacter*, and *Yersinia*; however, a significant enteric isolate was recovered and sent to

TABLE 38–1 Differential Diagnosis and Rationale for Inclusion (consideration)

Amebic colitis
Clostridium difficile colitis
Dysentery
   Shigella spp
   Salmonella spp
   Enteroinvasive *Escherichia coli* (EIEC)
   Yersinia enterocolitica
Hemorrhagic colitis due to *E. coli* O157:H7
Inflammatory bowel disease (IBD)

Rationale: It is important to carefully evaluate bloody diarrhea to rule out invasive bacterial pathogens because some should be treated and others may not need to be treated. In addition, empirical use of antidiarrheal medications is not recommended for most of these syndromes. Amebic colitis is usually associated with travel; *C. difficile* is associated with earlier antibiotic consumption. Dysentery would demonstrate fever along with bloody diarrhea. IBD, an idiopathic syndrome, is also important to consider, although it is distinctly unusual in this age group.

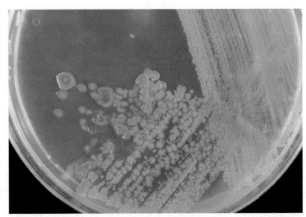

FIGURE 38-1 Growth of the pathogen on sorbitol-MacConkey agar medium. Note sorbitol-nonfermenting colonies that are pale colored. *(Courtesy of Dr. Paul Southern, Departments of Pathology and Medicine, University of Texas Southwestern Medical Center, Dallas, TX.)*

the State Public Health Laboratories for species confirmation.

ETIOLOGY

Enterohemorrhagic *Escherichia coli* (serotype *O157:H7*; EHEC)

MICROBIOLOGIC PROPERTIES

The *E. coli* O157:H7 is the most common pathogen isolated from bloody diarrhea, but other serotypes have also been associated with bloody diarrhea. Identification of *E. coli* O157:H7 is made on the basis of an **indole-positive, lactose-positive, sorbitol-nonfermenting isolate** (on sorbitol-MacConkey agar medium; Fig. 38-1) that is positive for agglutination by O157-specific antibodies. The **Shiga toxin-producing** strains of *E. coli* are confirmed by cytopathic effect on an assay using cell culture.

EPIDEMIOLOGY

In 1982, a previously unrecognized enteric pathogen, *E. coli* O157:H7, was associated with outbreaks of bloody diarrhea in persons eating hamburgers contaminated with the organism. The illness, subsequently called hemorrhagic colitis, was characterized by abdominal cramps and watery diarrhea that after a few days became streaked with blood or grossly bloody. More recently, outbreaks of enterohemorrhagic colitis have been linked to fecally contaminated drinking water and swimming pools. In addition to many well-described outbreaks, *E. coli* O157:H7 appears to be a common cause of sporadic infectious diarrhea, especially in the **northern states** and **Canada**, where isolation rates are similar to those of *Salmonella* and *Shigella*. **Ground beef** and **unpasteurized milk are the major vehicles** of infection, although unpasteurized apple cider (made from apples picked up from the ground presumably contaminated with bovine feces) has also been implicated. *E. coli* O157:H7 has been isolated in up to 6% of fecal samples from cattle and in up to 29% of raw ground beef. Other sources include Brussels sprouts, lettuce, and salami. Secondary person-to-person spread by direct fecal-oral transmission can occur, especially in day care centers or nursing homes. The organisms may also be transmitted from person to person as secondary spread. All ages are at risk; however, children younger than 5 years of age and the elderly are more likely to develop serious complications.

PATHOGENESIS

In colonic mucosa, EHEC strains produce virulence factors that allow these organisms to **attach and efface the brush border of the intestinal epithelium** (also seen in enteropathogenic *E. coli*; Table 38-2). Affected enterocytes exhibit a dramatic loss of microvilli and rearrangement of cytoskeleton elements, with a proliferation of filamentous actin beneath areas of intimate bacterial attachment. Colon pathology shows a pattern of injury similar to that seen on sigmoidoscopic examination in ischemic or infectious colitis. EHEC strains also produce one or more **phage-encoded (Shiga) toxins** that kill cells in culture and are cytotoxic for selected targets such as colonic and renal endothelial cells. Shiga toxins (Stx)-1 and Stx-2 are similar in structure to the classic Stx, produced by *Shigella dysenteriae* type I. Severe symptoms occur when local and systemic spread of a toxin from the gastrointestinal tract causes endothelial damage, leading to blood loss. Acute bloody diarrhea and abdominal cramps, with little or no fever, usually last 1 week.

> **NOTE** The **hemolytic uremic syndrome** (HUS) and **thrombotic thrombocytopenic purpura** (TTP) are the most serious complications of EHEC and occur overall in about 5% of cases, usually in the very young or very old. HUS occurs after binding of Stx1 and/or Stx2 to a glycolipid receptor molecule on the surface of endothelial cells in the kidney. TTP is believed to result from a combination of platelet effects: destruction, increased consumption, sequestration in the liver and spleen, and intrarenal aggregation. Table 38-2 describes the clinical, epidemiologic, and pathologic features of syndromes caused by other diarrheogenic *E. coli*.

TREATMENT

Most cases **do not require antimicrobial therapy**. Fluid and electrolyte replacement is important when diarrhea is watery or when there are signs of dehydration. The use of antibiotics to treat EHEC infections is controversial, and one hypothesis states that some antibiotics could increase the expression of Shiga toxin and contribute to kidney damage. HUS complications may require dialysis.

TABLE 38-2 Clinical, Epidemiologic, and Pathobiologic Features of Other *Escherichia coli*-Associated Diarrheal Illnesses

Characteristics	ETEC	EIEC	EPEC	EAEC
Clinical syndrome	Mild cholera-like illness in all ages	Dysentery (milder than shigellosis)	Infantile and childhood diarrhea	Persistent diarrhea in children
Stools				
Nature	Copious, watery	Scant, purulent	Copious, watery	Watery
Blood	Absent	Common	Absent	Absent
WBCs	Absent	Prominent	Minimal	Absent
Fever	Absent	Common	Uncommon	Absent
Complications	Serious dehydration	None	None	None
Major serotype	O:6	O:29	O26:H111	Nonclassic EPEC serotypes
Epidemiology	Endemic in developing countries; traveler's diarrhea	Endemic in developing countries; traveler's diarrhea	Endemic in developing countries; traveler's diarrhea	Endemic in developing countries; traveler's diarrhea
GI site	Small intestine	Large intestine	Small intestine	Small intestine
Primary pathogenic mechanism	Enterotoxins (heat-labile LT stimulates adenyl cyclase → cAMP↑)	Invasion of enterocytes	Attachment and effacement (loss) of the microvilli in mucosa	Cell signaling
Mucosal pathology	Intact, hyperemia	Inflammation, ulceration, necrosis	Effacing lesions	Intact

EPEC, enteropathogenic *E. coli*; ETEC, enterotoxigenic *E. coli*; EIEC, enteroinvasive *E. coli*; EAEC, enteroadherent *E. coli*.

OUTCOME

The patient was treated with intravenous fluids. He gradually recovered over the next 7 days.

PREVENTION

Useful measures may include the following: developing farm- and slaughterhouse-based methods to decrease contamination of meat; encouraging the use of irradiation to increase the safety of ground beef; identifying ways to prevent contamination of foods eaten raw (e.g., produce); educating the public to cook ground beef thoroughly, preferably using a digital instant-read thermometer; and conducting population-based surveillance for HUS and determining which serotype of Shiga toxin-producing *E. coli* was responsible for a particular illness.

FURTHER READING

Clarke SC, Haigh RD, Freestone PPE, Williams PH: Virulence of enteropathogenic *Escherichia coli*, a global pathogen. Clin Microbiol Rev 16:365, 2003.

Mahon BE, Griffin PM, Mead PS, Tauxe RV: Hemolytic uremic syndrome surveillance to monitor trends in infection with *Escherichia coli* O157:H7 and other Shiga toxin-producing *E. coli*.

Emerg Infect Dis 3:409, 1997.

Mead PS, Griffin PM: *Escherichia coli* O157:H7. Lancet 352:1207, 1998.

Nataro JP, Kaper JB: Diarrheagenic *Escherichia coli*. Clin Microbiol Rev 11:142, 1998.

A 31-year-old man returned to the United States in late summer from a 3-week-long **trip to Bangladesh.** On the second day after his return, he presented with sudden, severe, **profuse watery diarrhea.** In the emergency department, he passed a large, **watery stool with a rice-water appearance.** He vomited several times and became slightly sweaty. He complained of **muscle cramps and dizziness.**

He was on an **H₂-blocker drug** for ulcer disease. Otherwise, he had always maintained good health.

C A S E

39

PHYSICAL EXAMINATION

VS: T 37°C, P 124/min, R 28/min, BP 86/44 mmHg

PE: He was somewhat anxious; his pulse was rapid and weak.

LABORATORY STUDIES

Blood

Hematocrit: 49%

WBC: 8900/μL

Differential: Normal

Serum chemistries: Na 130 mmol/L, Cl 96 mmol/L, K 3.3 mmol/L, BUN 35 mg/dL, creatinine 1.4 mg/dL, glucose 204 mg/dL

Imaging

No imaging studies were done.

Diagnostic Work-Up

Table 39-1 lists the likely causes of illness (differential diagnosis). A clinical diagnosis of cholera was con-

A

B

FIGURE 39-1 *A,* Gram stain of the fecal specimen. Note the curved (*comma-shaped*), Gram-negative rods in the midst of many other normal flora of the colon. *B,* Culture of the pathogen on thiosulfate-citrate-bile-sucrose (TCBS) selective agar from fecal specimen. Note the sucrose-fermenting colonies of the isolate on TCBS agar are yellow in appearance. (*Courtesy of Dr. Paul Southern, Departments of Pathology and Medicine, University of Texas Southwestern Medical Center, Dallas, TX.*)

TABLE 39–1   Differential Diagnosis and Rationale for Inclusion (consideration)

Cholera
Cryptosporidiosis
Enterotoxic *Escherichia coli* (ETEC) diarrhea
Giardiasis
Viral gastroenteritis

Rationale: Watery diarrhea has many causes, including viruses, bacteria, and parasites. ETEC is the most common cause of watery diarrhea in travelers. However, severe, acute watery diarrhea that leads to rapid dehydration is characteristic of one of the above agents. No other illness causes such massive diarrhea. Certain areas of the world, such as South Asia, Africa, and South America, are highly endemic.

sidered based on symptoms and travel history. Investigational approach may include

- **Fecal examination** (microscopy). May confirm the absence of white cells (watery or secretory diarrhea)
- **Fecal cultures**. Routine enteric investigation and special enteric bacterial cultures need to be undertaken.
- In failed investigation, tests for enteric virus and ova and parasite examination for *Giardia* and *Cryptosporidium* may be undertaken.

COURSE

The patient was quickly recognized as being severely dehydrated and was immediately given a rapid infusion of intravenous fluids. Fecal specimen on special culture grew a significant pathogen that is part of an ongoing pandemic in the country of travel by this patient.

ETIOLOGY

Vibrio cholerae (cholera)

MICROBIOLOGIC PROPERTIES

The bacteria in the genus *Vibrio* are **curved (comma-shaped)**, **Gram-negative rods** (Fig. 39-1*A*). They are **highly motile**, with a single flagellum. They are nonspore-forming, **oxidase-positive**, **facultative anaerobes**. Special culture of fecal specimens on selective (thiosulfate-citrate-bile-sucrose) agar supports growth of *V. cholerae* while inhibiting the commensal colonic bacteria from fecal specimens (see Fig. 39-1*B*). The significant isolate is identifiable by biochemical tests and use of polyvalent antisera. Up to 141 types of LPS-associated somatic O antigens of *V. cholerae* are known. **Organisms that agglutinate in 0:1 anti-serum** usually **cause epidemics and pandemics of cholera**.

EPIDEMIOLOGY

Eight cholera pandemics have been reported since 1817 (Table 39-2). Six of the eight pandemics have swept out of the Ganges River Delta in the Indian subcontinent and were due to the O:1 classic biotype. **The seventh global pandemic in Asia, Africa, and Latin America,** which began in Sulawesi, Indonesia, in 1961, **has been ongoing** for more than four decades. **The O:1 classic biotype was replaced by O:1 biotype El Tor in the seventh pandemic of cholera**. Fortunately, virulence of El Tor is low, but the high carrier rate is problematic. The eighth pandemic, due to non O:1 (O:139 Bengal), was aborted in 1993, within a year of its emergence. Large bodies of fresh water contaminated by asymptomatic human carriers of *V. cholerae* are the habitats of the pathogen in the areas of the ongoing seventh pandemic. **Ingestion of contaminated water and food is the major mode of transmission**. Individuals with **achlorhydria, taking antacids or other drugs that reduce gastric acidity, are at risk** of developing cholera. Non-O:1 cholera vibrios inhabit the coastal waters of the United States (especially the Gulf coast) and cause diarrhea associated with the **consumption of raw shellfish**. Although there have been no recent outbreaks of cholera in the United States, sporadic cases (~5 to 10 per year) occur along the Gulf coast (mainly in Texas and Louisiana).

Other *Vibrio* species are natural inhabitants of brackish and salt water worldwide. *V. parahaemolyticus*, acquired by ingestion of contaminated seafood, causes intestinal disease ranging from watery diarrhea to frank dysentery in the United States. Most infections are associated with the ingestion of raw shellfish. *V. vulnificus* causes wound infection and sepsis, particularly in patients with cirrhosis.

TABLE 39–2	Cholera Pandemics: Origins, Years, and Other Findings		
Pandemic:	1st - 6th	7th (current)	8th (aborted)
Years:	1817-1960	1961 - to date	1992-1993
Origins:	Ganges Delta, Bangladesh	Indonesia	Madras, India
Bio/serotype:	Classic O:1	El Tor O:1	O:139 Bengal
Virulence:	High	Less	Moderate
Carrier rate:	Low	High	Unknown
Eradication:	Yes	Unlikely	Yes?

PATHOGENESIS

A potent cholera toxin (CTX), an enterotoxin produced by *V. cholerae*, causes the severe watery diarrhea of cholera. A very large inoculum of organisms is required for disease, except for patients with reduced gastric acidity. *V. cholerae* lacks a genetically controlled mechanism of acid resistance; the bacteria rely on large inoculum size. **Hypochlorhydria is a significant risk factor** for cholera. Individuals taking antacids or other drugs that reduce gastric acidity are at risk for cholera. *V. cholerae* reach the small intestine in sufficient numbers and multiply and colonize the small intestine via long filamentous pili (bundles). The synthesis of pili bundles is co-regulated with the synthesis of CTX by a *toxR* (sensor) gene product that regulates the virulence genes. CTX is an A-B type ADP-ribosylating enterotoxin. B pentamer binds to G_{M1} ganglioside, a glycolipid on the surface of jejunal epithelial cells that serves as the toxin receptor and facilitates the delivery of the A subunit to its target.

The functional A subunit **activates the adenyl cyclase cascade system** by irreversible transfer of an ADP-ribose subunit from NAD to membrane G_s protein, **thereby raising the intracellular concentrations of cyclic AMP (cAMP)** in the intestinal epithelial cells. **cAMP inhibits the absorptive sodium transport system in villus cells and activates the excretory chloride transport system in crypt cells, causing accumulation of sodium chloride in the lumen.** Watery diarrhea results from the passive movement of water into the lumen to maintain osmolality. Hypersecretion (fluid loss of 1 L/hr) of water and electrolytes (**"rice-water diarrhea"**) causes profound dehydration and associated symptoms and signs of cholera.

NOTE Of typical cases, 20% to 50% are fatal if untreated. In particularly severe cases, death may occur within hours.

TREATMENT

Fluid and electrolyte replacement is crucial, and treatment of cholera involves **intravenous and oral rehydration** therapy. The availability of **oral rehydration solution** has reduced mortality from more than 50% to less than 1%. Antibiotic therapy is of secondary value; **doxycycline** (drug of choice) or a fluoroquinolone (ciprofloxacin) can be used.

OUTCOME

The patient was given 4 L of fluid intravenously and then placed on an oral rehydration solution. He also received doxycycline. Stool volumes progressively diminished over 48 hours, and the patient was discharged home because there were no complications. He was able to manage himself with oral rehydration.

PREVENTION

Improved hygiene in endemic areas has great importance in the prevention of cholera, and disease control in endemic geographic areas includes filtration and chlorination of water systems, and health education. Killed cholera vaccine has limited value; it provides partial protection (for 50% of those vaccinated) of short duration (3 to 6 months). It is no longer recommended for travelers to endemic areas.

FURTHER READING

Besser RE, Feikin DR, Eberhart-Phillips JE, et al: Diagnosis and treatment of cholera in the United States. Are we prepared? JAMA 272:1203, 1994.

Lipp EK, Huq A, Colwell RR: Effects of global climate on infectious disease: The Cholera model. Clin Microbiol Rev 15:757, 2002.

Morris JG, Black RE: Cholera and other vibrioses in the United States. N Engl J Med 312:343, 1985.

Tauxe RV, Mintz ED, Quick RE: Epidemic cholera in the new world: Translating field epidemiology into new prevention strategies. Emerg Infect Dis 1:141, 1995.

Twenty-four people became **ill within 3 hours** after eating a meal at an office party. All had **nausea,** most had **vomiting,** and several had **crampy abdominal pain.** Three of the individuals sought medical care at the emergency department of a local hospital.

All 24 individuals had been in good health. The day before the office party, a food preparer had purchased a 17-pound **precooked packaged ham,** baked it at home at 400°F for 1.5 hours, and transported it to her work place, a large institutional kitchen, where she sliced the hot ham on a commercial slicer. The ham was **served cold at the party the next day.**

PHYSICAL EXAMINATION

Physical examination of one of the three patients (Patient X) is presented here.

VS: T 37°C, P 84/min, R 14/min, BP 136/80 mmHg

PE: Patient X appeared in mild distress due to abdominal pain. Abdomen was soft and nontender, with normal bowel sounds.

LABORATORY STUDIES (PATIENT X)

Blood

Hematocrit: 42%
WBC: Normal
Differential: Normal
Serum chemistries: Normal

Imaging

Abdominal x-ray was normal.

Diagnostic Work-Up

Table 40-1 lists the likely causes of outbreak (differential diagnosis). For Patient X (and others in the outbreak), a clinical diagnosis of food poisoning was considered based on **symptoms, onset,** and **incriminated food.** Illness in this outbreak had a rapid onset, within 3 hours after consumption of precooked ham, incriminating a preformed enterotoxin. Investigational approach and epidemiologic investigation may include

- **Search for food handlers** with skin infections, particularly of the hands

TABLE 40–1 Differential Diagnosis and Rationale for Inclusion (consideration)
Chemical poisoning
Food poisoning
Staphylococcus aureus
Bacillus cereus
Clostridium perfringens

Rationale: One of the most important factors to consider in cases of presumed food-borne illness is the time to onset of symptoms after eating the suspected food items. Organisms that cause an actual infection generally take at least 24-72 hours to cause symptoms. Those that produce pre-formed toxins lead to symptoms within hours. Food poisoning due to *S. aureus* and *B. cereus* usually has rapid onset (1-6 hours) of clinical symptoms. *C. perfringens* usually has a longer incubation period. Chemical poisoning would likely be a diagnosis of exclusion.

- **Cultures** of all purulent lesions and collection of nasal swabs from all food handlers

COURSE

Patient X was observed in the emergency department for several hours and given IV fluids until his vomiting subsided. One sample of leftover cooked ham, analyzed by latex agglutination to identify a bacterial toxin, was positive for enterotoxin A. Samples of stool or vomitus were not obtained from any of the ill persons. Cultures from nares or skin were not obtained from the food preparer.

ETIOLOGY

Staphylococcus aureus (food poisoning)

MICROBIOLOGIC PROPERTIES

Staphylococci are **Gram-positive cocci** that are arranged in irregular **grapelike clusters**, when viewed by light microscopy of Gram-stained cultures. These are non-motile and nonspore forming and are **resistant to high salt concentrations.** The normally yellowish colonies on blood agar are identified based on a coagulase-positive reaction. *S. aureus* produces multiple virulence factors, including **heat-stable enterotoxins** that cause food-borne intoxication.

EPIDEMIOLOGY

Food poisoning is one of the most common diseases in the United States, acquired by ingestion of contaminated food. The common theme that ties all food-borne illness together is the presence of an improper food-handling procedure, usually inadequate refrigeration or cooking. In an outbreak, recovery of large numbers ($>10^5$) of organisms per gram of epidemiologically incriminated food and detection of enterotoxin from the food item are useful for delineation of a common-source outbreak. Typing (or strain delineation) of all isolates is useful for epidemiology and is performed by DNA profiling (e.g., restriction fragment length polymorphism [RFLP] analysis).

PATHOGENESIS

Food poisoning usually results from the effects of a toxin that is preformed in the food (the organism no longer has to be alive). Enterotoxin-producing species are common causes of food-associated infections (Table 40-2). *S. aureus* is of human origin, from purulent discharges of an infected finger or eye, abscesses, acneiform facial eruptions, or nasopharyngeal secretions from the food preparers. Several enterotoxins of *S. aureus* are stable at boiling temperature, thus tolerating cooking conditions that kill the organisms that produced them. The case outbreak was caused by ingestion of precooked ham that was left at warm temperatures for prolonged periods, thereby allowing toxin production to occur. Preformed enterotoxins in incriminated food cause a rapid onset of mostly upper gastrointestinal symptoms.

TREATMENT

There is no specific therapy for food poisoning caused by organisms that produce preformed toxins. Intravenous fluids may be administered for patients with

TABLE 40–2   Clinical, Microbiologic, and Epidemiologic Characteristics of Food Poisoning

Organism	Mechanism	Incubation Period	Vehicles	Main Features
Staphylococcus aureus	Heat-stable enterotoxin	1-6 hours	Ham (meats), mayonnaise, custard	Nausea, vomiting (acute upper GI symptoms)
Bacillus cereus	Heat-stable enterotoxin (emetic)	1-6 hours	Reheated fried rice	Nausea, vomiting (acute upper GI symptoms)
	Heat-labile enterotoxin (diarrheogenic; produced in the gut	8-24 hours	Cream sauce	Watery diarrhea
Clostridium perfringens	Heat-labile enterotoxin	8-12 hours	Reheated meat dishes, gravy	Watery diarrhea

significant episodes of vomiting. The syndrome is self-limiting, usually within 12 to 24 hours.

OUTCOME

Patient X and two of his coworkers required intravenous fluids in the emergency department, but none were admitted to the hospital. Within 24 hours, all patients' symptoms had resolved.

PREVENTION

The most important preventive measure is proper preparation and handling (particularly storage) of food. Because the illness is not transmitted from person to person, hand washing is not useful once a food item has become contaminated.

FURTHER READING

Kluytmans J, van Belkum A, Verbrugh H: Nasal carriage of *Staphylococcus aureus*: Epidemiology, underlying mechanisms, and associated risks. Clin Microbiol Rev 10:505, 1997.
Shandera WX, Tacket CO, Blake PA: Food poisoning due to *Clostridium perfringens* in the United States. J Infect Dis 147:167, 1983.

Terranova W, Blake PA: *Bacillus cereus* food poisoning. N Engl J Med 298:143, 1978.

A **3-month-old** female infant was brought to the emergency department of a general hospital with a 5-day history of **decreased activity, decreased oral intake, upper airway congestion,** and **general irritability.**

There was **no history of fever or vomiting.** During the previous 2 weeks, she had been constipated, and twice **her mother had given her a tablespoon of honey** for treatment.

PHYSICAL EXAMINATION

VS: T 36.1°C, P 120/min, R 20/min, BP 90/65 mmHg

PE: The patient was **listless** and slightly pale but otherwise well nourished, with weight and height above the 50th percentile for age. Positive findings included moderately **dry oral mucosa,** upper airway congestion, a **sluggish pupillary response to light,** mild abdominal distention with hypoactive bowel sounds, and **significant hypotonia** (Fig. 41-1).

LABORATORY STUDIES

Blood

Hematocrit: Normal
WBC: Normal
Differential: Normal

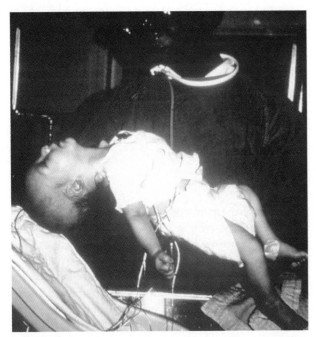

FIGURE 41-1 Severe hypotonia in this infant. *(Courtesy of Centers for Disease Control and Prevention, Atlanta, GA.)*

Blood gases: pO_2 65 mmHg, pCO_2 48 mmHg
Serum chemistries: Normal

Imaging

MRI brain scans were inconclusive.

Diagnostic Work-Up

Table 41-1 lists the likely causes of illness (differential diagnosis). A clinical diagnosis of sepsis (meningitis) was made, and botulism was considered. Investigational approach may include

- **Blood culture** to rule out blood-borne bacterial pathogen
- **Culture of CSF** to rule out bacterial meningitis
- **Detection of food-borne toxin** in stool, serum, or implicated food

COURSE

The patient was admitted to the pediatric intensive care unit and started on IV fluids and ceftriaxone for

---

TABLE 41–1 Differential Diagnosis and Rationale for Inclusion (consideration)

Botulism
Guillain-Barré syndrome
Meningitis
Myasthenia gravis
Sepsis
Tick paralysis

Rationale: Generalized weakness carries a broad differential, and specific features often distinguish the various etiologies. Sepsis would be expected to show a febrile response, as would meningitis, but these are nevertheless important to consider. Botulism, an afebrile illness, shows a characteristic clinical pattern. The other causes have their own clinical neurologic features, and tick paralysis should be associated with an engorged tick, usually on the scalp. Myasthenia gravis and Guillain-Barré syndrome are remote possibilities.

possible meningitis. During the 72 hours after admission, the patient was increasingly unable to handle her secretions and lost her gag reflex as well as her pupillary response. Her hypotonia increased markedly and she continued to have frequent episodes of apnea. Because of impending respiratory failure, she was electively intubated at this time. Stool and serum samples were sent to a reference laboratory to determine the presence of a toxin. A mouse lethality test became strongly positive, and the presence of a neurotoxin was identified by specific antisera.

> **NOTE** Empirical treatment for bacterial sepsis or meningitis is continued until an alternative diagnosis is established or cultures are negative, even though the patient may be afebrile.

ETIOLOGY

Clostridium botulinum (infant botulism)

MICROBIOLOGIC PROPERTIES

C. botulinum is a **Gram-positive, spore-forming rod** that is an obligate anaerobe. **A potent heat-labile neurotoxin produced by *C. botulinum* causes disease.** Organisms of types A through G are distinguished by the antigenic specificities of their toxins. Detection of toxin (types A, B, and E) in serum by bioassay in mice is definitive (sensitivity is low in infant botulism).

EPIDEMIOLOGY

Infant botulism is **the most frequently encountered form** of botulism seen in the United States (median 71 cases per year). **Ingestion of honey** and, to a lesser extent, corn syrup has been identified as a risk factor. Microbiologic surveys of honey have found that 4% to 25% of samples contain *C. botulinum* spores. Primary **adult botulism** occurs as food-borne clusters or sporadic single cases due to ingestion of preformed toxin in **home-canned fruit (bulging cans)**, sausage, and fish products with pH greater than 4.6. Occasionally, **wound botulism** (median three cases per year) occurs when *C. botulinum* **contaminates an injury**, especially associated with "skin popping" black tar heroin. Cases also occur after intranasal cocaine use, dental abscess, and major trauma.

The diagnosis of adult botulism must be suspected on clinical grounds in the context of an appropriate history.

> **NOTE** Botulinum toxin has been used as a biological warfare agent and has **bioterrorism potential** if it can be aerosolized by terrorists. It is the most potent toxin known.

PATHOGENESIS

Spores of *C. botulinum* are ingested in honey or syrup by the infants, and the organisms then germinate, colonizing the gastrointestinal tract and producing a potent neurotoxin. The spores are not able to germinate in older children or adults. The toxin is absorbed and circulates in the blood stream. The central nervous system is not involved. The toxin is transported to peripheral cholinergic nerve terminals, including neuromuscular junctions, postganglionic parasympathetic nerve endings, and peripheral ganglia. It attaches specifically to peripheral cholinergic synapses and is internalized inside the nerve cell in endocytic vesicles, from where it is translocated into the cytosol. The toxin, a zinc-metalloprotease, cleaves components of the neuroexocytosis apparatus, **irreversibly preventing release of acetylcholine from the neuromuscular junction**. Because of the specific cell type to which it binds, the result is **descending flaccid paralysis**.

> **NOTE** In adult (ingesting toxin in food) and wound botulism, the common symptoms are ptosis, dysphagia, dry mouth, blurred vision, dysarthria, bulbar weakness, symmetrical paralysis, diarrhea, and vomiting. The symmetrical, descending flaccid paralysis of motor and autonomic nerves, usually beginning with the cranial nerves, may progress rapidly. Death can result from respiratory failure. Recovery takes months (those who survive may have fatigue and shortness of breath for years).

TREATMENT

The standard treatment for severe botulism is **supportive therapy** with mechanical ventilation. Trivalent (A, B, E) **equine antitoxin serum**, which can be administered to neutralize toxin not yet internalized in neurons, should be given as soon as possible, even before diagnosis is confirmed. In infant botulism, antibiotics are contraindicated because these drugs may increase the levels of toxin in the gut by bacterial lysis. In adult food botulism, purgatives can be used to expel retained food. In wound botulism, wound débridement and antibiotics (penicillin, metronidazole) can be an effective modality for management.

OUTCOME

The patient remained intubated and mechanically ventilated for a total of 30 days. Ceftriaxone was discontinued after 72 hours, when all blood and CSF cultures were reported as negative. Pupillary response returned after 5 days. Although she was able to return home after 6 weeks, normal activity and strength did not return for several months.

PREVENTION

Infants should not be given honey. Proper food preparation and storage are effective preventive methods for prevention of adult botulism. Boiling food can inactivate the toxin, although the spores are much more difficult to destroy.

FURTHER READING

Arnon SS, Schechter R, Inglesby TV, et al: Botulinum toxin as a biological weapon: Medical and public health management. JAMA 285:1059, 2001.

Chia JK, Clark JB, Ryan CA, Pollack M: Botulism in an adult associated with food-borne intestinal infection with *Clostridium botulinum*. N Engl J Med 315:239, 1986.

Merson MH, Dowell VR, Jr: Epidemiologic, clinical and laboratory aspects of wound botulism. N Engl J Med 289:1005, 1973.

Midura TF: Update: Infant botulism. Clin Microbiol Rev 9:119, 1996.

Passaro DJ, Werner SB, McGee J, et al: Wound botulism associated with black tar heroin among injecting drug users. JAMA 279:859, 1998.

A **67-year-old** man presented with fever, abdominal cramping, and frequent **diarrhea** (six to nine bowel movements per day) for 4 days.

Three weeks before the current episode, he had undergone a hip replacement and was rehabilitating in an orthopedic unit. During that hospitalization, he developed a **nosocomial pneumonia** and was treated empirically with **cefuroxime** and **clindamycin.** He gradually improved and was discharged a week before his current presentation, with maintenance oral antibiotics, to recuperate at home. His wife had **no similar symptoms.**

C A S E

42

PHYSICAL EXAMINATION

VS: T 39°C, P 114 /min, R 18/min, BP 94/50 mmHg

PE: The patient appeared confused and very pale. He could not answer questions about his current condition. His skin showed decreased **turgor,** and his **oral mucosa** was **dry.**

LABORATORY STUDIES

Blood

Hematocrit: 45%

WBC: 12,800/μL

Differential: 71% PMNs, 24% lymphs

Blood gases: Normal

Serum chemistries: BUN 28 mg/dL, creatinine 1.5 mg/dL

Imaging

Sigmoidoscopy revealed erythematous and friable colonic mucosa.

Diagnostic Work-Up

Table 42-1 lists the likely causes of illness (differential diagnosis). A clinical diagnosis of antibiotic-associated diarrhea or colitis was considered based on diarrhea (more than five bowel movements per day), remark-

TABLE 42–1 Differential Diagnosis and Rationale for Inclusion (consideration)

Antibiotic-associated diarrhea or colitis (*Clostridium difficile*)

Bacterial enteritis (dysentery)

   Campylobacter

   Salmonellosis

Inflammatory bowel disease (IBD)

Irritable bowel syndrome (IBS)

Viral gastroenteritis

Rationale: Diarrhea has multiple etiologies, and specific clues are usually necessary in addition to microbiologic studies to determine a precise etiology. Prior antibiotic use is commonly associated with *C. difficile.* Bacterial and viral causes are certainly possible, but they are difficult to distinguish. Noninfectious causes, such as IBD and IBS, are somewhat less likely to manifest in the elderly but are also important to consider. Noninfectious causes are often associated with recurrent symptoms and not necessarily a single episode.

able sigmoidoscopy, and exposure to antibiotics. **Detection of toxins** in the diarrheal stool is the mainstay of delineation of the etiology.

COURSE

A stool specimen was sent to the laboratory, which within 24 hours yielded a positive test for a toxin.

ETIOLOGY

Clostridium difficile–associated diarrhea (CDAD)

MICROBIOLOGIC PROPERTIES

C. difficile is a strictly **anaerobic** bacterium. The organisms are **Gram-positive, spore-forming rods. Toxigenesis** is an important property of diarrheagenic *C. difficile*. These strains **produce two exotoxins: toxin A** and **toxin B**, which can be detected by ELISA (sensitive, specific, and simple).

EPIDEMIOLOGY

C. difficile is carried asymptomatically as **part of the large intestinal flora** of 50% of all healthy neonates during the first year of life. The carriage rate decreases to less than 4% in adults. This rate remains constant in the population. Among hospitalized adults who have received antibiotic therapy, carriage rates may be as high as 46% (particularly during outbreaks). The **primary** (index) **cases occur via endogenous mode in precolonized patients exposed to antibiotics. Secondary cases occur via exogenous transmission of spores** in the hospital environment and by the hands of health care attendants, causing nosocomial outbreaks. Antimicrobial agents of all classes and several anticancer chemotherapeutic agents have been incriminated as inciting agents of CDAD. **The most commonly incriminated antimicrobial agents are clindamycin, cephalosporins, and ampicillin.**

PATHOGENESIS

CDAD is toxin mediated. *C. difficile* is ordinarily suppressed by the normal colonic flora, preventing overgrowth. Broad-spectrum antibiotics suppress normal flora. Clindamycin, which inhibits growth or kills many different species of anaerobic bacteria in the colon, does not suppress *C. difficile*. The overgrowing vegetative organisms of *C. difficile* produce at least two toxins: toxin A and toxin B. Both toxins appear to act by the same mechanism, but toxin B is more potent. Both toxins exert their effects by binding to cellular GTP-binding proteins (in the Rho family within target cells). The toxins inactivate these proteins by glycosylation, dysregulating the action of the cytoskeleton in epithelial cells of the colonic mucosa, and causing depolymerization of actin. Break-up of actin filaments causes profound cytopathic effect, damaging the cellular lining of the bowel wall and causing **erythematous** and **friable colonic mucosa**, ulceration, and hemorrhagic necrosis.

TREATMENT

Treatment of CDAD begins with **discontinuation of the offending agent** and implementation of any necessary supportive measures. The preferred oral antimicrobial agent is **metronidazole**. The oral metronidazole therapy should be begun as soon as possible. Oral vancomycin is an alternative, but its use carries the risk of emergence of vancomycin-resistant enterococci and colonization, which may pose serious health risks. In addition, it is much more expensive. Relapse is common because *C. difficile* spores are resistant to many antibiotics, and continued antibiotic use can delay the return of normal flora that would inhibit growth of *C. difficile*.

NOTE Pseudomembranous colitis, a complication of CDAD, occurs in untreated acute cases and is characterized by **multiple elevated, yellowish white plaques (pseudomembranes) within the colon.** Toxic megacolon is a serious sequela of pseudomembranous colitis and may lead to sepsis due to perforation and polymicrobial infection of colonic flora.

OUTCOME

Specific antibiotic treatment for pneumonia was tapered off, and specific treatment for CDAD was begun with oral metronidazole in addition to hydration. The patient became afebrile within 36 hours, and his diarrhea resolved after 3 days. He was able to return home without any further problem and had no further episodes of diarrhea.

PREVENTION

Limiting use of broad-spectrum antibiotics is an important measure in reducing the risk of developing *C. difficile* colitis. Measures to control the spread of infection within the hospital include **hand washing, removing gloves** before attending another patient, enteric precautions, and isolating the index case or cohort patients with CDAD.

FURTHER READING

Bartlett JG: Antibiotic-associated diarrhea. N Engl J Med 346:334, 2002.

Knoop FC, Owens M, Crocker IC: *Clostridium difficile*: clinical disease and diagnosis. Clin Microbiol Rev 6:251, 1993.

Johnson S, Gerding DN: *Clostridium difficile*-associated diarrhea. Clin Infect Dis 26:1027, 1998.

Nath SK, Thornley JH, Kelly M, et al: A sustained outbreak of *Clostridium difficile* in a general hospital: Persistence of a toxigenic clone in four units. Infect Cont Hosp Epidemiol 15:382, 1994.

A **9-month-old baby** girl was brought to the emergency department of a local general hospital **during the winter** with a 2-day history of **vomiting, watery diarrhea,** and **fever.**

The patient had been **well until 24 hours before her presentation,** when she had experienced the acute onset of vomiting followed by multiple episodes of diarrhea. She refused to eat, and she drank very little fluid. Her parents were concerned about dehydration. The family had not traveled outside the United States recently, but the mother related that she had been leaving her baby in a **day care center** for 3 days a week for the past 3 months.

PHYSICAL EXAMINATION

VS: T 39.2°C, P 145/min, R 32/min, BP 90/44 mmHg

PE: The patient's **mucous membranes were dry,** and she was **listless** and **febrile.** Her neck was supple. Her chest was clear, and heart sounds were normal. Her abdomen was nontender. Bowel sounds were normal.

LABORATORY STUDIES

Blood

Hematocrit: 38%

WBC: 5300/μL

Differential: Normal

Serum chemistries: BUN 22 mg/dL, creatinine 1.2 mg/dL

Imaging

Chest x-ray was normal.

Diagnostic Work-Up

Table 43-1 lists the likely causes of illness (differential diagnosis). A clinical diagnosis of gastroenteritis was considered. Absence of fecal leukocytes (based on the examination of methylene blue-stained stool preparation) can rule out inflammatory diarrhea and bacterial causes. Investigational approach may include:

- **Routine enteric bacterial cultures**
- **Stool viral antigen test**
- In failed investigation, electron microscopy of enteric viruses and ova and parasite examination for protozoa

TABLE 43-1 Differential Diagnosis and Rationale for Inclusion (consideration)

Bacterial gastroenteritis
Dysentery: *Shigella* spp
Secretory diarrhea: enteropathogenic *Escherichia coli*
Protozoal diarrhea
   Cryptosporidium parvum
   Giardia lamblia
Viral gastroenteritis
   Astrovirus
   Enteric adenovirus
   Norovirus
   Rotavirus

Rationale: Diarrhea in infants is often due to viruses. In winter months, rotavirus is very common and can be severe. Dysentery can be associated with bloody stools. Protozoal infection usually relates to specific exposures and travel history, although it can also be associated with day care centers.

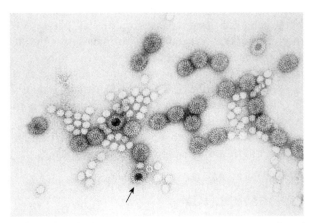

FIGURE 43-1 This electron micrograph reveals a number of 70-nm, nonenveloped virions and a number of unknown, 29-nm virion particles. Note the distinctive rim of radiating capsomeres of the virions (*arrow*). (*Courtesy of Centers for Disease Control and Prevention, Atlanta, GA.*)

COURSE

The patient was admitted to the hospital for intravenous hydration and further investigation. Her stool was sent for examination for fecal leukocytes, cultures for bacteria, electron microscopy for viruses, and tests for parasites. No fecal leukocytes were found in her stool. Microbiologic investigation of routine enteric pathogens was nondiagnostic, and the ova and parasite tests for protozoa and helminths were negative. An enzyme immunoassay-based stool antigen test was diagnostic.

ETIOLOGY

Rotavirus (infantile gastroenteritis)

MICROBIOLOGIC PROPERTIES

Rotaviruses are 70-nm **nonenveloped RNA viruses** in the family of Reoviridae. The viral nucleocapsid is composed of two concentric shells. The double-shelled particles, **shaped like wheels** with short spokes on an outer rim (Fig. 43-1), are the complete infectious virions. The virions have an icosahedral structure and lack an envelope. Two structural proteins on the outermost shell define the antigenic types of the virus. There are at least three major subgroups (A, B, and C) and nine serotypes of human rotaviruses. Rotaviruses have a **segmented genome**, with **11 segments of a double-stranded RNA genome**, yet, unlike influenza viruses, which also have segmented RNA genomes, **gene reassortment has played no role in genetic diversity of rotaviruses**, and the virus strains have remained stable.

Because rotavirus and other enteric viral agents (e.g., adenovirus, calicivirus, and astrovirus) grow poorly in cell culture, **assays that detect viral antigen in stool specimens have become the most widely used method of diagnosis**. Rotavirus antigen can be detected in the stool using enzyme immunoassay.

EPIDEMIOLOGY

Enteric viruses are responsible for the majority of cases of gastroenteritis worldwide. The leading viral pathogens are rotaviruses and noroviruses, followed by enteric adenoviruses, astroviruses, and coronaviruses. Rotaviruses are the **single most important cause of severe dehydrating diarrhea in infants** and children younger than 3 years of age **worldwide**. They are responsible for up to 10% of all diarrheal episodes among children younger than 5 years of age in the United States. Rotavirus infection has a **seasonal distribution** in the United States, with transmission occurring during **the winter months**. Rotaviruses are shed in high concentrations in the stools of infected children and are **transmitted by the fecal-oral route**, both through close person-to-person contact and through fomites. Elderly individuals who are immunocompromised or who have undergone hematopoietic or solid organ transplantation experience severe, prolonged, and sometimes fatal rotavirus diarrhea.

PATHOGENESIS

After the fecal-oral route of acquisition, the virus infects the mature villus tip cells of the small intestine. Viral invasion of the epithelial cells of the small intestine results in destruction of the mature absorptive cells. **Damaged cells on villi are replaced by immature crypt cells that cannot absorb carbohydrates or other nutrients efficiently, resulting in osmotic diarrhea**. The severity of rotaviral diarrhea is proportional to the extent of mucosal damage in the small intestine.

> **NOTE** Rotavirus infection frequently occurs in conjunction with respiratory tract symptoms, but the negative chest examinations and radiologic studies indicate that the virus does not spread and multiply in the respiratory tract. Local immune factors, such as IgA or interferon, may be important in protection against rotavirus infection. Reinfection in susceptible children in the age group 6 months to 3 years may occur owing to a different serotype of virus. The influence of cell-mediated immunity is likely related to recovery from infection and to protection against subsequent disease.

TREATMENT

There is no specific antiviral agent for this disease. **Symptomatic care** is directed at supportive measures, with particular attention to the prevention of dehydration through the use of intravenous hydration or oral rehydration therapy.

OUTCOME

The patient was rehydrated with IV fluids for 48 hours until the diarrhea improved significantly. She was discharged from the hospital after being able to tolerate fluids for 24 hours.

PREVENTION

An oral live attenuated rhesus-based rotavirus vaccine-tetravalent was licensed in the United States in 1998 for vaccination of infants. It was withdrawn a year later because of reports of intussusception, a form of intestinal obstruction, as an uncommon but serious side effect. Adherence to hand washing and barrier methods are important for preventing disease spread in day care and in the hospital setting. In view of the fecal-oral route of transmission, wastewater treatment and sanitation are important control measures in the community.

FURTHER READING

CDC: Foodborne Outbreak of group A rotavirus gastroenteritis among college students—District of Columbia, March-April 2000. JAMA 285:405, 2001.

Kapikian AZ: Viral gastroenteritis. In Evans A, Kaslow R (eds): Viral Infections in Humans: Epidemiology and Control. 4th ed. New York, Plenum,1997, p 285.

Midthun K, Kapikian AZ: Rotavirus vaccines: An overview. Clin Microbiol Rev 9:423, 1996.

Murphy TV, Gargiullo PM, Massoudi MS, et al: Intussusception among infants given an oral rotavirus vaccine. N Engl J Med 344:564, 2001.

Within **48 hours** of a college football game in Philadelphia, **158 students** with symptoms of gastrointestinal disease visited the university health service. The predominant symptoms included **nausea** in 99%, **vomiting** in 75%, **diarrhea** in 48%, and headache, fever, and myalgias. Marching band members, football players, and faculty and staff from both universities had similar symptoms. A total of several hundred individuals were afflicted with similar symptoms.

PHYSICAL EXAMINATION

Physical examination of one of the patients (Patient X) is presented here.

VS: T 38.6°C, P 96/min, R 16/min, BP104/60 mmHg

PE: A young female in mild distress due to **abdominal pain** and **nausea**; abdominal exam showed mild tenderness, and rectal exam revealed no blood.

LABORATORY STUDIES (PATIENT X)

Blood

Hematocrit: 38%
WBC: 7200/μL
Differential: Normal
Serum chemistries: Normal

Imaging

Abdominal x-ray was normal.

Diagnostic Work-Up

Table 44-1 lists the likely causes of illness of Patient X and the outbreak (differential diagnosis). Outbreaks like this are consistent with **acute nonbacterial gastroenteritis** given the **explosive nature** of the cases and the **high rates of nausea, vomiting**, and **diarrhea** in conjunction with the **low incidence of fever**. An enteric virus is the prime suspect in adult outbreaks, but patients' stools are still evaluated for bacterial and parasitic pathogens. Additional tests may include

- **Electron microscopy**
- **Stool antigen** for virus by enzyme immunoassay
- In failed investigation, **RT-PCR** of viral RNA targets, nucleotide hybridization probes.

TABLE 44–1 Differential Diagnosis and Rationale for Inclusion (consideration)
Acute nonbacterial gastroenteritis
Bacterial gastroenteritis (e.g., *Campylobacter jejuni*, *Salmonella* spp)
Caliciviruses
Norovirus (previously, Norwalk virus)
Protozoal diarrhea (e.g., *Giardia lamblia*, *Cryptosporidium parvum*)

Rationale: This is an explosive outbreak of gastroenteritis in adult individuals who were present at a football game. A common-source transmission of a highly communicable agent should be suspected. Viral etiologies are among the most common causes of these outbreaks. Unfortunately, these often cannot be differentiated from other pathogens based on clinical grounds alone, and even routine testing is inadequate. Although bacterial pathogens could be responsible, protozoal causes in this situation would be distinctly unusual.

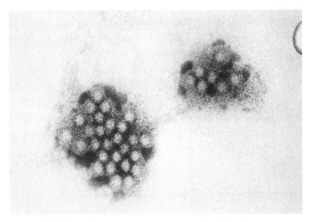

FIGURE 44-1 An electron micrograph of the 27- to 32-nm-sized viral particles from stool specimen. (*Courtesy of Centers for Disease Control and Prevention, Atlanta, GA.*)

COURSE

The patient was given IV fluids and antinausea medications and recovered after 2 days at home. No enteric bacterial or parasitic pathogens were found from cultures and light microscopic investigation of stool samples from Patient X or others. Immune electron microscopy revealed 27-nm virus-like particles in the stool samples of the patients (including Patient X), which helped delineate the etiology.

ETIOLOGY

Norovirus (epidemic nonbacterial gastroenteritis)

MICROBIOLOGIC PROPERTIES

Norovirus (NV) is a prototype strain of genetically and antigenically diverse single-stranded RNA viruses, previously called "small round-structured viruses," which are classified in the family of *Caliciviridae*. NV is named after the original strain "Norwalk virus," which caused an outbreak of gastroenteritis in a school in Norwalk, Ohio, in 1968. NV is a **spherical, nonenveloped, 27-nm virus** that has a **positive-sense, single-stranded**, polyadenylated **RNA**. Differentiation of NV from other similar viral pathogens (e.g., caliciviruses) can be made on the following clinical and epidemiologic grounds: (1) incubation period is 24 to 48 hours, (2) vomiting occurs in 50% or more of the cases, (3) duration of illness ranges from 12 to 60 hours, and (4) diarrheal illness rapidly spreads among many individuals. Assays based on cultivation are not available because the virus does not grow in cell culture. Unlike rotavirus, NV lacks distinctive morphology when viewed by EM. Most stools of infected individuals contain a very small number of virus particles. **Human antibodies** are required to **concentrate** and **visualize the NVs** (Fig. 44-1), which have a single structural protein with antigenicity.

EPIDEMIOLOGY

Epidemics of vomiting and diarrhea in adults and also children caused by norovirus are not uncommon. The clustered illnesses occur in families, communities, cruise ships, and nursing homes and other institutional settings. Transmission occurs around the year.

Noroviruses are transmitted primarily through the **fecal-oral route**, either by consumption of **fecally contaminated food or water** (e.g., eating raw or inadequately cooked shellfish, or drinking municipal or well water) or by direct person-to-person spread. Frequently, primary cases in an outbreak (like the one described here) result from exposure to a fecally contaminated vehicle (e.g., water or ice or food), whereas secondary and tertiary cases among contacts of primary cases result from person-to-person transmission.

PATHOGENESIS

Noroviruses are **highly contagious**, and it is thought that an **inoculum of as few as 10 viral particles may be sufficient to infect an individual.** The incubation period for norovirus-associated gastroenteritis in humans is usually between 24 and 48 hours, but cases can occur within 12 hours of exposure. Morphologic changes identified from biopsies of the mucosa of the proximal small intestine of patients with noroviral gastroenteritis include **shortening and atrophy of the villi, crypt hyperplasia, and infiltration of the lamina propria by polymorphonuclear and mononuclear cells**. The histologic alterations are seen in conjunction with carbohydrate malabsorption and decreased levels of some brush border enzymes.

Protective immune mechanisms are not known to provide long-term immunity, and therefore reinfection can occur.

TREATMENT

Treatment is symptomatic. **Oral rehydration** is clinically efficacious, but intravenous rehydration may be necessary when severe vomiting and diarrhea occur.

OUTCOME

The symptoms of the afflicted patients lasted approximately 12 to 48 hours, and a few older individuals were hospitalized briefly. However, all recovered completely. The explosive nature of this norovirus outbreak, in which a large number of young adults became ill within 48 hours, suggests that infection was acquired from a common source. Ninety-two percent of the students had purchased sodas with ice from the stadium concessionaire. In addition, the football team members developed similar symptoms and had also used ice, however, from a different vending machine. The ice consumed at the football game was traced to a manufacturer in southeast Pennsylvania whose wells had been flooded by water following a torrential rainfall. An increase in diarrheal illness had also been noted among residents along the creek who obtained their drinking water from private wells that were flooded.

PREVENTION

The norovirus is a highly communicable agent. The low infectious dose causes efficient transmission of the virus. Effective hand washing and disposal or disinfection of contaminated clothing and linens may decrease transmission. Good personal hygiene, careful food preparation, and purity of municipal drinking water and ice (from machines for sodas) are helpful in preventing transmission of norovirus gastroenteritis.

FURTHER READING

Blacklow NR, Greenberg HB: Viral gastroenteritis. N Engl J Med 325:252, 1991.

Graham DY, Jiang X, Tanaka T: Norwalk virus infection of volunteers: New insights based on improved assays. J Infect Dis 170:34, 1994.

Greenberg HB, Valdesuso J, Yolken RH: Role of Norwalk virus in outbreaks of nonbacterial gastroenteritis. J Infect Dis 139:564, 1979.

Griffin DW, Donaldson KA, Paul JP, Rose JB: Pathogenic human viruses in coastal waters. Clin Microbiol Rev 16:129, 2003.

Morse DL, Guzewich JJ, Hanrahan JP, et al: Widespread outbreaks of clam- and oyster-associated gastroenteritis. N Engl J Med 314:678, 1986.

A **54-year-old** woman presented to her family physician complaining of **abdominal pain** that had been **worsening for the past 2 weeks**. She stated that it **often improved immediately after meals or taking antacids.** She also noted occasional **heartburn** but denied fevers, nausea or vomiting, diarrhea, or bloody stools.

The patient was a schoolteacher who **worked** with a crowded class of grade seven students **under stressful conditions.** There were no other notable prior social or medical histories.

PHYSICAL EXAMINATION

VS: T 37°C, P 84/min, R 14/min, BP 110/70 mmHg

PE: Abdominal exam revealed mild **midepigastric tenderness** with no rebound. Rectal exam was normal; no blood on Hemoccult testing was noted.

LABORATORY STUDIES

Blood

Hematocrit: 38%

WBC: 5600/μL

Differential: Normal

Serum chemistries: Normal

Imaging

Not usually done unless worrisome symptoms are present.

Diagnostic Work-Up

Table 45-1 lists the likely causes of illness (differential diagnosis). A clinical diagnosis of peptic ulcer disease was considered. Investigational approach may include

- **Invasive tests**
  - **Histologic examination.** Endoscopic biopsy specimens sectioned, stained, and examined microscopically
  - **Culture.** Endoscopic biopsy specimens cultured on special agar under special conditions

- Noninvasive tests
  - **Fecal antigen** test to detect *Helicobacter pylori* antigen
  - **^{13}C urea-breath test**
  - **Serology** to detect *H. pylori*-specific IgG

TABLE 45–1 Differential Diagnosis and Rationale for Inclusion (consideration)

Appendicitis
Cholecystitis
Cholelithiasis
Crohn disease
Esophagitis
Gastroenteritis
Gastroesophageal reflux
Peptic ulcer disease

Rationale: Abdominal pain has an extremely broad differential diagnosis list. Certain features often help in determining which is the most likely etiology. Lower-right quadrant pain suggests appendicitis, whereas upper-right quadrant pain is suggestive of cholecystitis or cholelithiasis. The absence of diarrhea or emesis makes gastroenteritis and Crohn disease unlikely. Esophagitis or reflux disease would likely have chest pain as a prominent symptom. Pain associated with signs of acid hypersecretion suggests peptic ulcer disease. However, these are generalizations, and other factors, such as lab results and endoscopy, must also be considered.

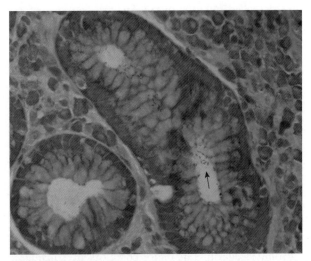

FIGURE 45-1 The Giemsa stain of a section of gastric biopsy, examined histologically. Note curved (or S-shaped) organisms (*arrow*) adjacent to gastric epithelial cells. (*Courtesy of Dr. Dominick Cavuoti, Department of Pathology, University of Texas Southwestern Medical Center, Dallas, TX.*)

COURSE

The patient received an H₂ blocker for her symptoms and also underwent serology testing, which was positive.

ETIOLOGY

H. pylori (peptic ulcer disease)

MICROBIOLOGIC PROPERTIES

The etiologic agent of peptic ulcer disease is a *Campylobacter*-like organism, *H. pylori*, which is a **Gram-negative curved rod**. It is a motile organism and is a profuse **urease producer**. Growth requires selective media with nutrient supplements (e.g., Skirrow agar), **microaerophilic** conditions, and incubation temperature of 37°C for up to 10 days.

> **NOTE** For diagnostic purposes, gastric mucosal biopsy specimens may be examined histologically, using Giemsa or Warthin-Starry stains; the S-shaped organisms are found adjacent to gastric epithelial cells or in the overlying mucus (Fig. 45-1). Specimens also can be tested directly for the presence of urease. The ^{13}C **urea-breath test** is carried out by having a patient drink a solution with ^{13}C-labeled urea. Urease in the stomach splits off labeled carbon dioxide, which is absorbed into the blood and expired from the lungs. Breath collections are tested for the presence of the radiolabel as an indirect marker for the presence of *H. pylori*.

EPIDEMIOLOGY

Human populations throughout the world are affected by *H. pylori* infection. In developing countries, more than 80% of adults are infected, whereas in the United States, as many as 30% of adults are infected with *H. pylori*. Humans and monkeys are the only known habitats of the pathogen. Fecally contaminated water causes more frequent human colonization in developing countries than in the developed countries. **Increasing age** and **poor socioeconomic conditions** are the most important risk factors for overgrowth and chronic infection. The major cofactors are considered to be excess acid, smoking, stress, and eating spicy foods.

PATHOGENESIS

Multiple virulence factors are necessary for *H. pylori* to colonize and persist in the stomach. The pathogenesis of *H. pylori* disease is complex, and prominent features associated with *H. pylori* include the following:

1. It is found only on gastric epithelium.
2. It **does not invade** cells.
3. It **elicits robust inflammation in stomach (chronic gastritis)** and immune response.
4. It is rarely spontaneously cleared.

A **cloud of ammonia (with buffering function)** produced by **urease activity (a major virulence factor)** protects the organisms from gastric acid. Bacterial flagella and mucinase allow the organisms to pass through the mucus layer (avoiding elimination). The bacteria are anchored at the intracellular junction of enteric cells to the epithelial cells by attachment pedestals. The **two other major virulence factors** are a **vacuolating cytotoxin, VacA,** and a group of genes termed the *cag* **pathogenicity island** (*cag* PaI). One of the identifiable markers of inflammation (in *cag* PaI) is the CagA protein. Mucosal levels of IL-8 (a potent chemoattractant) and other proinflammatory cytokines are significantly higher in patients carrying

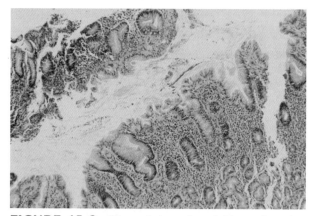

FIGURE 45-2 Histopathology of gastric biopsy of a patient with *Helicobacter pylori* infection. H & E stain. Note the marked inflammatory infiltrate in the submucosa, indicating chronic gastritis (most of the cells are mononuclear inflammatory cells, although one cannot tell this easily from this low magnification). There is also some erosion of the superficial layer of the gastric mucosa. *(Courtesy of Dr. Paul Southern, Departments of Pathology and Medicine, University of Texas Southwestern Medical Center, Dallas, TX.)*

cagA+ strains. Chronic gastritis is characterized by a **mononuclear inflammatory cell infiltration** (Fig. 45-2) **associated with neutrophilic infiltration into the lamina propria**. The amount of inflammation may range from minimal infiltration of the lamina propria, preserving intact glandular architecture, to severe, dense inflammation, with profound erosion of the gastric mucosa.

> **NOTE** Host response is predominantly an **IgG response** to *H. pylori* infection. The pathogen is, however, able to **evade elimination by virtue of its protected location** in the gastric mucosa. Reinfection or recurrence due to ineffective treatment is very common.

TREATMENT

Appropriate antibiotic regimens can now successfully eradicate gastrointestinal infection with *H. pylori*, permanently curing ulcers in a high proportion of patients. The most successful regimens are those using a combination of two antimicrobials and a proton-pump inhibitor (PPI) that is continued for between 2 and 4 weeks. The goal of treatment is eradication rather than temporary clearance. The most effective regimen is **metronidazole or tetracycline, or amoxicillin in addition to clarithromycin and omeprazole** (a PPI). Omeprazole increases the antibiotic effectiveness by increasing pH; it also suppresses the organism.

> **NOTE** In untreated cases, **duodenal ulcers** (due to elevated gastric acid secretion and/or reduced mucosal protection; almost 100% have persistent *H. pylori* infection) and gastric B-cell (**mucosa-associated lymphoid tissue [MALT]**) **lymphoma** (in a minority of colonized patients) are likely complications.

OUTCOME

The patient was treated with a combination of amoxicillin, clarithromycin, and omeprazole for 2 weeks. Her symptoms gradually resolved and did not recur.

PREVENTION

There are no known preventive measures, and there is a high prevalence of infection worldwide. No vaccine is available currently.

FURTHER READING

Dunn BE, Cohen H, Blaser MJ: *Helicobacter pylori*. Clin Microbiol Rev 10:720, 1997.

Munnangi S, Sonnenberg A: Time trends of physician visits and treatment patterns of peptic ulcer disease in the United States. Arch Intern Med 175:1489, 1997.

Suerbaum S, Michetti P: Medical progress: *Helicobacter pylori* infection. N Engl J Med 347:1175, 2002.

Uemura N, Okamoto S, Yamamoto S, et al: *Helicobacter pylori* infection and the development of gastric cancer. N Engl J Med 345:784, 2001.

A 36-year-old man presented to the emergency department of a general hospital with a 10-day history of intermittent **diarrhea** and **tenesmus,** with **blood** and **mucus** visible **in the stool.**

He had just returned from a working **trip to India,** where he had **visited a rural town** in the last week of his trip.

PHYSICAL EXAMINATION

VS: T 38.8°C, P 96/min, R 16/min, BP 130/80 mmHg

PE: Ill-appearing male in mild distress; abdominal exam revealed mild diffuse tenderness, and rectal exam was **positive for blood.**

LABORATORY STUDIES

Blood

Hematocrit: 44%
WBC: 11,600/μL
Differential: 72% PMNs, 20% lymphs
Serum chemistries: Normal

Imaging

Sigmoidoscopic examination revealed **multiple small hemorrhagic areas with ulcers.**

Diagnostic Work-Up

Table 46-1 lists the likely causes of illness (differential diagnosis). A clinical diagnosis of dysentery was considered. Investigational approach may include

- Enteric (bacterial) cultures
- Stool **antigen test** for amebic agent
- Microscopic (**ova and parasite**) examination

COURSE

The patient was admitted to the hospital for observation. Microscopic examination of his stool showed many WBCs and RBCs. Microscopic examination of fixed and stained stool specimens subsequently revealed a significant pathogen.

TABLE 46–1 Differential Diagnosis and Rationale for Inclusion (consideration)

Dysentery syndrome
  Entamoeba histolytica
  Enteroinvasive *Escherichia coli*
  Salmonella spp
  Shigella dysenteriae
  Yersinia enterolitica
Inflammatory bowel disease (IBD)

Rationale: The dysentery syndrome can be caused by multiple pathogens, and stool studies are required to definitively diagnose them. However, epidemiology (history of exposure) can be helpful. *E. histolytica* (amebic dysentery) and *S. dysenteriae* (bacillary dysentery), two of the most common colonic ulcerative diseases, are much more common in developing countries than in the Western hemisphere, and recent travel history should be obtained to rule out these diseases. IBD should always be considered, although after infectious etiologies have been ruled out.

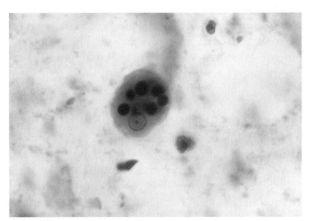

FIGURE 46-1 A trophozoite with ingested erythrocytes (trichrome stain). (*Courtesy of Centers for Disease Control and Prevention, Atlanta, GA.*)

ETIOLOGY

Entamoeba histolytica (amebic dysentery)

MICROBIOLOGIC PROPERTIES

E. histolytica exists in one of **two forms**: a **trophozoite** or a **cyst**. The trophozoite is fragile and cannot live outside the host because it is rapidly killed by poor environmental conditions. The cyst (the infective form) is resistant to killing by the low pH of the stomach. The **trophozoite (15 to 20 μm)** has a **single nucleus** with a central karyosome and uniformly distributed peripheral chromatin. The **cyst (12 to 15 μm)** is spherical with **four nuclei** with central karyosomes and fine, uniformly distributed peripheral chromatin.

The so-called "ova and parasite" examination is the recommended procedure for the recovery and identification of *E. histolytica* and other amebae in stool. **Light microscopy does not allow distinction between the invasive *E. histolytica* and lumen-dwelling *E. dispar* unless erythrophagocytosis** (the presence of ingested RBCs in trophozoites; Fig. 46-1) can be demonstrated. *E. histolytica* must also be distinguished from nonpathogenic amebic organisms that are intestinal colonizers of humans: *E. hartmanni*, *E. coli*, *E. polecki*, *Endolimax nana*, *Iodamoeba buetschlii*, and *Dientamoeba fragilis*. *D. fragilis* is the only intestinal parasite other than *E. histolytica* that is suspected of causing diarrhea. The presence of nonpathogenic amebae in the stool of a patient is strongly indicative of exposure to poor sanitation and fecal contamination of food or water. The presence of nonpathogenic amebae is also a warning of possible exposure to pathogenic *E. histolytica*.

TRANSMISSION

Humans are a natural reservoir. Cysts are passed in feces. Because of the protection conferred by their walls, cysts can survive days to weeks in the external environment and are responsible for disease transmission. Infection of suitable individuals occurs by **ingestion of mature cysts in fecally contaminated food, water,** or hands.

PATHOGENESIS

After the ingestion of cysts (the infective form) in contaminated food or water, excystation occurs in the small intestine, and trophozoites are released. Trophozoites migrate to the large intestine, where they adhere to intestinal mucosal cells by means of specific lectin-binding receptors. The amebic lectins facilitate adhesion to mucosa. In most patients trophozoites remain in the lumen as commensals. The trophozoites multiply by binary fission and produce cysts, which are passed in the feces.

In some patients the trophozoites invade the intestinal mucosa, with resultant pathologic manifestations. A number of virulence factors have been linked to invasion of colonic mucosa. Amebic **cytotoxins** enable the trophozoites to **invade the colon, with lysis of epithelial cells**. Cytotoxins also lyse PMNs, releasing hydrolytic enzymes that contribute to damage to colonic mucosa. Extracellular **cysteine proteinase** degrades collagen and elastin. Trophozoites of *E. histolytica* feed on neutrophils, monocytes, lymphocytes, cells of colonic mucosa, and other host materials, including RBCs, giving rise to amebic colitis. **Lesions in the colon range from nonspecific colitis with inflammatory cells and *E. histolytica* (Fig. 46-2) to flask-shaped ulcers,** and may extend through tissue planes.

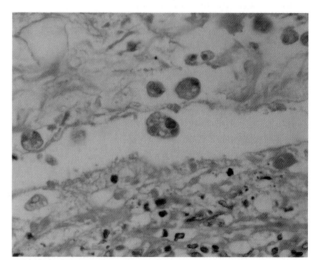

FIGURE 46-2 Histopathology of *Entamoeba histolytica* trophozoites in intestinal ulcer. *(Courtesy of Dr. Dominick Cavuoti, Department of Pathology, University of Texas Southwestern Medical Center, Dallas, TX.)*

TREATMENT

In general, antimotility agents are not recommended for patients with dysentery because of the possibility of an invasive pathogen. When a diagnosis is established, specific therapy may be started. Systemic therapy with **metronidazole** is used. For symptomatic patients iodoquinol, or paromomycin, or diloxanide furoate is used in conjunction with metronidazole to treat intraluminal infection. This combination therapy achieves complete elimination and cure.

NOTE In some patients with persistent colonization, the **trophozoites of *E. histolytica* invade the blood stream** and **reach** extraintestinal sites such as the liver, brain, and lungs. Unlike *E. dispar* (which is serum sensitive), the blood-borne trophozoites of *E. histolytica* are resistant to complement-mediated lysis and ascend the portal venous structures, producing hepatic infection. Liver infection is of grave concern because the cytolytic amebae irreversibly destroy hepatocytes, and **amebic liver abscess** may occur. Patients with amebic liver abscess typically present with a 1- to 2-week history of **fever, chills, leukocytosis, right upper quadrant abdominal pain,** and enlarged liver. Imaging studies (e.g., abdominal CT scan) and specific **serology confirm the diagnosis.**

OUTCOME

The patient was treated with metronidazole and paromomycin. He responded well and after 4 days was able to complete his therapy as an outpatient.

PREVENTION

Effective sanitation and clean water use during travel in endemic areas are the best methods of prevention. Travelers should avoid practices that may allow fecal-oral infection. Several vaccines to *E. histolytica* are under development and may provide the most effective means of disease control.

FURTHER READING

Espinosa-Cantellano M, Martínez-Palomo A: Pathogenesis of intestinal amebiasis: From molecules to disease. Clin Microbiol Rev 13:318, 2000.

Haque R, Huston CD, Hughes M, et al: Amebiasis: Current concepts. N Engl J Med 348:1565, 2003.

Ryan ET, Wilson ME, Kain KC: Illness after international travel: Current concepts. N Engl J Med 347:505, 2002.

Tanyuksel M, Petri WA, Jr: Laboratory diagnosis of amebiasis. Clin Microbiol Rev 16:713, 2003.

A 25-year-old man presented to a hospital clinic with a 2-week history of **sustained diarrhea** (three to five bowel movements per day), **nausea, flatulence,** and **lack of appetite.** He described his diarrhea as initially **watery,** and then greasy and **foul smelling.** He added that he had a **bloating** sensation. He did not have fever or chills.

The patient had been in good health. Four weeks previous to seeing his physician, he had visited Colorado for several days of **backpacking in the Rocky Mountains.**

PHYSICAL EXAMINATION

VS: T 37°C, P 82/min, R 14/min, BP 134/80 mmHg

PE: Abdomen was distended and mildly tender; no hepatosplenomegaly. Rectal exam was normal and Hemoccult negative.

LABORATORY STUDIES

Blood

Hematocrit: 46%

WBC: 6300/μL

Differential: Normal

Serum chemistries: BUN 22 mg/dL, creatinine 1.2 mg/dL

Imaging

No imaging studies were done.

Diagnostic Work-Up

Table 47-1 lists the likely causes of illness (differential diagnosis). History and physical examination do not contribute to microbiologic diagnosis. Investigational approach may include

- **Enteric bacterial cultures**
- **Stool ova and parasite** examination
- **Stool antigen** test for *Entamoeba histolytica* and *Giardia lamblia*
- In failed ova and parasite examination and stool antigen tests, examination of **upper endoscopy with biopsy** and **duodenal aspirate**

COURSE

Given his travel history, the patient was empirically given metronidazole after obtaining stool cultures. Fecal leukocytes and routine enteric culture investigation were negative. The patient's stools (three successive morning specimens) were sent to a

---

TABLE 47–1 Differential Diagnosis and Rationale for Inclusion (consideration)

Bacterial enteritis (e.g., *Salmonella, Campylobacter*)
Bacterial overgrowth
Inflammatory bowel disease
Irritable bowel syndrome
Protozoal diarrhea
   Cryptosporidium parvum
   Cyclospora cayetanensis
   Giardia lamblia

Rationale: Diarrhea is a very common clinical symptom, and history alone is usually not sufficient to narrow an etiology. Epidemiology is often useful, such as travel to highly endemic areas. Noninfectious etiologies are also possible, but are considered only after other causes are excluded. However, prolonged diarrhea, foul-smelling stools, and flatulence are particularly associated with a protozoal agent.

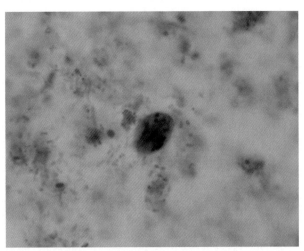

FIGURE 47-1 A cyst stained with trichrome stain. *(Courtesy of Dr. Dominick Cavuoti, Department of Pathology, University of Texas Southwestern Medical Center, Dallas, TX.)*

reference laboratory, where a protozoal agent was identified in stool specimens based on the microscopic examination of cysts.

ETIOLOGY

G. lamblia (giardiasis)

MICROBIOLOGIC PROPERTIES

G. lamblia (also known as *G. intestinalis* or *G. duodenalis*) is an **intestinal flagellate;** it exists in one of two forms: a trophozoite or a cyst. The **trophozoite** (the pathogenic form) is **pear-shaped** and has a convex dorsal surface, a flat **ventral surface with a sucking disk,** and **four pairs of flagella.** The sucking disk is composed of microtubules containing contractile proteins. Two anterior nuclei, each with a karyosome, give *G. lamblia* a facelike appearance. The trophozoites live freely in the upper small bowel lumen. Cyst is the infective form and is resistant to killing by the low pH of the stomach. The mature cyst is an oval structure that has four nuclei and is encased in a thin wall composed of *N*-acetylglucosamine.

Cysts are usually found in the feces of infected individuals (Fig. 47-1). **Identification of cysts is undertaken by using direct mounts of feces** as well as concentration procedures. If three successive feces specimens fail to demonstrate cysts, **duodenal fluid or duodenal biopsy may be examined to demonstrate trophozoites.** Alternative methods for detection include antigen detection by enzyme immunoassays and parasite detection by immunofluorescence. **Stool antigen test** is highly sensitive and specific for detection of *G. lamblia.*

TRANSMISSION

Humans are the common reservoir, and the asymptomatic carrier rate is high. Wild animals, such as beavers, and domestic animals, such as dogs, are also habitats. The mode of transmission is via **fecally contaminated water. Travelers** to the endemic countries and regions are **at risk. Hand-to-mouth transfer of cysts among infants** and **young children** occurs **in day care centers,** and the infection is readily spread to family members.

PATHOGENESIS

After the hardy cysts are ingested, excystation occurs in the small intestine, promoted by gastric acid. Trophozoites are released (each cyst produces two trophozoites), and they attach to brush border enterocytes. The mechanism of attachment involves (1) a ventral disk with contractile proteins; (2) flagella-mediated hydrodynamic forces; and (3) a receptor-ligand interaction mediated by lectin proteins. Trophozoites multiply by longitudinal binary fission, remaining in the lumen of the proximal small bowel (Fig. 47-2). Encystation occurs as the parasites transit toward the colon.

G. lamblia infection, initiated by antigen uptake into macrophages in Peyer patches, generates both an antibody and a cellular response. IgA can prevent adherence of organisms. *Giardia* produces IgA protease. T lymphocytes may contribute to crypt hyperplasia, which results in altered absorption. Trophozoites are neither invasive nor toxigenic; they **disrupt brush border by microvilli injury, causing villus atrophy** (via proteinase or mannose-binding lectin), resulting in **watery diarrhea.**

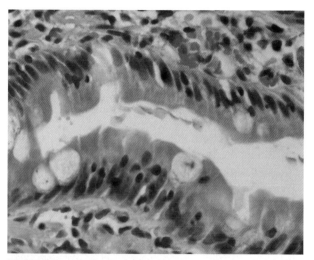

FIGURE 47-2 Histopathology of trophozoites of *Giardia lamblia* (*intestinalis*) in the lumen of the proximal small bowel in a patient with persistent diarrhea. *(Courtesy of Dr. Dominick Cavuoti, Department of Pathology, University of Texas Southwestern Medical Center, Dallas, TX.)*

TREATMENT

The drug of choice for symptomatic patients is metronidazole. It is given for 7 days and is effective in up to 95% of cases. Albendazole and paromomycin (in pregnancy) are alternatives, although they are less effective.

OUTCOME

The patient was treated with metronidazole for 7 days. He improved gradually over several days.

PREVENTION

Purification of water by filtration and other means that remove cysts is the primary method of prevention. However, boiling water is also effective. As with any other fecal-orally transmitted cause of gastroenteritis, regular hand washing is extremely important to prevent secondary spread of disease. Day care outbreaks can be especially problematic because not all children are symptomatic, but they may still transmit disease. If routine control measures are not effective, all children may have to be treated to stop an outbreak.

FURTHER READING

Adam RD: Biology of *Giardia lamblia*. Clin Microbiol Rev 14:447, 2001.

Faubert G: Immune response to *Giardia duodenalis*. Clin Microbiol Rev 13:35, 2000.

Gardner TB, Hill DR: Treatment of giardiasis. Clin Microbiol Rev 14:114, 2001.

Gunby P: Yet another way to contract giardiasis? JAMA 247:2078, 1982.

Wiesenthal AM, Nickels MK, Hashimoto KG, et al: Intestinal parasites in Southeast-Asian refugees. Prevalence in a community of Laotians. JAMA 244:2543, 1980.

A **35-year-old man with AIDS** presented to his primary care physician with a complaint of **several weeks of watery diarrhea.** Symptoms of low-grade fever, nausea, and anorexia were also present. In the past 2 weeks, the diarrhea had worsened considerably, and he had developed severe fatigue and weakness. He had also **lost 20 lbs.**

He had been **off his highly active antiretroviral therapy** (HAART) for several months due to intolerance.

PHYSICAL EXAMINATION

VS: T 37.4°C, P 112/min, R 18/min, BP 92/46 mmHg

PE: Cachectic, ill-appearing male in no acute distress. Abdominal exam revealed mild, diffuse tenderness.

LABORATORY STUDIES

Blood

Hematocrit: 28%

WBC: 2300/μL

Differential: 78% PMNs, 10% lymphs

CD4$^+$ cell count: 28/μL

Serum chemistries: BUN 38 mg/dL, creatinine 1.4 mg/dL, alkaline phosphatase 240 U/L

TABLE 48–1   Differential Diagnosis and Rationale for Inclusion (consideration)

Chronic enteritis
  C. difficile
  Mycobacterium avium-intracellulare
  Shigella sp.
Chronic protozoal infections
  Cryptosporidium parvum
  Cyclospora cayetanensis
  Giardia lamblia
  Isospora belli
  Microsporidium spp
Cytomegalovirus infection

Rationale: Diarrhea is the most common gastrointestinal manifestation of AIDS, occurring in 50-90% of patients. All of the enteric pathogens that cause diarrhea in the immunocompetent host may also cause diarrhea in AIDS patients. In patients with AIDS, however, the above organisms tend to produce a more virulent and protracted clinical course, and some are almost exclusively seen in AIDS patients. It is not usually possible to distinguish the likely cause of chronic watery diarrhea in AIDS patients based on clinical grounds alone. A thorough diagnostic evaluation is necessary owing to the broad range of organisms causing infection.

Imaging

No imaging studies were done.

Diagnostic Work-Up

Table 48-1 lists the likely causes of illness (differential diagnosis). Fecal WBCs can initially be examined to rule out inflammatory enteritis. Investigational approach for evaluation of diarrhea in a patient with AIDS may include

- **Enteric bacterial cultures**
- **Stool smear**

  - Ova and parasites examination
  - Acid-fast stain (*Cryptosporidium* and *Mycobacterium avium-intracellulare*)

- ***Clostridium difficile* toxin** to rule out antibiotic-associated colitis
- In failed investigation, small bowel mucosal biopsy for electron microscopic examination (microsporidia)

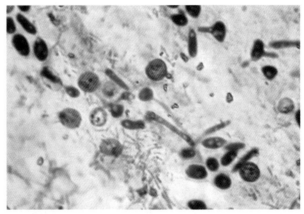

FIGURE 48-1   Oocysts in a fecal smear stained by the modified acid-fast method. *(Courtesy of Centers for Disease Control and Prevention, Atlanta, GA.)*

COURSE

The patient was admitted to the hospital and hydrated with intravenous fluids. Modified acid-fast staining of stool and examination by microscopy revealed small oocysts, which were diagnostic.

ETIOLOGY

Cryptosporidium parvum (cryptosporidiosis)

MICROBIOLOGIC PROPERTIES

C. parvum is an **intracellular coccidian** protozoal agent that exists as a small, 5-μm diameter oocyst containing four sporozoites. **Oocysts are acid-fast** (Fig. 48-1); they stain red with modified carbol fuchsin and mild acid wash. Auramine-rhodamine screening of stool sediment smears followed by **modified acid-fast (Ziehl-Neelsen) staining** is a sensitive and specific approach to identifying *Cryptosporidium* oocysts in stool (Fig. 48-1).

TRANSMISSION

This disease occurs worldwide. Humans, as well as cattle and other domestic animals, are habitats of *C. parvum*. **Zoonotic transmission** of *C. parvum* occurs through exposure to infected animals or **exposure to water contaminated by feces of infected animals**. People with **AIDS**, and renal or bone marrow transplant patients, are at risk for the disease. A unique feature of the life cycle of *C. parvum* is that the merozoites can undergo sexual reproduction (via production of gamonts) within the same human host to regenerate oocysts, which are infective on excretion, thus permitting direct and immediate fecal-oral transmission among men having sex with men. **Waterborne transmission of oocysts** accounts for infections in travelers and for common-source epidemics associated with drinking surface water and summer activities in recreational waters.

PATHOGENESIS

Humans are infected by ingesting the oocysts, which travel through the gut lumen to the small intestine, where they rupture, releasing the sporozoites. The sporozoites are motile forms, which adhere to the absorptive epithelial cells, lining the gastrointestinal tract. **The jejunum is the site most heavily infected**. The protozoa carry out their entire life cycle among the microvilli of the small intestine. The organisms focally disrupt the microvilli and slide in between the host cells, enveloping themselves in the host cell membranes in a process described as "intracellular but extracytoplasmic" (Fig. 48-2). The **parasites enter the intracellular asexual cycle. Diarrhea develops when intestinal absorption is impaired**, and secretion is also enhanced by the stimulation of prostaglandin production by intestinal epithelial cells. In immunocompetent individuals, the life cycle takes place only once or twice, resulting in a single episode of **mild diarrhea** that usually lasts 2 weeks or less. In patients with AIDS, the life cycle of the organisms is repeated many times, and it is associated with bouts of **protracted and severe watery diarrhea**.

TREATMENT

In immunocompromised hosts, particularly AIDS patients with CD4$^+$ cell counts below 200/mm^3, cryptosporidiosis can be life-threatening and must be treated aggressively in addition to **intravenous rehydration**. The **ideal treatment is restoration of immune function with HAART**. However, this can take many weeks to months. There is no reliable ther-

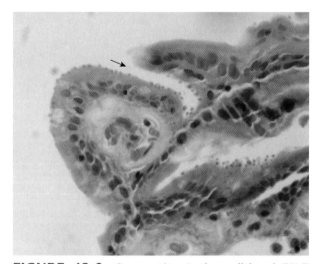

FIGURE 48-2 *Cryptosporidium* in the small bowel (H&E stain). Note the *Cryptosporidium* sporozoites (*arrow*) lining the brush border of a small bowel villus (sliding down the surface of the epithelium). They are described as "intracellular but extracytoplasmic." (*Courtesy of Dr. Dominick Cavuoti, Department of Pathology, University of Texas Southwestern Medical Center, Dallas, TX.*)

apy for cryptosporidiosis, although several antibiotics may have partial efficacy, including paromomycin, nitazoxanide, and azithromycin. Antidiarrheal agents may also provide additional temporary relief. No treatment other than rehydration is necessary in immunocompetent adults and children because the disease is self-limiting.

OUTCOME

The patient was given intravenous fluids, loperamide, and paromomycin with moderate improvement in symptoms. He was able to tolerate oral intake enough to return home after a week in the hospital. HAART was reimplemented in an effort to improve his immune function as well.

PREVENTION

The best approach to prevention of cryptosporidiosis in HIV-infected individuals is the maintenance of immune system function by using HAART because chronic cryptosporidiosis occurs only in severely immunocompromised individuals. Prophylactic clarithromycin, aimed at preventing mycobacterial infections in severely immunocompromised individuals, may also have a protective effect against cryptosporidiosis. Avoidance of tap water has been considered a good preventive approach in the AIDS community. Routine testing of drinking water using highly sensitive test methods is mandatory for all surface water utilities. Fine filtration methods (< 1 μm) are necessary to effectively remove *Cryptosporidium* oocysts.

FURTHER READING

CDC: Protracted outbreaks of cryptosporidiosis associated with swimming pool use—Ohio and Nebraska, 2000. JAMA 285:2967, 2001.

Clark DP: New insights into human cryptosporidiosis. Clin Microbiol Rev 12:554, 1999.

Hunter PH, Nichols G: Epidemiology and clinical features of *Cryptosporidium* infection in immunocompromised patients. Clin Microbiol Rev 15:145, 2002.

Marshall MM, Naumovitz D, Ortega Y, Sterling CR: Waterborne protozoan pathogens. Clin Microbiol Rev 10:67, 1997.

Xiao L, Fayer R, Ryan U, Upton SJ: *Cryptosporidium* taxonomy: Recent advances and implications for public health. Clin Microbiol Rev 17:72, 2004.

A **3-year-old girl** was brought to the emergency department of a general hospital following a 3-week history of **nausea,** poor appetite, and **abdominal pain.** She had not had any **bowel movements** for the last 2 days.

The patient was of Mexican origin and had recently moved from Mexico with her mother to South Texas.

PHYSICAL EXAMINATION

VS: T 37°C, P 110/min, R 20/min, BP 102/54 mmHg

PE: Young child in moderate distress due to abdominal pain. **Abdomen** was **distended** and mildly tender.

LABORATORY STUDIES

Blood

Hematocrit: 38%

WBC: 4500/μL

Differential: 62% PMNs, 23% lymphs, 12% eosinophils (**eosinophilia**)

Serum chemistries: Normal

Imaging

X-rays of her abdomen were consistent with **intestinal obstruction**.

Diagnostic Work-Up

Table 49-1 lists the likely causes of illness (differential diagnosis). Intestinal worm infection was considered based on clinical features and x-ray evidence. Diagnosis is confirmed by identification of ova and parasites by microscopy of trichrome- or iodine-stained concentrated fecal specimens.

TABLE 49–1   Differential Diagnosis and Rationale for Inclusion (consideration)
Appendicitis
Intestinal helminth infection
Ascaris lumbricoides
Schistosoma spp
Taenia spp
Trichuris trichiura
Small bowel obstruction from volvulus

Rationale: Abdominal symptoms with eosinophilia have a relatively limited differential, mainly parasitic infection. The various causes can be reliably determined only through stool examination for ova and parasites. Noninfectious causes may also cause similar symptoms but will not demonstrate eosinophilia.

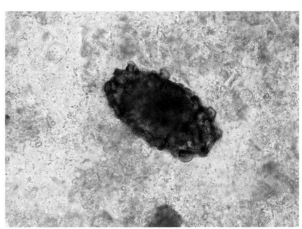

A   B

FIGURE 49-1   *A,* Adult worms. *B,* An unfertilized egg in feces. (*A, Courtesy of Dr. Paul Southern, Departments of Pathology and Medicine, University of Texas Southwestern Medical Center, Dallas, TX; B, Courtesy of Dr. Dominick Cavuoti, Department of Pathology, University of Texas Southwestern Medical Center, Dallas, TX.*)

COURSE

The patient was admitted to the hospital and put on intravenous hydration. Stool examination revealed characteristic nematode eggs.

ETIOLOGY

Ascaris lumbricoides (ascariasis)

MICROBIOLOGIC PROPERTIES

Ascaris lumbricoides is the **largest intestinal nematode parasite (roundworm) of humans** (20 to 35 cm in length; Fig. 49-1A). The **unfertilized eggs** are 45 to 75 μm in length and have a **characteristic "bumpy" coat** (see Fig. 49-1B). The eggs are **not infectious until a larva develops** inside (or eggs mature). The maturation of eggs requires a soil phase for approximately 30 days.

Microscopic identification of eggs in the stool is the most common method for diagnosing intestinal ascariasis. A direct wet mount examination of the specimen is adequate for detecting moderate-to-heavy infections. Formalin-fixed and permanently stained concentrated organisms are examined for morphology and identification.

TRANSMISSION

Ascariasis occurs in greatest frequency in moist regions of tropical countries where sanitary facilities are poor and the eggs have had adequate time to mature in the moist environmental conditions and become infective. In the United States, the clinical disease is most commonly detected in recent immigrants from developing countries. The prevalence and intensity of infections are usually highest in children between 3 and 8 years of age. Children can acquire **infection by ingestion of infective (larval) eggs** from soil contaminated by human feces. Uncooked produce contaminated with human feces may contain infective eggs (after maturation in soil).

PATHOGENESIS

After infective eggs are swallowed, the larvae hatch, invade the intestinal mucosa, and are carried via the portal, then systemic circulations to the lungs. The larvae mature further in the lungs (10 to 14 days) then penetrate the alveolar walls. During the **lung phase of larval migration**, pulmonary symptoms can occur, with cough, dyspnea, hemoptysis, and eosinophilic pneumonitis (on the basis of chest x-ray; known as **Loeffler syndrome**). Helminth infections induce T_H2 responses that are characterized by **eosinophilia**. T_H2 cells also induce antiparasite antibodies (of the IgE isotype) that express multiple effector functions in the immunity to helminths. Eosinophilia and elevated serum IgE levels are features of a variety of infections due to helminths.

The larvae ascend the bronchial tree to the throat and are swallowed. When they reach the small intestine, they develop into adult worms (see Fig. 49-1A). Adult worms live in the lumen of the small intestine. A female may produce eggs, which are passed with the feces. In young children, a large bolus of entangled worms in the small bowel can cause abdominal pain and **intestinal obstruction** (seen in the case patient). Most patients remain asymptomatic, however.

TREATMENT

Mebendazole and albendazole are the drugs of choice. Other effective agents include pyrantel and piperazine, although the latter can have significant toxicity.

> **NOTE** Aggressive treatment is very important in infected children with symptoms. If untreated, overwhelming infections, intestinal obstruction, and consequent inflammation of intestinal mucosa due to fecalith irritation may lead to bowel perforation, peritonitis, and sepsis (a rare but serious complication).

OUTCOME

The patient was given mebendazole and placed on intravenous fluids. Two days after the treatment, she passed many large roundworms. She responded to the treatment, and her symptoms disappeared in 5 days, with restoration of normal bowel movements. There was no recurrence of her illness.

PREVENTION

Infection with these worms is very common in developing countries. Adequate hygiene is important in reducing transmission of this parasite, but this is difficult due to its high prevalence in these areas.

FURTHER READING

Fuessl H: *Ascaris lumbricoides*. N Engl J Med 331:303, 1994.

Khuroo MS: Ascariasis. Gastroenterol Clin North Am 25:553, 1996.

Wiesenthal AM, Nickels MK, Hashimoto KG, et al: Intestinal parasites in Southeast-Asian refugees. Prevalence in a community of Laotians. JAMA 244:2543, 1980.

A 42-year-old man presented to a clinic with complaints of **3 weeks of worsening diarrhea, abdominal pain,** and **fevers.** He had noticed an **itchy rash** over his buttocks and groin area for the past 2 weeks. A **15-lb weight loss** was also noted.

A month before his symptoms started, he had returned from a 3-month **trip to El Salvador,** where he worked in a rural area.

PHYSICAL EXAMINATION

VS: T 38.8°C, P 84/min, R 16/min, BP 112/62 mmHg

PE: An **erythematous maculopapular rash** was present on his groin and buttock area.

LABORATORY STUDIES

Blood

Hematocrit: 43%

WBC: 10,200/μL

Differential: 52%PMNs, 15% lymphs, **24% eosinophils**

Serum chemistries: Normal

Imaging

No imaging studies were done.

Diagnostic Work-Up

Table 50-1 lists the likely causes of illness (differential diagnosis). Stool should be examined for leukocytes. Investigational approach may include

- **Enteric bacterial cultures**
- **Microscopic (ova and parasite) examination**
- **Toxin testing** for *Clostridium difficile*
- In failed investigation, **colonic biopsy** and histopathology to differentiate idiopathic ulcerative colitis and Crohn disease

COURSE

The patient was admitted to the hospital. A stool sample was sent to the state public health laboratory, where ova and parasite examination revealed nematode larvae.

TABLE 50–1 Differential Diagnosis and Rationale for Inclusion (consideration)
Amebic dysentery
Bacillary dysentery
Clostridium difficile colitis
Crohn disease
Helminth infection
Hookworms
Strongyloides stercoralis
Ascaris lumbricoides
Trichuris trichiura

Rationale: The unique feature regarding this case is the significant degree of eosinophilia. This limits the differential essentially to parasitic infections. Other etiologies can also be considered if initial work-up is negative, although these would be rare in the presence of eosinophilia. Noninfectious causes may be considered if infection has been ruled out.

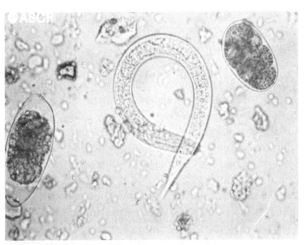

FIGURE 50-1 Rhabditiform larva and eggs from fecal specimen. *(Courtesy of Dr. Paul Southern, Departments of Pathology and Medicine, University of Texas Southwestern Medical Center, Dallas, TX.)*

ETIOLOGY

Strongyloides stercoralis (strongyloidiasis)

MICROBIOLOGIC PROPERTIES

The adult female *S. stercoralis* (2 to 3 mm in length) lives in the mucosa of the small intestine (duodenum and jejunum). The female lays eggs that hatch into uninfective larva (rhabditiform), which are excreted in feces. *Strongyloides* has the unique ability to replicate in the human host and produce infective larvae, which can cause reinfection (without leaving the same infected host and without any maturation). Microscopic (ova and parasite) examination of freshly passed feces from a symptomatic patient usually reveals **rhabditiform larvae** (Fig. 50-1). (Prominent genital primordium (*arrow*) is a diagnostic feature.) Adult worms in soil contaminated with feces may produce noninfective larvae that develop into free-living larvae that eventually develop into **filariform larvae (infective stage)**.

TRANSMISSION

Strongyloidiasis occurs worldwide, but the disease is most prevalent in humid regions of the tropical countries, where sanitation facilities are poor. One can acquire **infection by contact** with contaminated soil; infective **filariform larvae are able to penetrate intact skin**.

PATHOGENESIS

Filariform larvae in contaminated soil penetrate the human skin and are transported to the lungs, where they enter the alveolar spaces. In a heavy infection, pulmonary migration of the filariform larvae can cause pulmonary symptoms, including coughing, wheezing, and pulmonary inflammatory infiltrates, evident on chest x-ray. Larvae are carried through the bronchial tree to the pharynx, are swallowed, and then reach the small intestine. In the small intestine they become adult female worms. *Strongyloides* adult males do not exist. The **adult females inhabit the upper small intestine** (e.g., jejunum) where the **small worms burrow into the mucosa** (Fig. 50-2), causing **gastrointestinal symptoms** such as abdominal (midepigastric) pain (similar to peptic ulcer disease) and diarrhea

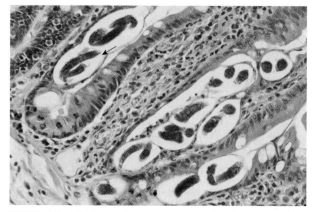

FIGURE 50-2   Section of jejunal mucosa in strongyloidiasis. Note the small (2 mm long) adult worms (*arrow*) of *Strongyloides stercoralis* in crypt in jejunal mucosa. (*Courtesy of Dr. Paul Southern, Departments of Pathology and Medicine, University of Texas Southwestern Medical Center, Dallas, TX.*)

(seen in the case patient). Heavy infections may lead to ulceration and sloughing of intestinal mucosa, with bloody diarrhea. **Blood eosinophilia** is generally present during the acute and chronic stages. Serum IgE is usually elevated (not tested in the case patient).

The female worms reproduce in the duodenojejunal mucosa by parthenogenesis and there deposit ova. The rhabditiform larvae (see Fig. 50-1) hatch in the mucosa and bore through the epithelium into the lumen, where they are normally passed in feces. A small number of rhabditiform larvae can mature into filariform larvae in the bowel. They cause reinfection by entering the body through the skin of the perianal area. Larval migration under the skin in the buttocks and groin areas causes **urticarial, raised, erythematous rashes,** the most common **dermatologic manifestation** of parasitic infections.

> **NOTE** In some patients with AIDS, infection can persist for the life of the host, as *Strongyloides* is able to complete its life cycle entirely within the human body, unlike most other parasites. Immunosuppression may result in hyperinfection with **dissemination of migrating larvae to other organs**, such as the liver, heart, kidneys, or CNS. This can be associated with translocation of Gram-negative bacteria from the GI tract, causing bacteremia (septic shock) and even meningitis.

TREATMENT

The drug of choice for the treatment of all forms of strongyloidiasis is **ivermectin**, with thiabendazole as an alternative. Albendazole has also been used. All patients who are infected should be treated, especially those at risk of disseminated strongyloidiasis.

OUTCOME

The patient was treated with ivermectin and gradually recovered.

PREVENTION

There is no vaccine available currently. Avoiding walking barefoot in highly endemic areas may reduce exposure, as this is a potential route of infection.

FURTHER READING

Genta RM: Global prevalence of strongyloidiasis: Critical review with epidemiologic insights into the prevention of disseminated disease. Rev Infect Dis 11:755, 1989.

Genta RM: Dysregulation of strongyloidiasis: A new hypothesis. Clin Microbiol Rev 5:45, 1992.

Keiser PB, Nutman TB: *Strongyloides stercoralis* in the immunocompromised population. Clin Microbiol Rev 17:208, 2004.

A 49-year-old woman presented with **high fever** and **chills, jaundice,** and upper abdominal pain for 3 days.

The patient was a **recent immigrant from Argentina.** One year before, she first noticed a **sensation of fullness in the right upper quadrant of her abdomen.** Her past medical history was unremarkable. In her country of origin, she had been healthy and active, **working in the field and breeding and raising sheepdogs.**

PHYSICAL EXAMINATION

VS: T 39.5°C, P 112/min, R 18/min, BP 102/60 mmHg

PE: The patient appeared acutely ill and was obviously jaundiced. **Right upper quadrant abdominal tenderness** was noted.

LABORATORY STUDIES

Blood

Hematocrit: 34%

WBC: 22,400/μL

Differential: 55% PMNs, 20% bands, 12% lymphs, 8% eosinophils

Serum chemistries: alkaline phosphatase 340 U/L; bilirubin 4.3 μmol/L

Imaging

A CT scan of the liver demonstrated **a large multiloculated cyst with bile duct dilation** (Fig. 51-1).

Diagnostic Work-Up

Table 51-1 lists the likely causes of the woman's illness (differential diagnosis). A clinical diagnosis of hepatic hydatid cyst (tapeworm) disease was considered based on the **imaging studies** (CT scans of liver). **Serologic tests** specific for a tapeworm may include

- **ELISA** to detect IgG in patient's serum (screening)
- **Immunoblot assay** to confirm the diagnosis

COURSE

Given the patient's clinical status, she was taken to the operating room. The liver lesion was completely removed surgically, and the biliary obstruction was relieved. The patient received an antihelminthic drug postsurgery. An ELISA and subsequent immunoblot test for antibodies against a tapeworm confirmed the specific etiology.

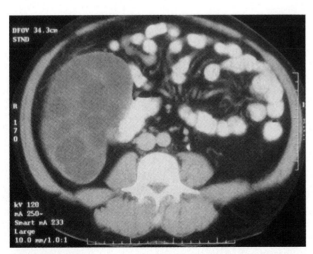

FIGURE 51-1 Large multiloculated cyst in liver of a patient, revealed by CT scan. *(Courtesy of Dr. Paul Southern, Departments of Pathology and Medicine, University of Texas Southwestern Medical Center, Dallas, TX.)*

TABLE 51–1 Differential Diagnosis and Rationale for Inclusion (consideration)
Cholangitis
Hydatid cyst (tapeworm) disease
Liver abscess
Viral hepatitis

Rationale: Clinical diagnosis is difficult; the history of exposure is helpful. The symptoms are suggestive of biliary tract infection, which may have many causes. Viral hepatitis would not produce lesions seen in the CT scan. The more common causes are listed above. Hydatid cyst is a possibility when appropriate epidemiology is present.

ETIOLOGY

Echinococcus granulosus (hydatid cyst disease or echinococcosis)

MICROBIOLOGIC PROPERTIES

The adult tapeworm, *E. granulosus*, is very small (3 to 6 mm long) and has only three proglottids (one immature, one mature, and one gravid). Gravid proglottid carries eggs. "Hydatid cysts" are found in mass lesions of infected patients with echinococcosis; a hydatid cyst shows multiple protoscolices (size approximately 100 μm), each of which is able to begin the life cycle as a "head of tapeworm" with typical hooklets (infective when evaginated; Fig. 51-2) for attachment to intestinal mucosa.

EPIDEMIOLOGY

Human echinococcosis (hydatid cyst disease) is caused by the larval stages of cestodes (tapeworms) of the genus *Echinococcus*, which are carried by canines. The highest prevalence is in sheep-raising areas of South America, Africa, Greece, the Middle East, Central Asia, and areas of Australia and New Zealand. Close contact with sheepdogs and household items soiled with dog feces facilitates transmission. The **worm resides in the small bowel of sheepdogs**. The sheepdogs become infected by ingesting cyst-containing organs of the infected sheep, which are intermediate hosts. After ingestion, the "protoscolices" evaginate (see Fig. 51-2), attach to the intestinal mucosa, and develop into adult stages in 32 to 80 days. Gravid proglottids release eggs that are passed in the dog feces.

PATHOGENESIS

Humans become **infected by ingesting eggs from food contaminated with infected dog feces**. In the intestine, oncospheres are released and invade and develop into cysts in various organs, predominantly **liver** (Fig. 51-3). Other sites are lung, brain, kidney, spleen, bone, and heart. Space-occupying effects in an involved organ yield symptoms. A **mass develops in the hepatic area**; **obstruction of the biliary duct** by the space-occupying mass can result in **jaundice** and **biliary duct dilation**, as in the case patient. Hydatid cyst (protoscolices, hooks of protoscolices) and the space-occupying mass produce **abdominal pain**.

> **NOTE** Leakage from hydatid cyst or rupture of mass may produce severe allergic reactions to echinococcus antigens, including anaphylaxis. Alveolar hydatid disease (AHD) is predominantly caused by a larval stage of *Echinococcus multilocularis,* a small tapeworm (1 to 4 millimeters) found in dogs and cats and passed to humans via contacts with soiled environment. Lung cysts are often asymptomatic and are discovered on routine chest x-ray (older cysts may calcify). Pulmonary hydatid cysts may rupture into the bronchial tree

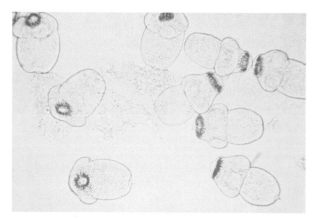

FIGURE 51-2 "Hydatid sand" found in the fluid tapped from the mass of this patient. Fluid aspirated from a hydatid cyst shows multiple protoscolices, each of which has typical hooklets. The protoscolices are normally invaginated, and evaginate when put in saline. *(Courtesy of Dr. Paul Southern, Departments of Pathology and Medicine, University of Texas Southwestern Medical Center, Dallas, TX.)*

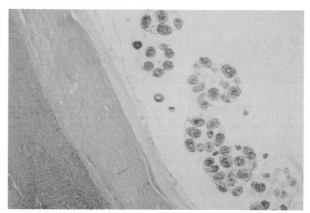

FIGURE 51-3 Hydatid cyst in liver. Note the cyst wall and contents of protoscolices from liver mass. *(Courtesy of Dr. Paul Southern, Departments of Pathology and Medicine, University of Texas Southwestern Medical Center, Dallas, TX.)*

and cause chest pain, cough, and hemoptysis. Rupture of the cysts can also produce fever, urticaria, eosinophilia, and anaphylactic shock, as well as cyst dissemination. If left untreated, infection with AHD can be fatal.

TREATMENT

Surgery is the most common form of treatment for hydatid cyst disease, although removal of the parasite mass is not usually 100% effective. The drug of choice for treatment of echinococcosis is **albendazole**. It is often given before surgery to reduce the risk of complications. After surgery, medication may be necessary to keep the cyst from recurring.

OUTCOME

The patient underwent additional surgeries but eventually recovered completely. She was given albendazole for 4 weeks. She was monitored with serial CT scans to ensure no further recurrences.

PREVENTION

In endemic areas, it is difficult to reduce exposure. Improved sanitation and the screening and treating of affected dog populations have been used with some success. Currently no vaccine for humans exists.

FURTHER READING

Baden LR, Elliott DD: Case 4-2003—A 42-year-old woman with cough, fever, and abnormalities on thoracoabdominal computed tomography. N Engl J Med 348: 447, 2003.

Eckert J, Deplazes P: Biological, epidemiological, and clinical aspects of echinococcosis, a zoonosis of increasing concern. Clin Microbiol Rev 17:107, 2004.

Zhang W, Li J, McManus DP: Concepts in immunology and diagnosis of hydatid disease. Clin Microbiol Rev 16:18, 2003.

A 41-year-old man presented with a 4-month history of worsening abdominal pain, diarrhea, nausea, and vomiting with blood. His abdominal pain was mainly in his right upper quadrant.

The man had **recently immigrated** to the United States **from Kenya.**

PHYSICAL EXAMINATION

VS: T 37.0°C, P 110/min, R 16/min, BP 136/80 mmHg

PE: An ill-appearing male in mild distress due to abdominal pain; **enlargement of the liver and spleen**, with mild tenderness, was noted.

LABORATORY STUDIES

Blood

Hematocrit: 32%

WBC: 9400/μL

Differential: 54% PMNs, 20% lymphs, **18% eosinophils**

Serum chemistries: AST 78 U/L, ALT 92 U/L, bilirubin 3.2 μmol/L, albumin 3.1 g/dL

Imaging

Abdominal ultrasound showed **enlarged liver and spleen with evidence of portal hypertension**.

Diagnostic Work-Up

Table 52-1 lists the likely causes of the man's illness (differential diagnosis). Diarrheal stool specimens may be positive for blood. Routine enteric investigation may be a good start to rule out enteric bacterial causes of diarrhea; however, further investigation for specific microbiologic diagnosis may include

- **Blood culture.** Can aid in ruling out typhoid fever or other blood-borne agents that cause febrile illness
- Ova and parasite examination of stool

COURSE

Stool specimens and blood cultures were obtained. The patient was started on appropriate empirical therapy. Characteristic fluke eggs were identified in the concentrated stool after trichrome staining.

TABLE 52–1 Differential Diagnosis and Rationale for Inclusion (consideration)
Dysentery syndrome
Entamoeba histolytica
Shigella dysenteriae
Infectious gastroenteritis (*Salmonella*)
Inflammatory bowel disease
Schistosomiasis
Typhoid fever
Viral hepatitis

Rationale: In a person from a superendemic area, infectious gastroenteritis must always be considered. Diarrhea is very common and is not very useful by itself for narrowing the differential diagnosis. Other symptoms, such as hepatosplenomegaly, are helpful because most common causes of diarrhea do not lead to this complication. Schistosomiasis is one of the causes that should be considered; it is very common in endemic areas. Typhoid fever is always a consideration in patients with fever and abdominal pain who have recently been in a developing country. Acute viral hepatitis should be considered in patients who have fever and right-upper-quadrant pain, although hematemesis would not be expected.

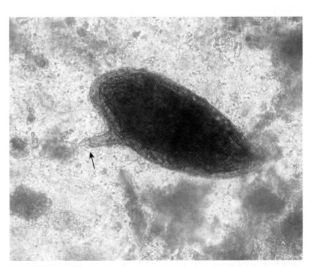

FIGURE 52-1 A parasite egg with a characteristic shape, and a prominent lateral spine *(arrow)* near the posterior end. *(Courtesy of Dr. Dominick Cavuoti, Department of Pathology, University of Texas Southwestern Medical Center, Dallas, TX.)*

ETIOLOGY

Schistosoma mansoni (schistosomiasis; bilharziasis)

MICROBIOLOGIC PROPERTIES

Schistosomiasis is caused by blood trematodes. The three main species infecting humans are *Schistosoma haematobium*, *Schistosoma japonicum*, and *Schistosoma mansoni*. *S. haematobium* lives in the venous plexus near the urinary bladder and ureters; *S mansoni* lives in the inferior mesenteric vein; and *S. japonicum* lives in the superior mesenteric vein of both the large and small intestines. Adult worms are small, 12 to 26 mm long and 0.3 to 0.6 mm wide (male and female remain paired together), varying with the different species. Adult worms mate and lay eggs. Identification of parasite eggs in stool is the most practical method for diagnosing schistosomiasis (should be requested). Size of eggs and the location of the spine (characteristic of schistosomal eggs) are useful in diagnosis of species (Fig. 52-1).

TRANSMISSION

S. mansoni is found in parts of South America and in the Caribbean, Africa, and the Middle East. *S. japonicum* is found in China and Southeast Asia, and *S. haematobium* is found in Africa and the Middle East. Approximately 200 million people are infected with schistosomes. There are 200,000 deaths per year caused by schistosomiasis. Of the parasitic diseases, schistosomiasis is second only to malaria in the number of people that die each year from infection. Human infection is transmitted by penetration of intact skin with infective cercariae, which are released from infected snails (intermediate hosts) in bodies of fresh water in the endemic countries and are free-swimming in those waters.

PATHOGENESIS

The life cycle of *S. mansoni* begins with eggs, shed with feces of a patient in the endemic areas. Under optimal conditions the eggs in bodies of fresh water hatch and release miracidia, which swim and penetrate specific snail intermediate hosts. The stages in the snail include two generations of sporocysts and the production of cercariae. Upon release from the snail, the **infective cercariae** swim, **penetrate the skin of the human host**, and shed their forked tail, becoming schistosomulae. The schistosomulae migrate through several tissues and stages to their residence in the veins. In the venous blood, adult male and female worms mate, and the female lays eggs 4 to 6 weeks after cercarial penetration. Adult worms rarely are pathogenic. The female adult worm lives for approximately 3 to 8 years and lays eggs throughout her life span. Adult worms reside in the lumen of mesenteric blood vessels for decades and resist all known immunologic effector mechanisms. The eggs are moved progressively toward the lumen of the intestine and are eliminated with feces.

Chronic schistosomiasis is due to immunologic reactions to *Schistosoma* eggs trapped in tissues. Multiple immune responses are induced by infection; however, the dominant immunologic effector mechanism involves IgE, mast cells, and eosinophils. As with other helminth infections, the hallmark of *Schistosoma* infection is the appearance of **eosinophilia** and hypergammaglobulinemia of the IgE isotype. Activation of the T_H2 subset of T cells (cellular immunity) in response to egg antigens in the liver is the primary pathogenic element in schistosomiasis. The T_H2 response contributes to a profound **granulomatous reaction** (antigen-specific T cells, macrophages, and eosinophils; Fig. 52-2), resulting in some of the symptoms of chronic disease. Manifestations of granulomatous

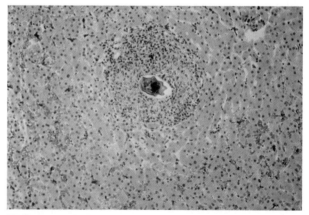

FIGURE 52-2   Histopathology of liver section. Note an egg of *Schistosoma mansoni* in liver with surrounding granulomatous reaction. (*Courtesy of Dr. Paul Southern, Departments of Pathology and Medicine, University of Texas Southwestern Medical Center, Dallas, TX.*)

inflammation in liver include **portal hypertension with hematemesis** and **hepatosplenomegaly** (seen in the case patient). In the latter stages of the disease, fibroblasts, giant cells, and B lymphocytes predominate, and the significant pathologic changes at that time, **collagen deposition** and **fibrosis**, result in liver damage that may be only partially reversible.

TREATMENT

The drug of choice of schistosomiasis is **praziquantel**, used for infections caused by all *Schistosoma* species (including *S. haematobium*; Table 52-2). Oxamniquine has been effective in treating infections caused by *S. mansoni* in areas where praziquantel is less effective.

OUTCOME

The patient received praziquantel and recovered over the course of a few weeks. His liver function improved but remained slightly abnormal.

PREVENTION

In endemic areas, there is no highly effective way to reduce exposure. Travelers should avoid contact with fresh water in endemic areas. Currently no vaccine exists.

TABLE 52–2 Other Important Clinical Features of *Schistosoma* Infections

Syndromes	Clinical Features
Allergic dermatitis	Skin penetration of **cercariae** produces an **allergic dermatitis** due to sensitization from prior exposure.
Katayama fever (KF)	KF (**acute schistosomiasis**) occurs 4-6 weeks after infection, at the time of the initial egg release.
	Manifestations include fever, cough, abdominal pain, diarrhea, hepatosplenomegaly, and eosinophilia. ***Schistosoma japonicum*** (found in Asia) can cause a severe form of KF, which can be fatal.
	A serum sickness-like illness is caused by egg **antigen stimuli** and a resultant immune complex formation.
Chronic schistosomiasis of bladder	Similar to *Schistosoma mansoni* infection of liver, ***Schistosoma haematobium*** eggs trapped in the bladder trigger granuloma formation at the lower end of the ureters, obstructing urinary flow.
	The urinary schistosomiasis manifests as cystitis with hematuria. Diagnosis is on the basis of detection of characteristic eggs in voided urine.
	The advancing infection is associated with scarring and deposition of calcium in the bladder wall, sometimes progressing to bladder cancer.

FURTHER READING

Elliott DE: Schistosomiasis. Pathophysiology, diagnosis, and treatment. Gastroenterol Clin North Am 25:599, 1996.

Lucey DR, Maguire JH: Schistosomiasis. Infect Dis Clin North Am 7:635, 1993.

MMWR: Acute schistosomiasis with transverse myelitis in American students returning from Kenya. MMWR 33:445, 1984.

Mostafa MH, Sheweita SA, O'Connor PJ: Relationship between schistosomiasis and bladder cancer. Clin Microbiol Rev 12:97, 1999.

Ross AGP, Bartley PB, Sleigh AC, et al: Schistosomiasis: Current concepts. N Engl J Med 346:1212, 2002.

Rothenberg ME: Mechanisms of disease: Eosinophilia. N Engl J Med 338:1592, 1998.

A 23-year old man presented to the emergency room of a general hospital with a 5-day history of **fever, jaundice with dark yellow urine,** and **pale colored stools.** He also complained about malaise, fatigue, abdominal pain, intermittent nausea, and vomiting, and he noted loss of appetite, to the point that even the **sight of food made him nauseated.**

He denied a history of intravenous drug use and he had no sexual contact for the previous 2 months. Five weeks ago, he attended a family reunion.

PHYSICAL EXAMINATION

VS: T 38.4°C, P 94/min, R 14/min, BP 124/80 mmHg

PE: Physical examination revealed an **icteric** patient with **hepatomegaly** but **no evidence of spleno-megaly.**

LABORATORY STUDIES

Blood

Hematocrit: 42%
WBC: 8200/μL

TABLE 53–1 Differential Diagnosis and Rationale for Inclusion (consideration)

Alcoholic liver injury
Amebic liver abscess
Bacterial liver abscess
Drug-induced hepatic injury
Gallstone hepatitis
Other viral causes (cytomegalovirus [CMV], Epstein-Barr virus [EBV], yellow fever, Lassa virus)
Viral hepatitis
   Hepatitis A virus (HAV)
   Hepatitis B virus (HBV)
   Hepatitis C virus (HCV)

Rationale: Hepatitis is a relatively common clinical syndrome associated with hepatocyte injury, jaundice, and elevation of liver enzymes (ALT and AST >500 IU/L). The major infectious causes of hepatitis are hepatitis viruses (HAV, HBV, or HCV), where elevation of ALT is more pronounced than is AST. Alcoholic liver injury does not usually lead to ALT or AST levels >500 IU/L, and AST is higher than ALT. The other listed agents may not cause such extreme elevations of ALT and AST, but they can also occur with similar symptoms. Sometimes risk factors can shed light on one etiology over others. Liver abscess can be detected with abdominal imaging and is not usually associated with significant jaundice. Drug-induced hepatic injury is not uncommon and is often associated with very high transaminases. Gallstone hepatitis may cause similar symptoms owing to blockage of bile but may not show marked increase in liver enzymes. Infection with CMV and EBV can be associated with elevated liver enzymes, but no significant jaundice.

Differential: Normal
Serum chemistries: **ALT 1240 U/L, AST 1090 U/L,** bilirubin 14 μmol/L

Imaging

No imaging studies were done.

Diagnostic Work-Up

Table 53-1 lists the likely causes of illness (differential diagnosis). A clinical diagnosis of viral hepatitis was considered. The viral hepatitides resemble each other clinically and can be differentiated only with the aid of specific virology lab tests. Assessment of risk factors is also important.

- **Serology.** Serologic diagnosis is based on **hepatitis panel** and should include testing for hepatitis (A, B, and C) viral antigens and antibodies.
- **RT-PCR.** HAV and HCV RNA can be detected in the blood. HAV RNA can be detected in the stool of most persons during the acute phase of infection by using PCR.

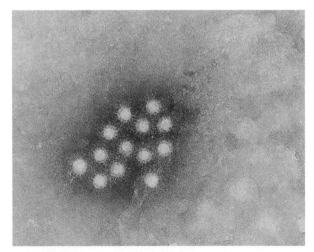

FIGURE 53-1 Electron micrograph of virion particles from fecal specimen. *(Courtesy of Centers for Disease Control and Prevention, Atlanta, GA.)*

COURSE

The patient was given IV fluids for mild dehydration. A viral serology work-up was positive for the IgM subclass of antibody and yielded the diagnosis.

ETIOLOGY

Hepatitis A virus (HAV)

MICROBIOLOGIC PROPERTIES

Hepatitis A is caused by HAV, **a small (27 to 32 nm**; Fig. 53-1) **RNA virus** (genus: *Hepatovirus*; family: Picornaviridae). It is a single-stranded RNA (polyadenylated) genome with (+) polarity (RNA polymerase is carried by the virus). The virus has an **icosahedral capsid without a lipid** envelope; it is resistant to heat (60°C for 20 minutes) and is stable in the environment. Only one serotype has been observed among HAV isolates collected from various parts of the world. HAV is the only hepatitis virus that can be cultivated in cell cultures, and it can be cultivated only with difficulty.

EPIDEMIOLOGY

Hepatitis A occurs worldwide (sporadic and epidemic). In the developing countries, where environmental sanitation is poor, infection is common and occurs at an early age. **Fecal-oral transmission** is the major mode (the virus is shed in feces of infected persons). In the United States, common-source outbreaks have been related to water and food contaminated by infected food handlers. Shellfish have been implicated in HAV outbreaks in Atlantic and Gulf Coasts. Shellfish concentrate the virus during filter feeding and are common sources of HAV infection. **Travelers to high or moderately endemic countries, children in day care centers**, and homosexual individuals are at risk for development of infection. Parenteral transmission (e.g., by way of injection drug use) is rare, because the viremia is short.

PATHOGENESIS

The average incubation period for hepatitis A is 30 days, with a range of 10 to 50 days (mean: 30 days, which is **shorter** than other viral hepatitides). In infected persons, HAV replicates in the liver, causing selective infection of hepatocytes, the parenchymal cells of the liver. The virus can be detected in bile, and it is shed in the stool. Most infections are asymptomatic. In symptomatic cases, HAV **hepatic pathology is indistinguishable (microscopically) from other viral hepatitides**. It is associated with a pattern of injury to and destruction of hepatocytes, a result of the host immunologic response to the virus, as **HAV-specific cytotoxic T cells destroy virus-infected hepatocytes**.

Viremia follows soon after infection and persists through the period of liver enzyme (alanine aminotransferase [ALT]) elevation (Fig. 53-2). Peak infectivity occurs during the 2-week period before the onset of jaundice or elevation of liver enzymes, when the concentration of virus in stool is highest. The concentration of virus in stool declines after jaundice appears. **IgM HAV Ab is generally present 5 to 10 days before the onset of symptoms** and is no longer detectable in the vast majority of patients 6 months later. Following the immune response, the virus is rapidly cleared, and the patient is no longer infectious. IgG HAV Ab, which also appears early in the course of infection, remains detectable for the lifetime of the individual and confers lifelong protection against reinfection. **Chronic infection does not occur** following HAV infection, and **no chronic carrier state** is recognized.

TREATMENT

No specific treatment is available. Management is mostly supportive.

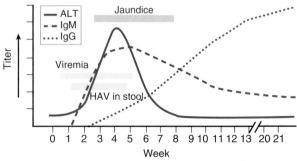

FIGURE 53-2 Course of HAV infection. Note the titers of liver enzymes (ALT) and HAV serologic markers (IgM and IgG) from patient's blood.

OUTCOME

The patient did not receive any specific treatment, and he gradually recovered over the course of the next 3 months.

PREVENTION

Since transmission is fecal-oral, proper hygiene is important for prevention. However, **passive immunization** with immunoglobulin given less than 2 weeks after exposure can prevent 80% to 90% of cases and is also used in cases in which travel to endemic areas will occur in less than 2 weeks from immunization. **A killed vaccine** is also available, with long-lasting immunity (up to 10 to 30 years). It can be given to children as young as 2 years old. Two injections (IM) given 6 to 12 months apart is the recommended schedule. Travelers to endemic areas should also receive the vaccine.

FURTHER READING

CDC: Hepatitis A vaccination programs in communities with high rates of hepatitis A. MMWR 46:4, 1997.

Cuthbert JA: Hepatitis A: Old and new. Clin Microbiol Rev 14:38, 2001.

Hadler SC, Webster HM, Erben JJ, et al: Hepatitis A in day-care centers: A community-wide assessment. N Engl J Med 302:1222, 1980.

Hutin YJF, Pool V, Cramer EH, et al: A multistate, foodborne outbreak of hepatitis A. N Engl J Med 340:595, 1999.

Krugman S, Giles JP: Viral hepatitis: New light on an old disease. JAMA 212:1019, 1970.

Stapleton JT: Host immune response to hepatitis A virus. J Infect Dis 171(Suppl):S9, 1995.

A 27-year-old woman was seen in a clinic for complaints of **fevers,** chills, headache, malaise, anorexia, and abdominal pain for several days. She stated she came today because she noticed that her **eyes had turned yellow,** and she had developed a very bothersome **generalized itching.**

She admitted to using **IV drugs** for the past few years, saying she frequently shared **needles with friends.** She had had one sexual partner in the past year.

PHYSICAL EXAMINATION

VS: T 38.8°C, P 104/min, R 16/min, BP 112/70 mmHg

PE: She was in mild distress due to pruritus. **Scleral icterus** and **jaundice** were present. Her **liver was enlarged** and mildly tender.

LABORATORY STUDIES

Blood

Hematocrit: 31%
WBC: 9400/μL

TABLE 54–1 Differential Diagnosis and Rationale for Inclusion (consideration)
Alcoholic liver injury
Amebic liver abscess
Bacterial liver abscess
Drug-induced hepatic injury
Gallstone hepatitis
Other viral causes (cytomegalovirus [CMV], Epstein-Barr virus [EBV], yellow fever, Lassa virus)
Viral hepatitis
Hepatitis A virus (HAV)
Hepatitis B virus (HBV)
Hepatitis C virus (HCV)

Rationale: Hepatitis is a relatively common clinical syndrome associated with hepatocyte injury, jaundice, and elevation of liver enzymes (ALT and AST >500 IU/L). The major infectious causes of hepatitis are hepatitis viruses (HAV, HBV, or HCV), where elevation of ALT is more pronounced than AST. Alcoholic liver injury does not lead to ALT or AST levels >500 IU/L and AST is higher than ALT. The other listed agents may not cause such elevations of ALT and AST, but they can also occur with similar symptoms. Sometimes risk factors can shed light on one etiology over others. Liver abscess should be detected with abdominal imaging and is not usually associated with significant jaundice. Noninfectious etiologies should be ruled out. Drugs, particularly acetaminophen, can also cause such elevations in transaminases. Gallstone hepatitis may cause similar symptoms due to blockage of bile but may not show marked increase in liver enzymes. Infection with CMV and EBV can be associated with elevated liver enzymes but without significant jaundice.

Differential: 32% PMNs, **52% lymphs**, 5% atypical lymphs

Serum chemistries: **ALT 2730 U/L, AST 2390 U/L,** bilirubin 6.8 μmol/L

Imaging

Ultrasound showed **hepatomegaly.**

Diagnostic Work-Up

Table 54-1 lists the likely causes of illness (differential diagnosis). For suspected viral hepatitis, **serology** is the mainstay of specific diagnosis. The **hepatitis serology panel** should include testing of hepatitis (A, B, and C) viral antigens and antibodies.

The potential for diagnostic confusion requires careful evaluation of clinical and serologic findings and of risk factors.

COURSE

Antibodies to HAV and HCV were absent, but hepatitis B virus surface antigen, HB_sAg, was detected in the patient's serum. The clinical suspicion of HBV infection was confirmed by hepatitis serology based on the significantly high level of the IgM HB_cAb marker.

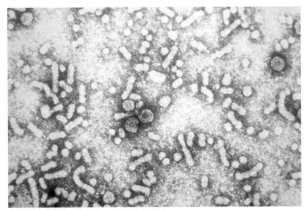

FIGURE 54-1 Virus particles in blood. *(Courtesy of Centers for Disease Control and Prevention, Atlanta, GA.)*

ETIOLOGY

Hepatitis B virus (HBV)

MICROBIOLOGIC PROPERTIES

HBV is **a 42-nm, double-shelled DNA virus** (genus: *Orthohepadnavirus*; family: Hepadnaviridae). The genome is a 3.2-kb circular, double-stranded, incomplete DNA. The viremic patient's blood may present three different particles under electron microscopy (Fig. 54-1). The larger (42-nm) rounded bodies are **full assembly infectious virus** particles with a lipid envelope. The two smaller particles are a spherical (27-nm) nucleocapsid core and tubular or filamentous (22-nm) particles of surface proteins (Hb_sAg). The envelope protein expressed on the outer surface of the full virion and on the smaller spherical and filamentous particles is called **hepatitis B surface antigen** (HB_sAg). The smaller spherical and tubular structures carrying HB_sAg are thought to be produced in blood for decoy functions and evasion of the immune system. HB_sAg is the primary component of the hepatitis B vaccine; this antigen **induces a protective, neutralizing antibody** that provides long-term protection against HBV infection. The inner core of the virus contains hepatitis B core antigen (HB_cAg), hepatitis B$_e$ antigen (HB_eAg), a single molecule of partially double-stranded DNA, and DNA-dependent DNA polymerase (with reverse transcription function). The three antigens, **HB_sAg, HB_cAg, and HB_eAg,** and the selected, **specific antibodies, constitute hepatitis B diagnostic markers**. HBV does not grow in cell cultures, but HBV DNA can transfect surrogate cells, producing viral proteins.

EPIDEMIOLOGY

Transmission in humans occurs via percutaneous or **parenteral** (IV drug abuse, needles) inoculation, by permucosal exposure to fluids (**sexual contact**), or by **perinatal** (maternal–neonatal; vertical) transmission. In the United States, the most important route of HBV transmission is by sexual contact (heterosexual or homosexual) with an infected person. Direct parenteral inoculation of HBV by needles during injecting drug use is also an important mode of transmission, as is transmission of HBV by other percutaneous exposures, including tattooing, ear piercing, and acupuncture, and by needle sticks or other injuries from sharp instruments sustained by medical personnel; however, these exposures account for only a small proportion of reported cases in the United States. In addition, transmission can occur perinatally from a chronically infected mother to her infant, most commonly by contact of an infant's mucous membranes with maternal blood at the time of delivery.

PATHOGENESIS

The incubation period for hepatitis B ranges from 45 to 210 days for jaundice. After entry, **virus becomes blood borne, producing sustained viremia**. The virus replicates in hepatocytes with minimal cytopathic effects. Replication continues for relatively long periods without causing liver damage. **Virus-specific cytotoxic T lymphocytes** are responsible for clinical manifestations, and for the eventual resolution of infection. The onset of acute disease is generally insidious. Most **acute HBV infections** in adults result in complete recovery. The first serologic marker to appear following acute infection is HB_sAg, which can be detected as early as 1 or 2 weeks and as late as 11 or 12 weeks (Fig. 54-2*A*) after exposure to HBV. In persons who recover, HB_sAg is no longer detectable in serum after an average period of about 3 months. Anti-HBs becomes detectable during convalescence after the disappearance of HB_sAg in patients who do not progress to chronic infection. The presence of anti-HBs following acute infection generally indicates recovery and immunity from reinfection. HB_eAg is transiently detectable in patients with acute infection. IgM class antibody to hepatitis B core antigen (**IgM HB_cAb**) in serum is generally detectable at the time of clinical onset and declines to subdetectable levels within 6 months (see Fig. 54-2*A*). IgG HB_cAb persists indefinitely as a marker of past infection. A **diagnosis of acute HBV infection** can be made on the basis of the detection of **IgM HB_cAb** (Table 54-2).

> **NOTE** At the progression of infection, copies of the HBV genome integrate into the hepatocyte chromatin and remain latent. This **"hide and infiltrate" strategy** of the virus makes its systemic presence more prolonged, leading to chronicity. In patients with **chronic HBV infection**, both HB_sAg and IgG HB_cAb remain persistently detectable, generally for life, whereas HB_eAg is variably present. The **presence of HB_sAg for 6 months or more** is generally **indicative of chronic infection** (Fig. 54-2*B*). In addition, **a negative test for IgM HB_cAb**

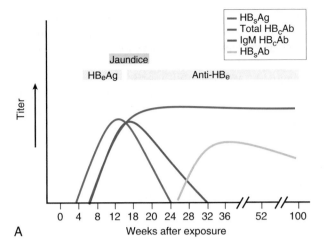

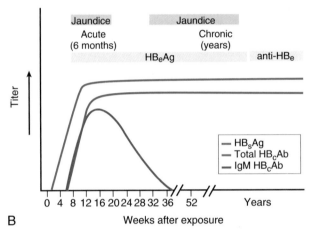

FIGURE 54-2 *A,* Course of HBV infection: Acute HBV infection with recovery. Note the titers of serologic markers (viral antigens, and IgM and IgG from patient's blood. *B,* Progression to chronic HBV infection. Note the titers of HBV serologic markers (surface antigen, and IgM and IgG) from patient's blood.

together with a **positive test for HB$_e$Ag** in a single serum specimen usually **indicates** that an individual has **chronic HBV infection.** Chronicity manifests with two stages: chronic persistent (healthy carrier) and chronic active hepatitis (symptomatic, sick patients). **Chronic persistent hepatitis** manifests with a variable course (but milder), hepatospleno-megaly (may be asymptomatic; see Table 54-2), and abnormal liver function tests; **the presence of HB$_s$Ag and IgG HB$_c$Ab** (HB$_e$Ag is nondetectable) may remain for the lifetime. **Chronic active hepatitis** manifests with jaundice, a variable course, cirrhosis, liver failure, and death; it carries a risk of hepatocellular carcinoma. A symptomatic HB$_s$Ag-carrying patient is highly infectious, and there are **significantly high levels of HB$_e$Ag** in serum (see Fig. 54-2*B*). The presence of HB$_e$Ag in serum correlates with higher titers of HBV and greater infectivity. Risk factors for **hepatocellular carcinoma** are older age at onset of HBV infection, and alcohol-induced liver injury, Down syndrome, chronic lymphocytic leukemia, and immunodeficiency.

Hepatitis D virus (delta agent) is found only in patients also infected with HBV. It can cause fulminant disease in these patients. It is particularly common in Africa, South America, and the Mediterranean.

TREATMENT

There is no specific antiviral treatment for acute hepatitis B infection. Management is mostly supportive. Treatment for chronic viral hepatitis B includes interferon-α, lamivudine, or adefovir. Other antiviral agents are being tested, but many direct antiviral therapies have been disappointing or toxic.

TABLE 54–2	Hepatitis B Serologic Markers*					
	HB$_s$Ag	IgG HB$_s$Ab	IgM HB$_c$Ab	Total HB$_c$Ab	HB$_e$Ag	IgG HB$_e$Ab
Acute infection	+	−	++	+	+	−
Chronic persistent carrier (asymptomatic)	+	−	−	+	−	+
Chronic active hepatitis (symptomatic; infectious)	+	−	−	+	++	−
Past exposure to HBV	−	+	−	+	−	−
Successful immunization	−	++	−	−	−	−

*Detectable levels of serologic markers are noted in the diagnostic grid.

OUTCOME

The patient gradually recovered over the following 2 months and did not require specific antiviral therapy.

PREVENTION

Avoiding risk factors such as multiple sexual partners and IV drug use with needle sharing can reduce the risk for transmission. Needle sticks in the hospital, also an important source of infection, can be reduced by using proper needle disposal techniques. A recombinant vaccine licensed for use in the United States is highly effective and safe. It is recommended for those at risk and even for newborns. The vaccine is also useful for preventing infection in a recently exposed individual. An older, plasma-derived vaccine is no longer in use in the United States. Hepatitis B immune globulin (HBIG) is recommended for post-exposure prophylaxis and is most effective when administered soon after exposure, preferably within 48 hours. Sexual partners of infected patients should receive a single dose of hepatitis B immune globulin and the HBV vaccine. There is no vaccine for hepatitis D.

FURTHER READING

Blumberg BS: Australia antigen and the biology of hepatitis B. Science 197:17, 1977.

CDC: Hepatitis B vaccination of adolescents—California, Louisiana, and Oregon, 1992-1994. MMWR 43:605, 1994.

Dienstag JL, Schiff ER, Wright TL: Lamivudine as initial treatment for chronic hepatitis B in the United States. N Engl J Med 341:1256, 1999.

Lau JY, Wright TL: Molecular virology and pathogenesis of hepatitis B. Lancet 342:1335, 1993.

Marcellin P, Chang TT, Lim SG: Adefovir dipivoxil for the treatment of hepatitis B e antigen-positive chronic hepatitis B. N Engl J Med 348:808, 2003.

A 48-year-old man presented to an internist for a physical exam, since he had not been to a physician for many years. He was asymptomatic, but routine chemistry panel showed **elevated transaminases.**

He denied IV drug use, but on further questioning, he did recall that he **received a blood transfusion 25 years** before, after an appendectomy.

PHYSICAL EXAMINATION

VS: T 37°C, P 72/min, R 12/min, BP 124/76 mmHg
PE: Physical examination was unremarkable.

LABORATORY STUDIES

Blood

Hematocrit: 45%
WBC: 6200/μL
Differential: Normal
Serum chemistries: **AST 124 U/L, ALT 145 U/L**
Urine bilirubin: Normal

Imaging

Liver was slightly enlarged but otherwise normal.

Diagnostic Work-Up

Table 55-1 lists the likely causes of an elevation of liver enzymes (differential diagnosis). For suspected infection, **serology** is the mainstay of specific diagnosis. The **hepatitis panel** should include testing of viral antigen and antibodies.

COURSE

The patient was referred to a gastroenterologist, who performed a liver biopsy and determined that bridging fibrosis was present. The serologic assays for the hepatitis markers yielded evidence of a chronic viral infection.

TABLE 55–1   Differential Diagnosis and Rationale for Inclusion (consideration)

Alcoholic liver injury
Drug-induced hepatic injury
Gallstone hepatitis
Other viral causes (cytomegalovirus [CMV], Epstein-Barr virus [EBV], yellow fever, Lassa virus)
Viral hepatitis
   Hepatitis A virus (HAV)
   Hepatitis B virus (HBV)
   Hepatitis C virus (HCV)

Rationale: Elevations in hepatic transaminases can be due to many causes. Hepatitis is a relatively common clinical syndrome associated with hepatocyte injury, jaundice, and elevation of liver enzymes (ALT and AST >500 IU/L). It may be asymptomatic. The major infectious causes of hepatitis are hepatitis viruses (HAV, HBV, or HCV), where elevation of ALT is more pronounced than AST. Alcoholic liver injury does not lead to ALT or AST levels >500 IU/L and AST is higher than ALT. Sometimes risk factors can shed light on one etiology over others. Many drugs can also cause asymptomatic elevations in transaminases. Gallstone hepatitis may cause similar symptoms due to blockage of bile but may not show marked increase in liver enzymes. Infection with CMV and EBV can be associated with elevated liver enzymes but without significant jaundice.

ETIOLOGY

Hepatitis C virus (HCV)

MICROBIOLOGIC PROPERTIES

HCV belongs to the genus *Hepacivirus* and the family **Flaviviridae**. HCV is an **enveloped virus (50 nm) with an icosahedral capsid, containing a single-stranded, polyadenylated, positive-sense (9.4-kb) RNA** genome. The replication of genome requires viral RNA polymerase. At least six different genotypes and more than 90 subtypes of HCV exist (based on nucleotide sequence differences), with genotype 1 being the most common in the United States. Because variability between different HCV isolates within a genotype or subtype may not vary sufficiently (a few percent difference in nucleotide level) to define a distinct genotype, these intragenotypic differences are referred to as **quasispecies**. Unlike HBV, infection with one HCV genotype or subtype does not protect against reinfection or superinfection with other HCV strains.

EPIDEMIOLOGY

Humans are the only habitat of HCV. Exposures known to be associated with HCV infection in the United States are **parenteral** (percutaneous; e.g., **IV drug use**, needle sticks, tattooing, transfusion or organ transplantation from an infected donor before routine testing of blood products began), **permucosal** (sex with infected partner), and **perinatal** (infected pregnant woman to fetus). Most infections are due to **illegal injection drug use**, accounting for 60% of cases in the United States. Transfusion-related cases are also common in individuals who received blood before routine testing was instituted. Sexual transmission is also possible, although it is not considered a major mode of transmission.

PATHOGENESIS

The incubation period ranges from 14 to 180 days (average: 6 to 7 weeks). The pathogenesis is similar to HBV infection. HCV infection resembles HBV infection in its course and severity. Persons with newly acquired (acute) HCV infection typically are either asymptomatic or have a mild clinical illness (80% of

persons have no signs or symptoms), in which case it is indistinguishable from other forms of acute viral hepatitis. The elevations in serum ALT levels are the most characteristic feature (Fig. 55-1*A*) and precede anti-HCV seroconversion. After **acute HCV infection**, 15% to 25% of persons with normal immune status appear to **resolve** their infection without sequelae, as defined by sustained absence of HCV RNA. In some persons, ALT levels normalize, suggesting full recovery (see Fig. 55-1*A*), but this is frequently (in 75% to 85% of infected persons) followed by ALT elevations that indicate **progression to chronic disease** (see Fig. 55-1*B*). Of persons with chronic HCV infection, 60% to 70% have persistent or

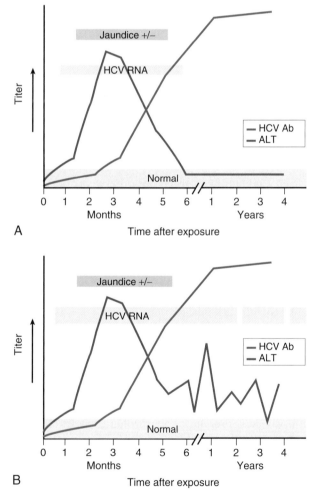

FIGURE 55-1 *A*, Course of HCV infection: acute infection with recovery. Note the titers of liver enzymes (ALT) and serologic marker (IgG) from patient's blood. *B*, Course of HCV infection: acute infection with progression to chronic infection. Note the titers of liver enzymes (ALT) and serologic markers (IgG) from patient's blood.

fluctuating ALT elevations, indicating active liver disease. A **chronic carrier state** is common with HCV infection (> 50%).

Tests for the diagnosis of HCV infection are those that **measure HCVAb** and include enzyme immunoassays (EIAs) and a supplemental recombinant immunoblot assay (RIBA). These tests detect anti-HCV in greater than 97% of infected persons but do not distinguish between acute, chronic, or resolved infection. The diagnosis of HCV infection also can be made by qualitatively **detecting HCV RNA** using gene amplification techniques (e.g., RT-PCR).

TREATMENT

Supportive care is important. Interferon and ribavirin are two drugs licensed for the treatment of chronic HCV infection. Interferon can be taken alone or in combination with ribavirin. Combination therapy, using a pegylated form of interferon and ribavirin, is currently the treatment of choice and appears to provide the best responses.

OUTCOME

The gastroenterologist decided that therapy with interferon and ribavirin was appropriate, and the patient began therapy with this regimen.

PREVENTION

There is no vaccine to prevent hepatitis C. Avoidance of specific risks is the best method of prevention.

FURTHER READING

Alter MJ, Margolis HS, Krawczynski K, et al: Natural history of community-acquired hepatitis C in the United States. N Engl J Med 327:1899, 1992.

Conry-Cantilena C, VanRaden M, Gibble J, et al: Routes of infection, viremia, and liver disease in blood donors found to have hepatitis C virus infection. N Engl J Med 334:1691, 1996.

Cuthbert JA: Hepatitis C: Progress and problems. Clin Microbiol Rev 7:505, 1994.

Henderson DK: Managing occupational risks for hepatitis C transmission in the health care Setting. Clin Microbiol Rev 16: 546, 2003.

Muir AJ, Bornstein JD, Killenberg PG, et al: Peginterferon Alfa-2b and ribavirin for the treatment of chronic hepatitis C in blacks and non-Hispanic whites. N Engl J Med 350:2265, 2004.

Zein NN: Clinical significance of hepatitis C virus genotypes. Clin Microbiol Rev 13:223, 2000.

Nervous System Infections

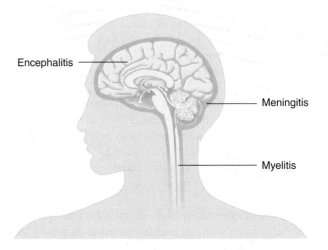

Encephalitis

Meningitis

Myelitis

INTRODUCTION TO NERVOUS SYSTEM INFECTIONS

Infections of the **central nervous system** (CNS) are uncommon because several anatomic barriers exist that effectively prevent organism entry. The most obvious barriers are the scalp and skull, which protect the brain from external elements. Three meninges surround and protect the brain and spinal cord. The outermost is the dura mater, a thick membrane that is adherent to the periosteum of the skull. The next two are the arachnoid, which loosely surrounds the brain, and the pia mater, which is contiguous with the external surface of the brain. Finally, the blood-brain barrier (BBB), which consists of tight junctions in the endothelial cells, prevents organisms from entering the CNS.

CNS infections can be categorized depending on where they occur, and they are usually grouped as meningitis, encephalitis, brain abscess, or myelitis (infection of the spinal cord). The **subarachnoid space** is the site where **meningitis develops**. Epidural abscesses are limited in occurrence because of the tight attachment of the dura to the periosteum. **Infections of the brain parenchyma** may be focal, such as abscesses, or more diffuse, such as **encephalitis**. Whereas some of these syndromes may have a subacute or chronic presentation, most infections of the CNS are relatively acute and demand prompt diagnostic and therapeutic attention.

The **signs and symptoms of meningitis** are variable and are **dependent on the age of the patient**. Anorexia, nausea, and vomiting are common at any age. The clinical presentation of meningitis in infants and the elderly is often nonspecific and may consist primarily of altered mental status, whereas in older children and adults, a **classic triad of fever, headache**, and **nuchal rigidity** is the common manifestation.

The **most common cause of meningitis is viruses**, which are associated with the **aseptic meningitis** syndrome. Enteroviruses account for the majority of the cases, although mumps, herpes, and HIV are also common causes. Typical symptoms are **fever, headache, neck stiffness,** and **photophobia**, which may be accompanied by upper respiratory symptoms. Beyond the neonatal period, viral meningitis is rarely fatal, and no specific therapy currently exists.

Life-threatening meningitis is caused primarily by a small group of bacterial pathogens. These include *Streptococcus pneumoniae* (pneumococcus), *Neisseria meningitidis* (meningococcus), *Listeria mono-*

cytogenes, and *Streptococcus agalactiae* (Group B streptococcus). These **pathogens affect different age groups** with some overlap (Table IV-1). *Haemophilus influenzae* type b (Hib) was the most common cause of bacterial meningitis until the introduction of the highly effective Hib vaccine, which reduced the incidence of infection by more than 90%. Before the introduction of effective vaccines against Hib, more than 65% of cases of bacterial meningitis occurred in children younger than 5 years of age.

As a conceptual model for all human infectious diseases, bacterial meningitis results when the virulence factors of a CNS pathogen overwhelm host defense mechanisms. The "successful" meningeal pathogen must sequentially colonize the host, invade the subarachnoid space, and trigger an inflammatory host reaction. The development of acute meningitis progresses through a series of stages:

1. Colonization of a meningeal pathogen in the nasopharynx, followed by primary multiplication
2. Invasion of the mucosa
3. Entry of the encapsulated bacteria into the bloodstream and evasion of phagocytic killing, causing bacteremia
4. Crossing the choroid plexus and entry of the pathogen into the subarachnoid space and CSF
5. Bacterial multiplication and induction of inflammation
6. Progression of inflammation with associated pathophysiologic alterations via the mediation of proinflammatory cytokines
7. Development of edema and intracranial pressure, leading to clinical signs and symptoms of bacterial meningitis

The bacterial capsule, the primary virulence factor, facilitates survival of organisms in the blood stream, operating against circulating antibodies, complement-mediated bacterial killing, and phagocytosis by PMNs. **Encapsulation is the common evasive mechanism among meningeal pathogens** (see Table IV-1). In nonimmune individuals (who cannot opsonize bacteria) or in individuals with underlying risks (e.g., sickle-cell disease or asplenia), the alternative complement pathway is the main host defense against encapsulated pathogens. Bacterial meningitis is an infection that occurs in a space lacking any host defenses (specific antibody and complement) necessary for efficient phagocytosis.

Bacterial meningitis is classically characterized by **fever, severe headache,** and **neck stiffness**, but an **altered mental status** (which is **uncommon in viral**

TABLE IV–1   Age-Specific Microbial Causes of Acute Meningitis and Microbial Virulence

Affected Group	Bacterial Agents	Major Virulence*
Immunocompetent Patients		
Newborns (0 to <3 months)	Group B streptococcus	Type III capsule
	Escherichia coli	K1 antigen capsule
	Listeria monocytogenes	Invasin
Infants to young adults (3 months to <18 years)	*Streptococcus pneumoniae*	Capsular polysaccharides (23 major types)
	Neisseria meningitidis	Capsular polysaccharides (serogroups A, B, C, Y, W-135)
	Haemophilus influenzae type b (rare)	Type b capsule
Ages 18 to 50 years	*Streptococcus pneumoniae*	See above
	Neisseria meningitidis	See above
Older adults (>50 years)	*Streptococcus pneumoniae*	See above
	Listeria monocytogenes	
	Gram-negative bacilli:	
	Klebsiella pneumoniae	Capsule
	Pseudomonas aeruginosa	Microcapsule
Any age (head trauma; cranial injury or surgery; cerebrospinal fluid shunt)	*Staphylococcus aureus*	Microcapsule
	Streptococcus pneumoniae	See above
	Anaerobic bacteria	Mixed infection
Immunosuppressed patients with impaired cellular immunity	*Listeria monocytogenes*	As above
	Gram-negative bacilli:	
	Klebsiella pneumoniae	As above
	Pseudomonas aeruginosa	As above
AIDS Patients	*Cryptococcus neoformans*	Capsule

*Note the common string of virulence associated with the major pathogens.

meningitis) is **frequently seen** as the bacterial infection progresses. On physical examination, alterations in the level of consciousness are highly variable. Most children are described as irritable or lethargic. A stiff neck is reported in 60% to 80% of children and in more than 90% of adults. Kernig and Brudzinski signs are indicative of meningeal irritation. A bulging fontanelle in an infant may suggest increased intracranial pressure. Petechiae or purpura are noted in more than 50% of adults with meningococcal meningitis. Seizures are noted in as many as 20% to 30% of patients before admission or early in the course of the illness.

The presence of immunosuppression (e.g., AIDS) should arouse the suspicion of meningitis occurring in patients presenting primarily with fever and alterations in consciousness. *Mycobacterium tuberculosis* and fungi, particularly *Cryptococcus neoformans* and *Coccidioides immitis*, can also cause meningitis in these patients (in addition to normal individuals) and are generally associated with a more indolent course. Neck stiffness is often absent.

Bacterial meningitis demands rapid diagnosis and therapeutic intervention. The **most common and useful diagnostic test is a lumbar puncture,** where a small (10 to 20 mL) amount of **cerebrospinal fluid (CSF) is removed and analyzed**. If the diagnosis of bacterial meningitis is a serious consideration, lumbar puncture should be performed promptly. **Lumbar puncture is not done in cases of focal mass lesions**, such as abscesses, due to the risk of brain herniation from increased intracranial pressure. Blood culture findings are also useful, as they are often positive. It is important to remember that apart from culture, analysis of CSF does not lead to a definitive diagnosis, and there is considerable overlap in the CSF picture in viral and bacterial meningitis.

Some generalizations are useful to remember. Opening pressure is almost always elevated in bacterial meningitis. **The WBC count is typically >1000/μL in bacterial meningitis**, less in viral meningitis (Table IV-2). **Protein levels are higher in bacterial meningitis** (usually >100 mg/dL) and **glucose levels are lower** (usually <40 mg/dL) than in

TABLE IV–2   Typical CSF Abnormalities in Meningitis

Parameter	Normal	Viral	Bacterial
CSF opening pressure	<180 mm H_2O	>180 to <250 mm H_2O	>250 mm H_2O
WBCs	<10/μL	200-600/μL	>1000/μL
PMNs	None	<50%	>50%
Protein	<30-40 mg/dL	<100 mg/dL	>100 mg/dL
Glucose	>50 mg/dL	>50 mg/dL (or 2/3 of serum)	<40 mg/dL
Gram stain	Negative	Negative	Positive (80%)

viral meningitis. Conversely, the finding of no WBC in the CSF makes the diagnosis of bacterial meningitis highly unlikely, although this may be seen particularly in neonates. **Gram stain is positive in most cases of bacterial meningitis** and has important implications for therapy, as seen in Table IV-3. Gram stain is helpful in identifying the meningeal pathogen in 60% to 90% of cases of bacterial meningitis. If the CSF is cloudy, smears should be obtained from fresh, uncentrifuged fluid. If the CSF is clear, smears should be obtained from the cyto-centrifuged sediment. Specific antigen tests are available for *Streptococcus agalactiae*, *Streptococcus pneumoniae*, and *Neisseria meningitidis*, although their critical utility has come into question recently.

Radiologic imaging, such as head CT or MRI, is usually not helpful in making the diagnosis of acute meningitis, although head CT is often used to exclude mass lesions prior to lumbar puncture. *Mycobacterium tuberculosis* and certain fungi can produce meningeal enhancement that can be seen on MRI studies. Analysis of CSF in fungal meningitis is often similar to analysis of CSF in viral meningitis. *Cryptococcus* can usually be diagnosed by India ink stain, although a highly sensitive antigen test (cryptococcal latex agglutination [crypto LA]) is available. Other fungi often do not grow from CSF cultures; serology is therefore needed to make the diagnosis.

Meningitis can progress rapidly, within minutes to hours. **Aggressive management is critical to improving outcomes** for patients with bacterial meningitis. For this reason, patients who present with an acute meningitis syndrome should be started on empirical antibiotics as soon as possible. If lumbar puncture is delayed, blood culture is obtained, then antibiotics begin, as delays in therapy are associated with worsened outcomes.

A basic tenet of the treatment is rapid killing of the invading meningeal pathogen by a potent antimicrobial drug in the CSF. Maximal bactericidal activity is accomplished when the achievable concentration of antibiotic reaches 10-fold or greater than the minimal bactericidal concentration for the meningeal pathogen. The characteristics of the drug and the integrity of the BBB primarily influence the penetration of an antibiotic into CSF. During meningitis, the integrity of the BBB is altered, resulting in increased permeability and enhanced CSF penetration of selected antibiotics.

Therapy of bacterial meningitis has also evolved, based on changing resistance patterns of common pathogens (see Table IV-3). The most important consideration is for *S. pneumoniae*, which has developed high-level resistance to penicillin and, in recent years, to third-generation cephalosporins as well. All isolates remain susceptible to vancomycin, however, and this is often given in addition to cephalosporins until

TABLE IV–3   Choice of Antibiotics Based on the Gram Stain of CSF

Organism	Gram Stain (Presumptive Species)	Antibiotics
Cocci	Gram positive (*Streptococcus pneumoniae*)	Vancomycin plus broad-spectrum cephalosporin (e.g., cefotaxime)
	Gram negative (*Neisseria meningitidis*)	Broad-spectrum cephalosporin (e.g., cefotaxime)
Bacilli	Gram positive (*Listeria monocytogenes*)	Ampicillin or penicillin G plus aminoglycoside (gentamicin)
	Gram negative (*Escherichia coli* or *Pseudomonas aeruginosa*)	Broad-spectrum anti-pseudomonal cephalosporin (e.g., ceftazidime) plus aminoglycoside (gentamicin)

susceptibilities are known. The administration of high-dose steroids (dexamethasone) to adults improves survival, particularly in cases of pneumococcal meningitis. They must be given just before or at the time of antibiotics for maximum benefit.

Encephalitis, in contrast to meningitis, is **characterized by fever, headache**, and an early **alteration of mental status,** which may be seen as erratic behavior and confusion, progressing to lethargy and coma in advanced cases. This is due to diffuse involvement of the brain parenchyma. Viruses are frequent causes of encephalitis, although many other organisms may produce similar syndromes. Enteroviruses are the most common cause of viral encephalitis and occur primarily in late summer to fall. Mosquito-borne viral encephalitis, such as St. Louis, eastern and western equine encephalitis, and LaCrosse and West Nile encephalitis are also seen in the summer months.

Herpes virus encephalitis, the most frequent cause of sporadic viral encephalitis, is seen year round, as it is generally due to reactivation of latent virus. It is one of the few treatable viral causes of encephalitis and so it is important to diagnose. Rabies virus causes a uniformly fatal encephalitis that is often characterized by hydrophobia, which is practically diagnostic of this disease. Animal bites are commonly associated with rabies, but there are cases without any bites, particularly with bat exposure.

The diagnosis of encephalitis rests on clinical history and a combination of CSF analysis with serology or culture. An elevated WBC count is almost invariably seen, although not usually as high as in bacterial meningitis. CSF protein is usually elevated, although glucose is usually normal. Serology is useful for viral and some nonviral causes, and for herpes simplex virus (HSV), PCR has become a useful method for diagnosis. Electroencephalogram is often abnormal, but it is nonspecific. Head CT is not usually helpful, but MRI can reveal the more subtle changes in brain parenchyma that are seen early in the disease course. HSV encephalitis, specifically, is associated with temporal lobe disease, particularly hemorrhagic lesions. Other uncommon pathogens (e.g., rabies) should be sought based on appropriate risk factors and other clinical findings.

Therapy for encephalitis is not available for the vast majority of viral causes, except for HSV, varicella-zoster virus, and possibly cytomegalovirus. Most nonviral causes of encephalitis are treatable, and include rickettsial infections, brucellosis, syphilis, toxoplasmosis, Whipple disease, and amoeba, among others.

Brain abscess, in contrast to encephalitis, is a focal infection of the brain's parenchyma. Its primary symptoms are fever, headache, and neurologic deficits, arising from mass effect of the abscess. Seizures are also seen. Bacteria, particularly streptococci, are by far the most common etioloy and are asscociated with a variety of risk factors, including poor dentition, otitis media, trauma, and immunosuppression. Fungi can also cause brain abscess, especially in immunocompromised patients. The diagnosis of brain abscess has been revolutionized by the advent of computed tomography of the head, which allows rapid, precise localization of lesions. Determining a precise etiology by brain biopsy and culture is critical in instituting effective therapy, which usually must be continued for several weeks. Head CT is also invaluable in monitoring the lesions during and after therapy.

Infections of the **peripheral nervous system** include tetanus and botulism, caused by extremely potent toxins produced by *Clostridium tetani* and *Clostridium botulinum*, respectively. Tetanus is characterized by painful spasms of muscles, which may progress to generalized rigidity. Tetanus is usually associated with traumatic wounds contaminated with soil, and it may involve only the limb affected by the wound. Therapy is generally supportive, and tetanus is easily prevented by a highly effective vaccine. Botulism is characterized by flaccid paralysis and, ultimately, respiratory failure. It is usually acquired as a preformed toxin from improperly stored food, although wounds may also be a source. Infants in particular can acquire the disease by ingesting the organism, often found in honey, and are unable to eradicate the spores before toxin is produced (a detailed discussion of infant botulism can be found in Section III).

In summary, CNS infections are uncommon, and most do not have specific treatments. The most important issue is recognition of treatable causes, especially bacterial meningitis and HSV encephalitis. Empirical therapy is the standard of care because prompt treatment leads to improved outcomes for these rapidly progressive and often fatal infections.

FURTHER READING

Attia J, Hatala R, Cook DJ, Wong JG: Does this adult patient have acute meningitis? JAMA 282:175, 1999.

Gold R: Epidemiology of bacterial meningitis. Infect Dis Clin North Am 13:515, 1999.

Quagliarello VJ, Scheld WM: Treatment of bacterial meningitis. N Engl J Med 336:708, 1997.

van der Flier M, Geelen SPM, Kimpen JLL, et al: Reprogramming the host response in bacterial meningitis: How best to improve outcome? Clin Microbiol Rev 16:415, 2003.

Questions on the case (problem) topics discussed in this section can be found in Appendix F. Practice question numbers are listed in the following table for students' convenience.

Self-Assessment Subject Areas (book section)	Question Numbers
Central and peripheral nervous system infections (Section IV)	14-17, 31-34, 38-43, 68-70, 114-115, 145, 199-201, 204, 209, 211-216, 220-224, 228, 250

A **20-year-old** white male college student was brought to an emergency department in early **January** with a **12-hour history** of high **fever, chills, and severe headache.** Soon after arriving at the ED, he vomited twice. He looked confused and was highly agitated. He was admitted to the hospital, and within two hours he developed **purpuric skin lesions.** He had received all appropriate immunizations and was otherwise healthy.

PHYSICAL EXAMINATION

VS: T 39.9°C, P 124/min, R 38/min, **BP 71/54** mmHg

PE: The patient was unable to answer questions or follow commands. **Neck stiffness** was present, and a **purpuric rash** (Fig. 56-1), mainly on his extremities, was noted.

LABORATORY STUDIES

Blood

Hematocrit: 40%

WBC: 2,400/μL

Differential: 55% PMNs, 20% bands

Platelets: 24,000/μL

Blood gases: pH 7.28, pCO_2 34 mmHg, pO_2 84 mmHg

Serum chemistries: BUN 24 mg/dL, creatinine 1.6 mg/dL, fibrinogen 80, D-dimers greater than 8

Imaging

A head CT scan was normal.

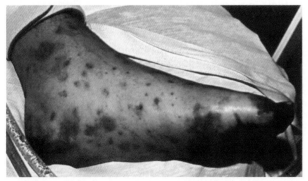

FIGURE 56–1 Severe purpuric rash on the legs of this patient.

Diagnostic Work-Up

Table 56-1 lists the likely causes of illness (differential diagnosis). Lumbar puncture is performed if the patient does not have papilledema or lateralizing neurologic findings. Laboratory examination of the CSF usually confirms the presence of meningitis (viral or bacterial). Investigation approach should include

- Gram stain and culture of CSF
- Blood cultures
- Direct antigen detection in problem scenarios

COURSE

Blood cultures and CSF were collected, and a high dose of IV cefotaxime and vancomycin were administered on an empirical basis. Analysis of the patient's CSF demonstrated 1200 WBCs/μL (91% PMNs), protein concentration of 280 mg/dL, and glucose concentration of 29 mg/dL. Gram stain of CSF showed many PMNs and Gram-negative diplococci. Blood and CSF cultures grew a significant isolate with the same Gram characteristics.

TABLE 56–1 Differential Diagnosis and Rationale for Inclusion (consideration)
Bacterial meningitis
Haemophilus influenzae
Listeria monocytogenes
Neisseria meningitidis
Streptococcus pneumoniae
Rocky Mountain spotted fever
Sepsis

Rationale: The presentation of acute meningitis is one of the most dramatic in medicine. There are relatively few organisms commonly associated with this syndrome. The age of the patient is important in determining the most likely organisms. In this case, *S. pneumoniae* and *N. meningitidis* are the most common causes, and the purpuric rash is highly characteristic of one of these pathogens. Rocky Mountain spotted fever can also cause a very similar clinical picture, especially the rash. The other causes listed do not generally present with a characteristic rash.

ETIOLOGY

Neisseria meningitidis (meningococcal meningitis)

MICROBIOLOGIC PROPERTIES

N. meningitidis is **Gram negative** and is found as cocci, occurring in pairs (**diplococci**) or in tetrads. Diplococci may show flattening of their adjacent sides, giving a "coffee bean" shape (Fig. 56-2). A thin capsule, forming a halo, may be apparent. The surface **capsule** and outer membrane-bound **lipo-oligosaccharide (LOS)-associated endotoxin** are **virulence factors**. The common capsular poly-saccharide-specific serogroups of the encapsulated pathogen are A, B, C, Y, and W-135 (encompassing 95% of strains isolated worldwide).

Blood and CSF cultures require lysed blood agar (chocolate) or sheep blood agar media and CO_2 for growth. *Neisseria meningitidis* isolates are **oxidase positive**. Acid is produced from **fermentation of glucose** and **maltose**, but not from lactose, which aids in identification of *N. meningitidis* isolates. Clinical isolates may be agglutinated by specific standard antisera for serogrouping and epidemiologic investigation of outbreaks.

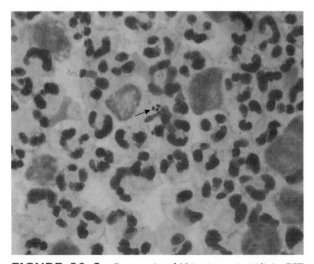

FIGURE 56–2 Gram stain of *Neisseria meningitidis* in CSF from this patient with purulent meningitis. Note Gram-negative diplococci (*arrow*) in association with many PMNs. (*Courtesy of Dr. Dominick Cavuoti, Department of Pathology, University of Texas Southwestern Medical Center, Dallas, TX.*)

EPIDEMIOLOGY

Human carriers are the habitat of the pathogen. **Person-to-person transmission occurs via respiratory droplets**. Systemic infections due to *N. meningitidis* occur worldwide as sporadic cases, school or college or military institution–based outbreaks, and community epidemics. *N. meningitidis* is the **leading cause of bacterial meningitis** in **children** and **young adults** in the United States, where meningococcal meningitis tends to occur in outbreaks (predominantly due to serogroup B). The incidence of meningococcal meningitis peaks in the late winter and early spring. Large-scale epidemics of meningococcal meningitis (due to serogroups A and C) in the meningitis belts of sub-Saharan Africa affect mostly adults. Children younger than 5 years of age without protective antibodies are at risk, with the attack rates being highest in children 3 to 12 months of age. **A congenital deficiency of the terminal complement components** (C5 to C9) predisposes individuals to meningococcemia. Both close contacts of infected individuals (e.g., family members) and individuals with **functional or anatomic asplenia** are also at risk.

PATHOGENESIS

The incubation period varies from 2 to 10 days (mean 3 to 4 days). Sequential steps in the pathogenesis of bacterial meningitis involve

1. Nasopharyngeal colonization by bacteria
2. Bacterial invasion of and survival within blood stream
3. Penetration of blood-brain barrier and egress into CSF
4. Local release of inflammatory cytokines in CSF
5. Adhesion of leukocytes to brain endothelium and diapedesis into CSF
6. Exudation of albumin through opened intercellular junctions of meningeal venules
7. Brain edema, increased intracranial pressure, and altered cerebral blood flow

Invasive disease follows nasopharyngeal colonization by the organisms. Meningococci attach selectively to the microvilli of nonciliated columnar epithelium via specific receptors. Meningococcal pili allow colonization of the nasopharynx, and **IgA protease enhances colonization** of the upper respiratory tract. Organisms enter and pass through the cell to the submucosa. The **polysaccharide capsule is antiphagocytic**, protecting bacteria from phagocytosis in the blood stream. The CNS invasion mechanism is unclear; CNS injury is due to inflammatory response, mediated by the proinflammatory cytokines (tumor necrosis factor [TNF] and interleukin [IL]-1), triggered by LOS-associated endotoxin. Intense pyogenic inflammation occurs in the CSF and the meninges. Cytokine mediators induce fever; brain edema with increased intracranial pressure causes severe headache and stiff neck. Patients may also be nauseated and vomit.

In some patients, **meningococcal infections involve both the meninges and the blood stream**, where meningococci shed LOS-containing "membrane blebs" as they grow. Release of LOS-associated endotoxin into the blood stream leads to thrombocytopenia, which is associated with disseminated intravascular coagulation (DIC) with skin manifestations (**purpuric rash**). Patients rapidly become **hypotensive**.

In others, the infection **involves only the blood stream** (**meningococcemia**), with the organisms multiplying in the blood stream extremely quickly and reaching blood titers higher than any other bloodborne pathogen. Patients with fulminant meningococcemia usually have extremely high blood levels of TNF and IL-1. Hemorrhagic skin lesions (**petechiae and purpura**) are present in most cases of meningococcemia. The fulminant form of meningococcemia is **a rapidly progressing lethal disease**; death is due to shock and cardiac failure.

> **NOTE** Mental retardation, deafness, and hemiparesis are possible permanent sequelae of meningococcal meningitis, whereas the morbidity of meningococcemia includes arthritis and the loss of limbs, digits, and skin, caused by ischemic necrosis. Acute **adrenal gland failure** in children younger than 12 years of age due to bleeding into the adrenal gland caused by severe meningococcal infection is associated with **Waterhouse-Friderichsen syndrome**. The patients present with profound shock. It is fatal if not treated immediately.

Less frequent clinical forms, such as arthritis, pericarditis, and pneumonia due to *N. meningitidis*, are known to occur—either as complications or as primary clinical diseases.

TREATMENT

Penicillin G is the treatment of choice for meningococcal infections. Alternative antibiotics are cefotaxime or ceftriaxone and chloramphenicol. The antimeningococcal activity of penicillin G is excellent; it can penetrate inflamed meninges. Resistance due to β-lactamase is very rare in the United States but has been reported in other countries.

OUTCOME

The patient was admitted to the ICU and cefotaxime was continued until cultures were positive. He was then switched to IV penicillin G. He slowly improved over the next several days, but it was another 3 weeks until he had recovered enough to be discharged from the hospital.

PREVENTION

Polyvalent **vaccine against capsular polysaccharide groups A, C, Y, and W135** is available for children older than 2 years of age. Vaccine against type B capsule is not available. Type B is a sialic acid polymer, which is an endogenous surface component of host RBC; therefore, purified serogroup B polysaccharide fails to elicit human bactericidal antibodies. Vaccination is useful for protection of travelers to hyperendemic areas and for individuals at increased risk.

Prophylaxis for individuals exposed to patients with meningitis is recommended. Rifampin is effective and has a long history of use, but oral ciprofloxacin or ceftriaxone IM as single doses are also highly effective and probably better tolerated.

FURTHER READING

Harrison LH, Dwyer DM, Maples CT, Billmann L: Risk of meningococcal infection in college students. JAMA 281:1906, 1999.

Rosenstein NE, Perkins BA, Stephens DS, et al: Meningococcal disease: Medical progress. N Engl J Med 344:1378, 2001.

van Deuren M, Brandtzaeg P, van der Meer JWM: Update on meningococcal disease with emphasis on pathogenesis and clinical management. Clin Microbiol Rev 13:144, 2000.

Vienne P, Ducos-Galand M, Guiyoule A, et al: Unusual meningococcal infections. Clin Infect Dis 37:1639, 2003.

Wilder-Smith A, Goh KT, Barkham T, et al: Hajj-associated *N. meningitidis* W135. Clin Infect Dis 36:679, 2003.

A **3-week-old** baby boy was brought to the emergency department with a **24-hour** history of **fever, poor feeding, irritability,** and a **seizure** that occurred just before arriving at the ED.

He was **born preterm (32 weeks' gestation) with very low birth weight** (1490 g) after a normal **vaginal delivery.**

PHYSICAL EXAMINATION

VS: T 38.2°C, P 142/min, R 32/min, **BP 90/42** mmHg

PE: Male infant who was irritable, with **nuchal rigidity.** Neurologic exam was otherwise nonfocal.

LABORATORY STUDIES

Blood

Hematocrit: 36%

WBC: 21,000/μL

Differential: 71% PMNs, 19% bands, 10% lymphs

Blood gases: pH 7.24, pCO_2 36 mmHg, pO_2 89 mmHg

Serum chemistries: Normal

Imaging

Head CT was normal.

Diagnostic Work-Up

Table 57-1 lists the likely causes of illness (differential diagnosis). Lumbar puncture can be performed if the patient does not have papilledema or lateralizing neurologic findings. Bacterial or viral cultures from the mother may provide valuable information. Investigational approach may include

- **CSF analysis**
- **Blood cultures**
- **Gram stain and cultures of CSF**
- **Direct detection of bacterial capsular antigens** in CSF and urine in pretreated problem scenarios
- **Polymerase chain reaction** for amplification of DNA of herpes simplex virus or varicella-zoster virus from CSF

TABLE 57–1   Differential Diagnosis and Rationale for Inclusion (consideration)

Bacterial meningitis
  Escherichia coli
  Streptococcus agalactiae
  Listeria monocytogenes
  Streptococcus pneumoniae
  Neisseria meningitidis
  Haemophilus influenzae
Viral meningitis
  Herpes simplex virus type 2
  Varicella-zoster virus (VZV)
  Enteroviruses
Sepsis (bacterial)
Congenital infections
  Toxoplasma gondii
  Cytomegalovirus

Rationale: In infants younger than 1 month, there are relatively few common causes of meningitis. *E. coli*, *S. agalactiae*, and *S. pneumoniae* are the most common agents in infants; the remaining are much more common in adults. The viral causes are also less common, and sepsis in general would not be expected to typically cause seizures, although sepsis is much more common than meningitis. The other congenital infections usually do not manifest with such severity.

COURSE

The patient was hospitalized. Blood was drawn, and a lumbar puncture was performed. Examination of the CSF was remarkable for WBCs 2340 mm³ (89% PMNs), protein 180 mg/dL, and glucose 15 mg/dL. A Gram stain of the CSF showed Gram-positive cocci in chains. The patient was given ampicillin and gentamicin as empirical therapy until final culture results were known. The cultures of blood and CSF were diagnostic.

ETIOLOGY

Streptococcus agalactiae (group B streptococcus [GBS]) type III meningitis

MICROBIOLOGIC PROPERTIES

S. agalactiae is **Gram-positive**, occurring as **cocci in chains** (Fig. 57-1*A*). The cell surface of *S. agalactiae* exhibits Lancefield **group B carbohydrate antigen**, thus the organism is known as group B streptococcus (GBS). Lancefield antigenic classification is based on precipitin reactions with homologous antiserum, and it divides the streptococci into serogroups A to H and K to V. Group B streptococci causing human infections are **encapsulated** by one of nine antigenically distinct polysaccharides. Type III capsular type is associated with invasive infections in newborns.

CSF cultures are obtained in symptomatic infants with nonspecific symptoms of meningitis. Blood-borne GBS (bacteremia) can be detected by blood cultures, which are moderately sensitive and highly specific. *S. agalactiae* form glistening, gray-white colonies with a narrow zone of β-**hemolysis** on sheep blood agar medium. The isolates are **catalase negative** and **bacitracin resistant**. The isolates on blood agar medium can also be identified by CAMP test (see Fig. 57-1*B*). Direct antigen detection permits identification of bacterial antigen in urine or in CSF. In infants whose mothers received antimicrobial drugs before delivery, this becomes an important diagnostic tool (moderately sensitive).

EPIDEMIOLOGY

GBS infections are the **leading bacterial cause of disease and deaths among newborns** in the United States. Approximately 19,000 cases occur annually in the United States. Asymptomatic carriage in gastrointestinal and genital tracts is common. Up to 40% of pregnant women harbor GBS in the genital tract. Approximately one-half of colonized mothers transmit the organism to their infant. Maternal risks associated with serious disease in newborns are

- Delivery at less than 37 weeks of gestation
- Premature rupture of membranes at less than 37 weeks of gestation
- Rupture of membranes 18 or more hours before delivery

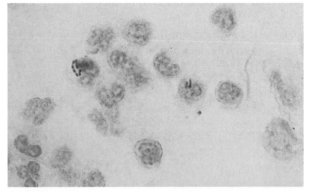

A

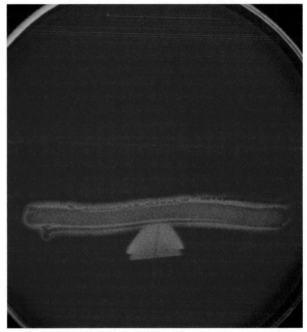

B

FIGURE 57–1 *A*, Gram stain of *Streptococcus agalactiae* (GBS). *B*, CAMP test of *Streptococcus agalactiae* (GBS). Note an "arrowhead-shaped" zone of synergistic hemolysis of sheep RBCs produced by CAMP factor (a phospholipase) of GBS and β-hemolysin of *Streptococcus aureus* (*horizontal streak*). Remainder of GBS streak (*vertical*) shows lighter hemolysis without the synergistic effect of β-hemolysin. *(Courtesy of Dr. Paul Southern, Departments of Pathology and Medicine, University of Texas Southwestern Medical Center, Dallas, TX.)*

- Prior delivery of an infant with GBS disease
- Heavy maternal colonization (vaginal inoculum >10^5 CFU/mL)
- Deficient maternal antibody to the capsular type of GBS at delivery

Although acquisition of the organism by the neonate is efficient, the rate of subsequent clinical disease is quite low (1% to 2%).

PATHOGENESIS

GBS is an **invasive pathogen that causes sepsis, pneumonia, and meningitis in newborns** depending on the age at onset (Table 57-2).

It is best known as the most common cause of neonatal infection, particularly in the setting of prematurity and prolonged rupture of the membranes (termed **early onset disease**). **Pathogenesis is dictated by maternal colonization with a GBS strain with enhanced virulence.** During vaginal delivery, the skin and mucous membranes (e.g., nasopharynx) of the infant are exposed to the bacteria, and during the first 24 to 48 hours after birth, babies become colonized with the bacteria. **A late onset infection** due to GBS is usually acquired by contact with a colonized mother, or by contact with nursery personnel.

The **type III capsular polysaccharide** of GBS is the **major virulence factor**, which interferes with ingestion by phagocytes. Immunity is mediated by antibody to the capsular polysaccharide and is serotype specific. Infection in the infant is more common in the absence of maternal antibody to group B streptococci. The capsule is antiphagocytic, allowing the organisms to disseminate and multiply in blood and CSF due to lack of opsonization and, therefore, subsequent phagocytic killing by PMNs.

TREATMENT

Penicillin G and ampicillin are the drugs of choice. Alternatives include ampicillin, vancomycin, and third-generation cephalosporins. Gentamicin is often added for synergy in severe infections.

OUTCOME

After a rocky hospital course and a 14-day course of high-dose ampicillin and gentamicin, the patient was discharged home.

PREVENTION

The large number of women who carry this organism and the fact that neonatal sepsis is a rare event make a preventive approach to this problem difficult. GBS isolates remain universally susceptible to penicillin and ampicillin, the first-line agents for intrapartum antibiotic prophylaxis. The administration of IV ampicillin at the onset of and throughout labor to women who are colonized with GBS interrupts transmission. All "at risk" patients, including vaginal and rectal colonizers of GBS, should receive ampicillin or penicillin G (intrapartum maternal administration). A conjugated vaccine composed of type III capsular polysaccharide for the target risk group of pregnant women is under development.

TABLE 57–2 Important Clinical Features of *Streptococcus agalactiae* Infections

Syndromes	Clinical Features
Early-onset neonatal disease (24 hours—7 days after birth)	Pneumonia and bacteremia are common. Case fatality rate (CFR) in untreated patients is 50%.
Late-onset disease (1 week—before 3 months after birth)	Characterized predominantly by meningitis (seen in the case patient). Significant bacteremia is also seen in some patients. CFR for meningitis is 25%; long-term morbidity such as hearing loss, global brain injury, and mental retardation may also occur.
Maternal disease	Manifests as bacteremia (especially with cesarean section), chorioamnionitis, postpartum endometritis, or urinary tract infection.
Disease in nonpregnant adults	May manifest as skin and soft tissue infection, bacteremia without an identifiable focus, pneumonia, septic arthritis, endocarditis, or meningitis. Host risk factors are diabetes mellitus, age >65 years, liver disease, or immune compromise.

FURTHER READING

CDC: Prevention of perinatal group B streptococcal disease: A public health perspective. MMWR 45:1, 1996.

Jackson LA, et al: Risk factors for group B streptococcal disease in adults. Ann Intern Med 123:415, 1995.

Kaufman D, Fairchild KD: Clinical microbiology of bacterial and fungal sepsis in very-low-birth-weight infants. Clin Microbiol Rev 17:638, 2004.

Schuchat A: Epidemiology of group B streptococcal disease in the United States: Shifting paradigms. Clin Microbiol Rev 11:497, 1998.

Schuchat A: Group B streptococcus. Lancet 353:51, 1999.

Schwartz B, Schuchat A, Oxtoby MJ, et al: Invasive group B streptococcal disease in adults. A population-based study in metropolitan Atlanta. JAMA 266:1112, 1991.

Weisner AM, Johnson AP, Lamagni TL, et al: Invasive GBS in England and Wales. Clin Infect Dis 38:1203, 2004.

A **64-year-old** woman was brought to the emergency department with 5 days of **fever, headache,** and **confusion.**

She had had diarrhea for 2 days that resolved a few days before her current symptoms began. She had a **history of rheumatoid arthritis,** for which she had been taking **prednisone daily for the past several months.**

PHYSICAL EXAMINATION

VS: T 38.8°C, P 102/min, R 18/min, BP 134/78 mmHg

PE: The patient was unable to answer simple questions and appeared agitated; exam revealed no **nuchal rigidity** and **no focal neurologic deficits.**

LABORATORY STUDIES

Blood

Hematocrit: 33%

WBC: 12,400/μL

Differential: 65% PMNs, 8% bands, 24% lymphs

Blood gases: pH 7.34, pCO_2 33 mmHg, pO_2 78 mmHg

Serum chemistries: Normal

Imaging

Head CT was normal. Brain MRI showed **meningeal enhancement** but no focal lesions.

Diagnostic Work-Up

Table 58-1 lists the likely causes of illness (differential diagnosis). Bacterial meningitis was suspected. Lumbar puncture should be performed. The investigational approach may include

- **CSF analysis**
- **Gram stain and cultures of CSF**
- **Blood cultures**

COURSE

Blood was drawn, a lumbar puncture was performed, and the CSF was collected. The patient was put on an empirical regimen of intravenous vancomycin, cefotaxime, and ampicillin. CSF examination was remarkable with WBCs 620/μL (89% PMNs), glucose 40 mg/dL, and protein 175 mg/dL. Gram stain of

TABLE 58–1 Differential Diagnosis and Rationale for Inclusion (consideration)
Bacterial meningitis
Streptococcus pneumoniae
Listeria monocytogenes
Haemophilus influenzae
Neisseria meningitidis
Encephalitis (e.g., HSV-1, arthropod-borne viruses)
Viral meningitis (e.g., enteroviruses)

Rationale: In adults, the pathogens primarily responsible for bacterial meningitis are different from those causing meningitis in neonates, although *Listeria* is common to both. Although *S. pneumoniae* is the most common etiology for all adults, *Listeria* is one of the more common agents, particularly in the immunocompromised, and in patients more than 50 years old. With altered mental status, encephalitis should always be considered as well. Aseptic (viral) meningitis is quite common but is self-limited and is generally a diagnosis of exclusion.

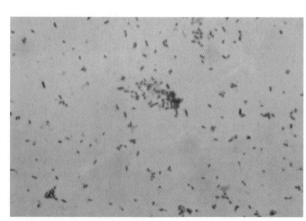

FIGURE 58–1 Gram stain of the isolate from a culture of CSF. *(Courtesy of Dr. Paul Southern, Departments of Pathology and Medicine, University of Texas Southwestern Medical Center, Dallas, TX.)*

CSF revealed moderate numbers of PMNs, but no organisms. Culture of blood and CSF grew a significant bacterial pathogen.

ETIOLOGY

Listeria monocytogenes (*Listeria* meningitis)

MICROBIOLOGIC PROPERTIES

L. monocytogenes is **Gram positive**, occurring generally as **small** (nonspore-forming) **rods with rounded ends** (Fig. 58-1). The organisms are facultatively anaerobic. Blood and CSF cultures grow well on blood agar medium, producing a **narrow zone of β-hemolysis**. These bacteria also grow in cold temperature (4°C). The organisms are **motile**, with a characteristic **"tumbling motility"** in wet mounts at room temperature (22°C). The organisms produce acid from glucose. The somatic "O" and flagellar "H" antigens determine the serotypes. Only a few serotypes cause human diseases.

EPIDEMIOLOGY

L. monocytogenes is typically responsible for **food-borne illness in adults** and **meningitis in newborns** (Table 58-2). In the United States, an estimated 2500 persons become seriously ill with listeriosis each year; of these, 500 die. *L. monocytogenes* is found in **soil and animal reservoirs**. Processed meat and dairy products (e.g., soft cheeses) may support the growth of *Listeria* and have caused outbreaks. The major **risk groups** are **pregnant women, fetuses, neonates, immunocompromised patients** (increased incidence in AIDS), and the **elderly**. Transmission occurs via ingestion of contaminated food products (in adults).

PATHOGENESIS

L. monocytogenes is an **intracellular pathogen** and, therefore, is able to cause systemic infections in patients having defective cell-mediated immunity. The bacteria induce host cells to engulf them via the mediation of a *Listeria* protein called "internalin." Intracellular survival and spread of the bacteria are

TABLE 58–2 Important Clinical Features of *Listeria monocytogenes* Infections

Syndromes	Clinical Features
Early-onset (<5 days) disease in newborns	Many pregnant women carry *L. monocytogenes* asymptomatically in their large bowel and vagina. Vertical transmission can occur from mother to newborn during passage through an infected birth canal or because of an ascending infection through ruptured amniotic membranes. **Sepsis** or **meningitis** occurs in newborns.
Late-onset (>5 days) disease in newborns	Poor personal hygiene and poor nursing care by the colonized mother may transmit the pathogen to her newborn. **Meningitis** is the primary presentation.
Adult (food-borne) listeriosis	**Pregnant women** may experience a mild, flu-like illness. In these patients with febrile illness, a blood culture is the most reliable way to find out if the patient's symptoms are due to listeriosis. **Fetal loss** may occur in pregnant women following transplacental transmission of the invasive organism to the fetus. **Elderly** and **immunocompromised** patients present with bacteremia and **meningitis**. *L. monocytogenes* is the most common etiologic agent of bacterial meningitis in transplant patients. **Immunocompetent** persons may experience acute febrile gastroenteritis (fever, nausea, vomiting, and diarrhea)

critically important for pathology. All virulent strains produce **listeriolysin O** (a β-hemolysin) and **phospholipases**, which permit the bacteria to escape from the membrane-bound vacuole (phagosome) into the cytoplasm, where they rapidly multiply. Bacteria stimulate nucleation and rearrangement of host cell actin. Cell-to-cell spread occurs via pseudopods that extend into adjacent host cells; cellular **actin "tails"** and listerial phospholipase activity are needed for spread. Immunity to intracellular bacteria relies on T cell-mediated activation of macrophages by lymphokines. Activated macrophages can kill the bacteria and control an ongoing infection. In an uncontrolled infection, microabscesses and granulomas contribute to pathology.

TREATMENT

Penicillin or **ampicillin**, usually in combination with an **aminoglycoside**, is the drug of choice. Alternatives include trimethoprim/sulfamethoxazole or vancomycin.

OUTCOME

The cultures turned positive on the third hospital day, and the patient was switched to high-dose ampicillin and gentamicin for a total of 21 days, with gradual recovery.

PREVENTION

There are no vaccines. People at high risk (e.g., pregnant women, immunocompromised individuals) should avoid eating raw or partially cooked food of animal origin, soft cheeses, and unwashed raw vegetables. Regular hand washing should be encouraged.

FURTHER READING

Drevets DA, Leenen PJM, Greenfield RA: Invasion of the central nervous system by intracellular bacteria. Clin Microbiol Rev 17:323, 2004.

Frye DM, Zweig R, Sturgeon J, et al: Outbreak of febrile gastroenteritis. Clin Infect Dis 35:943, 2002.

Hof H, Nichterlein T, Kretschmar M: Management of listeriosis. Clin Microbiol Rev 10:345, 1997.

Southwick FS, Purich DL: Mechanisms of disease: Intracellular pathogenesis of listeriosis. N Engl J Med 334:770, 1996.

Vázquez-Boland JA, Kuhn M, Berche P, et al: *Listeria* pathogenesis and molecular virulence determinants. Clin Microbiol Rev 14:584, 2001.

During a 4-week period in **August** in a rural county in the Midwest, a total of **29 persons** (between the **ages of 9 and 15 years**) had a rapid onset of **fever, headache, stiff neck,** and **photophobia.** Some patients had had diarrhea for a few days preceding the headache.

PHYSICAL EXAMINATION

Physical examination of one of the individuals (Patient X) is presented here.

VS: T 39.2°C, P 112/min, R 16/min, BP 130/84 mmHg

PE: An ill-appearing male in moderate distress due to headache; he had some photophobia and mild **nuchal rigidity**, but negative Kernig sign (flexion of the neck when the knee is flexed).

TABLE 59–1 Differential Diagnosis and Rationale for Inclusion (consideration)

Arboviruses
Aseptic (viral) meningitis
   Enteroviruses
   Herpes simplex virus type 2
   Mumps
Aseptic (nonviral) meningitides (e.g., fungal, tuberculous, parasitic, or syphilitic disease)
Bacterial meningitis
   Haemophilus influenzae
   Listeria monocytogenes
   Neisseria meningitidis
   Streptococcus pneumoniae
Neoplastic meningitis
Noninfectious inflammatory diseases (e.g., sarcoid, Behçet disease)

Rationale: Meningitis should be considered. The major categories of meningitis (e.g., bacterial causes) should first be excluded owing to their high mortality. Often the initial distinction between bacterial and viral meningitis is difficult. Viral meningitis should be considered when an outbreak occurs in the summer months, especially in children younger than 15 years of age. This seasonal predilection of some viruses (e.g., enteroviruses) can provide a valuable clue to diagnosis. Arboviral meningitis should be considered when clusters of meningitis cases occur in a region proximal to marshy lands during the summer. Aseptic (nonviral) infectious meningitides may not yield culture-positive diagnosis. Noninfectious causes are uncommon and are usually considered after other etiologies have been excluded.

LABORATORY STUDIES (PATIENT X)

Blood

Hematocrit: 45%
WBC: 8200/µL
Differential: Normal
Serum chemistries: Normal

Imaging

Head CT was normal.

Diagnostic Work-Up

Table 59-1 lists the likely causes of illness (differential diagnosis). Lumbar puncture is performed for diagnostic investigation. The most important laboratory test in the diagnosis of meningitis is examination of the CSF. In viral meningitis, a typical profile demonstrates lymphocytic pleocytosis and bacteria-free CSF. The investigational approach for specific microbiologic diagnosis may include

- **Gram stain** and **cultures of CSF**
- **Virus isolation** from CSF, stool, and oropharyngeal secretions using cell cultures

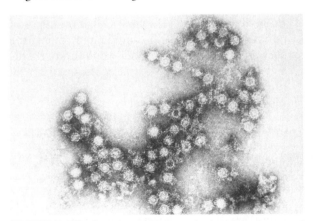

FIGURE 59–1 Electron micrograph of virions concentrated from stool specimen. (*Courtesy of Centers for Disease Control and Prevention, Atlanta GA.*)

- **Blood cultures** to detect any blood-borne bacterial agent
- **RT-PCR** of CSF in negative cultures for bacteria and viruses

COURSE

CSF and blood cultures were obtained. Aseptic meningitis was suspected based on CSF markers: WBC 54/μL, 92% lymphs; glucose 65 mg/dL; protein 30 mg/dL. Empirical therapy was started pending bacterial cultures. After 3 days, all patients (including Patient X) had negative bacterial cultures of blood and CSF. Preliminary identification of the cause of the outbreak was made following virus-isolation studies.

ETIOLOGY

Echovirus type 9 (enteroviral meningitis)

MICROBIOLOGIC PROPERTIES

Echovirus belongs to the enteroviruses (in the Picornaviridae family), which are small (20 to 30 nm), ether-resistant, nonenveloped viruses that exhibit cubic symmetry (Fig. 59-1). The RNA genome is single stranded and (+) sense (it can serve as mRNA), and it is 7.2 to 8.4 kb in size. RNA is translated into one large polypeptide that is cleaved by proteases into many small proteins. The picornaviruses infecting humans are classified in three major groups. Table 59-2 lists disease associations of the major enterovirus types.

- **Polioviruses** (three serotypes). All three types of polioviruses have been eliminated from the Western Hemisphere, as well as Western Pacific and European regions, by the widespread use of vaccines.
- **Nonpolioviruses** (61 serotypes)

  - **Coxsackie** A (23 serotypes) and B (6 serotypes)
  - **Echoviruses** (28 serotypes)
  - **New enteroviruses** (4 serotypes)

- **Hepatovirus** (hepatitis A virus)
- **Rhinoviruses** (115 types). The "common cold" viruses are the most common viral infectious agents in the United States and worldwide.

TABLE 59–2   Selected Picornaviruses and the Clinical Syndromes

Virus	Disease (Virus Types)
Polioviruses (types 1-3)	Aseptic meningitis (types 1-3)
	Paralysis and encephalitic diseases (types 1-3)
Coxsackievirus group A (A1-A22, A24)	Acute hemorrhagic conjunctivitis (type 24 variant)
	Hand-foot-mouth disease (types 5, 10, 16)
	Aseptic meningitis (types 1, 2, 4-7, 9, 10, 14, 16, 22)
	Myopericarditis (types 4, 16)
Coxsackievirus group B (B1-B6)	Pleurodynia (types 1-5)
	Pericarditis, myocarditis (types 1-5)
	Aseptic meningitis (types 1-6)
	Severe systemic infection in infants, meningoencephalitis and myocarditis (types 1-5)
	Exanthem, hepatitis, diarrhea (type 5)
Echoviruses (1-7, 9, 11-27, 29-33)	Aseptic meningitis (many serotypes)
	Exanthem (types 1-9, 11, 14, 16, 18, 19, 25, 30, 32)
	Hand-foot-mouth disease (19)
	Pericarditis and myocarditis (types 1, 6, 9, 19, 22)
	Upper and lower respiratory illnesses (types 4, 9, 11, 20, 22, 25)

Enteroviruses are acid stable and have a low density. Rhinoviruses are acid labile; they multiply better at 33°C and have a higher density than do enteroviruses. In cell cultures of enteroviruses, cytopathic effect (CPE) on the infected, fibroblastic cells is usually evident within 36 hours. A virus isolate can be identified by neutralization of CPE. Culture-negative CSF specimens from aseptic meningitis are subjected to RT-PCR testing for type-specific detection of viral isolates.

EPIDEMIOLOGY

Aseptic (viral) meningitis occurs both sporadically and in outbreaks, and greater than 90% of cases with an identified cause are associated with enteroviruses. Enteroviruses cause an estimated 10 to 15 million or more symptomatic infections a year in the United States. Enteroviral illnesses and aseptic meningitis typically demonstrate **a seasonal pattern**, with the **highest incidence during the summer and fall**

months. The community outbreak in the study problem occurred during the usual season of greatest enterovirus activity. Transmission of enteroviruses usually is **person-to-person**, through either the **fecal-oral** or oral-oral routes. Enteroviruses can be readily isolated from waste waters and occasionally from recreational waters.

PATHOGENESIS

The portal of entry of this virus is the mouth; multiplication takes place in the **oropharynx** and the **small intestine**. By 3 to 5 days after exposure, virus can be recovered from blood, throat, and feces. At this time, symptoms of minor illness may appear, or the infection may remain asymptomatic. **Viremia** is present for a few days. Virus spreads by the way of the blood stream to the CNS, where the viruses multiply further and induce an inflammatory response. The CSF is, however, clear and bacteria free (aseptic), and the cells are mainly lymphocytes, although PMNs may be present in the early stages. Viruses are isolated from the CSF in less than 50% of aseptic meningitis cases. **Aseptic meningitis is a milder disease than is bacterial (septic) meningitis**, with headache, fever, and general illness, as in bacterial meningitis, but with less neck stiffness, and **hospitalization** is not required in most cases.

TREATMENT

Specific antiviral drugs are not available, although some are in development. Treatment is symptomatic and involves reducing pain and muscle spasm and maintaining respiration and hydration. Complete recovery generally takes place.

OUTCOME

Echovirus type 9 was the predominant agent isolated from stool specimens of 16 case patients, including the Patient X. The same virus serotype was also isolated from CSF of 19 case patients, including the Patient X. It was determined that on opening day of the facility, a swimmer became ill and vomited into the pool; subsequently other swimmers became ill. All patients recovered spontaneously with no complications.

PREVENTION

No vaccine is currently available for the nonpolio enteroviruses. Aseptic meningitis outbreaks caused by enteroviruses, such as the one in this outbreak, underscore the importance of public health messages that emphasize the role personal hygiene plays in interrupting the transmission of enteroviral infections.

FURTHER READING

Centers for Disease Control and Prevention: Enterovirus surveillance—United States, 1997-1999. MMWR 49:913, 2000.
Jeffery KJM, Read SJ, Peto TEA: Diagnosis of viral infections of the central nervous system: Clinical interpretation of PCR results. The Lancet 349:313, 1997.

Ramers C, Billman G, Hartin M, et al: Impact of a diagnostic cerebrospinal fluid enterovirus polymerase chain reaction test on patient management. JAMA 283:2680, 2000.

In July, a **61-year-old homeless man** was brought to the hospital by paramedics with complaints of **fever, malaise,** and **worsening headache** for the past 3 days. He also noted **nausea, vomiting,** and diarrhea. The paramedics who treated him stated that he appeared somewhat confused.

It had been a **hot summer,** and the **mosquito population had been hard to control.**

PHYSICAL EXAMINATION

VS: T 40°C, P 108/min, R 18/min, BP 138/78 mmHg

PE: The patient was not able to answer more than simple questions, and he frequently seemed to drift into sleep. The neck was supple, but neurologic exam was difficult due to poor cooperation. Slight **tremors of the face and extremities** were noted.

LABORATORY STUDIES

Blood

Hematocrit: 42%

WBC: 14,400/μL

Differential: Normal

Platelets: 210,000/μL

Serum chemistries: Sodium 128 mmol/L (other markers were normal)

Imaging

Chest x-ray was normal; head CT scan was normal (there was no evidence of hydrocephalus).

Diagnostic Work-Up

Table 60-1 lists the likely causes of illness (differential diagnosis). Diagnosis of encephalitis on clinical grounds alone is difficult. There are not enough specific findings on physical examination. A high index of suspicion in the summer months in the appropriate geographic areas may support a presumptive diagnosis of arboviral encephalitis. Lumbar puncture is performed, and CSF is examined to rule out bacterial and viral meningitis. The CSF of patients with arboviral encephalitis may reveal lymphocytic leukocytosis, with a normal protein and a normal glucose level. Further investigational approach may include

- **Cultures** of neurotropic virus in CSF
- **Direct cryptococcal antigen** in CSF

TABLE 60–1 Differential Diagnosis and Rationale for Inclusion (consideration)

Arthropod-borne encephalitides
  California/LaCrosse
  Eastern equine
  St. Louis
  Western equine
Aseptic (nonviral) meningitides (e.g., tuberculous, cryptococcal, parasitic, or syphilitic disease)
Aseptic (viral) meningitis
Bacterial meningitis
Brain abscess
Herpes simplex virus (HSV) type 1 encephalitis
Rabies encephalitis

Rationale: Encephalitis should be considered on the basis of altered mental status, in addition to fever and headache. The various causes of encephalitis may be distinguished by specific epidemiologic factors (e.g., outdoor loitering in mosquito-infested areas). Brain abscess usually exhibits focal neurologic deficits on presentation. HSV-1 is important to consider because it is one of the few treatable causes of encephalitis. HSV-1 usually causes temporal lobe lesions and may be associated with hemorrhagic CSF. Meningitis of various causes can also manifest similarly, although mental status changes are typically not prominent until later in the course. Rabies has specific clinical features (i.e., hydrophobia) that often distinguish its presentation.

- **Viral serology**
  - Presence of specific **IgM** in acute-phase serum or CSF
  - Elevated **arboviral IgG** between early and late specimens of serum
- **PCR amplification** of DNA from CSF to rule out a neurotropic virus (e.g., HSV-1, rabies)

COURSE

A lumbar puncture was performed, and the CSF was clear. CSF markers were remarkable for WBCs 120/μL (85% lymphs and 15% PMNs), RBCs 5/μL, protein 80 mg/dL, and glucose 60 mg/dL. Gram stain of the CSF was negative. Viral and bacterial cultures from CSF did not grow any organisms. The cryptococcal antigen test result was negative. The patient remained in this state for 72 hours. Thereafter, his level of consciousness improved slightly and he became more easily aroused. Nevertheless, he did not regain full consciousness for another 48 hours. Positive serology for a viral encephalitis was subsequently diagnostic.

ETIOLOGY

St. Louis encephalitis virus

MICROBIOLOGIC PROPERTIES

Arthropod-borne **(arbo-) viruses are a leading cause of viral encephalitis** in the world. Arboviruses that cause human encephalitis are members of three virus families: the Bunyaviridae (genus *Alphavirus*), Togaviridae, and Flaviviridae. Members of the family of Bunyaviridae are spherical or pleomorphic (80 to 120 nm) enveloped viruses. The genome is made up of a triple-segmented, circular, single-stranded, negative-sense (ambisense) RNA. Virion particles contain three circular, helically symmetrical nucleocapsids about 2.5 nm in diameter and 200 to 3000 nm in length. The members of the two other main families of arboviruses (Togaviridae [Table 60-2] and Flaviviridae [Table 60-3]) have enveloped virions (45 to 60 nm in diameter) that contain a single-stranded RNA genome of positive polarity. Flaviviruses (e.g., St. Louis virus, associated with encephalitis in the case patient) also include yellow fever virus and dengue fever viruses. Most members are transmitted by bloodsucking arthropods.

EPIDEMIOLOGY

The incidence of arboviral encephalitis varies markedly with geographic regions of the world and also local low-level endemicity in the United States, depending on ecologic factors. **Arboviruses usually exist** in woodland habitat, passing between tree hole

TABLE 60–2	Principal Medically Important Togaviruses (Alphaviruses)			
Virus	**Clinical Syndrome**	**Properties**	**Hosts**	**Distribution**
Eastern equine encephalitis (EEE)	Encephalitis	EEE complex	Birds	Americas
Western equine encephalitis (WEE)	Encephalitis	WEE complex	Birds	North America
Venezuelan equine encephalitis (VEE)	Febrile illness, encephalitis	VEE complex (several subtypes)	Rodents, horses	Americas

TABLE 60–3	Principal Medically Important Flaviviruses			
Virus	**Antigenic Clinical Syndrome**	**Properties**	**Hosts**	**Distribution**
Dengue (DEN)	Febrile illness, rash; **hemorrhagic shock syndrome**	DEN group four viruses: DEN-1, 2, 3, 4	Humans	Tropics, worldwide
Yellow fever (YF)	**Hemorrhagic fever,** hepatitis (jaundice)	YF group	Primates, humans	Africa, South America
St. Louis encephalitis	Encephalitis	JE group	Birds	Americas
Japanese encephalitis (JE)	Encephalitis	JE group	Pigs, birds	India, China, Japan, Southeast Asia

mosquitoes and vertebrate hosts (e.g., birds and rodents) in the summer months, when the mosquito activity is maximal. The vector also uses trash-filled drainage systems and artificial containers (e.g., dumped tires) in urban areas. In the United States, St. Louis virus is transmitted between mosquitoes and birds. *Culex tarsalis* is the vector, causing transmission of St. Louis virus in residents of the rural western and central United States. The more urbanized mosquito species (e.g., *Culex pipiens*) are responsible for cases in cities of the central and eastern United States. The mosquitoes feed on viremic birds (or rodents) then infect humans. All age groups are at risk, but severity of disease increases with age.

PATHOGENESIS

Arboviruses gain access to humans via the bite of an infective mosquito. Viruses contained in the saliva of mosquitoes are introduced into the capillary bed as the mosquito's proboscis penetrates the skin and enters the endothelial cell of the capillary wall. The viruses localize in the vascular endothelium of the reticuloendothelial system, replicating in the endothelial cells. A primary viremia (unlike in HSV-1 and rabies) is induced as the viruses are liberated from these infected cells. After the viruses enter the circulation, they may localize primarily in the endothelial cells lining the small vessels of the brain or the choroid plexus. Damage to brain or cerebellar tissues is due largely to vascular involvement because virus-antibody complexes may trigger a complement activation process, leading to disseminated intravascular coagulation (DIC) in the brain.

NOTE If a protective antibody titer develops, infection usually resolves, but the immune response is usually delayed in eastern equine encephalitis and Japanese encephalitis. Laboratory diagnosis of human arboviral encephalitis may require positive identification using IgM- and IgG-based assays, with a fourfold increase in IgG titer between acute and convalescent serum samples. Rapid serologic assays such as IgM ELISA (based on monoclonal antibody capture) and IgG ELISA may be employed soon after infection. Early in infection, IgM antibody is more specific, whereas later in infection, IgG antibody is more reactive.

Complications of St. Louis viral encephalitis include cranial nerve palsies, hemiparesis, and convulsions, with case-fatality rates as high as 20% among patients older than the age of 60. Older patients who recover from the disease may have sequelae such as difficulties in concentration and memory, asthenia, and tremor. The clinical presentations of the different arboviral encephalitides are often indistinguishable.

TREATMENT

Specific treatment is not available. Management is supportive.

OUTCOME

The patient gradually improved over the course of the next several weeks with no residual neurologic deficit.

PREVENTION

There are no human vaccines against the arboviral encephalitides. A killed vaccine is available to protect horses from eastern equine encephalitis. Preventive measures for human disease include avoidance of mosquito-infested areas, control of vectors, and use of protective clothes and mosquito repellents. Additional measures include careful surveillance, including searching for missed cases and the presence of vectors, and necropsy investigation of dead animals.

FURTHER READING

Calisher CH: Medically important arboviruses of the United States and Canada. Clin Microbiol Rev 7:89, 1994.

Deresiewicz RL, Thaler SJ, Hsu L, Zamani AA: Clinical and neuroradiographic manifestations of eastern equine encephalitis. N Engl J Med 336:1867, 1997.

McJunkin JE, de los Reyes EC, Irazuzta JE, et al: LaCrosse encephalitis in children. N Engl J Med 344:801, 2001.

Solomon T: Flavivirus encephalitis: Current concepts. N Engl J Med 351:370, 2004.

Solomon T, Dung NM, Vaughn DW, et al: Neurological manifestations of dengue infection. Lancet 355:1053, 2000.

The family of a 56-year old man became concerned when they noted **changes in his personality** that began gradually over several days, making him more **irritable** and **confused.** He also developed **fevers** and **headache** and progressively became unable to perform daily activities. He was brought to the emergency department of a local general hospital by his wife after she noticed a **left-sided weakness** and then witnessed him having a **seizure.**

He had no other medical problems and was on no medications.

PHYSICAL EXAMINATION

VS: T 38°C, P 86/min, R 16/min, BP 124/80 mmHg

PE: On exam, he was confused and unable to coherently answer questions. Pupils were equal and reactive, and his neck was supple. Neurologic exam revealed **left arm and leg weakness,** but further exam was difficult due to his lack of cooperation.

LABORATORY STUDIES

Blood

Hematocrit: 44%

WBC: 8600/μL

Differential: Normal

Blood gases: Normal

Serum chemistries: Normal

Imaging

A **head MRI** showed an irregular area of **hemorrhagic necrosis in the right temporal lobe.**

Diagnostic Work-Up

Table 61-1 lists the likely causes of illness (differential diagnosis). A clinical diagnosis of viral encephalitis was considered, based on symptoms such as fever, headache, altered mental state, and seizure. A definitive diagnosis of viral encephalitis relies on

- **Virus isolation** from CSF or brain
- **Polymerase chain reaction** (PCR)-based amplification of viral DNA from CSF

TABLE 61–1 Differential Diagnosis and Rationale for Inclusion (consideration)

Aseptic meningitis
Bacterial meningitis
Brain abscess
Cryptococcal meningoencephalitis
Rabies encephalitis
Stroke
Viral encephalitides
  Arthropod-borne viruses
  Cytomegalovirus
  Herpes simplex virus (HSV) type 1

Rationale: The presentation is highly suggestive of encephalitis, with predominantly mental status changes rather than meningeal signs typical of bacterial or viral meningitis. The focal neurologic signs and seizures may also suggest a mass lesion (brain abscess) or stroke. The causes of viral encephalitis can be difficult to distinguish on clinical grounds, and other epidemiologic features are often helpful in making a diagnosis.

COURSE

Lumbar puncture showed CSF pleocytosis (WBCs 70/μL with 82% lymphs, RBCs 1500/μL), a normal glucose concentration, and an elevated protein concentration. Additional tests were ordered, and the patient was started on empirical therapy. A CSF PCR test for a neurotropic virus was positive.

ETIOLOGY

Herpes simplex virus (HSV) type 1 (herpes simplex encephalitis [HSE])

MICROBIOLOGIC PROPERTIES

All herpesviruses are structurally similar, with an **icosahedral nucleocapsid, a linear ds-DNA genome**, and a **lipoprotein envelope**. Virus-infected cells reveal the presence of **multinucleated giant cells**, although the findings are not specific for the type of herpesvirus. Oral herpes simplex virus (HSV-1) is distinguished from venereal herpesvirus (HSV-2) by antigenicity. **Latency** after primary infection is the **characteristic feature of all herpesviruses, including HSV-1**.

EPIDEMIOLOGY

HSE accounts for about 10% of all encephalitis cases and is the most common cause of sporadic fatal encephalitis, which occurs in about 1 per 250,000 to 500,000 persons per year in the United States. Infection with HSV-1 is acquired earlier in life than is HSV-2. Asymptomatic HSV-1 infection is very common, as evidenced by the high prevalence of HSV-1 antibodies in the general population. **Transmission** can result **from close contact** (kissing) with persons who are shedding the virus or on whose mucosal surfaces the virus is replicating. **Oral lesions** (gingivostomatitis) and **pharyngitis** (seen **in children and young adults**) are the most frequent clinical manifestations of first-episode HSV-1 infection. Recurrent **herpes labialis** is the most frequent clinical manifestation of reactivation HSV-1 infection **in young adults**.

PATHOGENESIS

Primary HSV-1 infection is often subclinical (without clinically apparent lesions). The HSV-1 **virus remains in the body throughout an exposed person's entire life**. Trigeminal ganglion or autonomic nerve roots are the anatomic location of latency of HSV-1. Reactivation may occur in certain older adults, and direct neuronal transmission of the replicating virus primarily occurs from a peripheral site to the brain via the trigeminal or olfactory nerve. The virus finds a new niche in the temporal lobe of brain. The **cytolytic virus causes direct damage (necrosis) to the brain parenchyma**. Virus-specific T-cell responses are also mounted and contribute to **perivascular inflammation**, resulting in **hemorrhage**, distributed in an irregular fashion **throughout the right temporal lobe**. Manifestation of encephalitis is associated with deeper disturbance of brain function, with **confusion and seizures**.

> **NOTE** The accurate diagnosis of herpes simplex encephalitis is important, because effective antiviral therapy is available. Virus is present in brain and CSF. The most sensitive method for diagnosis of HSV encephalitis is the detection of HSV DNA in CSF by PCR.

TREATMENT

Effective therapy requires early use of high-dose acyclovir for a total of 21 days. Case fatality rate is decreased from 70% to 30% in treated cases, although many are left with residual neurologic and cognitive deficits.

OUTCOME

The patient was started on acyclovir, which was continued for 21 days. He had a prolonged hospital course and continued in rehab for several months. He had persistent mild left-sided weakness.

PREVENTION

There is no vaccine for HSV currently. Whereas prophylactic therapy with a variety of agents is effective in preventing recurrent mucosal disease, this is not a practical method to prevent development of encephalitis because this is a sporadic and rare complication.

FURTHER READING

Jones C: Herpes simplex virus type 1 and bovine herpesvirus 1 latency. Clin Microbiol Rev 16:79, 2003.

Sauerbrei A, Eichhorn U, Hottenrott G, Wutzler P: Virological diagnosis of herpes simplex encephalitis. J Clin Virol 17:31, 2000.

Whitley RJ: Viral encephalitis. N Engl J Med 323:242, 1990.

Whitley RJ, Cobbs CG, Alford CA, et al: Diseases that mimic herpes simplex encephalitis. Diagnosis, presentation, and outcome. JAMA 262:234, 1989.

A 45-year-old white homosexual man was brought by his partner to the emergency department of a general hospital because of **fever, severe headache, nausea,** vomiting, and **mental status changes** that had been **progressive** over the course of the **past 2 weeks.**

The patient had been diagnosed with HIV infection 2 years before and was not currently on antiretroviral therapy.

PHYSICAL EXAMINATION

VS: T 38.5°C, P 106/min, R 18/min, BP 110/62 mmHg

PE: At the ED, the patient was lethargic and disoriented. On exam, **nuchal rigidity** was noted as well as a **positive Kernig sign** (flexion of the neck when the knee is flexed).

LABORATORY STUDIES

Blood

Hematocrit: 34%

WBC: 3100/μL

Differential: 70% PMNs, 12% lymphs

CD4+ cell count: 42/μL

Blood gases: Normal

Serum chemistries: Normal

Imaging

Chest x-ray was normal. A head CT scan was normal.

Diagnostic Work-Up

Table 62-1 lists the likely causes of the patient's illness (differential diagnosis). A clinical diagnosis of meningitis was considered. However, microbiologic diagnosis based on clinical evidence alone is difficult. Investigational approach may include

- **Gram stain** and **acid-fast stain** of CSF and respiratory specimens
- **India ink** stain in CSF or **cryptococcal (latex agglutination)** antigen test in CSF and serum
- **Cultures of CSF** and **blood** for bacteria, fungi, and mycobacteria

TABLE 62–1 Differential Diagnosis and Rationale for Inclusion (consideration)

Bacterial meningitis
  Mycobacterium tuberculosis
  Streptococcus pneumoniae
  Treponema pallidum
Brain abscess
Fungal meningitis
  Coccidioides immitis
  Cryptococcus neoformans
  Histoplasma capsulatum
Viral meningitis (e.g., herpes simplex viruses)

Rationale: Classic bacterial meningitis usually has a more acute and severe presentation than is described in this case. In patients with AIDS, *Cryptococcus neoformans* is the most common cause of meningitis. Tuberculosis is also an important consideration, because it is much more common in AIDS patients than in other populations. Other fungi are uncommon causes of meningitis. Brain abscess will often have associated focal neurologic findings. Syphilis is always a consideration, especially in AIDS, although it often manifests more indolently. Viral causes would not be expected to last this long. Other causes should always be considered, even though clinically, those possibilities are difficult to distinguish.

- In failed tests above
  - CSF VDRL
  - Serologic tests for other blood-borne pathogens included in the differential diagnosis

COURSE

Blood and CSF were collected for microbiology investigation. The CSF examination revealed WBC 80/μL (32% PMNs and 66% lymphs), protein 68 mg/dL, and glucose 46 mg/dL. An India ink preparation of the CSF was diagnostic.

ETIOLOGY

Cryptococcus neoformans (cryptococcal meningitis)

MICROBIOLOGIC PROPERTIES

C. neoformans is a yeast-like fungus (**not dimorphic**), with an **oval, budding yeast cell** (4 to 6 μm). A **thick gelatinous capsule** frequently surrounds each yeast cell. Light microscopy of **India ink-treated CSF** from a symptomatic patient usually reveals the **budding cells, surrounded by a refractile, sharply demarcated capsule** (Fig. 62-1*A*) and is diagnostic. Capsular antigens of the isolates have been classified into serotypes A through D. A **latex agglutination** test can demonstrate cryptococcal **polysaccharide antigen in CSF** and **serum**. Blood, CSF, and a biopsy of the skin lesions are cultured for confirmatory diagnosis, in failed India ink and latex agglutination tests. The fungus **grows well** on most mycologic media (e.g., **Sabouraud-dextrose agar**) at 37°C within 2 to 5 days, and the highly mucoid, creamy-white colonies (Fig. 62-1*B*) are identified by microscopic morphologies and biochemical reactions (e.g., urease positive, phenol oxidase positive) and on the basis of oxidation of certain sugars and KNO_3.

EPIDEMIOLOGY

C. neoformans var. *neoformans* has been isolated from the soil worldwide, usually in association with **bird (pigeon) droppings**. A less common etiologic agent, *C. neoformans* var. *gattii*, has been isolated from eucalyptus trees in tropical and subtropical regions. Sporadic cases occur in all parts of the world. Mainly adults are infected (0.4 to 1.3 cases per 100,000 in the general population). Infection is acquired via **inhalation of airborne fungus**. This is usually an **opportunistic infection** and is not transmitted from person to person. Immunocompromised persons, especially **patients with AIDS**, are **at risk for development of meningitis**, although normal individuals can be affected as well.

PATHOGENESIS

The primary *C. neoformans* infection of lung in most exposed individuals is asymptomatic. Some patients may develop pneumonitis, with fever, chills, cough, and shortness of breath, a result of granulomatous inflammation. Abnormal chest x-ray and positive sputum and blood cultures are diagnostic. *C. neoformans* **causes a deep mycosis in AIDS patients with low**

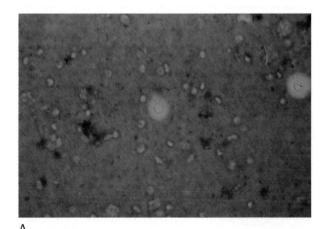
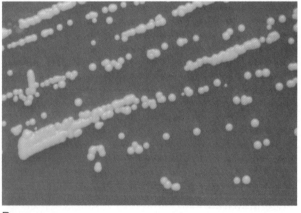

A   B

FIGURE 62-1   *A*, This photomicrograph depicts *Cryptococcus neoformans* from CSF of a patient with meningitis, using a light India ink staining preparation. Note the thick capsule around the two budding yeasts. The cells in the background are the inflammatory cells from CSF. *B*, Culture of *Cryptococcus neoformans* on Sabouraud-dextrose agar medium. Note the highly mucoid, creamy-white, glossy colonies (resembling pasty bacterial colonies). (*A, Courtesy of Dr. Paul Southern, Departments of Pathology and Medicine, University of Texas Southwestern Medical School, Dallas, TX; B, Courtesy of Lisa Forrest, Department of Microbiology, University of Texas Southwestern Medical Center, Dallas, TX.*)

CD4⁺ counts (<100). *C. neoformans* silently spreads from the lung (where the infection may be asymptomatic) to the blood stream. Hematogenous dissemination to the CNS and other organs (e.g., skin) can occur during a primary lung infection or during reactivation of infection years later. **Cryptococci cross the blood-brain barrier and accumulate in the perivascular areas of cortical gray matter** and other areas of the CNS. *C. neoformans* does not elaborate any toxin. **Without the control by cell-mediated immunity, the encapsulated cryptococci multiply and accumulate within the brain parenchyma** (with little necrosis or organ dysfunction), resulting in **macroscopically visible gelatinous pathology.** The inflammatory response to infection is minimal. CNS symptoms include gradual onset with worsening headache and fevers over weeks. More severe cases are associated with altered mental status. CT or MRI head scans are rarely positive for mass lesions in cryptococcal meningoencephalitis.

> **NOTE** Skin involvement after hematogenous dissemination may cause acneiform (nodular) or molluscum-like lesions, ulcers, or subcutaneous tumor-like masses. Cryptococci can be visualized in these lesions by Gomori methenamine silver staining.

TREATMENT

Amphotericin B remains the standard treatment for acute meningitis or pneumonia. The addition of flucytosine (5-FC) is often recommended as combination therapy. **Fluconazole** has been approved for treatment of acute cases but is usually reserved **for maintenance therapy.** Treatment is life long because of frequent relapses. Therapy may be discontinued if sufficient immune restoration occurs from the use of HAART. Itraconazole is used sometimes but is less effective because of poor penetration into the CSF.

OUTCOME

The patient received IV amphotericin B and 5-FC for 2 weeks, with gradual recovery of his mental status. He required some physical therapy and remained in the hospital for a total 3 weeks. He was given fluconazole indefinitely to prevent further relapses.

PREVENTION

Improvement of a patient's immune system through treatment of HIV infection (HAART therapy) is the most important measure of prevention in AIDS patients.

FURTHER READING

Chuck SL, Sande MA: Infections with *Cryptococcus neoformans* in the acquired immunodeficiency syndrome. N Engl J Med 321:794, 1989.

Minamoto GY, Rosenberg AS: Fungal infections in patients with acquired immunodeficiency syndrome. Med Clin North Am 81:381, 1997.

Pappas PG, Perfect JR, Cloud GA, et al: Cryptococcosis in human immunodeficiency virus-negative patients in the era of effective azole therapy. Clin Infect Dis 33:690, 2001.

Perfect JR, Casadevall A: Cryptococcosis. Infect Dis Clin North Am 16:837, 2002.

A 30-year-old white woman was brought to the emergency department of a local hospital with a 2-week history of progressively **severe headache, nausea,** and **vomiting;** several **seizures** had occurred over the past 2 days.

She had been HIV positive for 3 years and had been diagnosed with AIDS a year before the current episode. She had been on HIV therapy, but was **currently failing her regimen.** She was also on aerosolized pentamidine because of a Bactrim allergy. Her brother, who brought her to the ED, could not recall any history of seizures.

PHYSICAL EXAMINATION

VS: T 38°C, P 86/min, R 14/min, BP 104/70 mmHg

PE: She was in moderate distress due to the headache but was able to answer questions. Mild right-sided weakness was apparent on exam.

LABORATORY STUDIES

Blood

Hematocrit: 34%
WBC: 7400/µL
Differential: Normal

CD4+ counts: 62/µL
Blood gases: Normal
Serum chemistries: Normal

Imaging

Head MRI revealed ring-enhancing lesions in the left parietal lobe and right frontal lobe (Fig. 63-1).

Diagnostic Work-Up

Table 63-1 lists the likely causes of illness (differential diagnosis). A clinical diagnosis of encephalitis was considered based on the evidence of brain lesions. Investigational approach may include

- **Cultures of blood and CSF** for bacteria and *Cryptococcus neoformans*
- **Serology.** A significantly high titer of IgG specific for *Toxoplasma gondii* in a single serum may support the diagnosis of *Toxoplasma* encephalitis.

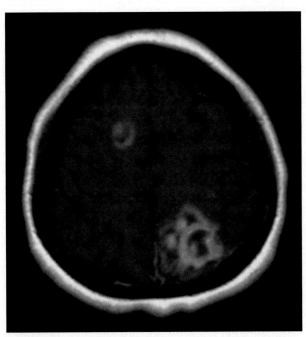

FIGURE 63–1 Head MRI scan of this patient with seizure illness. Note ring-enhancing lesions in the left parietal lobe and right frontal lobe in this patient. *(Courtesy of Dr. Daniel Skiest, Department of Medicine, University of Texas Southwestern Medical Center, Dallas, TX.)*

TABLE 63–1 Differential Diagnosis and Rationale for Inclusion (consideration)
Cryptococcus neoformans
Listeria monocytogenes
Mycobacterium tuberculosis
Nocardia asteroides
Primary CNS lymphoma
Progressive multifocal leukoencephalopathy (PML): JC virus
Toxoplasma gondii

Rationale: Focal brain lesions are due to only a few causes in patients with AIDS. *T. gondii* and CNS lymphoma—the most common causes—can usually be differentiated based on lab and radiologic findings, although clinically, they may manifest similarly. The first three organisms are more likely to cause meningitis, as opposed to mass lesions with neurologic deficits. The other causes are much less common, and they may or may not exhibit ring enhancement. PML does not usually cause mass lesions.

- In failed investigation, the following tests can be undertaken:
  - **Histology** of brain biopsy
  - Detection of genetic material by **PCR** in CSF and biopsy

COURSE

The patient was admitted to the hospital for further evaluation and lab investigation. Lumbar puncture revealed normal opening pressure, and CSF examination showed WBCs 45/µL, protein 78 mg/dL, glucose 64 mg/dL. Bacterial, fungal, and viral cultures of CSF were all negative. Based on results of a positron emission tomography (PET) scan and serology tests, the patient was empirically started on therapy.

ETIOLOGY

Toxoplasma gondii (*Toxoplasma* encephalitis)

MICROBIOLOGIC PROPERTIES

T. gondii is an obligate intracellular parasite that exists in two forms. The proliferative form, **banana-shaped tachyzoites**, is seen in tissues (e.g., CSF; Fig. 63-2) in the active stage (postreactivation) of a chronic infection or a primary infection. The resting cyst form, **slow-growing bradyzoites**, is found in muscle and brain during (asymptomatic) chronic infection.

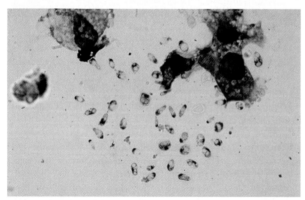

FIGURE 63–2 *Toxoplasma gondii* tachyzoites in CSF. Note the banana-shaped organisms. *(Courtesy of Dr. Paul Southern, Departments of Pathology and Medicine, University of Texas Southwestern Medical Center, Dallas, TX.)*

EPIDEMIOLOGY

Toxoplasmosis occurs worldwide but is absent from areas where there are no cats, such as isolated Pacific Islands. Cats are the definitive host, as sexual stages occur only in cats. Oocysts produced in the cat intestine are excreted in feces and develop into the infectious form. Cattle and pigs feeding on cat feces are intermediate hosts. Humans are infected by (1) **ingesting pseudocysts in undercooked meat** (mutton, pork, beef) or (2) accidental **ingestion of material contaminated with cat feces containing oocysts. AIDS patients** and immunocompromised individuals are **at risk** for infection. **Transplacental crossing of the blood-borne parasite** from a pregnant woman to the fetus causes congenital infection. Congenital infection rate in the United States is 1 to 3 cases per 1000 live births.

PATHOGENESIS

Following ingestion of tissue pseudocysts or oocysts, sporozoites are released in the intestinal tract. Sporozoites transform into invasive tachyzoites in the enterocytes. The invasive tachyzoites penetrate the intestinal mucosa and disseminate into various organs. Specialized structures at the anterior end of the *T. gondii* cell aid in invasion of the cells in target tissues (e.g., brain). Invasion is receptor mediated. Inside the cell, the **intracellular organism** is surrounded by a membrane-bound vacuole called the phagosome. The organisms inhibit the fusion of lysosome with phagosome. Inside the phagosome, the organisms differentiate into tachyzoites and begin to divide rapidly. Infection with *T. gondii* mounts both humoral and cell-mediated immune (CMI) responses. In immunocompetent individuals, both humoral and CMI responses slow division of tachyzoites. Macrophages are activated with production of interferon-γ, and parasite-specific cytotoxic T lymphocytes (CTLs) of the CD8+ phenotype are produced. Activated macrophages and the CTLs kill extracellular organisms and tachyzoite-infected cells. Dormant bradyzoites develop, however, and remain sequestered in the brain matrix as long as the immune system remains active.

In patients with AIDS, *T. gondii* is the most common cause of **intracerebral mass lesions** and is thought to be caused by reactivation of chronic infection. **In patients with AIDS with CD4+ cell** count less than 200, **bradyzoites reactivate** and transform into tachyzoites, which evade killing and continue to multiply. The **replicating tachyzoites rupture the brain cells, resulting in focal necrosis**. Histopathology of tissue sections may demonstrate inflammatory cells as well as numerous tachyzoites of *T. gondii*. MRI brain scan or, better still, a PET scan may aid the diagnosis. A **significant high level (≥1:256) of IgG for *T. gondii*, however, strongly suggests the diagnosis** (a base level IgG in the absence of *Toxoplasma* encephalitis in the AIDS population is 1:16). Table 63-2 lists the other important clinical features of *T. gondii* infections in other risk groups.

TREATMENT

Sulfadiazine and **pyrimethamine** with or without leucovorin are the drugs of choice. The treatment in patients with AIDS does not eradicate infection, but it reduces severity and length of infection. In pregnancy, treatment of the mother may reduce the incidence of congenital infection and reduce sequelae in the infant; therefore prompt and accurate diagnosis is important. Most infants with subclinical infection at birth will subsequently develop signs or symptoms of congenital toxoplasmosis unless the infection is treated.

OUTCOME

The patient was started on pyrimethamine and sulfadiazine with leucovorin, based on a PET scan and a positive serology for *T. gondii*. She gradually improved over the course of a week, and her dose was reduced after 2 weeks of therapy. Repeat MRI showed the lesions had significantly decreased in size.

PREVENTION

Individuals with HIV who are at risk for toxoplasmosis are advised to avoid inadequately cooked meat or contact with cat feces. Susceptible pregnant women are advised to avoid cats or cat-soiled items, and those at risk for infection require serial testing. It is recommended that an initial test early in the first trimester be performed, with retesting at 20 to 22 weeks to identify women who have undergone seroconversion during the first half of pregnancy. This allows identification of infected fetuses sufficiently early to start effective antibiotic therapy.

TABLE 63–2 Other Important Clinical Features of *Toxoplasma gondii* Infections

Syndromes	Clinical Features
Toxoplasmosis in pregnant women	Seronegative pregnant women acquiring infection may present with **mononucleosis** or **flu-like symptoms**. The most common manifestations are nontender lymphadenopathy, fatigue, fever, headache, malaise, and myalgia. **Immunity screen of pregnant or childbearing women is important** and is based on high IgG titer (newborn is protected via passive immunity from mother).
Congenital toxoplasmosis	**Congenital toxoplasmosis (birth defects)** results from an acute primary infection of the fetus by *T. gondii* acquired by a seronegative woman during pregnancy. The incidence and severity of congenital toxoplasmosis vary depending on trimester during which infection was acquired. The infected infants usually are **asymptomatic at birth** but later manifest a wide range of signs and symptoms, including chorioretinitis, epilepsy, and psychomotor retardation. Retinochoroiditis is one of the common manifestations. **The brain may be damaged, causing hydrocephalus** and calcification. A significant titer of *Toxoplasma*-specific IgM in the cord blood of the newborn is indicative of infection. If IgG is measured, the titer in the cord blood of the infected newborn maintains a high level or becomes elevated with time (half-life of maternal IgG is 20 days).

FURTHER READING

Denkers EY, Gazzinelli RT: Regulation and function of T-cell-mediated immunity during *Toxoplasma gondii* infection. Clin Microbiol Rev 11:569, 1998.

Dubey JP, Lindsay DS, Speer CA: Structures of *Toxoplasma gondii* tachyzoites, bradyzoites, and sporozoites and biology and development of tissue cysts. Clin Microbiol Rev 11:267, 1998.

Guerina NG, Hsu H-W, Meissner HC, et al: Neonatal serologic screening and early treatment for congenital *Toxoplasma gondii* infection. N Engl J Med 330:1858, 1994.

Luft BJ, Brooks RG, Conley FK, et al: Toxoplasmic encephalitis in patients with acquired immune deficiency syndrome. JAMA 252:913, 1984.

Wallace MR, Rossetti RJ, Olson PE: Cats and toxoplasmosis risk in HIV-infected adults. JAMA 269:76, 1993.

A 28-year-old Hispanic man was brought to the emergency department of a general hospital for **severe headaches** and **two generalized seizures.**

He had first noticed headaches, which had been getting more frequent, several weeks before. He denied fevers or chills. The patient was a **recent immigrant from Mexico.** There was **no previous history of seizures.**

PHYSICAL EXAMINATION

VS: T 37°C, P 83/min, R 14/min, BP 136/80 mmHg
PE: Young male in moderate distress due to headache; neurologic exam was normal.

LABORATORY STUDIES

Blood

Hematocrit: 45%

WBC: 7200/µL

Differential: 52% PMNs, 23% lymphs, **12% eosinophils**

Blood gases: Normal

Serum chemistries: Normal

Imaging

A **CT scan** of his brain revealed **an intracranial calcified cyst,** and further imaging with MRI confirmed the presence of similar lesions, some with a scolex visible (Fig. 64-1).

Diagnostic Work-Up

Table 64-1 lists the likely causes of illness (differential diagnosis). The presumptive diagnosis should be on the basis of clinical picture and epidemiologic information. The likelihood of exposure during foreign travel, and characteristic findings on CT or MRI scans, are adjuncts to clinical diagnosis. Lumbar puncture and peripheral blood collection are an essential beginning of investigation. Microbiologic investigation may include

- **Gram stain** and **acid-fast stain** of CSF
- **Cultures of CSF** and **blood** for bacteria, fungi, mycobacteria
- **Antibody detection** (e.g., immunoblot assay of IgG in serum or CSF specific for the invasive stage of a parasite)

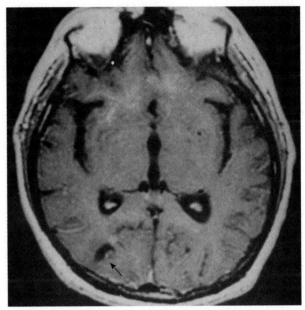

FIGURE 64–1 Head MRI scan. Note an intracranial cyst and a scolex (*arrowhead*) in the right occipital lobe. (*Courtesy of Dr. Paul Southern, Departments of Pathology and Medicine, University of Texas Southwestern Medical Center, Dallas, TX.*)

TABLE 64–1 Differential Diagnosis and Rationale for Inclusion (consideration)
Brain abscess
Craniopharyngioma
Cryptococcal meningoencephalitis
Medulloblastoma
Neurocysticercosis
Toxoplasmosis
Tuberculoma

Rationale: Intracerebral mass lesions have many possible causes (common ones are listed above). Homogeneous masses may be malignancies, and ring enhancement is classically associated with brain abscess. Cystic masses may be malignant, but they are also classically associated with neurocysticercosis. Other infectious causes of intracerebral masses include toxoplasmosis, tuberculoma, and cryptococcosis.

- In failed tests above
  - Cryptococcal antigen in CSF and serum
  - CSF VDRL
  - Serologic tests for others included in differential diagnosis

COURSE

The patient was admitted to the hospital for observation and diagnostic investigation. A lumbar puncture was performed. The CSF analysis revealed a WBC count of 35/μL, with lymphocytic and eosinophilic pleocytosis, slightly elevated protein (62 mg/dL), and a normal glucose level. A positive serologic investigation specific for a parasite yielded the diagnosis.

ETIOLOGY

Taenia solium (neurocysticercosis)

MICROBIOLOGIC PROPERTIES

The adult pork tapeworm has an attachment organ (**scolex**; Fig. 64-2) with suckers or grooves, and the area behind the scolex (neck region) is linked to a long chain of segments (proglottids). Mature worms contain male and female reproductive organs (hermaphroditic). Gravid worms contain eggs in the uterus.

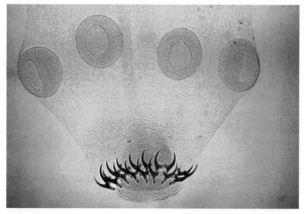

FIGURE 64–2 Scolex of *Taenia solium*. Note four suckers, and two rows of hooks. (*Courtesy of Centers for Disease Control and Prevention, Atlanta, GA.*)

EPIDEMIOLOGY

Cysticercosis occurs worldwide; the disease is particularly frequent in the rural regions of the developing countries, where sanitary conditions permit pigs to have access to human feces. Prevalence is highest in parts of **Africa**, Latin America, Southeast Asia, and Eastern Europe. Pigs are the intermediate hosts for *T. solium*. **Ingestion of uncooked or inadequately cooked meat with *T. solium* larvae may cause the development of human cysticercosis** in various forms (subcutaneous, ocular, and CNS).

PATHOGENESIS

In the human intestine, the cysticercus from uncooked meat develops over 2 months into an adult tapeworm. The adult tapeworm attaches to the small intestine by its scolex and remains alive in the small intestine for many years. Adults produce proglottids, which mature, become gravid, detach from the tapeworm, and migrate to the anus or are passed in the stool. Eggs are also passed with feces and can survive for days to months in the environment.

Ingested eggs hatch in the intestine. The embryo penetrates the mucosa and is carried via the circulation to a site of development (e.g., brain). The embryo grows into a cysticercus, which contains a single inverted scolex (Fig. 64-3). The cysticercus is

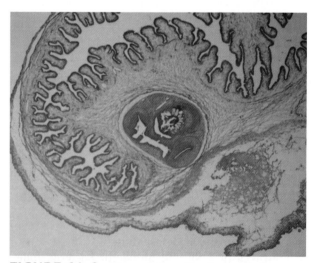

FIGURE 64–3 Histopathology of cysticercosis in low power field. Note a single *Taenia solium* larva in cysticercus (a fluid-filled cyst), which is about 1 to 2 cm in diameter and contains an inverted scolex. (*Courtesy of Dr. Dominick Cavuoti, Department of Pathology, University of Texas Southwestern Medical Center, Dallas, TX.*)

deposited in the cerebral parenchyma. A **mass effect of the larval cyst on the brain parenchyma causes seizures** (seen in the case patient). The **appearance of a scolex on an MRI scan** (see Fig. 64-1) **supports the diagnosis of neurocysticercosis**. Serologic evaluation of **IgG** specific for *T. solium* usually **confirms the clinical diagnosis of cysticercosis**. The CDC immunoblot assay is based on detection of IgG antibody (in serum or CSF) to purified structural glycoprotein antigens from the larval cysts of *T. solium*.

NOTE Unless large numbers of cysts are present, the body's immune system does not act to destroy the *T. solium* organisms, and cysts can live for many years undetected. Cysticerci may also be found in subcutaneous tissues and eye. The organisms cause stationary **subcutaneous masses,** which must be distinguished from lipomas. **Ocular cysticercosis** may occur when aqueous or vitreous humor interferes with vision. Neurologic symptoms may also arise when the encysted worm dies and the host mounts an inflammatory response. **Intraventricular cysts cause hydrocephalus.**

TREATMENT

Asymptomatic cysts and easily controllable seizures probably do not require treatment. Hydrocephalus from intracranial hypertension may require CSF shunting procedures. Medical treatment is possible using **praziquantel** or albendazole, but there is not universal agreement on the indications for the use of these drugs. Corticosteroids may be needed to reduce brain swelling and edema. Antiepileptic drugs are needed to control seizures.

OUTCOME

The patient received praziquantel, anti-inflammatory medications to reduce his brain swelling, and antiepileptic drugs to control his seizures. Serology results confirmed neurocysticercosis. The patient gradually recovered over the course of the next 30 days.

PREVENTION

Avoiding consumption of raw and undercooked pork and beef is the most effective way to prevent these infections. Transmission may be reduced by improving sanitary conditions in the endemic countries. Proper inspection of meat is also important.

FURTHER READING

Del Brutto OH, Rajshekhar V, White AC, Jr: Proposed diagnostic criteria for neurocysticercosis. Neurology 57:177, 2001.
Garcia HH, Gonzalez AE, Evans CAW: *Taenia solium* cysticercosis. Lancet 362:547, 2003.
Garcia HH, Pretell EJ, Gilman RH, et al: A trial of antiparasitic treatment to reduce the rate of seizures due to cerebral cysticercosis. N Engl J Med 350:249, 2004.

A 31-year-old white man was brought in to the emergency department (ED) of a general hospital with the complaints of fever and **visual hallucinations.**

Four days before his arrival at the ED, the patient developed malaise and back pain while **working on a roadside clean-up crew.** The next day, he sought medical attention, complaining of **muscle pains, vomiting,** and abdominal cramps, which were treated with acetaminophen.

PHYSICAL EXAMINATION

VS: T 37.8°C, P 116/min, R 28/min, BP 168/104 mmHg

PE: The patient was alert, with increased tone in the right forearm and **hyperesthesia over the entire right side of the body.** During the next 12 hours, he **became increasingly agitated and less well oriented.** His condition worsened, and he **developed hydrophobia,** because even the sight of water set off severe spasms of the neck and chest. **Hypersalivation** and **wide fluctuations in body temperature** and **blood pressure** were also noted.

LABORATORY STUDIES

Blood

Hematocrit: 46%
WBC: Normal
Differential: Normal
Blood gases: Normal
Serum chemistries: Normal

Imaging

A chest x-ray was normal; no abnormal findings were found on brain CT.

Diagnostic Work-Up

Table 65-1 lists the likely causes of illness (differential diagnosis). A definitive history of an animal bite could not be established for this patient. Investigational approach may include:

- **DFA staining** of nuchal skin **biopsy** sections
- **RT-PCR** assay of CSF, saliva, or nuchal skin biopsy
- In failed diagnosis, **viral serology**

COURSE

Blood was drawn, a lumbar puncture was performed, and CSF was collected. Analysis of the CSF showed WBCs 57/μL (90% lymphs), protein 88 mg/dL, and glucose 66 mg/dL. The patient was admitted to the hospital, and his condition deteriorated. Six days later,

TABLE 65–1 Differential Diagnosis and Rationale for Inclusion (consideration)
Arboviral encephalitis
Guillain-Barré syndrome
Herpes simplex virus type-1 encephalitis
Lyme meningoencephalitis
Poliomyelitis
Rabies

Rationale: Altered consciousness is the sine qua non of encephalitis in the setting of other signs of infection, such as fever, headache, and neurologic signs. Some degree of weakness would be expected with Guillain-Barré or polio. However, the findings of agitation and hydrophobia are specific for rabies encephalitis. If these are not present, it is difficult to distinguish the various causes listed above.

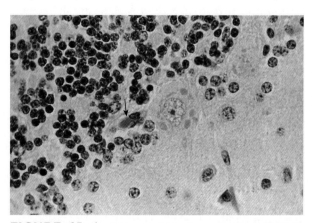

FIGURE 65–1 Histopathologic changes in biopsy (H & E) stain. Negri body," an eosinophilic intracytoplasmic mass, made up of a finely fibrillar matrix and rabies virus particles (*arrow*) is seen. (*Courtesy of Centers for Disease Control and Prevention, Atlanta, GA.*)

he was intubated and heavily sedated. Samples for testing included a nuchal skin biopsy, which tested positive for neurotropic virus by DFA test. Saliva and skin subsequently tested positive by RT-PCR assay.

ETIOLOGY

Rabies virus (rabies encephalitis)

MICROBIOLOGIC PROPERTIES

Rabies virus belongs to the Rhabdovirus family. A **bullet-shaped nucleocapsid** is surrounded by a lipoprotein envelope. **Envelope glycoproteins are arranged in knob-like structures**. This virus has ssRNA with negative polarity and contains RNA-dependent RNA polymerase. The virus grows in tissue culture but does not destroy cells (no cytopathic effect is visualized).

EPIDEMIOLOGY

Worldwide, an estimated 35,000 to 40,000 persons die of rabies annually, almost all in developing countries. Rabies exists in two epidemiologic forms. **Urban rabies** is transmitted by dogs, whereas **sylvatic rabies** is a disease of wild carnivores and **bats**. Some of the most common animal species that transmit rabies are coyotes, foxes, raccoons, and bats. Cats and dogs are important as well, given the frequent contact with people—especially in areas, such as developing countries, where vaccination is not routine. Transmission to humans is by a bite or scratch from an infected animal with virus-laden saliva. Person-to-person transmission is also possible, via saliva or aerosolized froth from an infected person. On rare occasions, transplantation of infected tissues may cause disease. In June 2004, the Centers for Disease Control and Prevention confirmed diagnoses of **rabies in four recipients** (in Alabama, Arkansas, Oklahoma, and Texas) **of transplanted organs from their common donor**, who was found subsequently to have serologic evidence of rabies infection. Infection with rabies virus likely occurred via neuronal tissue contained in the transplanted organs because rabies virus is not spread hematologically.

PATHOGENESIS

The **incubation period for rabies virus is usually** 3 to 8 weeks, although it can be as short as 9 days or as long as 7 years depending on the severity of wound, site of wound, and its distance from the brain. The virus enters cells of the mucous membranes at the bite site and within minutes multiplies locally. The virus buds from the cells but does not lyse them. Chronic infection is established when the progeny virus infects the sensory neurons in the peripheral nervous system. **Viral glycoproteins bind to acetylcholine receptors, contributing to neurovirulence of the rabies virus**. The virus moves up by axonal transport to the CNS and then multiplies in the CNS. There is no viremic stage. Once in the brain, the **virus replicates exclusively within the gray matter**. It then **travels down the peripheral nerves to the salivary glands** and other body tissues, including lungs, kidney, and skin. Secondary viral replication in mucinogenic acinar cells of salivary glands facilitates secondary transmission via saliva.

There is little cytopathic effect and almost no cellular infiltration. The most **characteristic pathologic finding of rabies in the CNS** is the formation of **intracytoplasmic inclusions**, called "**Negri bodies**" (Fig. 65-1), **found in neurons** (most frequently in the pyramidal cells of Ammon's horn and in the Purkinje cells of the cerebellum). They can be detected by immunofluorescence staining of biopsy specimens, which is more sensitive than histology. **Virus-specific cytotoxic T cells, induced by the viral glycoprotein, kill the virus-infected neurons, causing brain damage**, and **encephalitis develops**. The disease progresses to paralysis, and painful spasms of swallowing muscles lead to avoidance (fear) of water (**hydrophobia**). Convulsions follow, and the final stage is coma. The **disease is essentially 100% fatal**; death is due to respiratory failure. The few reported cases of survival are all associated with immediate postexposure prophylaxis.

TREATMENT

There is **no antiviral therapy** for a patient with rabies. Only supportive care is available.

OUTCOME

After removal of all sedatives, the patient showed no purposeful movement, and there was loss of brain stem reflexes. He died 2 weeks later.

PREVENTION

Individuals at high risk (e.g., veterinarians, park rangers, Peace Corps members) should receive **pre-exposure immunization**. Human diploid cell rabies vaccine (HDCV), an inactivated vaccine prepared from virus grown in human diploid cell culture, is an approved vaccine. Three-step **postexposure prophylaxis** is available for the exposed individuals and their contacts:

1. Treatment of the bite wound or contact with salivary secretions: immediate and thorough cleaning with soap and water
2. Passive immunization with rabies antiserum, provided by administration of human rabies immune globulin, which neutralizes the virus at the bite wound site
3. Active immunization with antirabies HDCV vaccine

The decision to give postexposure prophylaxis (i.e., step 2 and 3) depends on a variety of factors, such as area of endemicity (e.g., Texas communities bordering with Mexico), type of animal (all wild animal attacks require immunization), or whether an attack by a domestic animal was spontaneous or provoked.

FURTHER READING

Hemachudha T, Laothamatas J, Rupprecht CE: Human rabies: A disease of complex neuropathogenetic mechanisms and diagnostic challenges. Lancet Neurol 1:101, 2002.

Jackson AC, Warrell MJ, Rupprecht CE, et al: Management of rabies in humans. Clin Infect Dis 36:60, 2003.

Moran GJ, Talan DA, Mower W, et al: Appropriateness of rabies postexposure prophylaxis treatment for animal exposures. JAMA 284:1001, 2000.

A 9-day-old female **newborn** was taken to a hospital by her parents, who reported a 10-hour history of an **inability to nurse and difficulty in opening her jaw.**

The patient's parents had noticed a **foul-smelling discharge from her umbilical cord** during the preceding 2 days.

PHYSICAL EXAMINATION

VS: T 38°C, P 130/min, R 36/min, BP 94/48 mmHg

PE: The newborn had **trismus, opisthotonus** (Fig. 66-1), and **hyper-responsiveness** to external stimuli. The umbilical cord was covered with dry clay, which when retracted revealed **a foul-smelling yellow-green discharge**.

LABORATORY STUDIES

Blood

Hematocrit: 38%
WBC: 12,400/μL
Differential: 85% PMNs, 10% lymphs
Blood gases: pH 7.34, pCO_2 46, pO_2 75
Serum chemistries: Normal

Imaging

Chest x-ray was normal; head CT scan or MRI was not done.

FIGURE 66–1 Opisthotonus in a newborn. *(Courtesy of Centers for Disease Control and Prevention, Atlanta, GA.)*

Diagnostic Work-Up

Table 66-1 lists the likely causes of illness (differential diagnosis). Tetanus is a **clinical diagnosis**. Electromyography may help in uncertainty, and **cultures of umbilical stump** wound may be necessary to determine the qualitative nature and extent of wound infection.

COURSE

The patient was admitted to the hospital. Culture from the umbilical cord grew several anaerobic organisms (*Clostridium perfringens*, *C. sporogenes*) and *Staphylococcus*, *Streptococcus*, and *Bacillus* spp.

TABLE 66–1   Differential Diagnosis and Rationale for Inclusion (consideration)
Dystonic reaction to dopamine blocker
Meningitis (bacterial or viral)
Neonatal tetanus
Neurologic disorder (infectious or noninfectious cause)
Sepsis
Strychnine poisoning

Rationale: A clinical diagnosis of tetanus should be considered. The characteristic symptoms and signs of this syndrome in neonates have a very limited differential. Dystonic reactions this severe would not be common and are caused by a limited number of drugs. Strychnine poisoning, a second possibility, manifests similarly and should always be ruled out. In neonates and infants, there may be limited clinical signs from which to make a diagnosis. Fever alone is enough to warrant a complete sepsis work-up in a baby younger than 1 month old. The presence of paralysis may suggest a neurologic disorder, such as tetanus. Meningitis would be expected to cause lethargy rather than paralysis.

ETIOLOGY

C. tetani (tetanus)

MICROBIOLOGIC PROPERTIES

Culture of polymicrobic infection rarely grows the causative agent, *C. tetani*. The organism is a **Gram-positive anaerobic rod**. It forms **terminal spores**. Spores germinate in the wound sites, producing an extremely potent exotoxin, tetanospasmin.

EPIDEMIOLOGY

C. tetani spores are **ubiquitous in the soil environment**. The bacteria are found in abundance in the **GI tract of animals** (e.g., horse). **Transmission occurs via injury and trauma, where there is wound contamination with soil.** In the United States, most cases of tetanus are caused by an acute injury (e.g., puncture wound, laceration). Neonatal tetanus is a major cause of death in some developing nations and is not uncommon in the United States. Because of her family's philosophic beliefs, the mother of the case patient had never been vaccinated. A licensed "direct-entry" midwife attended her throughout her pregnancy. The parents, under the supervision of the midwife, took care of the newborn. For umbilical cord care, the parents applied a "Health and Beauty Clay" (bentonite) powder provided by the midwife. The family lived in a rural area, in a house adjacent to a horse pasture. Although the newborn and her mother stayed primarily indoors, the family's dog often ran between the house and the pasture.

PATHOGENESIS

Spores of *C. tetani* from soil environment contaminate the wound. Once the toxin is synthesized, it moves from the contaminated site to peripheral motor neuron terminals, enters the axon, and travels to the spinal cord within 2 to 14 days. When the toxin reaches cell bodies in the brain stem and spinal cord, it diffuses to terminals of neurons. The blood-brain barrier prevents direct entry of toxin into the central nervous system. **Tetanospasmin is a Zn++-dependent endo-**peptidase, which cleaves synaptobrevin, a vesicle-associated membrane protein critical to the "docking and fusion" apparatus of neuroexocytosis. The **toxin blocks the release of the inhibitory neurotransmitters glycine and gamma-aminobutyric acid by the descending inhibitory neurons, leaving lower motor neurons uninhibited and resulting in muscle rigidity**. Autonomic system involvement also results in a hypersympathetic state caused by the uninhibited release of adrenal catecholamines. Localized or cephalic tetanus may occur initially, followed by **generalized tetanus, manifesting as trismus ("lockjaw")**, risus sardonicus (sardonic smile) in adults, abdominal rigidity, **opisthotonus**, severe pain, and respiratory compromise. Trismus is caused by tetanic spasm of the masseter muscles, preventing opening of mouth. Continuous stimulation contributes to spastic paralysis. Patients die of respiratory failure due to chest muscle collapse.

TREATMENT

The most important consideration for management of a patient with tetanus is to **maintain the airway and intubate** if necessary. A feeding tube should be placed, as high caloric demands are required. Benzodiazepines are used as needed to control spasms and rigidity. Beta-blockers such as labetalol may be used to control sympathetic hyperactivity. **Passive immunization** with human tetanus immune globulin should be administered in addition to **active immunization with tetanus toxoid**. **Metronidazole** is the drug of choice because penicillin may act as a GABA antagonist.

OUTCOME

The newborn was treated with tetanus immune globulin and IV metronidazole for 10 days. She was discharged 19 days later, with no apparent neurologic sequelae.

PREVENTION

Initially, a full series (3 injections) of tetanus toxoid (adult Td, child DPT) is needed, with a booster every 10 years. The elderly, especially women, often need a

Td booster as part of general medical maintenance. A fresh tetanus-prone wound in a patient without an adequate immunization history (>5 years since booster) warrants administration of human tetanus immune globulin (HTIG) in addition to a tetanus toxoid booster at the initial encounter. Interestingly, a case of tetanus does not render the patient immune, so tetanus toxoid series or boosters must be given throughout life.

FURTHER READING

Bowie C: Tetanus toxoid for adults—too much of a good thing. Lancet 348:1185, 1996.

CDC: Tetanus among injecting-drug users—California, 1997. MMWR 47:149, 1998.

CDC: Shortage of tetanus and diphtheria toxoids. MMWR 49:1029, 2000.

SECTION V

Skin, Wound, and Multisystem Infections

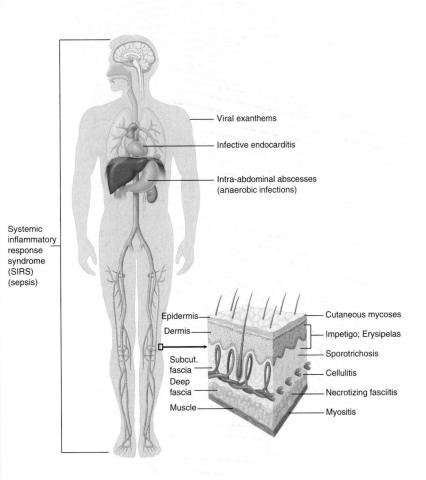

Viral exanthems

Infective endocarditis

Intra-abdominal abscesses (anaerobic infections)

Systemic inflammatory response syndrome (SIRS) (sepsis)

Epidermis

Dermis

Subcut. fascia

Deep fascia

Muscle

Cutaneous mycoses

Impetigo; Erysipelas

Sporotrichosis

Cellulitis

Necrotizing fasciitis

Myositis

INTRODUCTION TO SKIN, WOUND, AND MULTISYSTEM INFECTIONS

The skin is constantly exposed to and colonized with a variety of organisms from the environment. It is an effective barrier against invasion by most pathogens. Skin is divided into three distinct layers: the **epidermis, dermis,** and **subcutis/fascia.** Any or all of these layers may be involved by infectious agents causing pathology. **Folliculitis** is an infection of the hair follicles that consists of small, erythematous, often pruritic, lesions. Local therapy (warm compresses and topical antibiotics) is often sufficient treatment. **Furuncles** and **carbuncles** usually develop from folliculitis, with carbuncles involving deeper tissues and manifesting with systemic symptoms, but rarely bacteremia. Antibiotic therapy and occasionally surgical therapy are indicated. **Impetigo** is a superficial infection involving the epidermis, usually due to β-hemolytic group A streptococcus (BHGAS). It presents as vesicles that have a purulent discharge and crusting, and it is highly contagious. Large vesicles, or bullae, can be seen with infection due to *Staphylococcus aureus*. There are generally no systemic symptoms. Resolution leaves no scars, as deeper tissues are not involved.

Erysipelas is an acute inflammation of the dermis, primarily involving lymphatic vessels. It is usually caused by BHGAS. Fever and leukocytosis may occur, and the affected area characteristically has sharp borders ("butterfly" wing rash on the face). **Cellulitis** involves all layers of skin to the subcutaneous tissue and typically causes fever and leukocytosis and occasionally bacteremia. Although BHGAS and *S. aureus* are the most common etiologies, a variety of other organisms may be responsible, such as *Erysipelothrix* spp (following exposure to saltwater fish or other meats), *Aeromonas* spp (following trauma while swimming in fresh water), or *Vibrio* spp (following trauma while in salt water). When BHGAS or *S. aureus* is the suspected cause of cellulitis, an appropriate penicillin is generally adequate, and surgery is not usually indicated unless abscesses are present.

An **abscess** is a localized collection of purulent material (pus), and its formation usually involves efforts by host defenses to localize, or "wall off," potentially spreading, pyogenic infection. The pathogenesis of abscess involves (1) a characteristic location (e.g., bone marrow in osteomyelitis); (2) a primary process (e.g., injury) as the predisposing host factor; and (3) an extension of (common or uncommon) resident flora beyond anatomic boundaries to a deeper tissue plane.

Osteomyelitis is often seen with diabetic or ischemic foot ulcers, decubitus ulcers, skull/facial bones, and contaminated open fractures. **Intra-abdominal infection** due to anaerobes manifests as diverticulitis, ischemic bowel, liver abscess, pylephlebitis (septic portal vein thrombosis), occasionally biliary tract infection, and postsurgical infections. It may also be the primary presentation of a ruptured appendix (appendicitis), secondary peritonitis (**retroperitoneal abscess**), or inflammatory bowel disease. Intra-abdominal abscesses are uncommon and are almost invariably mixed infections, with enteric Gram-negative rods and anaerobic bowel flora (*Bacteroides fragilis*) predominating. Broad-spectrum antibiotics with anaerobic coverage and surgical drainage are required. **Soft tissue infections** due to **anaerobic bacteria** include infected diabetic ulcers, decubitus ulcers, bite wounds, burn wounds, infected subcutaneous cysts, and life- and limb-threatening infections of deep fascia and muscle (e.g., **necrotizing fasciitis**, clostridial myonecrosis [**gas gangrene**]).

Gangrene is an advanced stage of cellulitis that has led to significant tissue necrosis. A strong clue to diagnosis is gas in the soft tissues. Typical organisms are streptococci, mixed infection with anaerobes, and clostridial infection (classic gas gangrene). Cultures often yield a mixture of organisms that include *Bacteroides fragilis, Prevotella, Porphyromonas, Fusobacterium,* group A or other streptococci, staphylococci, and Gram-negative aerobic or facultative rods. **Fournier gangrene** is mixed aerobic-anaerobic necrotizing cellulitis of the scrotum. It is usually a rapidly progressive infection, and wide surgical débridement in addition to antibiotics is required.

Necrotizing fasciitis (NF) is a rare but life-threatening infection of subcutaneous tissues, most often caused by BHGAS. NF is a focus of spreading streptococcal infection that causes bacteremia and systemic inflammatory response syndrome (SIRS), resulting in widespread desquamation of the skin, shock, multiorgan failure, and death in more than 30% of cases. SIRS is caused by specific **exotoxins** produced at the site of infection that have systemic effects. More than 50% of patients have experienced a recent minor trauma, surgery, or varicella infection. Since the mid 1980s, aggressive life-threatening group A streptococcal invasive infections have increased in incidence, although they are sporadic in nature. These types of infections are rare (3 to 5 per 100,000), but they can occur in healthy adult individuals of an age not normally targeted by BHGAS (20 to 45 years). If NF is present, the infection may have begun with a break in the skin, with invasion or deep blunt trauma (hematoma) that extends from superficial skin layers down through fascia and muscle. Aggressive surgical intervention is required to prevent the spread of infection, in addition to antibiotics.

Head and neck infections due to anaerobes include chronic otitis media, chronic sinusitis, peritonsillar abscess, odontogenic infections, and deep fascial space infections (retropharyngeal and prevertebral space abscesses). Uncommon anaerobic abscesses are (1) peritonsillar abscess with *Fusobacterium necrophorum* infection of the lateral pharyngeal space, with septic jugular vein thrombosis and metastatic abscesses in other organs (Lemierre syndrome); (2) submandibular space infection, usually mandibular odontogenic, with mixed aerobes/anaerobes; swelling in the floor of mouth and in the neck, drooling, trismus, dyspnea (Ludwig angina); and (3) acute necrotizing ulcerative gingivitis caused by *Fusobacterium* and fusospirochetosis ("trench mouth").

Fungi often cause **superficial infections** of the skin, hair, and nails. Most of these infections are caused by a group of fungi called dermatophytes. They infect the epidermis only because they are unable to survive at body temperature of 37°C. Common fungi are *Microsporum, Epidermophyton,* and *Trichophyton.* Multiple syndromes are recognized, based on the infecting fungus and the site of infection. Topical therapy may be adequate, but systemic antifungals are also highly effective. Sporotrichosis, caused by *Sporothrix schenckii,* is a specific **subcutaneous infection**, resulting from minor trauma. This has been classically associated with rose thorns, but most patients do not give this history.

Several viral infections, including **measles, rubella,** and **varicella-zoster (chickenpox),** cause febrile exanthematous illnesses, manifesting with characteristic diffuse rashes. Effective vaccines are available for all these infections, and they are uncommonly seen in areas with high rates of immunization. Measles and rubella both produce a widespread maculopapular rash, although patients with measles tend to be more systemically ill. In addition, measles is associated with lesions, called Koplik spots, on the buccal mucosa that precede the rash and are considered pathognomonic. Chickenpox produces an initially vesicular rash that progresses to pustular lesions, in varying stages at any one time. This is in contrast to smallpox, in which the lesions are generally all at the same stage. Cytomegalovirus, a very common infection in childhood, usually causes asymptomatic infections. It can, however, be a significant problem in immunocompromised patients, particularly transplant patients and those with AIDS,

in whom it causes multiorgan disease. Typical involvement is in the eye (retinitis), GI tract (esophagitis and colitis), nervous system (radiculitis), and lung (pneumonia). Therapy is difficult, and preventive measures are used when possible.

Multisystem infectious diseases often involve bloodstream infections, which explains their effects on several organ systems simultaneously. Fever is practically universal for all disseminated infections, and mortality for the blood stream infections (especially caused by bacteria) is high if untreated. The classic example of blood stream infections is **infective endocarditis** (IE), an infection of the endocardial surface of the heart, characterized by continuous **bacteremia** (bacteria in the blood stream). Most cases of bacteremia are not due to endocarditis. Distinguishing endocarditis from bacteremia may be difficult. Endocarditis may be associated with (1) new or changing murmur(s); (2) embolic phenomena; (3) peripheral manifestations; and (4) vegetations seen by echocardiography. IE can lead to infection of other deep tissues, including lungs, bones and joints, the CNS, and the renal system. Common organisms are viridans streptococci, *S. aureus,* and enterococci.

Arthropod-borne diseases are common and often have multisystem involvement. **Malaria** is the most common arthropod-borne disease worldwide, affecting hundreds of millions of patients a year. There are four main species, with *Plasmodium falciparum* being the most virulent. It typically causes severe disease, involving the kidneys, liver, spleen, and CNS. Mortality is high, particularly in cases with cerebral involvement. **Rocky Mountain spotted fever (RMSF)** is the most severe of the tick-borne infections, with a mortality rate of 30%. It is caused by an intracellular Gram-negative bacillus, *Rickettsia rickettsiae.* The characteristic rash is usually petechial and often involves the palms and soles, one of few infections to do so. Contrary to what the epithet suggests, most cases of RMSF now occur in the mid-eastern states, and it is uncommon in the Rocky Mountains. Early therapy with doxycycline is most effective. **Lyme disease** is caused by *Borrelia burgdorferi,* a spirochete transmitted by ticks. The initial manifestation is a rash, called erythema migrans, followed by arthritis, myocarditis, or meningitis in untreated cases. Early therapy effectively prevents complications and should be considered in patients presenting in endemic areas.

Leptospirosis, caused by another spirochete, *Leptospira interrogans,* is typically associated with animal exposure. Leptospirosis leads to bacteremia, with fevers and liver and spleen involvement, as well as renal and CNS disease in severe cases. Many other spirochetes exist, and some are transmitted by ticks, as in Lyme disease. **Pasteurellosis**, a zoonosis, is seen with cat or dog exposure. Cat or dog bites often lead to cellulitis or septic arthritis. It can also cause pneumonia or meningitis, with high mortality.

In summary, most skin and soft-tissue infections are caused by either group A streptococcus or *S. aureus*. Involvement of deeper tissue layers is associated with higher mortality, including toxic shock syndrome. Viral infections, usually seen in childhood, may be difficult to distinguish based on the presenting rash alone. Systemic infections, following the blood meals by infective vectors such as mosquitoes and ticks, are transmitted through the blood stream and generally involve multiple organs. Careful epidemiologic (exposure) history is needed to elicit specific risk factors for each of these infections.

FURTHER READING

Bowler PG, Duerden BI, Armstrong DG: Wound microbiology and associated approaches to wound management. Clin Microbiol Rev 14:244, 2001.
Brook I: Abscesses. In Brogden KA, Guthmiller JM (eds): Polymicrobial Diseases. Washington DC, ASM Press, 2002, p 153.
Hotchkiss RS, Karl IE: The pathophysiology and treatment of sepsis. N Engl J Med 348:138, 2003.

Questions on the case (problem) topics discussed in this section can be found in Appendix F. Practice question numbers are listed in the following table for students' convenience.

Self-Assessment Subject Areas (book section)	Question Numbers
Skin, wound, cardiovascular, and multisystem infections (Section V)	11-13, 27-30, 58, 75-77, 85-88, 105-107, 129-131, 137-139, 142-143, 147-149, 152, 163-165, 176-178, 181, 190, 202, 203, 225-227, 229-233, 248, 254

An **18-year-old male** college student was brought to the emergency department of a hospital with the complaints of **fever, chills,** and **pain while walking.**

The patient had experienced progressively **spreading boils on his left leg** for the past week. The **boils** were **painful** and **tender,** and he had **fever.** He had not sought medical attention earlier, hoping the infection would resolve spontaneously.

PHYSICAL EXAMINATION

VS: T 39.4°C, P 112/min, R 18/min, BP 124/70 mmHg

PE: The lower **left leg** was **swollen** and tender. The **overlying skin** was **warm and red** with multiple necrotic-appearing lesions (Fig. 67-1). The knee joint was normal with full range of motion.

LABORATORY STUDIES

Blood

ESR: **98 mm/h**

C-reactive protein: 18 mg/L unit

Hematocrit: 42%

WBC: 16,400/µL

Differential: 68% PMNs, 10% bands, 16% lymphs

Serum chemistries: Normal

Imaging

X-rays of the left femur showed **soft-tissue edema** without any abnormalities of the bone. **Bone scans** of the same area showed **intense uptake in the proximal femur**.

Diagnostic Work-Up

Table 67-1 lists the likely causes of illness (differential diagnosis). A clinical diagnosis of acute osteomyelitis was considered. Microbiologic diagnosis of acute osteomyelitis is often made by CT-guided aspiration. Investigational approach to delineating the etiology may include

- **Gram stain** and **cultures** of wound aspirate (usually positive in 75% of cases)
- **Blood culture** (results are positive in 50% of children with acute osteomyelitis)

TABLE 67–1 Differential Diagnosis and Rationale for Inclusion (consideration)
Cellulitis
Gas gangrene (*Clostridium perfringens* and other anaerobic bacteria)
Mycobacterium tuberculosis (Pott disease)
Neoplasms
Osteomyelitis
Pseudomonas aeruginosa
Salmonella species
Staphylococcus aureus
Streptococcus pyogenes

Rationale: Skin infections may be deep seated with an abscess (which exhibits fluctuance of the area), or mainly cellulitis (superficial redness and warmth). Gas gangrene would have crepitance on palpation due to the presence of gas. Skin cancers may manifest with similar lesions, but usually in older individuals. Sarcomas may manifest with skin or soft-tissue lesions. TB classically causes osteomyelitis of the vertebrae, called Pott disease, although other areas may be seen; overlying cellulitis is uncommon in such cases. *S. aureus* is the most common cause of skin and soft-tissue infections. Other causes are much less common and cannot be distinguished clinically, although the history of a traumatic wound makes some of the above organisms (e.g., *Salmonella*) less likely. *Pseudomonas* is often associated with IV drug users, and *Salmonella* with patients with sickle cell disease.

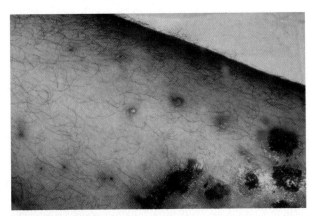

FIGURE 67-1 Multiple necrotic-appearing lesions. Note that the overlying skin is red. (*Courtesy of Dr. Paul Southern, Departments of Pathology and Medicine, University of Texas Southwestern Medical Center, Dallas, TX.*)

COURSE

Blood cultures and wound aspirate were obtained. Empirical management with antibiotics was initiated. Cultures of blood and wound aspirate yielded a significant Gram-positive bacterial pathogen.

ETIOLOGY

Staphylococcus aureus (staphylococcal osteomyelitis)

MICROBIOLOGIC PROPERTIES

Species belonging to the genus *Staphylococcus* are **Gram-positive cocci** that occur individually, in pairs, and in irregular **grapelike clusters** (Fig. 67-2*A*). They are nonmotile, nonspore forming, and **catalase positive**. The cell wall contains peptidoglycan and teichoic acid. These bacteria are resistant to temperatures as high as 50°C, to high salt concentrations, and to drying. **Colonies on sheep blood agar are usually large** (6 to 8 mm in diameter), smooth, and translucent; colonies of most strains are β-**hemolytic** and **pigmented**, ranging in color from cream-yellow to orange (see Fig. 67-2*B*). The ability to clot plasma (**coagulase activity**) is generally an accepted criterion for the identification of *S. aureus*. One such factor, bound coagulase, also known as clumping factor, reacts with fibrinogen to cause organisms to aggregate (a widely used key diagnostic test in the hospital laboratory). Another factor, extracellular coagulase, reacts with prothrombin to form thrombin, which can convert fibrinogen to fibrin (the tube or slide coagulase test is also used in the diagnostic laboratory).

EPIDEMIOLOGY

Humans are the primary habitat. The major site of **colonization** is the **anterior nares** (20% to 30% of the general population carry *S. aureus*). Approximately 60% of persons have intermittent carriage, with the organisms being present not only in the nasopharynx but also on the surface of the skin. The organisms may be found on skin, in skin glands, and in mucous membranes of humans. Most community-acquired infections are **endogenous**. Exogenous infections may arise from persons with a draining lesion or any purulent discharge. Transmission is through **contact** either with a person who has a purulent lesion or with an asymp-

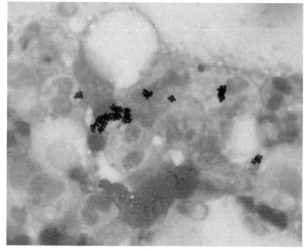

A

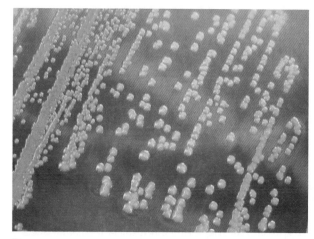

B

FIGURE 67-2 *A*, Gram stain of *Staphylococcus aureus* from a skin abscess. *B*, Culture of *Staphylococcus aureus*. (*A, Courtesy of Dr. Dominick Cavuoti, Department of Pathology, University of Texas Southwestern Medical School, Dallas, TX; B, Courtesy of Lisa Forrest, Department of Microbiology, University of Texas Southwestern Medical Center, Dallas, TX.*)

tomatic (usually nasal) carrier. The incubation period is variable and indefinite; onset of the disease occurs 4 to 10 days after acquisition of the organism. Susceptibility is greatest among **newborns** and the **chronically ill**. Elderly, debilitated individuals, drug abusers, and those with diabetes mellitus, cystic fibrosis, and neoplastic disease are also at risk, as are burn patients. Patients with Chédiak-Higashi disease, who have a primary defect in lysosomal enzyme release, present with recurrent pyogenic infections due to *S. aureus*. *S. aureus* is also an important hospital-acquired pathogen, associated with **surgical wound infec-**

tions. The hands are the most important instruments for transmitting infection.

PATHOGENESIS

Osteomyelitis is an acute infection of the bone, caused as a result of hematogenous dissemination or local extension from a skin infection and subsequent seeding of the bacteria within the bone. Skin colonization increases the likelihood that staphylococci will find and enter the deeper tissues through a mucosal breach, beginning the process of infection. **Fibronectin-binding protein** is responsible for **colonization** of organisms in skin breaks. Staphylococci are equipped with virulence factors (Table 67-2) that facilitate the following:

1. Invasion across mucosal barriers
2. Adherence to materials in the extracellular matrix
3. Evasion or neutralization of host defenses
4. Destruction of host tissues

Although **coagulase** is not a toxin, it probably plays some role in the pathogenesis of staphylococcal infections. The typical infection caused by *S. aureus* involves a pyogenic exudate or an abscess, which is a discrete localized collection of inflammatory cells and bacteria. The rapidly growing and highly vascular metaphysis of growing bones in older children is the predominant site of pathology, resulting in osteomyelitis. Strains of *S. aureus* produce many other extracellular, biologically active substances, including hemolysin, hyaluronidase, nuclease, protease, and a plasminogen activator. Their roles in the pathogenesis of pyoderma, osteomyelitis, and other infections (Table 67-3) remain obscure.

TREATMENT

A **penicillinase-resistant penicillin** (e.g., nafcillin) is the drug of choice. So-called nafcillin-resistant *S. aureus* (widely known as methicillin-resistant *S. aureus* [**MRSA**]) is treated with **vancomycin.** Since the 1980s (when MRSA emerged in the United States) vancomycin has been the last uniformly effective antimicrobial agent available for treating serious *S. aureus* infections. Synergistic combination with an aminoglycoside is often used in severe systemic infection. Clindamycin, Bactrim, and tetracycline are often useful choices in less severe infections. Widespread use of vancomycin, since the emergence of MRSA, has contributed to the emergence of vancomycin-intermediate resistant *S. aureus* (**VISA**) and vancomycin-resistant *S. aureus* (VRSA). This trend is worrisome, and development of new antibiotics is needed. A recent addition to antistaphylococcal antibiotics is linezolid, the first of a new class of antibiotics with broad-spectrum Gram-positive activity, including against MRSA.

OUTCOME

The patient responded to intravenous nafcillin and gradually improved. He received 4 weeks of antibiotics, and the lesions healed completely.

PREVENTION

There are no specific measures that are useful in preventing staphylococcal infections in children in the community. Infections in the hospital usually result from organisms carried by the patient or derived from other patients. Antistaphylococcal soaps (e.g., **chlorhexidine**) are used to block infection and patient-to-patient transmission, particularly before procedures or surgery. **Mupirocin** (Bactroban) cream (in the nares) and **oral rifampin** may be used by carriers or patients with noncomplicated infections to

TABLE 67–2   Virulence Factors of *Staphylococcus aureus*

Virulence Factors	Functions
Catalase	Reduces killing by phagocytes
Protein A	Binds to Fc-IgG, inhibiting complement fixation and phagocytosis
Coagulase	Binds prothrombin and causes the conversion of fibrinogen to fibrin, allowing plasma to clot; may facilitate production of a protective layer around bacteria, providing protection from phagocytes and antibiotics
Hemolysins	Lyse red blood cells and other target cells
Hyaluronidase	Hydrolyzes hyaluronic acid, a mucopolysaccharide; its action may facilitate the spread of the organisms through the extracellular matrix by destroying connective tissue ground substance
Panton-Valentine leukocidin	Active against WBCs; damages PMNs, monocytes, and macrophages

TABLE 67–3   Important Clinical Features of Other Infections due to *Staphylococcus aureus*

Syndromes	Clinical Features
Pyogenic Infections (abscesses)	
Localized infections of the skin	**Folliculitis** (boil) is an abscess that develops in a hair follicle. Multiple interconnected abscesses, or furuncles, develop into large lesions called **carbuncles.**
	Staphylococci may spread in the subcutaneous or submucosal tissue and cause diffuse relapsing infections that show little evidence of host immune reaction. **Chronic furunculosis** results when immune response against the pyogenic organisms is impaired. This occurs owing to inherited deficiency of phagocytic killing activities of leukocytes.
	Exfoliatin-producing strains cause bullous **impetigo**. Impetigo is limited to the epidermis and manifests as a bullous, crusted, or pustular eruption of the skin.
	Cellulitis is a spreading inflammation of cellular or connective tissue, demonstrating an erythematous area with ill-defined margins. A rapidly developing cellulitis of the subcutaneous tissues of the leg presents with large bullae and scabs.
Surgical wound infections (nosocomial)	Most common wound infections that are acquired in hospitals
	Often difficult to treat because many *S. aureus* isolates are resistant to β-lactams (MRSA)
Bacteremia/Endocarditis	Bacteremic spread of infection to the heart causes an acute endocarditis, most commonly in drug abusers.
	Endocarditis may also result from a secondary infection by *S. aureus* following varicella-zoster virus infection.
Toxin-Associated Diseases	
Toxic shock syndrome (TSS)	Can manifest as menstrual or nonmenstrual (wound infection) forms
	Following wound infection, **a staphylococcal exotoxin (known as toxic shock syndrome toxin 1, TSST-1), produced by certain strains of *S. aureus*, can enter the blood stream** (staphylococci do not cross the mucosal barrier into the blood stream; bacteremia is rare).
	TSST-1 is a **superantigen**; it **binds directly** to major histocompatibility **(MHC) class II** proteins without intracellular processing. This complex **interacts with** the **T-cell receptor, causing release** of large amounts of **interleukin (IL)-2,** inducing **IL-1** and **TNF** activities. These proinflammatory cytokines mount a **systemic inflammatory response,** causing many of the signs and symptoms of toxic shock, a **multiorgan dysfunction** syndrome.
Staphylococcal scalded-skin syndrome	Also known as Ritter disease; usually occurs in children less than five years of age
	Exfoliatin, a staphylococcal exotoxin, causes sloughing of skin in scalded-skin syndrome.

eliminate the nasal carrier state and reduce transmission. Contact precautions (**hand washing** and using **gowns** and **gloves**) are to be followed in nosocomial infections (hospitalized cases), especially those due to MRSA. Investigation of outbreaks is undertaken using epidemiologic methods and pulse-field gel electrophoresis of restriction enzyme-digested DNAs of isolates.

FURTHER READING

Archer GL: *Staphylococcus aureus*: A well-armed pathogen. Clin Infect Dis 26:1179, 1998.

Archer GL, Climo MW: *Staphylococcus aureus* bacteremia—consider the source. N Engl J Med 344:55, 2001.

Bagger JP, Zindrou D, Taylor KM: Postoperative infection with methicillin-resistant *Staphylococcus aureus* and socioeconomic background. Lancet 363:706, 2004.

Chambers HF: Methicillin resistance in staphylococci: Molecular and biochemical basis and clinical implications. Clin Microbiol Rev 10:781, 1997.

Czachor J, Herchline T: Bacteremic nonmenstrual staphylococcal toxic shock syndrome associated with enterotoxins A and C. Clin Infect Dis 32:E53, 2001.

Lowy FD: *Staphylococcus aureus* infections: Medical progress. N Engl J Med 339:520, 1998.

von Eiff C, Becker K, Machka K, et al: Nasal carriage as a source of *Staphylococcus aureus* bacteremia. N Engl J Med 344:11, 2001.

A 24-year-old man was brought to a local hospital emergency department because of **severe pain and swelling** that had developed in his left thigh that day. The **pain had progressed rapidly;** before seeking treatment he developed a **high fever** and became extremely weak and was **unable to walk** without assistance.

The patient had always been in good health, but the day before this illness, a minor injury occurred to his leg while he was playing soccer. He noted that it started as a small area of **redness at the site of injury on his left thigh,** but in the **last several hours** it appeared **more grayish.**

CASE 68

PHYSICAL EXAMINATION

VS: T 40°C, P 138/min, R 24/min, BP 70/40 mmHg

PE: A young male in moderate distress due to pain who appeared to be acutely ill. The **left thigh** was **dusky and purplish, swollen,** and **tense.** Pulses in that leg were diminished.

LABORATORY STUDIES

Blood

C-reactive protein: 16 mg/L

ESR: 120 mm/h

Hematocrit: 42%

WBC: 28,000/μL

Differential: 40% PMNs, 28% bands, 21% lymphs

Blood gases: pH 7.18, pCO_2 30 mmHg, pO_2 88 mmHg (room air)

Serum chemistries: HCO_3 12 mmol/L, BUN 30 mg/dL, Cr 1.6 mg/dL

TABLE 68–1 Differential Diagnosis and Rationale for Inclusion (consideration)

Anaerobic (*Clostridium perfringens*) infection (polymicrobic)

Gram-negative bacteria (including *Pseudomonas aeruginosa, Vibrio vulnificus*)

Staphylococcus aureus

Streptococcus pyogenes

Rationale: The presentation of rapidly progressive soft-tissue infection with signs of acute toxicity is characteristic of necrotizing fasciitis, which in immunocompetent patients is classically caused by *S. pyogenes,* much less commonly by other organisms listed above (including anaerobic bacteria; e.g., *Clostridium perfringens*). *Vibrio* species often cause severe soft-tissue infections in patients with cirrhosis. *S. aureus* commonly causes soft-tissue infections, but usually not this severe. *P. aeruginosa* does not commonly cause infections in immunocompetent individuals without other risk factors (e.g., diabetes, hospitalization).

Imaging

CT showed **edema of the soft tissues** and possible compression of the vessels.

Diagnostic Work-Up

Table 68-1 lists the likely causes of illness (differential diagnosis). Investigational approach may include

- **Gram stain** and **cultures of wound** aspirate and surgical specimens (aerobic and anaerobic)
- **Blood cultures**

COURSE

The patient was admitted to the hospital and received an intravenous antibiotic and an immediate surgical

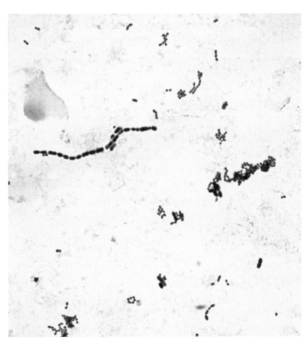

FIGURE 68-1 Gram stain of blood culture. *(Courtesy of Dr. Rita Gander, Department of Pathology, University of Texas Southwestern Medical Center, Dallas, TX.)*

consultation. He underwent surgical exploration for possible necrotizing fasciitis and subsequently received a total fasciotomy of his left thigh, extending into his pelvis. Initial blood cultures were positive for a significant pathogen. Facultative anaerobic cultures of tissue specimens obtained at surgery grew the same pathogen that grew in blood cultures.

ETIOLOGY

Streptococcus pyogenes (streptococcal necrotizing fasciitis)

MICROBIOLOGIC PROPERTIES

Streptococci are **Gram-positive cocci** that grow **in chains in body fluid** (e.g., blood cultures; Fig. 68-1). Facultative anaerobic growth of *S. pyogenes* on blood agar medium yields small colonies with clear, sharp **β-hemolysis on blood agar** culture. The bacteria are **catalase negative** and **bacitracin sensitive** in a diagnostic disc susceptibility test. The cell surface of *S. pyogenes* exhibits a carbohydrate antigen (Lancefield group A). The fibrillar M protein that extrudes from the cell membrane as a layer of protein fuzz is a major cell-surface antigen with more than 100 serotypes. β-hemolytic group A streptococci (BHGAS) are extremely virulent bacteria. Consequently, detection and reporting of these organisms from all types of specimens are emphasized in the clinical laboratory.

EPIDEMIOLOGY

Humans are the only habitat of BHGAS. **Direct contact** with patients or carriers is the major mode of transmission. BHGAS may be recovered from the normal skin for 1 to 2 weeks before skin lesions develop. Particular **streptococcal M types** (e.g., **M1, M3**) **have predilection for skin**. General susceptibility is common worldwide.

PATHOGENESIS

The pathogenic BHGAS enter via wounds (e.g., trauma from soccer injury) or other breaks in the integrity of the epidermis. BHGAS have in its arsenal a number of virulence factors (e.g., hyaluronic acid capsule, M-protein, lipoteichoic acid [LTA], enzymes,

and toxins), which contribute to the **multifactorial pathogenesis of abscesses**. One of the first steps in pathogenesis is the ability of BHGAS to adhere to epithelial cells of the mucosa at the site of wound. The LTA-M protein complex dominates in binding to epidermis via fibronectin. BHGAS multiply in tissues and **resist phagocytosis** by using **M-proteins**, which interfere with the alternative complement pathway and opsonization. Although many M types can lead to invasive disease, M1 and M3 are by far the most prominent. Wound infection is usually associated with a thin exudate, and streptococci may spread rapidly in the subcutaneous tissue and into the **dermis and fascia**. Spread **occurs when BHGAS rapidly advance into** the deeper layers of the skin owing to hyaluronidase and other hydrolytic enzymes. The host mounts **an acute pyogenic inflammation at the provocation of bacterial products**. Bacterial cytotoxins (e.g., hemolysins O and S) damage host cell membranes. Tissue ischemia is usually evident clinically in necrotizing soft-tissue inflammation (necrotizing fasciitis). The bacteria can also invade the blood stream, causing bacteremia.

> **NOTE** In all age groups, *S. pyogenes* causes a variety of clinical syndromes and complicated illnesses (Table 68-2).

TREATMENT

This organism is highly sensitive to **penicillin G**. Erythromycin and clindamycin are useful in case of penicillin-allergic patients. The addition of clindamycin to penicillin is often recommended in necrotizing fasciitis to reduce the production of toxins contributing to the pathogenesis of the tissue necrosis and sepsis.

OUTCOME

The therapy was switched to high dose IV penicillin and clindamycin for 2 weeks. The patient required additional surgical débridements and was transferred to the regular floor after spending almost 3 weeks in the ICU. He eventually left the hospital after 3 months of wound care and rehabilitation.

PREVENTION

Cases such as the one just described are not amenable to prevention, given their highly sporadic nature and

TABLE 68–2 Important Clinical Features of Skin and Wound Infections due to *Streptococcus pyogenes*

Syndromes	Clinical Features
Localized abscesses	**Impetigo** is confined to the epidermis; invasion is through minor trauma. Clusters of superficial vesicles break out to form a raw, weeping surface; characteristic yellow crusts are often the main features on presentation.
Spreading abscesses	**Erysipelas** involves the dermal lymphatics and is characterized by a spreading area of erythema and edema with rapidly advancing, well-demarcated edges (**"butterfly-wing"** rash characteristically seen on the face), pain, and systemic infections, including fever and lymphadenopathy.
	Cellulitis involves the subcutaneous fat layer and often develops at a site of previous trauma or skin lesion; the involved skin becomes tender, red, warm, and swollen. Patients frequently have an abrupt onset of malaise, fever, chills, and headache.
	Necrotizing fasciitis (described in this case) and **myositis** are infections of the deeper subcutaneous tissues, fascia, and muscle, characterized by extensive, rapidly spreading necrosis and gangrene of the skin and surrounding structures.
Surgical/obstetric (nosocomial) infections	Surgical and obstetric patients are predisposed to hospital-acquired infection because broken cutaneous or mucosal barriers facilitate invasive infection after exposure.
	Health care workers can shed the organisms into the immediate environment despite proper gowning and gloving. The common sites of asymptomatic carriage of BHGAS are anus, vagina, skin, and oropharynx.
Streptococcal toxic shock syndrome (STSS)	Following wound infection due to BHGAS, when local host defenses are inadequate to control progression of infection, invasive streptococci enter the blood stream (bacteremia is common, in contrast with *S. aureus*).
	Selected strains (from M1 and M3 serotypes) produce streptococcal pyrogenic exotoxins (Spe). SpeA, a toxin functionally similar to scarlet fever toxin A, is a **superantigen**. It allows simultaneous binding to MHC class-2 molecules and T-cell receptors, leading to the direct stimulation of T cells and IL-2 production. In response, there is a dramatic increase in production of cytokines IL-1 and TNF, and a **systemic inflammatory response occurs**.
	Fever is the most common presenting sign, although some patients present with hypothermia secondary to shock. Patients also have tachycardia and hypotension. Necrotizing fasciitis or myositis is found in 50-70% of patients, requiring surgical débridement or amputation.
	Patients with STSS require prompt antibiotic therapy using empirical antibiotic coverage but intensive care is urgently needed to manage the complications of cardiovascular collapse. Patients with invasive BHGAS disease have significantly lower serum levels of protective antibodies against M-protein and superantigens, compared with other streptococcal infections. Intravenous immunoglobulin therapy of patients with STSS has improved prognosis.
Acute post-streptococcal glomerulonephritis (APSGN)	APSGN, a nonpyogenic complication, is similar in pathogenesis to rheumatic fever but is more commonly associated with skin infections in children.
	Pathogenesis involves immune complex deposition and inflammation of the target tissue. A toxin associated with a distinct nephritogenic strain (e.g., M-type 6) activates plasminogen to produce plasmin, which may in turn promote complement deposition and glomerular hypercellularity. Complement components usually decrease in serum during the acute phase. Elevated titers of IgG against streptolysin O may be diagnostic.

the high prevalence of carriage in the community. Infections in the hospital usually result from organisms carried by the patient or derived from other patients. Patient-to-patient transmission needs standard precautions. Infection-control measures (**hand washing** and **gloving**) should be followed. Occasionally, outbreaks of invasive BHGAS may require specific screening and prophylaxis measures.

FURTHER READING

Bisno AL, Stevens DL: Streptococcal infections of skin and soft tissues: Current concepts. N Engl J Med 334:240, 1996.

Bryant AE: Biology and pathogenesis of thrombosis and procoagulant activity in invasive infections caused by group A streptococci and *Clostridium perfringens*. Clin Microbiol Rev 16:451, 2003.

CDC: Invasive *Streptococcus pyogenes* after allograft implantation—Colorado, 2003. JAMA 291:174, 2004.

Cunningham MW: Pathogenesis of group A streptococcal infections. Clin Microbiol Rev 13:470, 2000.

Darenberg J, Ihendyane N, Sjölin J, et al: Intravenous immunoglobulin G therapy in streptococcal toxic shock syndrome. Clin Infect Dis 37:333, 2003.

Davies HD, McGeer A, Schwartz B, et al: Invasive group A streptococcal infections in Ontario, Canada. N Engl J Med 335:547, 1996.

A 66-year-old white man underwent surgery for **colon carcinoma,** and 2 days later he experienced **severe pain at the surgical wound site.** Within several hours, local **edema** and **tenderness** developed at the wound, as well as a **thin, brownish discharge.**

Prior to surgery and this episode, he had always maintained good health. His social history was unremarkable.

PHYSICAL EXAMINATION

VS: T 37.6°C, P 136/min, R 26/min, BP 80/52 mmHg

PE: The patient appeared very ill. The surgical wound site exhibited **discoloration of skin** and **hemorrhagic bullae.** There was a **serosanguineous discharge** from the infected wound. The affected muscles showed **failure to bleed,** and there was **extensive gas in the soft tissues.**

LABORATORY STUDIES

Blood

ESR: 36 mm/h

Hematocrit: 34%

WBC: 15,400/μL

Differential: 60% PMNs, 23% bands, 15% lymphs

Blood gases: pH 7.23, pCO_2 30 mmHg, pO_2 90 mmHg

Serum chemistries: creatinine 2.3 mg/dL, LDH 310 U/L, bilirubin 2.8 μmol/L

Imaging

No imaging studies were done.

Diagnostic Work-Up

Table 69-1 lists the likely causes of illness (differential diagnosis). Both needle aspirate of pus and tissue biopsy are appropriate for anaerobic cultures. Investigational approach may include

- **Gram strain.** Typical bacterial morphologies are suggestive of anaerobes.
- **Cultures** of pus or wound aspirate on selective and nonselective media

TABLE 69–1   Differential Diagnosis and Rationale for Inclusion (consideration)

Anaerobic infection (*Clostridium perfringens;* polymicrobic)

Gram-negative bacterial (e.g., *Escherichia coli, Klebsiella* spp) infection

Mixed infection

Staphylococcus aureus

Streptococcus pyogenes

Rationale: The medical history and findings on physical examination should arouse suspicion of anaerobic infection. Predisposing factors (e.g., solid tumor), foul odor of lesion or drainage, gas or discoloration in tissue, and tissue necrosis, gangrene, or abscess point toward a narrowed differential diagnosis. Gas gangrene is commonly caused by clostridial species (e.g., *C. perfringens*). However, Gram-negative bacteria, which are much more common with surgical infections, may, rarely, cause gas in mixed infections of soft tissue. Gram-positive organisms (e.g., *Strep. pyogenes, Staph. aureus*) also commonly cause postsurgical infections, but usually not gas gangrene.

- **Aerobic culture.** "Sterile pus" (no growth) indicates anaerobes
- Selective **anaerobic** cultures and species identification

COURSE

The patient was taken to a surgical unit, where his wound was débrided for cleaning. He was treated with intravenous antibiotics (directed at the anaerobic and facultative anaerobic organisms). A Gram stain of the wound aspirate revealed sparse PMN leukocytes and many large, Gram-positive rods together with a mixture of Gram-negative rods and Gram-positive cocci. Anaerobic cultures were diagnostic.

ETIOLOGY

Clostridium perfringens (gas gangrene)

MICROBIOLOGIC PROPERTIES

The important clostridial species, *C. perfringens*, *C. septicum*, *C. novyi*, and *C. ramosum*, are **large, box-shaped Gram-positive rods** (Fig. 69-1*A*) and are **anaerobic**, **spore-forming** bacteria. Spores are rarely seen in clinical specimens. The organisms are catalase negative and are unable to deactivate H_2O_2 or $O_2^{\cdot}$ (superoxide), which are toxic to the bacteria. The organisms grow only in the deep tissues of the body with low redox pontential. Many clinically significant anaerobes are aerotolerant (2% to 8% oxygen), in part because they produce superoxide dismutase. **Anaerobes fail to grow on solid media in 10% CO_2 in air** (18% oxygen). Clostridia grow slowly (48 to 72h) under anaerobic conditions. Blood agar culture of *C. perfringens* from an anaerobic cellulitis lesion yields round, grayish-white colonies, with an opaque center surrounded by a **double zone of β-hemolysis** (see Fig. 69-1*B*). Toxigenic strains cause disease; a significant link to clinical disease is the toxigenicity of the *C. perfringens* isolate. Nagler reaction for positive lecithinase on egg-yolk agar, neutralized by antitoxin, is the positive proof for a toxigenic isolate.

EPIDEMIOLOGY

Clostridium species are part of the normal human or animal flora, and vegetative cells are found in the colon. *C. perfringens* organisms cause disease only when they leave their normal niche (colon) and make their way to a new location (as occurred in this case). Most infections are caused by a mixture of bacteria (2 to 10 different species; anaerobes and aerobes or facultative anaerobes of colonic origin). Any host abnormality that causes **vascular stasis** elevates the **risk** for anaerobic infection. **Carcinoma, diabetes mellitus**, colonic (fecalith) obstruction, treatment with immunosuppressive agents, and chemotherapy for malignancy are recognized **predisposing factors for endogenous anaerobic infection**. Clostridial spores are found in soil. **Exogenous infection is caused via soil contamination of a deep wound from trauma.**

A

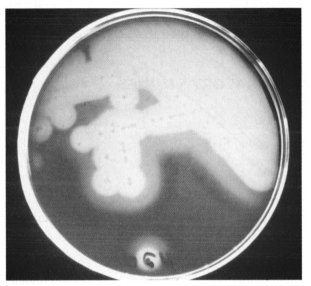

B

FIGURE 69-1 *A,* Gram stain of needle aspirate of anaerobic cellulitis lesion revealing *Clostridium perfringens*. Note absence of PMN exudates. *B,* Culture of *Clostridium perfringens* grown on anaerobic blood agar medium. Note a characteristic double-zone β-hemolysis. *(Courtesy of Dr. Paul Southern, Departments of Pathology and Medicine, University of Texas Southwestern Medical Center, Dallas, TX.)*

PATHOGENESIS

This is a typical case of cellulitis with progression to gas gangrene, and the usual culprit is *C. perfringens*. Bowel surgery to remove a solid tumor allows anaerobes and facultative anaerobes from the colon access to tissues. The facultative anaerobes assist the primary anaerobes to proliferate outside their normal sites of habitation, creating a growth environment with low oxygen potential. Under the right conditions, *C. perfringens* can invade and multiply in essentially any tissue. At the new location, toxigenesis occurs. **Clostridial α-toxin is cytolytic owing to activity of phospholipase C activity on cell membranes**. Other catabolic enzymes produced by *C. perfringens* yield gas in tissues, producing crepitance. Systemically, the patient develops fever, sweating, and low blood pressure. Muscle grows black and gangrenous.

> **NOTE** Clostridial infection often disseminates and causes blood stream infection (bacteremia). Isolation of *Clostridium septicum* from blood cultures suggests colon cancer or leukemic cecitis. *C. perfringens* and *C. novyi* cause soft-tissue inflammation and fatal bacteremia in injecting drug users.

TREATMENT

Surgical drainage and **débridement** are essential to control the infection initially. **Antibiotic therapy** relies on drugs with generally good activity against anaerobes, with some variation in susceptibility. Most isolates are susceptible to penicillin, and it is considered the drug of choice. However, other antibiotics are also effective and have a broader spectrum: metronidazole (anaerobes only), clindamycin, piperacillin/tazobactam, imipenem/cilastatin, or meropenem. Antibiotic susceptibility testing is difficult and is indicated only for certain isolates from serious infections (from normally sterile sites).

OUTCOME

The patient received penicillin and clindamycin and gradually recovered over the course of the next 2 weeks. His recovery was attributed to successful treatment of polymicrobial infection.

PREVENTION

There is no vaccine for *C. perfringens*. Gas gangrene is best prevented by immediate and thorough irrigation and débridement of traumatic wounds and appropriate antibiotic prophylaxis before surgery. Antibiotics used should target anaerobes as well as enteric Gram-negative bacteria.

FURTHER READING

Bryant AE: Biology and pathogenesis of thrombosis and procoagulant activity in invasive infections caused by group A streptococci and *Clostridium perfringens*. Clin Microbiol Rev 16:451, 2003.

Stevens DL, Tweten RK, Awad MM: Clostridial gas gangrene: Evidence that alpha and theta toxins differentially modulate the immune response and induce acute tissue necrosis. J Infect Dis 176:189, 1997.

A 27-year-old man was brought to the emergency department of a hospital with **high, spiking fevers, severe diffuse pain over the lower abdomen,** and **loss of appetite.**

Two weeks earlier he first noticed mild abdominal pain and **anorexia,** which gradually progressed to include fevers and night sweats. The day before admission his abdominal pain became severe, and his fever became constant.

PHYSICAL EXAMINATION

VS: T 39.5°C, P 118/min, R 20/min, BP 92/50 mmHg

PE: Examination revealed **tenderness in the right lower quadrant of his abdomen,** with **rebound tenderness** (voluntary guarding or rigidity).

LABORATORY STUDIES

Blood

ESR: 68 mm/h
Hematocrit: 42%
WBC: 23,400/μL
Differential: 50% PMNs, 24% bands, 15% lymphs
Serum chemistries: Normal

Imaging

CT scan showed an intra-abdominal fluid collection consistent with an **abscess in the right lower quadrant**.

Diagnostic Work-Up

Table 70-1 lists the most likely causes of illness (differential diagnosis). A presumptive diagnosis of intra-abdominal abscess may be based on persistent systemic signs (fever and elevated WBC count) and local signs and symptoms (diffuse abdominal pain). Further imaging studies are warranted to determine the location of pathology. Microbiologic diagnosis is often made by CT-guided aspiration or surgical explorations and specimen collection. Investigational approach may include

- **Gram strain** of peritoneal fluid
- **Aerobic and anaerobic cultures** of peritoneal fluid
- **Blood cultures**

TABLE 70–1 Differential Diagnosis and Rationale for Inclusion (consideration)
Inflammatory bowel disease with associated abscess
Postsurgical abscess
Ruptured appendix with polymicrobic abscess
Rationale: In a young adult, an intra-abdominal abscess is often due to a ruptured appendix. Preceding symptoms such as diarrhea with or without blood may suggest inflammatory bowel disease, especially if these symptoms have occurred in the past. In patients with prior abdominal surgery, complications such as abscesses may also occur.

COURSE

The patient was admitted to the hospital and an admission abdominal CT revealed a retroperitoneal abscess. He was taken to the surgery unit, where at surgery 300 mL of purulent peritoneal fluid was

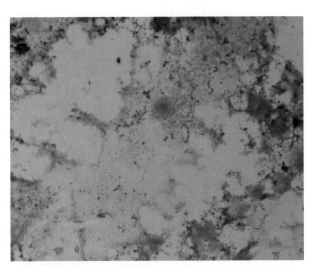

FIGURE 70-1 Gram stain of peritoneal fluid. (*Courtesy of Dr. Dominick Cavuoti, Department of Pathology, University of Texas Southwestern Medical Center, Dallas, TX.*)

drained. Blood cultures were negative. Anaerobic cultures yielded a significant pathogen. Facultative anaerobic cultures were positive for three different Gram-negative species of the Enterobacteriaceae family.

ETIOLOGY

Bacteroides fragilis (polymicrobic infection)

MICROBIOLOGIC PROPERTIES

B. fragilis is a **Gram-negative, nonspore-forming rod** (Fig. 70-1) that **constitutes 1% to 2% of the normal flora of the gastrointestinal tract of humans**. Table 70-2 lists an array of nonspore-forming anaerobic organisms and the diseases they cause. *B. fragilis* is an anaerobe that exhibits O_2 sensitivity in laboratory cultures (although in vivo the organisms are somewhat aerotolerant); O_2-sensitive colonies on selective culture media are identified biochemically and by examination of short chain fatty acid production with the aid of gas-liquid chromatography. A **thick (polysaccharide) capsule** is found at the outermost layer of the bacteria (see Fig. 70-1). Lipopolysaccharide (LPS) associated with the Gram-negative cell wall is not toxic (unlike LPS of other Gram-negative bacteria, such as *Fusobacterium*); this is because the lipid A component of *B. fragilis* LPS lacks phosphate groups on the glucosamine residues, and the number of fatty acid chains linked to the amino sugars is, therefore, reduced.

EPIDEMIOLOGY

Endogenous infection in this case resulted from the spillage of bacteria into the peritoneum from the perforated colon due to appendix rupture. *B. fragilis* and coinfecting Gram-negative bacteria are common residents of the human colon. Other important origins of intra-abdominal infections include ruptured gall-

TABLE 70–2	Diseases Caused By Nonspore-Forming Anaerobic Bacteria	
Anaerobe	**Microbiology**	**Common Syndromes**
B. fragilis	Gram-negative rods	Intra-abdominal abscesses (described in this case)
B. melaninogenicus		Oral, dental, pleuropulmonary infections
B. gingivalis		Oral, dental, pleuropulmonary infections
B. bivius		Pelvic infections
Prevotella spp *Porphyromonas* spp (pigmented, often fluorescent)	Gram-negative rods	Pleuropulmonary infections: aspiration of oral or gastric contents, especially with severe periodontal disease (high counts of anaerobes), resulting in aspiration pneumonia, lung abscess, or pleural empyema. May be accompanied by viridans streptococci or, in a hospital setting, by Gram-negative aerobes or staphylococci.
Fusobacterium nucleatum	Gram-negative rods	**Vincent angina**: "trench mouth"—acute necrotizing ulcerative gingivitis. **Ludwig angina**: submandibular space infection, usually mandibular odontogenic, with mixed aerobes/anaerobes; swelling in the floor of mouth and in the neck, drooling, trismus, dyspnea.
Fusobacterium necrophorum	Gram-negative rods	**Lemierre syndrome**: peritonsillar abscess with *F. necrophorum* infection of lateral pharyngeal space, with septic jugular vein thrombosis and metastatic abscesses in other organs. High mortality rate.
Peptostreptococcus anaerobius	Gram-positive cocci	Blood infections (rare)
Actinomyces spp	Gram-positive (filamentous) rods	Actinomycosis (described in an earlier section)

bladder, perforated peptic ulcer, ruptured intestinal diverticulum, acute pancreatitis, chronic peritoneal dialysis surgery, and trauma.

PATHOGENESIS

The presence of an abscess means a break in normal barriers occurred allowing normal flora to create a localized pocket of infection. Multiple species of anaerobic bacteria of colonic origin can cross an inflamed or ruptured colonic mucosa, but most die, and only the aerotolerant ones survive in the higher redox potential of the vascularized peritoneum and grow in the peritoneal fluid. The **synergistic presence of a facultative anaerobe is required to cause disease**. For instance, *Klebsiella pneumoniae*, found in peritoneal fluid, grows well at the early stage of an infection and utilizes oxygen, thereby lowering the **redox potential** of the tissue environment and **allowing the anaerobic B. fragilis to grow**.

B. fragilis is capable of causing disease and is generally able to tolerate exposure to a minute level of oxygen in human tissues. *B. fragilis* is especially suited to causing intra-abdominal infection and abscess due to the presence of a thick polysaccharide capsule, which promotes pyogenic inflammation and abscess development. The first line of host defense against invasion in a sterile site (e.g., peritoneum) is **rapid mobilization of PMNs** attracted by **interleukin-8**. However, **highly encapsulated bacteria resist phagocytosis and continue to grow**. Host control of bacterial proliferation and containment (abscess formation) is mediated by both humoral (e.g., C3 components) and cellular immune factors (e.g., tumor necrosis factor [TNF]). C3 components of the alternative complement pathway deposit on bacterial surfaces, opsonizing them. TNF upregulates the expression of intercellular adhesion molecule 1 (ICAM-1). When PMNs in the peritoneum are unable to eradicate the unwanted bacteria, they **adhere to ICAM-1-expressing cells,** and an **abscess develops, thus "walling off" the infection**.

TREATMENT

Surgical drainage should accompany antibiotic therapy. Both facultative anaerobes and strict anaerobes should be targeted. Oral anaerobes are often susceptible to penicillin and are not usually resistant. Gut anaerobes (particularly *B. fragilis*) are sensitive only to metronidazole, cefoxitin or cefotetan, clindamycin, imipenem, or β-lactam/β-lactamase inhibitor combinations (i.e., piperacillin/tazobactam). Aminoglycosides may be combined with one of these agents for synergistic activity, primarily for resistant Gram-negative rods, although their activity may be diminished in an anaerobic environment.

OUTCOME

The patient underwent an emergency surgical drainage and was in the ICU for 2 days before being transferred to a regular ward for an additional week of IV antibiotics. He was then discharged home with close follow-up and an additional oral course of antibiotics until his abdominal wound began to heal.

PREVENTION

There is no specific way to prevent intra-abdominal abscess. No vaccines exist for anaerobic bacteria, and their ubiquitous presence means that they are commonly involved in abscesses involving the oral cavity and GI tract.

FURTHER READING

Brook I: Abscesses. In Brogden KA, Guthmiller JM (eds): Polymicrobial Diseases. Washington DC, ASM Press, 2002, p 153.

Goldstein EJC: Intra-abdominal anaerobic infections: Bacteriology and therapeutic potential of newer antimicrobial carbapenem, fluoroquinolone, and desfluoroquinolone therapeutic agents. Clin Infect Dis 35:S106, 2002.

Studd RC, Stewart PJ: Intraabdominal abscess after acupuncture. N Engl J Med 350:1763, 2004.

Styrt B, Gorbach SL: Recent developments in the understanding of the pathogenesis and treatment of anaerobic infections. N Engl J Med 321:298, 1989.

An **8-year-old** white boy was brought to the dermatology clinic for the examination of **raised lesions on his head.** He was otherwise healthy.

The lesions were **first noticed a month before as erythematous areas that enlarged and coalesced over a period of several weeks.** He had received a **dog** for his birthday a few months before, which was apparently without illness.

PHYSICAL EXAMINATION

VS: T 37°C, P 78/min, R 16/min, BP 126/78 mmHg

PE: On examination, the **lesions were raised nodules in the center and scaling at the peripheries,** covering a large area on his head (Fig. 71-1). There were also a few small pustules on the scalp.

LABORATORY STUDIES

Blood

Hematocrit: 43%
WBC: 7800/μL
Differential: Normal
Blood gases: Normal
Serum chemistries: Not done

Imaging

No imaging studies were done.

Diagnostic Work-Up

Table 71-1 lists the likely causes of lesions (differential diagnosis). A clinical diagnosis of ringworm was considered. Investigational approach may include

- **Wood's light**
- **Direct mount** microscopy
- **Culture** of skin scrapes to demonstrate fungus

COURSE

An examination of the scalp using Wood's (low-wavelength ultraviolet) lamp was fluorescence positive. A scraping of a lesion from the active area was sent for direct microscopy and culture. KOH/Calcofluor preparation of the specimen demonstrated hyphal elements, and fungal culture of the scraped lesion using Sabouraud agar medium subsequently grew a white mold.

TABLE 71–1 Differential Diagnosis and Rationale for Inclusion (consideration)

Abscesses
   Staphylococcus aureus
   Streptococcus pyogenes
Allergic rashes
Parasitic rashes
Ringworm lesions (dermatophytoses)
   Microsporum spp
   Epidermophyton spp
   Trichophyton spp
Viral exanthem lesions (e.g., varicella-zoster virus)

Rationale: Allergic reactions can produce significant lesions, which may be localized, making them difficult to distinguish from infections. Parasitic infections usually have geographic exposure history. Ringworm infections, caused by dermatophytes, are common in children and distinctly unusual in adults. Classic clinical features can help in presumptive diagnosis, but occasionally a bacterial superinfection due to *Staphylococcus* or *Streptococcus* can confound the clinical diagnosis. Viral exanthems are not usually confined to the scalp, but occur more commonly as part of a systemic illness.

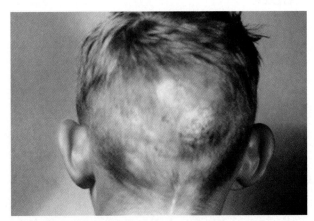

FIGURE 71-1 Raised lesions on the scalp covering a large area in the back of the head on this patient. *(Courtesy of Centers for Disease Control and Prevention, Atlanta, GA.)*

ETIOLOGY

Microsporum canis (ringworm; tinea capitis)

MICROBIOLOGIC PROPERTIES

Dermatophytic molds are **monomorphic fungi**. Occurring as a **keratinophilic** mold, dermatophytes are characterized by **colorless conidia** (spores). Conidia of dermatophyte molds are found as

● **Arthroconidia—infective form**
● **Macroconidia** and **microconidia—diagnostic form**

Large numbers of closely related fungi in the genera of *Microsporum*, *Epidermophyton*, and *Trichophyton* are pathogenic. Important species of dermatophytes are:

1. *Microsporum canis* (diagnostic form is a mixture of **spindle-shaped macroconidia** [Fig. 71-2*A*] and microconidia)
2. *Epidermophyton floccosum* (diagnostic form is characteristic **dumbbell-shaped macroconidia** only)
3. *Trichophyton rubrum, T. mentagrophytes, T. tonsurans, T. verrucosum, T. schoenleinii, T. violaceum* (diagnostic form is **microconidia** only)

Some species of *Microsporum* cause hairs to fluoresce a greenish-yellow color when exposed to a UV light (365 nm; Wood light). KOH (10%) preparations are made and observed for branching septate hyphae in skin and nail scrapings. Arthroconidia are sought in hairs, and it is noted whether they are outside the hair shaft (ectothrix invasion) or inside (endothrix invasion). The dermatophytes grow on Sabouraud dextrose agar (SDA) medium at pH 5.2. Chloramphenicol and cycloheximide are used in agar to make it selective by inhibiting bacteria and saprophytic molds. On SDA plate, *E. floccosum* colonies are slow growing and greenish-brown or khaki colored, having a suede-like surface. Colonies of most members of *Trichophyton* are pale, yellowish-brown or reddish-brown from the reverse side of the plate. Colonies of most members of *Microsporum* appear white from the top of the plate (see Fig. 71-2*B*).

EPIDEMIOLOGY

The dermatophytes are worldwide in distribution, but some species are strictly limited to certain geograph-

A

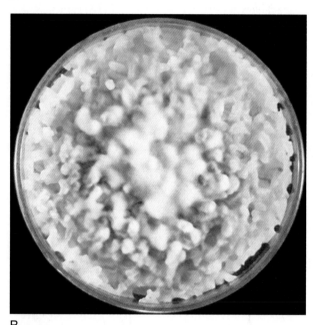

B

FIGURE 71-2 *A*, Lactophenol cotton blue stain of *Microsporum canis*. Note the presence of characteristic spindle-shaped macroconidia (which aid in identification of the species). *B*, Culture of *Microsporum canis* growing on boiled polished rice grains. *(Courtesy of Centers for Disease Control and Prevention, Atlanta, GA.)*

ical areas. **Habitats are human, animal, and soil**. Ringworm infections are transmitted by **direct contact** with lesions or by contact with materials (e.g., infected scales and hairs), animals, or soil. The following three modes of transmission are found worldwide:

1. **Person to person. Anthropophilic** species (e.g., *T. tonsurans, E. floccosum*) are found solely on humans; they commonly show reduced conidiation.

2. **Animal to human. Zoophilic** species (e.g., *M. canis*) occasionally infect humans, but they are predominantly **found on animals** (e.g., cats and dogs).
3. **Environment to humans. Geophilic** species (*Microsporum fulvum*) are those **found in soil.**

Infected materials have been shown to remain infectious for periods greater than 1 year. The dermatophytoses are contagious, while all other mycoses are not.

PATHOGENESIS

The term **tinea** (means worm in Latin) is used descriptively because of the common presence of **serpentine** (snake like) **and annular lesions** that occur on skin, making it appear that a worm is burrowing at the margin. **Nodular** and **vesicular lesions** are also noted in tinea diseases. Table 71-2 describes a variety of tinea lesions based on the anatomic locations of the body.

Dermatophytes invade keratinized tissue (e.g., hair, nails, any area of the skin) but are restricted to the dead cornified layer of the epidermis. They rarely invade living epidermis/dermis, and dissemination is very rare because the dermatophytes are unable to grow at body temperature (37°C). The arthroconidia of dermatophytes adhere to keratinocytes and germinate. Humid or moist skin provides a highly favorable environment for the establishment of a fungal infection. Slow-growing organisms cause chronic infection of keratinized tissues. Their ability to produce **keratinases and proteinases enables them to invade keratinized tissues**. The infection and clinical manifestations of tinea lesions are initiated by an **immunologically mediated reaction to the fungal antigens that diffuse to the epidermis**.

> **NOTE** Superficial fungi infect the outer areas of the body, especially skin and hair. **Superficial mycoses** involve only the keratin-containing layers of skin or hair. **Pityriasis (tinea) versicolor,** caused by *Malassezia furfur,* manifests as hyperpigmented or hypopigmented macular lesions that scale readily. Diagnosis is on the basis of microscopic examination of KOH-treated skin scrapings, demonstrating fungal elements as the characteristic "spaghetti and meatballs."

TABLE 71–2   Classification of Tinea Infections Based on Anatomic Locations

Cutaneous Infections	Location of Pathology and Clinical Features
Tinea capitis	Infection of the **scalp** caused by dermatophytes, which thrive in warm, moist areas
	Microsporum canis, a zoophilic dermatophyte often found in cats and dogs, is a common cause of tinea capitis in humans (described in this case).
	Poor hygiene, prolonged moist skin, and minor skin or scalp injuries predispose children to tinea capitis. Disease may be seen as an epidemic, particularly affecting inner city children.
Tinea corporis	Ringworm infection on the body (**trunk** and extremities)
	Caused predominantly by the genus *Trichophyton*
Tinea pedis	A fungal infection of the **feet** (athlete's foot), principally involving the toes (web space between the fourth and fifth toes) and soles, characterized by pruritus, erythema, scaling, and occasionally edema
	Athlete's foot can be caused by the fungi *Epidermophyton floccosum* or by members of the *Trichophyton* genus.
	Individuals with poor hygiene, who wear tennis shoes and endure prolonged wetting of the skin (i.e., sweating during exercise), are susceptible to infections.
Tinea barbae	Infection around the **bearded area** of men occurring as a follicular inflammation or as chronic cutaneous granulomatous lesions
Tinea faciei	Infection of the **face**, but not including infection of the bearded areas
Tinea unguium	Infection of **nails beds**, caused predominantly by *Trichophyton* spp (**onychomycosis**)

TREATMENT

The dermatophytes can be treated by prescription or nonprescription drugs. Prescription drugs are griseofulvin (an antibiotic that concentrates in keratinized areas of the body), terbinafine, itraconazole, fluconazole, and ketoconazole (used systemically or topically). Nonprescription drugs are undecylenic acid, tolnaftate, or miconazole (applied topically and usually effective, although recurrences are common).

OUTCOME

The patient received an 8-week regimen of griseofulvin and recovered completely.

PREVENTION

There are no specific preventive measures.

FURTHER READING

Behr M, Lewis TP, Barone GW, et al: Tinea barbae: Man and boxer. N Engl J Med 339:272, 1998.

Elewski BE: Onychomycosis: Pathogenesis, diagnosis, and management. Clin Microbiol Rev 11:415, 1998.

Noble SL, Forbes RC, Stamm PL: Diagnosis and management of common tinea infections. Am Fam Physician 58:163, 1998.

Wagner DK, Sohnle PG: Cutaneous defenses against dermatophytes and yeasts. Clin Microbiol Rev 8:317, 1995.

Weitzman I, Summerbell RC: The dermatophytes. Clin Microbiol Rev 8:240, 1995.

A 58-year-old man presented with a **3-week history of progressive, mildly painful skin lesions on his left arm** that had begun with an erythematous lesion on his left thumb. A **reddish streak** was apparent along these lesions. The patient did not have any fevers or chills.

He enjoyed **working in his garden** but he did not recall any specific injury.

PHYSICAL EXAMINATION

VS: T 37°C, P 74/min, R 16/min, BP 138/82 mmHg

PE: Multiple erythematous lesions were seen on his left arm from his **thumb to his elbow**, associated with **lymphangitic streaking** (Fig. 72-1). There was a dry, shallow, **ulcerated lesion** on his thumb.

LABORATORY STUDIES

Blood

Hematocrit: 44%
WBC: 9000/µL
Differential: Normal
Blood gases: Normal
Serum chemistries: Normal

Imaging

No imaging studies were done.

Diagnostic Work-Up

Table 72-1 lists the likely causes of illness (differential diagnosis). A clinical diagnosis of a subcutaneous infection was considered. Investigational approach may include

- **Gram** and **acid-fast stain** of biopsy of a freshly opened skin lesion or pus
- **Fungal stain** and **microscopy** of biopsy
- **Cultures** of biopsy specimen or pus

  - Routine bacterial cultures
  - Fungal cultures on Sabouraud dextrose agar (SDA)

COURSE

A biopsy of a skin lesion was sent for Gram stain, direct fungal stain, acid-fast stain, and routine bacterial, mycobacterial, and fungal cultures. Gram

TABLE 72–1 Differential Diagnosis and Rationale for Inclusion (consideration)
Boils due to pyogenic bacteria
Staphylococcus aureus
Streptococcus pyogenes
Cat-scratch disease
Cutaneous manifestations of AIDS
Sporotrichosis (*Sporothrix schenckii*)
Nocardiosis (*Nocardia* spp)
Mycobacterial infections of the skin (*Mycobacterium marinum*)
Tularemia

Rationale: There are few organisms that manifest with such lesions in a subacute manner. *Sporothrix schenckii* is the most common cause, followed by *Nocardia* and atypical mycobacterial infections. Some of the other infections, notably cat-scratch disease and tularemia, usually produce significant systemic symptoms and have appropriate exposure histories. Pyogenic infections (e.g., *S. pyogenes*, *S. aureus*) present more acutely and perhaps with systemic symptoms. AIDS patients may have lesions due to Kaposi sarcoma or bacillary angiomatosis that manifest indolently as well.

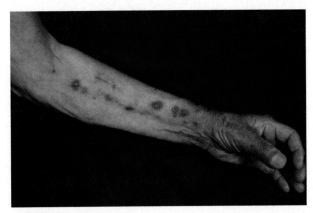

FIGURE 72-1 Multiple erythematous lesions on the left arm, associated with lymphangitic streaking, were seen in this patient.

stain was negative. A fungal stain of a biopsy specimen on direct microscopy revealed yeast-like organisms. Routine bacterial cultures were negative. The fungal culture was diagnostic.

ETIOLOGY

Sporothrix schenckii (sporotrichosis)

MICROBIOLOGIC PROPERTIES

S. schenckii is a **dimorphic fungus**; the mold form occurs in compost soil environment and on laboratory media at room temperature, and the yeast form occurs in infected subcutaneous tissue and nutrient-rich media at 37°C. Demonstration of the organisms in tissue or skin biopsy may be difficult because numbers are not numerous. Infected subcutaneous tissue may occasionally reveal subglobose-to-ovoid **(cigar-shaped) yeasts** (3 to 5 μm in diameter; Fig. 72-2*A*). When cultured on SDA at 22°C, *S. schenckii* grows as a **darkly pigmented mold** (Fig. 72-2*B*). Teased aerial growth stained with lactophenol blue usually demonstrates slender hyphae and **microconidia in floral (daisy) arrangements**.

EPIDEMIOLOGY

Sporotrichosis is reported from all parts of the world, and *S. schenckii* is commonly found in **sphagnum moss, hay**, and soil. Individuals handling thorny plants, sphagnum moss, or baled hay may contract sporotrichosis. The **organisms enter the subcutaneous tissue through punctures from thorns**, barbs, pine needles, or wires, causing infection. Therefore, sporotrichosis is an occupational disease of gardeners (**"rose handler's disease"**), horticulturists, and farmers. Outbreaks occur among children playing in, and adults working with, baled hay. Zoonotic transmission has been described in isolated cases or in small outbreaks.

PATHOGENESIS

Sporotrichosis may develop when the **skin is punctured or abraded by thorns or other vegetation contaminated with fungal spores**. The organisms establish themselves in the skin and produce a localized infection in the surrounding underlying tissue. The initial erythematous papulonodular lesions evolve into either smooth or verrucose painless nodules of about 3 cm that may ulcerate and drain. The fungus

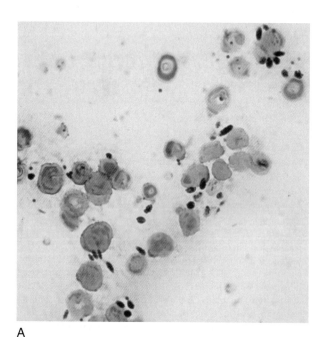

A

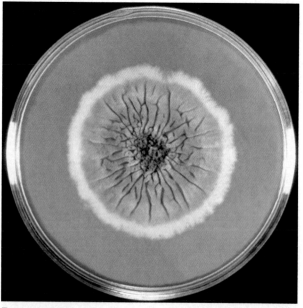

B

FIGURE 72-2 *A*, Direct tissue stain of *Sporothrix schenckii*. Note the subglobose to ovoid (cigar-shaped) yeasts. *B*, Culture of *Sporothrix schenckii* on SDA at 22°C. (*A, Courtesy of Dr. Rita Gander, Department of Pathology, University of Texas Southwestern Medical Center, Dallas, TX; B, Courtesy of Centers for Disease Control and Prevention, Atlanta, GA.*)

spreads from the initial lesion along lymphatic channels (although the lymph nodes themselves are not usually involved), forming the **chain of nodular lesions** that typifies **the lymphocutaneous form** of the disease. The lesions are **suppurating granulomas, which are composed of histiocytes and giant cells surrounded by neutrophils, lymphocytes, and plasma cells**. Subcutaneous lesions are characterized by chronic, ulcerated areas that are hard, lumpy, and crusted. These lesions periodically exude fluid and do not heal.

TREATMENT

Sporotrichosis is treatable medically, although surgical intervention may, rarely, be indicated. Orally administered saturated solution of potassium iodide cures mild subcutaneous disease. Oral **itraconazole** is highly effective and better tolerated in lymphocutaneous and osteoarticular disease.

Inhalation of the organisms may also cause a chronic, cavitary pneumonia, which is clinically and radiographically indistinguishable from tuberculosis, and disseminated infection is rare. In severe extracutaneous disease (e.g., fungal pneumonias), amphotericin B is usually administered as the primary therapy.

OUTCOME

The patient was treated with itraconazole for 6 months with gradual resolution of the lesions.

PREVENTION

Thorough débridement and cleansing of wounds may reduce the risk of developing disease, especially when soil has been inoculated into the skin. However, no specific measures are useful in preventing disease.

FURTHER READING

Al-Tawfiq JA, Wools KK: Disseminated sporotrichosis and *Sporothrix schenckii* fungemia as the initial presentation of human immunodeficiency virus infection. Clin Infect Dis 26:1403, 1998.

Coles FB, Schuchat A, Hibbs JR, et al: A multistate outbreak of sporotrichosis associated with sphagnum moss. Am J Epidemiol 136:475, 1992.

Kauffman CA, Hajjeh R, Chapman SW: Practice guidelines for the management of patients with sporotrichosis. Clin Infect Dis 30:684, 2000.

A group of 28 college students traveled to India. Several of the students had **not received childhood immunization** because of nonmedical exemptions. **Six of these students became sick with a febrile rash disease** while they were in India. The index patient returned to the United States early despite recommendations not to do so by the local public health authority. During his travel, he had **fever, cough, conjunctivitis,** and **coryza,** and within 24 hours of his arrival he developed a **rash** that began on his forehead and **spread to his trunk** and **extremities.**

PHYSICAL EXAMINATION

Physical examination of the index patient (Patent X) is presented here.

VS: T 38.8°C, P 104/min, R 16/min, BP 126/68 mmHg

PE: An ill-appearing patient with **Koplik spots** on the buccal mucosa (Fig. 73-1); a **red, blotchy rash** on **his trunk** and **extremities** was also noted.

LABORATORY STUDIES (PATIENT X)

Blood

Hematocrit: 42%
WBC: 4200/μL
Differential: 70% PMNs, 8% lymphs, 18% monos
Platelets: 110,000/μL
Blood gas: Normal
Serum chemistries: Normal

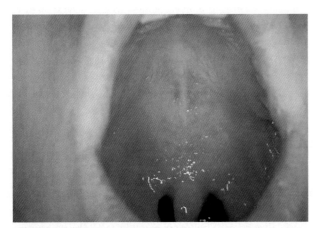

FIGURE 73-1 Koplik spots on palate of a patient with a febrile rash disease. *(Courtesy of Centers for Disease Control and Prevention, Atlanta, GA.)*

Imaging

No imaging was done.

Diagnostic Work-Up

Table 73-1 lists the likely causes of illness (differential diagnosis). A diagnosis was made on the basis of exanthematous fever and epidemiologic grounds (e.g., lack of immunization and exposure history). Investigational approach for confirmation of clinical diagnosis (if needed) may include

- **Isolation of virus** from throat (or nasopharyngeal swab) and urine
- **Serologic tests**
  - Virus-specific **IgM** to indicate acute infection
  - Elevation of **IgG** titer (>fourfold increase)

TABLE 73–1 Differential Diagnosis and Rationale for Inclusion (consideration)

Erythema infectiosum (parvovirus B19)
Measles
Meningococcemia
Roseola (HHV-6)
Rubella
Scarlet fever (*Streptococcus pyogenes*)
Typhoid fever
Varicella

Rationale: Many viral infections manifest with fevers, coryza, and a rash. Measles and rubella should be considered in such a patient who has not had prior immunizations. Except for the Koplik spots, the presentation would not be distinguishable from a variety of other infections. Petechial rashes are more common with meningococcemia. Streptococcal scarlet fever and typhoid fever do not manifest with cough and coryza; rashes (e.g., in scarlet fever: red papillae on tongue, in typhoid fever: rose spots on torso) are characteristic. Varicella would manifest with a vesicular rash. Leukopenia is also common with many of these infections, as is thrombocytopenia.

COURSE

Patient X's symptoms progressed for 4 days, then his rash began to resolve. A few days later, he developed a bacterial pneumonia for which he was briefly hospitalized and treated with antibiotics. Subsequently, clinical diagnosis was confirmed serologically.

ETIOLOGY

Measles (rubeola)

MICROBIOLOGIC PROPERTIES

The measles virus is a Morbillivirus in the family Paramyxoviridae. Virions are pleomorphic. The internal **nucleocapsid** has **linear, single-stranded, non-segmented, negative-sense RNA** (16 to 20 kb in size). Both the nucleocapsid and hemagglutinin are formed in the cytoplasm. The **enveloped virus has only one serotype**. The vaccine-type virus is differentiated from the wild-type measles virus by gene sequencing. Eight genotypes of measles virus are found worldwide.

EPIDEMIOLOGY

Although measles is not endemic in the United States, Canada, and the countries in Europe because of high levels of vaccine usage, it remains a common disease in many developing countries of the world, including India. Virus habitat is humans. **Respiratory droplets transmit the virus**, as does direct contact with nasal or throat secretions of infected persons. It is highly contagious.

In the United States, **measles still occurs among unvaccinated school populations and college students**, and, rarely, among vaccinated school populations. Small outbreaks continue to occur owing to importations, with limited domestic spread. The unvaccinated students in this case were infected during travel in India during an epidemic.

PATHOGENESIS

The virus infects the epithelial cells lining the upper respiratory tract. Following primary multiplication, the virus enters the blood stream and infects reticulo-endothelial cells, where it replicates again. Following secondary multiplication in the blood stream (causing viremia), the virus infects many types of white blood cells (primarily monocytes) and spreads to the skin and the respiratory tract. A biologic characteristic of viral infection in the target cells is the production of multinucleated giant cells with inclusion bodies in the nucleus and cytoplasm. Infection of the upper and lower respiratory tracts contributes to **cough, conjunctivitis**, and **coryza**, and, in rare cases, croup, bronchiolitis, and pneumonia. The incubation period is 7 to 8 days preceding rash onset, and infection is highly contagious from the first respiratory symptoms until 4 days after rash onset. After the febrile prodrome stage, **virus-specific T cells attack virus-infected vascular endothelial cells** of dermal capillaries and play an important role in the development of Koplik spots (pathognomonic) and skin rash. Antigen-antibody complex mediated vasculitis, coupled with necrosis of epithelial cells, may also contribute to febrile exanthem illness. After a few days of uncomplicated measles, fever subsides and the rash fades. Capillary leakage at the height of illness is revealed by transient purpura (i.e., post-measles staining) in the distribution of rash.

TREATMENT

No specific treatment is available.

> **NOTE** Acquired humoral and cell-mediated immunity after illness is life long. Unvaccinated older persons may still be susceptible to measles. In patients with defective cell-mediated immunity (e.g., HIV), measles can progress to **giant cell pneumonia** and CNS complications, which include **subacute sclerosing panencephalitis** (SSPE—a fatal disease) manifesting with brain inflammation several years after primary infection. In normal individuals, SSPE occurs at a rate of 1 per 1000 cases of measles.

OUTCOME

Over the course of the next 2 weeks the patient recovered completely.

PREVENTION

This illness can be prevented by childhood immunization with measles-containing vaccine, a trivalent vaccine (MMR) containing live, attenuated mumps, measles, and rubella viruses. It is contraindicated in the immunosuppressed, in pregnant women, and in other high-risk individuals. There is a requirement in the United States that school-aged children receive 2 doses of MMR, administered at 12 to 15 months of age and preschool, rendering 95% immunity. Immunization by live, attenuated virus vaccine, if given within 72 hours of exposure, may provide protection.

FURTHER READING

Gustafson TL, Lievens AW, Brunell PA, et al: Measles outbreak in a fully immunized secondary-school population. N Engl J Med 316:771, 1987.

Moss WJ, Monze M, Ryon JJ, et al: Measles and HIV. Clin Infect Dis 35:189, 2002.

Wood DL, Brunell PA: Measles control in the United States: Problems of the past and challenges for the future. Clin Microbiol Rev 8:260, 1995.

A **3-year-old** boy was brought to a clinic for complaints of **low-grade fever, swollen lymph nodes,** and a **rash for 3 days.**

Because of his family's religious beliefs, he **had not had any vaccinations.** A **close family friend** had also been **sick with a similar illness** 2 weeks earlier.

PHYSICAL EXAMINATION

VS: T 38°C, P 118/min, R 20/min, BP 112/64 mmHg

PE: Young child in no acute distress. A **diffuse maculopapular rash** was present **over the trunk and extremities** (Fig. 74-1); no lesions were present in the mouth. Cervical lymphadenopathy was also noted.

LABORATORY STUDIES

Blood

Hematocrit: 38%
WBC: 8200/μL
Differential: Normal
Platelets: 95,000/μL
Blood gases: Normal
Serum chemistries: Normal

Imaging

No imaging studies were done.

Diagnostic Work-Up

Table 74-1 lists the likely causes of illness (differential diagnosis). A diagnosis can be made on clinical and

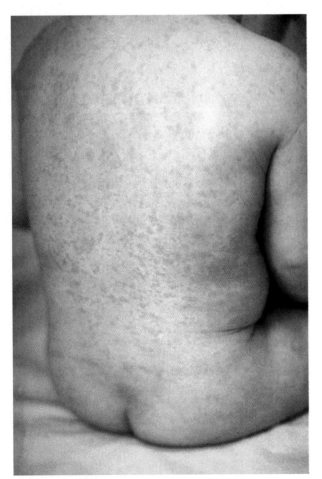

FIGURE 74-1 Rash on skin of a child's back. *(Courtesy of Centers for Disease Control and Prevention, Atlanta, GA.)*

TABLE 74–1 Differential Diagnosis and Rationale for Inclusion (consideration)

Disseminated bacterial infections
Meningococcemia
Rocky Mountain spotted fever (RMSF)
Scarlet fever (*Streptococcus pyogenes*)
Typhoid fever (*Salmonella typhi*)
Viral exanthem illnesses
   Roseola (HHV-6)
   Measles
   Erythema infectiosum (parvovirus B19)
   Rubella
   Varicella (VZV)

Rationale: Childhood viral infections can be difficult to distinguish from one another. If the patient is quite ill, a bacterial etiology may be suspected, although with RMSF and typhoid fever, there is often a relevant exposure history. Rubella is rare in vaccinated children, but it certainly may be considered in those not immunized. It is similar to measles, but with rubella, patients are less sick, and Koplik spots are not present. The other viral infections (HHV-6 and parvovirus B-19) are also common and may be seen in fully immunized children because there are no specific vaccines against these viruses.

epidemiologic grounds, although laboratory confirmation is preferred. Laboratory investigational approach to delineating the etiology of such infection may include

- **Isolation of virus** from throat (or nasopharyngeal swab) and urine
- **Serology**
  - Virus-specific **IgM** to indicate acute infection
  - Elevation of **IgG** titer (>fourfold increase in paired sera)

COURSE

A clinical diagnosis was made. The patient was sent home and parents told to give symptomatic therapy.

ETIOLOGY

Rubella (German measles)

MICROBIOLOGIC PROPERTIES

Rubella virus belongs to the Togavirus family. The virus has a **central icosahedral nucleocapsid core** and is covered externally by a lipid-containing **envelope**. The genome is a **single-stranded RNA**, and **surface spikes on the envelope contain hemagglutinin**. Only **one antigenic (sero) type** of rubella virus exists.

Virus can be **cultivated from throat, nasopharynx,** and **urine.** Interference assay using a specific enterovirus is a reliable method of diagnosis.

EPIDEMIOLOGY

In 2005, the CDC announced that rubella had been eradicated in the United States. Importation of the virus can still occur, however. Disease is common in countries where mass immunization programs have been unsuccessful. Humans are the only natural host of the virus. Susceptibility to **contagious infection** generally occurs after loss of transplacentally acquired maternal antibodies. Infection in unvaccinated individuals is usually **transmitted by droplet spread or by direct contact** with patients. If a seronegative woman contracts rubella in early pregnancy, there can be a high rate of fetal damage or birth defects, known as **congenital rubella syndrome** (Table 74-2).

PATHOGENESIS

Rubella virus enters and infects the nasopharynx and lungs, attaching to and invading the respiratory epithelium. It then spreads hematogenously (primary viremia) to regional and distant lymphatics and replicates in the reticuloendothelial system. This is followed by a febrile stage (secondary viremia), which occurs 6 to 20 days after infection, spreading the virus to other tissues and to the skin. **Virus-specific T cells attack virus-infected vascular endothelial cells** of dermal capillaries and play an important role in the development of skin rash. Antigen-antibody complex mediated **vasculitis may also contribute to febrile exanthem illness**. Rash follows the prodromic febrile stage and lasts for 3 days.

TREATMENT

There is no specific antiviral therapy for rubella.

OUTCOME

The patient recovered after 10 days without any complications.

PREVENTION

A single dose of live, attenuated rubella virus vaccine elicits a significant antibody response in approximately 98% to 99% of susceptible individuals. Because it is a live vaccine, it should not be given to immunocompromised patients. The vaccine induces a respiratory mucosal IgA response, thus interrupting the spread of virulent virus by nasal carriage. To protect a fetus from exposure to rubella virus, maternal immunity should be demonstrated prior to pregnancy, either by serologic testing or by proof of immunization. For immunity screen in pregnant women, greater than or equal to 1:8 titer of IgG antibody indicates immunity and consequent protection of the fetus.

TABLE 74–2 Clinical Features of Rubella Virus Infections

Syndromes	Clinical Features
Rubella (German measles)	In an unvaccinated individual, rubella is a mild, febrile disease, manifesting as a diffuse punctate and maculopapular rash. Distribution is similar to that of measles, but the lesions are less intensely red. Children usually present few or no constitutional symptoms.
Congenital rubella syndrome (CRS)	Fetuses infected early (at the first trimester of pregnancy) are at greatest risk. Congenital malformations and even fetal death may occur. CRS occurs in up to 90% of infants born to women who acquire rubella during pregnancy.
	Fetal infection occurs transplacentally during the maternal viremic phase and the mechanism of fetal damage may be one of (1) tissue necrosis without inflammation, or (2) direct cytopathic effects, or (3) apoptotic cell death.
	In **pregnant women,** a clinical diagnosis of rubella is often inaccurate, so laboratory confirmation is important. Acute rubella infection in pregnant women can be confirmed by a fourfold rise in specific antibody titer between acute- and convalescent-phase serum specimens by ELISA. Amniocentesis and rubella culture may reveal whether there is rubella virus in amniotic fluid (in case of current infection).
	The **diagnosis of CRS in the newborn** may be supported by the presence of specific IgM antibody in a single specimen taken between 2 weeks and 3 months of age.

FURTHER READING

Banatvala JE, Brown DWG: Rubella (A Review). Lancet 363:1127, 2004.

Lee J-Y, Bowden DS: Rubella virus replication and links to teratogenicity. Clin Microbiol Rev 13:571, 2000.

Plotkin SA, Katz M, Cordero JF: The eradication of rubella. JAMA 281:561, 1999.

Schluter WW, Reef SE, Redd SC, Dykewicz CA: Changing epidemiology of congenital rubella syndrome in the United States. J Infect Dis 178:636, 1998.

A **55-year-old** woman presented to her family physician with a 3-day history of **burning and pain over her left forearm.** The symptoms were rather abrupt in onset, and she had not experienced anything like this before. Over the previous 2 days, **several vesicles had developed in a band-like distribution on her arm,** and new ones were erupting daily. On the day she went to see her family doctor, the lesions had started turning purplish. At no time did she have any fever, but her appetite was markedly reduced.

She had previously been healthy except for childhood illnesses of measles, chickenpox, and mumps. The patient worked as a caregiver to elderly individuals. She had **no known allergies.** She did smoke one pack of cigarettes per day.

PHYSICAL EXAMINATION

VS: T 37.5°C, P 112/min, R 16/min, BP 116/64 mmHg

PE: On examination, the patient was in mild distress due to the pain in her arm. It was readily evident that there was a **vesicular rash over the left arm** (Fig. 75-1). **Numerous vesicles measuring 2 to 3 mm**, **with a hemorrhagic base**, were present in a dermatomal distribution. One or two of the vesicles had crusted over.

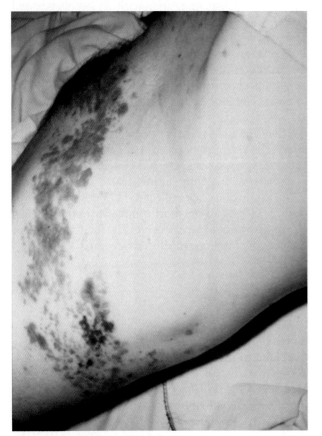

FIGURE 75-1 Lesions on the arm of the case patient. *(Courtesy of Dr. Daniel Skiest, Department of Medicine, University of Texas Southwestern Medical Center, Dallas, TX.)*

LABORATORY STUDIES

Blood

Hematocrit: 38%
WBC: Normal
Differential: Normal
Serum chemistries: Normal

Imaging

No imaging studies were done.

Diagnostic Work-Up

Table 75-1 lists the likely causes of illness (differential diagnosis). A clinical diagnosis was considered based on the distinctive crops of rashes. HSV infection may, in rare cases, interfere with the diagnosis of zoster.

TABLE 75–1   Differential Diagnosis and Rationale for Inclusion (consideration)
Contact dermatitis
Enteroviral rash
Herpes simplex virus type 1
Herpes zoster (shingles)
Measles
Rubella
Varicella (chickenpox)

Rationale: The presence of a rash with systemic symptoms often means a viral illness. However, contact dermatitis can also cause severe reactions. In children, common viruses are usually responsible (see above). However, localized lesions are not commonly seen with systemic viral infections and should prompt search for an allergic or local reaction. Herpes zoster should always be considered in localized rashes, particularly if the rash occurs in an adult and has a vesicular component.

In difficulty with clinical diagnosis, investigational approach may include

- Tzanck smear of the base of a lesion
- Demonstration of **viral antigen** in fluid by direct fluorescent antibody **(DFA)**
- **Cultivation of virus** from the skin lesions and other body sites

COURSE

A clinical diagnosis was made, and the patient was started on therapy with a mild narcotic for pain relief. The area was kept covered until all lesions had crusted.

ETIOLOGY

Varicella-zoster virus (VZV; zoster or shingles)

MICROBIOLOGIC PROPERTIES

VZV is a member of the herpesvirus group. The virus has an **envelope** (Fig. 75-2) and a **double stranded-DNA** genome. Antigenically it is different from other herpesviruses because it has a **single serotype.** VZV is **cultivable in tissue culture**; presumptive identification of the virus may be made based on the **characteristic** appearance of eosinophilic intranuclear inclusions and multinucleated giant cells in the

infected human fetal diploid kidney cells (apparent at 3 to 7 days). **Latency** is the key biologic feature of the virus, as is found in other viruses in the herpesvirus family.

EPIDEMIOLOGY

VZV causes **two distinct diseases: varicella (chickenpox in children)** and **herpes zoster (shingles in adults)**. Chickenpox is a febrile rash disease, manifesting as a sudden onset of slight fever, mild constitutional symptoms, and a skin eruption. The skin eruption is maculopapular for a few hours, progressing rapidly to vesicular and then to pustular for 3 to 4 days, and finally crusting over. The vesicles collapse on puncture. Lesions commonly occur in successive crops, with several stages of maturity present at the same time. They tend to be more abundant on scalp, high in the axilla, and on mucous membranes of the mouth and upper respiratory tract. Humans are the only reservoir of the virus. In temperate zones, chickenpox occurs most frequently in late winter and early spring. Transmission of varicella is by the inhalation of airborne respiratory aerosol from a patient 1 to 2 days before the onset of rash, which makes this infection highly contagious. There is a universal susceptibility among the population in general. Nonimmune children are particularly affected and have a benign course. Nonimmune neonates, immunodeficient individuals, and oncology patients are at increased risk for poor prognosis. Children with acute leukemia are also at increased risk of severe disseminated disease with high mortality.

    Zoster occurs sporadically as a reactivation of latent VZV and occurs most commonly **in individuals above 50 years of age**. Zoster can occur any time of the year, as it is a reactivation syndrome.

PATHOGENESIS

In susceptible children, VZV primarily infects the mucosa of the upper respiratory tract and spreads via the blood to the skin, where the typical vesicular rash occurs. Incubation period is 2 to 3 weeks (15 days median). After primary varicella infection, VZV may infect and become dormant in the dorsal root ganglia (sensory nerve roots) for life. In latency, the virus persists in a noninfectious form, and there may be intermittent periods of reactivation and shedding. The latent virus can be reactivated due to medications,

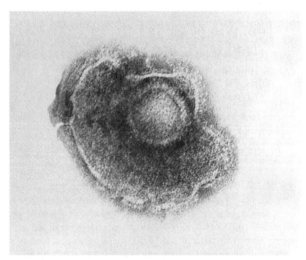

FIGURE 75-2 Electron micrograph of a varicella-zoster virus. *(Courtesy of Centers for Disease Control and Prevention, Atlanta, GA.)*

illness, or malnutrition, or from the natural decline in immune function with aging, causing **zoster** (shingles). Upon reactivation, the virus migrates down the sensory nerve to the skin, causing the characteristic painful dermatomal rash. Patients experience pain in the distribution of the rash (postherpetic neuralgia). The reactivated virus causes the vesicular skin lesions and nerve pain of zoster due to damage to the peripheral nerves. The passage of the virus progresses down the axon to mucocutaneous sites; local spread and replication occur to form clusters of blister-like lesions in a strip-like pattern on one side of the body. **Multinucleated giant cells** with intranuclear inclusions are seen in the base of the skin lesions. Histopathology may also reveal hemorrhage, edema, and lymphocytic infiltration.

NOTE In complicated reactivation of VZV in immunocompromised patients, there may be **hematogenous spread** of the virus to the lung, causing **interstitial pneumonia,** and to the CNS, causing **encephalitis.** These complications can also occur in immunocompetent adults. Primary varicella is the second most common **viral antecedent for Reye syndrome** after influenza B in children.

TREATMENT

Acyclovir (ACV) is generally considered the antiviral agent of choice for varicella and zoster. ACV shortens the duration of symptoms and pain of zoster in the normal older patient. Therapy with oral ACV is recommended for disease in normal hosts over 50 years of age to shorten the duration of zoster-associated pain. Corticosteroids may also be helpful in older adults with zoster. Successors to ACV, famciclovir and valacyclovir, are also effective and may be taken less often than acyclovir.

OUTCOME

The patient completed a course of acyclovir, and her lesions began healing within a week. Over the course of several weeks, her pain resolved completely.

PREVENTION

A live attenuated varicella virus vaccine (Varivax) has been licensed for use in the United States and is recommended for all children older than 12 months. Immunization with the live attenuated vaccine significantly enhances VZV-specific IgG antibodies and T-cell immunity. Varicella-zoster immune globulin (VZIG) is effective in preventing disease if given within 96 hours after exposure of the virus in high-risk individuals (e.g., patients receiving cancer chemotherapy and/or radiotherapy).

FURTHER READING

Arvin AM: Varicella-zoster virus. Clin Microbiol Rev 9:361, 1996.
Cohen JI: Varicella-zoster virus. Infect Dis Clin North Am 10:457, 1996.
Cohen JI, Brunell PA, Straus SE: Recent advances in varicella-zoster virus infection. Ann Intern Med 130:922, 1999.

Gilden DH, Kleinschmidt-DeMasters BK, LaGuardia JJ, et al: Neurologic complications of the reactivation of varicella-zoster virus: Medical progress. N Engl J Med 342:635, 2000.

A 62-year-old man was seen in his family physician's office for complaints of low-grade **fevers, night sweats,** and **fatigue for 3 weeks.**

The patient had a **history of a heart murmur** but had never undergone extensive evaluation. He had been in generally excellent health with normal exercise tolerance. Approximately 6 weeks ago he underwent an uncomplicated **extraction of an impacted wisdom tooth** but received **no antibiotics** prior to the procedure.

CASE
76

PHYSICAL EXAMINATION

VS: T 38°C, P 104/min, R 14/min, BP 130/82 mmHg

PE: The patient was alert; a rough, diamond-shaped systolic murmur—heard loudest in the upper left sternal border—was noted. A **"splinter hemorrhage" in the nail** of his right index finger and **conjunctival petechiae** were also noted. His spleen was palpable as well.

LABORATORY STUDIES

Blood

ESR: 80 mm/h

Hematocrit: 36%

WBC: 10,500/µL

Differential: 74% PMNs, 18% lymphs

Serum chemistries: Normal

Urinalysis: WBC 5 to 10/µL, RBC 10 to 20/µL, no bacteria

Coagulation panel: Normal

Imaging

A chest x-ray was within normal limits.

Diagnostic Work-Up

A clinical diagnosis of infective endocarditis (IE) was considered, based on the symptoms and peripheral manifestations. Table 76-1 lists the likely causes of IE (differential diagnosis). Investigational approach may include

- **Echocardiography (echo).** Useful when endocarditis is suspected; transesophageal echo is more sensitive than is transthoracic echo
- **Blood cultures**
- **Serology** for noncultivable agents

TABLE 76–1   Differential Diagnosis and Rationale for Inclusion (consideration)

Enterococci

Fungi

Gram-negative aerobic bacilli

HACEK organisms

  Haemophilus

  Actinobacillus

  Cardiobacterium

  Eikenella

  Kingella

Miscellaneous bacteria (*e.g., Chlamydia, Coxiella,* and *Bartonella*)

Oral bacteria (e.g., viridans streptococci)

Staphylococcus aureus

Staphylococcus epidermidis

Streptococcus pneumoniae

Rationale: The clinical syndrome presented above is nonspecific and may be seen with a variety of infectious as well as noninfectious causes. Endocarditis is an important consideration, especially in older individuals and in those with specific risk factors. *S. aureus,* viridans streptococci, *S. epidermidis,* enterococci, and *S. pneumoniae* are commonly encountered.

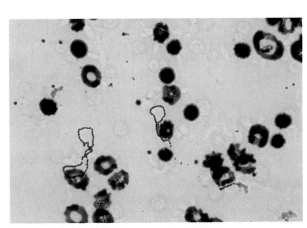

FIGURE 76-1   Gram stain of blood culture.

COURSE

The patient was admitted to the hospital, and blood cultures were drawn. **Echocardiogram** revealed a bicuspid aortic valve **with an 8-mm mobile vegetation** on the anterior cusp. Blood cultures were diagnostic.

ETIOLOGY

Viridans streptococci (native-valve endocarditis)

MICROBIOLOGIC PROPERTIES

Viridans streptococci are **Gram-positive cocci** that occur in pairs or **chains** (Fig. 76-1). The important microbiologic features are α-**hemolytic** (colonies on blood agar medium are surrounded by a partial zone indicating incomplete lysis of red blood cells and a greenish discoloration) or nonhemolytic (grayish) growth, and **resistance to Optochin** (*Streptococcus pneumoniae*, also α-hemolytic, is sensitive to Optochin). Many species exist as commensals (human **oral flora**), and the list of important species includes *Streptococcus mitis (mitior)*, *S. oralis*, *S. gordonae*, *S. sanguis*, *S. anginosus (milleri)*, *S. salivarius*, *S. mutans*, and nutritionally variant (vitamin B_6–dependent) species.

Multiple blood cultures are necessary to demonstrate **continuous bacteremia** in IE. Viridans group streptococci are difficult to identify in the laboratory, and only for patients with serious infections—particularly endocarditis—are viridans group streptococci identified to the species level.

EPIDEMIOLOGY

Viridans streptococci are **normal inhabitants of the oral cavity** and GI tract. These organisms enter the blood stream following any procedure (e.g., dental extraction or vigorous tooth brushing) that causes submucosal damage.

PATHOGENESIS

Infective endocarditis, an infection of the endocardial surface of the heart, usually occurs as a result of transient bacteremia in the setting of a preexistent valve abnormality. IE can be divided into two main categories (with somewhat distinct association with the causal agent)

- **Acute bacterial endocarditis.** Usually caused by highly virulent *Staphylococcus aureus* (in patients with intravenous drug use [IVDU]) and, less commonly, *Streptococcus pyogenes* and *Streptococcus pneumoniae*. Pathology includes large vegetations and occasionally abscesses. There is generally a fulminant course.
- **Subacute bacterial endocarditis** (SBE). Pathology includes smaller vegetations on abnormal valves. SBE is generally associated with a more indolent course and is classified further as

  - **Native-valve** endocarditis (often caused by viridans streptococci, enterococci, or HACEK [*Haemophilus, Actinobacillus, Cardiobacterium, Eikenella, Kingella*] organisms). It is a sequela of dental procedures, as seen in the case patient.
  - **Prosthetic valve** endocarditis (often caused by *Staphylococcus epidermidis*). A patient may have a history of an uneventful replacement of the aortic valve with a mechanical prosthesis because of severe aortic stenosis, for instance.

Regarding etiologic agents of endocarditis, streptococci in general and viridans streptococci in particular predominate in cases of **SBE**, which usually has a more indolent course than does the acute form. In patients with **prosthetic heart valve** endocarditis, staphylococcal infections predominate. Prosthetic valve endocarditis arising a few months after valve surgery is generally the result of intraoperative contamination of the prosthesis. Coagulase-negative staphylococcus (*S. epidermidis*) is more commonly seen than is coagulase-positive *S. aureus*, which is more common in acute endocarditis in intravenous drug use.

Native-valve IE generally occurs on abnormal valves (i.e., rheumatic heart disease due to early, recurrent *Streptococcus pyogenes* infection) or in instances when **turbulence to the endothelial surface of the heart** leads to minor trauma. This can lead to deposition of fibrin and platelets, so-called nonbacterial thrombotic endocarditis (NBTE). Transient bacteremia then leads to **seeding of lesions with adherent bacteria** (e.g., oral flora in vigorous dental irrigation). Fibronectin-binding proteins on viridans streptococci facilitate adherence to thrombi, whereas surface clumping factors on more virulent *S. aureus* facilitate adherence to intact endothelium or exposed subendothelial tissue. Organisms become entrapped in the growing platelet-fibrin vegetation and proliferate, forming microcolonies. Pathology in other organs is a **result of deposition of circulating immune complexes** to deposited bacterial antigens.

There are no highly specific clinical and laboratory features (bacteremia or blood culture positivity) of IE. An elevated ESR and anemia are commonly seen in SBE. Endocarditis in patients with **IVDU (heroin addiction)** is predominantly a **right-sided** disease, caused by *S. aureus* and involving the **tricuspid valve**. Infection may lead to pulmonary septic emboli but may otherwise have symptoms similar to subacute bacterial endocarditis. In subacute bacterial endocarditis, prosthetic valves and native valves are not easily differentiated. In all cases of endocarditis, the **most common symptom is fever**, followed by malaise, night sweats, fatigue, weight loss, and arthralgias. Signs include **Janeway lesions** (small erythematous lesions on palms or soles), **Osler nodes** (painful raised lesions on finger and toe pads; often transient), **Roth spots** (retinal lesions), and **splinter hemorrhages** on the nail bed.

TREATMENT

Use of bactericidal drug(s) is considered essential. Successful outcome usually requires 4 to 8 weeks of intravenous therapy, depending on the organism and type of valve; prosthetic valves require longer courses of therapy. Antibiotic choice is based on the susceptibility of the blood isolate. Viridans streptococci can usually be treated with penicillin G with or without an aminoglycoside. Surgery may be indicated for failure of medical therapy or development of heart failure, intracardiac abscess, or multiple systemic emboli.

OUTCOME

The patient was treated with 4 weeks of intravenous penicillin G and had a complete recovery.

PREVENTION

In certain high-risk groups, such as those with known abnormal heart valves, or mitral valve prolapse with regurgitation, antibiotic prophylaxis is indicated after procedures that are frequently associated with transient bacteremia. These include dental procedures, GI endoscopy with biopsy, and urologic procedures. Guidelines available from the American Heart Association detail which prophylaxis to use in specific situations. Although these are based on expert opinion and not on controlled trials, they are widely used and recommended.

FURTHER READING

Brouqui P, Raoult D: Endocarditis due to rare and fastidious bacteria. Clin Microbiol Rev 14:177, 2001.
Moreillon P, Que Y-A: Infective endocarditis. Lancet 363:139, 2004.
Morris AJ, Drinkovic D, Pottumarthy S, et al: Gram stain, culture, and histopathological examination findings for heart valves removed because of infective endocarditis. Clin Infect Dis 36:697, 2003.
Mylonakis E, Calderwood SB: Infective endocarditis in adults: Medical progress. N Engl J Med 345:1318, 2001.

A 36-year-old man was brought to the hospital for complaints of persistent **high fever, dry cough,** and **worsening shortness of breath** for a week.

He had been diagnosed with **acute myelogenous leukemia 6 months before,** and he **received an allogeneic bone marrow transplant 6 weeks before.** He was taking Bactrim for prophylaxis of pneumocystis pneumonia.

PHYSICAL EXAMINATION

VS: T 39°C, P 122/min, R 36/min, BP 120/82 mmHg

PE: Thin male in moderate respiratory distress; lungs with crackles diffusely in both lungs.

LABORATORY STUDIES

Blood

Hematocrit: 32%

WBC: 3200/μL

Differential: 50% PMNs, 32% lymphs, 12% monos

Blood gases: pH 7.48, pCO$_2$ 30 mmHg, pO$_2$ 58 mmHg

Serum chemistries: BUN 24 mg/dL, creatinine 1.8 mg/dL

Imaging

A chest x-ray revealed bilateral interstitial infiltrates.

Diagnostic Work-Up

Table 77-1 lists the likely causes of illness (differential diagnosis). Specific diagnosis is difficult owing to the high frequency of asymptomatic and reactivation infections in patients with transplants. Investigational approach for specific microbiologic diagnosis may include a combination of the following tests:

- **Gram stain**, **acid-fast stain**, and **silver stain (histopathology)** of bronchoalveolar lavage (BAL) or bronchoscopic biopsy specimens
- **Virus isolation** from BAL
- **CMV antigen detection** from serum
- **CMV DNA detection** by PCR

COURSE

A bronchoscopy was done, and transbronchial biopsy and bronchoalveolar lavage specimens were sent for histopathologic and cytologic examination. Empirical antibiotics were started pending cultures. Routine

TABLE 77–1   Differential Diagnosis and Rationale for Inclusion (consideration)

Adenoviral acute respiratory distress syndrome
Aspergillosis
Cytomegalovirus (CMV)
M. avium-intracellulare
Pneumocystis pneumonia

Rationale: Because of their immunocompromised state, transplant patients are at high risk for developing a variety of unusual infections due to bacteria and mycobacteria. Patients with bone marrow transplants are the most immunosuppressed; those with kidney transplants are the least immunosuppressed. A variety of clinical syndromes may be seen, depending on the organ that has been transplanted. Aspergillosis is often seen with prolonged neutropenia and carries a high mortality. Pneumonitis is often caused by *Pneumocystis jiroveci* in an immunocompromised patient, but Bactrim is a reliable prophylactic drug. CMV is among the most common viral causes of disease in these patients and should always be considered, but adenovirus is also seen.

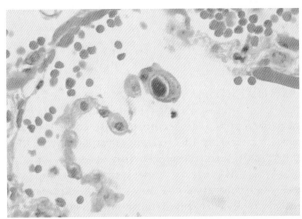

FIGURE 77-1   Histopathology of lung. Note intranuclear inclusion with surrounding halo in a pneumocyte. (*Courtesy of Centers for Disease Control and Prevention, Atlanta, GA.*)

bacteriologic and mycobacteriologic work-up were all negative. Virus culture was positive for an agent typical of this patient population.

ETIOLOGY

Cytomegalovirus (CMV pneumonitis)

MICROBIOLOGIC PROPERTIES

CMV, a member of the herpesvirus group (which has **double-stranded DNA, a protein capsid, and a lipoprotein envelope**), includes herpes simplex virus types 1 and 2, varicella-zoster virus, and Epstein-Barr virus. CMV is a β-herpesvirus that is lymphotropic and has the largest genome of any herpesvirus. **Antigenically, CMV is different from other herpesviruses**, as the virus is associated with only a single serotype. CMV replicates in the nucleus of the target cells, and the **infected cells develop into cytomegalic (giant) cells**. Like other herpesviruses, CMV can progress to latency after primary infection; reactivation causes either lytic or productive infection.

Multiple modalities for laboratory diagnosis of CMV disease should always be used if possible. CMV can be **cultured** from specimens obtained from **urine**, throat swabs, and tissue samples to detect active infection. Antigen detection is often used as a sensitive marker for CMV infection, and CMV PCR is emerging as a useful tool as well.

EPIDEMIOLOGY

CMV is found universally throughout all geographic locations and socioeconomic groups. It infects between 50% and 85% of adults in the United States by 40 years of age and is even more widespread in developing countries and in areas of lower socioeconomic conditions. Transmission of CMV occurs from **person to person and requires close contact with a person excreting the virus** in saliva, urine, or other body fluids. Virus is acquired early in life via saliva in children. CMV can be sexually transmitted and can also be transmitted via breast milk, transplanted organs, and, rarely, blood transfusions. The major **risk groups are transplant patients receiving immuno-suppressive drugs, AIDS patients, and fetuses**. Patients with bone marrow transplant, organ transfer, or blood transfusion receive the virus from the donors, and it can also be transmitted by vertical transmission from mother to fetus.

PATHOGENESIS

CMV is the most **common cause of post-transplant infection**, for both solid organ and bone marrow transplants. For most healthy donors who acquire CMV after birth, there are few symptoms and no long-term health consequences. **Multiple cell types and various organs** (including lungs, as seen in the case patient) **are targeted by the virus** and become the sites of viral latency. CMV, like other herpesviruses, shares a **characteristic ability to remain dormant** within the body over a long period. CMV also enters the latent state in leukocytes, including lymphocytes. The virus is reactivated when cell-mediated immunity is compromised due to immunosuppressive therapy, causing interstitial pneumonia, among other clinical syndromes (Table 77-2). Characteristic microscopic changes include **enlarged cells with intranuclear and intracytoplasmic inclusions**, which **contain viral particles, in the target organ**. Histopathology of transbronchial lung biopsy from patients with CMV pneumonitis may reveal pneumocytes containing characteristic **cytomegalic inclusions with a surrounding halo** ("owl's eye"; Fig. 77-1). Often there is a cellular infiltrate in proximity to the CMV-containing cells consisting of lymphocytes and plasma cells. Patients also present with altered graft function or rejection or with systemic evidence of a "viral" illness.

TREATMENT

Ganciclovir with or without anti-CMV immunoglobulin is effective in treating pneumonitis in immunocompromised patients (e.g., transplant patients). Ganciclovir can be used for CMV esophagitis and retinitis in AIDS patients or other immunocompromised patients. **Foscarnet** is useful **in ganciclovir-resistant strains**. There is often no specific duration of therapy; it is usually continued until significant immunosuppression has resolved.

TABLE 77–2 Other Important Clinical Features of Cytomegalovirus Infections

Syndromes	Clinical Features
Congenital infection	One of the **common congenital infections** (Toxoplasmosis, Rubella, Cytomegalovirus, Herpes [HIV-1, HBV] viruses, Syphilis, and Listeria [ToRCH³eS-List]); a consequence of either primary or reactivation infection of the mother Infants with CMV infection have signs and symptoms of severe generalized infection, especially involving the CNS, eye, and liver. Major manifestations are mental retardation, spasticity, eye abnormalities, hearing defects, and hepatosplenomegaly. Clinical diagnosis **in the newborn** is confirmed by virus isolation. In the first week of life of the newborn, virus shedding can be identified in the throat and in urine and other body fluids.
Infectious mononucleosis	Primary infection in late childhood or early adulthood, manifesting as a heterophile antibody-negative (EBV-negative) infectious mononucleosis (IM), occasionally with prolonged fever and a mild hepatitis Appearance of atypical lymphocytes in the peripheral blood is comparable to IM caused by EBV.
Infectious colitis, esophagitis, (opportunistic infections in AIDS)	Colitis can manifest with diarrhea, fever, abdominal pain, cramps, or bloating. Diagnosis requires colonoscopy with biopsy of ulcerative lesions. Biopsy of the ulcer shows the typical CMV histopathologic changes (inclusions). Esophagitis typically manifests with odynophagia (painful swallowing). Biopsy of the ulcer may show the typical CMV histopathologic changes (inclusions).
Retinitis (in AIDS)	Although CMV may infect virtually any organ in HIV-infected patients, retinitis is the most common clinical manifestation, accounting for 60-71% of cases. Common symptoms include floaters, flashes, blurred vision, decreased visual acuity, and scotomata (blind spots).

OUTCOME

The patient was admitted to the hospital, and bronchoscopy showed positive culture for CMV. His condition deteriorated and he was intubated. He was also started on ganciclovir with good, albeit slow, response. He was discharged after a 4-week hospital stay.

PREVENTION

No vaccine is available. In CMV-negative patients and in immunocompromised patients, transfusion of blood from a CMV-positive donor should be avoided. Many centers use prophylactic ganciclovir to reduce the risk of CMV infection.

FURTHER READING

Goodrich J, Khardori N: Cytomegalovirus: The taming of the beast? Lancet 350:1718, 1997.
Meijer E, Boland GJ, Verdonck LF: Prevention of cytomegalovirus disease in recipients of allogeneic stem cell transplants. Clin Microbiol Rev 16:647, 2003.
Revello MG, Gerna G: Diagnosis and management of human cytomegalovirus infection in the mother, fetus, and newborn infant. Clin Microbiol Rev 15:680, 2002.
Sia IG, Patel R: New strategies for prevention and therapy of cytomegalovirus infection and disease in solid-organ transplant recipients. Clin Microbiol Rev 13:83, 2000.
Skiest DJ: Cytomegalovirus retinitis in the era of highly active antiretroviral therapy (HAART). Am J Med Sci 317:318, 1999.
Whitcup SM: Cytomegalovirus retinitis in the era of highly active antiretroviral therapy. JAMA 283:653, 2000.

One **summer** evening, a **63-year-old** white man came to the emergency department of a nearby hospital presenting with a **6-day history of fever,** moderate headache, **generalized myalgia, arthralgias,** and fatigue. He had noticed a **rash under the armpit** that day that had **spread rapidly,** prompting him to seek medical attention.

He had lived in Connecticut and had recently moved to a cottage in a **wooded area outside of a small town.** He noted multiple tick bites after his daily walks in the woods. He had otherwise maintained good health.

PHYSICAL EXAMINATION

VS: T 38.8°C, P 102/min, R 14/min, BP 134/82 mmHg
PE: An expanding erythematous skin lesion (**erythema migrans** was noted under the axilla that had a central area of clearing (Fig. 78-1).

LABORATORY STUDIES

Blood

Hematocrit: 43%
WBC: 7400/µL
Differential: Normal
Blood gases: Normal
Serum chemistries: AST 156 U/L, ALT 189 U/L

Imaging

No imaging studies were done.

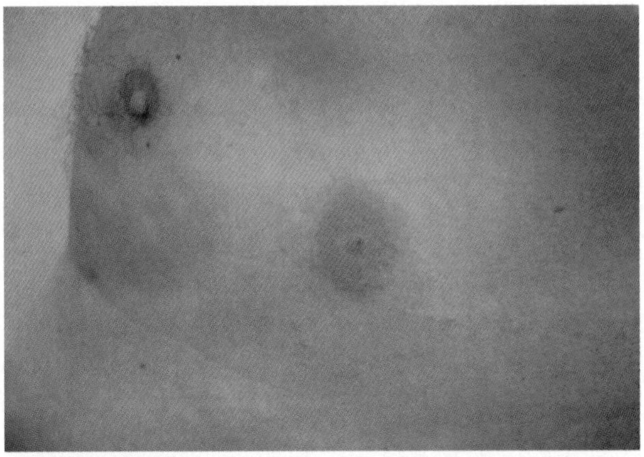

Diagnostic Work-Up

Table 78-1 lists the likely causes of illness (differential diagnosis). A clinical diagnosis was considered based on protracted fever, characteristic rash, and history of exposure, in an area where this disease is known to occur. Investigational approach for confirmation of microbiologic diagnosis may include

- **Serologic tests**. Two-step ELISA followed by Western blot techniques
- In failed investigation, **PCR** of target DNA

COURSE

Acute-phase serum was collected. Based on the clinical picture, therapy was begun for the most likely infectious agent.

FIGURE 78-1 Expanding macular rash with central clearing in a patient with tick bite. *(Courtesy of Dr. Paul Southern, Departments of Pathology and Medicine, University of Texas Southwestern Medical Center, Dallas, TX.)*

TABLE 78–1 Differential Diagnosis and Rationale for Inclusion (consideration)

Aseptic meningitis
Ehrlichiosis
  Ehrlichia chaffeensis (human monocytic ehrlichiosis)
  Ehrlichia equi (human granulocytic ehrlichiosis)
Lyme disease (*Borrelia burgdorferi*)
Rheumatoid arthritis
Rickettsia rickettsii
Septic arthritis

Rationale: Symptoms and signs are nonspecific and can be associated with several diseases, listed above. The fever, muscle aches, and fatigue can easily suggest viral infections, such as influenza or infectious mononucleosis, or aseptic meningitis if headache is a prominent symptom. Joint pain can be mistaken for other types of arthritis, such as rheumatoid arthritis or septic arthritis with the history of fevers. Tick bites are associated with several infections, including ehrlichiosis, RMSF, and Lyme disease. The first two infections usually manifest with generalized rashes. Characteristic expanding rash and history of tick bite in a geographic area are supportive of a diagnosis of Lyme disease.

ETIOLOGY

Borrelia burgdorferi (Lyme disease [LD])

MICROBIOLOGIC PROPERTIES

The Lyme agent is a spirochete, *B. burgdorferi sensu stricto*. *B. burgdorferi* is a **flexible, motile bacterium** (Fig. 78-2). It can be grown in a special liquid medium, called Barbour, Stoenner, Kelly (BSK) medium. Routine cultures of clinical specimens are typically negative. Culture of the organism from the tick vector is usually positive. Spirochetal outer-surface proteins A and B [OspA and OspB] are prominent antigens.

Serologic diagnosis is currently based on two-step ELISA followed by Western blot techniques. The ELISA serologic test has a relatively high false positivity rate and must be interpreted with caution, because the antibody cross-reacts with other spirochetes. The Western blot is more specific and confirmatory. By **PCR**, *B. burgdorferi* genetic material has been detected in urine, blood, synovial fluid, CSF, skin, and other tissues. Urine PCR, the target of which is a specific region of the flagellin gene, may allow monitoring of the efficiency of therapy in patients with early Lyme borreliosis.

EPIDEMIOLOGY

Lyme disease is the most common vector-borne infection in the United States. Black-legged ticks (***Ixodes scapularis***; Fig. 78-3) **transmit *B. burgdorferi* to

FIGURE 78-3 *"Black-legged ticks," Ixodes scapularis.* From left to right: The adult male, adult female, and nymph. *(Courtesy of Centers for Disease Control and Prevention, Atlanta, GA.)*

humans in the northeastern** United States (from Massachusetts to Maryland) and in the **north-central** United States (Wisconsin and Minnesota). On the Pacific Coast, the bacteria are transmitted to humans by black-legged ticks of a different species, *Ixodes pacificus*. Maintenance of endemic foci requires a reservoir of infection in rodent hosts. People of all ages are susceptible in the summer months during outdoor activities in the endemic areas. Larval ticks take blood meals during the early summer months and the nymphs transmit *B. burgdorferi* to humans. White-tailed deer support the growth and maintenance of the adult stage of *I. scapularis* and are the key to the tick's survival.

PATHOGENESIS

Spirochetes live in the tick midgut (where OspA and OspB are expressed), and they become "activated" during the blood meal. They migrate to salivary glands, from which they are inoculated into host (the whole process takes 48 to 72 hours). Cutaneous migration of spirochetes outward from a tick bite site causes erythema migrans. No exotoxins, enzymes, or other virulence factors have been linked to the spirochetal agent; vascular changes are induced by local effects of proinflammatory cytokines (including tumor necrosis factor and interleukin-1) on the vessel wall.

The immune response to *B. burgdorferi* is a delayed response, with several weeks needed for the development of seropositivity, so antibody (IgG) tests at presentation may yield false-negative results. The humoral response may elicit the production of auto-

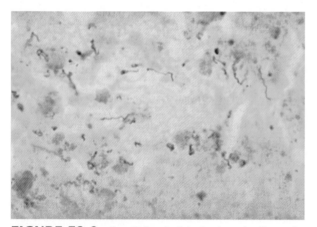

FIGURE 78-2 *Borrelia burgdorferi* spirochetes in silver stain. *(Courtesy of Centers for Disease Control and Prevention, Atlanta, GA.)*

TABLE 78–2 Other Clinical Features of *Borrelia burgdorferi* Infections (Lyme disease [LD])

Syndromes	Clinical Features
Early disseminated LD (stage 2)	**Musculoskeletal:** arthralgias, migratory arthritis, frank arthritis **Neurologic:** meningitis, cranial neuropathy (especially facial palsy), and peripheral neuropathy **Cardiac:** least common form. Usually manifests as conduction system abnormalities, usually heart block
Late disseminated LD (stage 3: persistent infection)	The **late symptoms** and signs (or complications) of LD may not appear until weeks, months, or years after a tick bite. **Musculoskeletal** abnormalities include **arthritis**, which is most likely to appear as intermittent bouts of pain and swelling, usually in one or more large joints, especially the knees. **Neurologic** abnormalities can include numbness, pain, Bell palsy (paralysis of the facial muscles, usually on one side), and a chronic encephalomyelitis.

antibodies capable of recognizing and damaging normal host tissues (e.g., arthritis or meningitis).

> **NOTE** After several days or weeks, spirochetes disseminate from the site of inoculation by cutaneous, lymphatic, and vascular routes to many different sites (disseminated infections; see Table 78-2). The role of autoimmunity in immune pathogenesis is suggested by the association of chronic disease (arthritis) with certain HLA types and the failure of some cases to respond to antibiotics.

TREATMENT

Doxycycline or amoxicillin (for children) is the drug of choice. In a few patients who are treated for LD, symptoms of persisting infection may continue or recur, making additional antibiotic treatment necessary. Varying degrees of permanent damage to joints or the nervous system can develop in patients with late chronic LD. Typically, these are patients in whom LD was unrecognized in the early stages or for whom initial treatment was unsuccessful. Disseminated infections (arthritis or meningitis; see Table 78-2) can be treated with intravenous ceftriaxone.

OUTCOME

The patient responded well to doxycycline, with resolution of all signs and symptoms. Serologic testing for LD by enzyme immunoassay and Western blot confirmed the diagnosis.

PREVENTION

Personal protective measures, such as use of insect repellent and routine tick checks in the endemic areas, are key components of primary prevention. Removing infected ticks within 24 hours of attachment can reduce the likelihood of transmission. Most exposed people can be successfully treated with antibiotic therapy when diagnosed in the early stages of LD. A single dose of doxycycline immediately after a tick bite in endemic areas can prevent LD. Recently, LD vaccine was developed that used recombinant *B. burgdorferi* lipidated OspA as the immunogen. However, the vaccine was withdrawn from the market, reportedly because of poor sales.

FURTHER READING

Hayes EB, Piesman J: How can we prevent Lyme disease? N Engl J Med 348:2424, 2003.
Stanek G, Strle F: Lyme borreliosis. Lancet 362:1639, 2003.
Steere AC, Sikand VK, Schoen RT, Nowakowski J: Asymptomatic *B. burgdorferi* infection. Clin Infect Dis 37:528, 2003.

Wang TJ, Liang MH, Sangha O, et al: Coexposure to *Borrelia burgdorferi* and *Babesia microti* does not worsen the long-term outcome of Lyme disease. Clin Infect Dis 31:1149, 2000.

A 9-year-old boy was brought to the hospital by his parents with complaints of repeating **intense chills and daily high fever for 4 days.** The parents said that when **his fevers would abate, he would become drenched in sweat and feel exhausted and drained.** The parents also reported diarrhea, nausea, and abdominal pain. On the day of admission the patient was noted to be lethargic and difficult to arouse. A **generalized seizure** was witnessed in the emergency department.

The family had **immigrated to the United States from West Africa** 3 weeks before the onset of the current illness.

PHYSICAL EXAMINATION

VS: T 40°C, P 140/min, R 28/min, BP 82/40 mmHg

PE: Thin male minimally responsive to verbal commands. Pupils were reactive and neck was supple. Conjunctiva was pale, and abdominal exam showed **hepatosplenomegaly.**

LABORATORY STUDIES

Blood

Hematocrit: 18%

WBC: 16,300/µL

Differential: 50% PMNs, 20% bands, 15% lymphs

Platelets: 42,000/µL

Blood gases: pH 7.28, pCO$_2$ 30 mmHg, pO$_2$ 64 mmHg

Serum chemistries: Glucose 40 mg/dL, BUN 45 mg/dL, AST 240 U/L, ALT 310 U/L, LDH 820 U/L, creatinine 2.6 mg/dL

Imaging

Head CT was unremarkable.

Diagnostic Work-Up

Table 79-1 lists the likely causes of illness (differential diagnosis). Investigational approach for specific microbiologic diagnosis may include

- **Lumbar puncture and CSF examination** to rule out bacterial meningitis
- **Blood cultures** to detect blood-borne bacteria
- **Thick and thin smear** for blood-borne parasites
- In failed diagnosis, virus-specific serology for the listed infections

COURSE

The patient was admitted and required mechanical ventilation for impending respiratory failure. Lumbar

TABLE 79–1 Differential Diagnosis and Rationale for Inclusion (consideration)

African trypanosomiasis (sleeping sickness)
Aseptic meningitis
Babesiosis
Bacterial meningitis
Dengue fever
Leptospirosis
Malaria
Typhoid fever

Rationale: A diagnosis should be aggressively sought in patients who present with severe neurologic symptoms and fever. It is always important to rule out bacterial meningitis initially. Epidemiology is important for a patient to determine the possible history of exposure in an area endemic for a variety of infections. Typhoid fever and parasitic infections should be considered. Many of the above infections are geographically limited, so a good history of travel is important as well. Babesiosis is found in the northeastern and upper midwestern U.S., and leptospirosis is associated with animal exposure. Dengue, malaria, and trypanosomiasis are all endemic in Africa, with the latter two infections typically causing periodic fever.

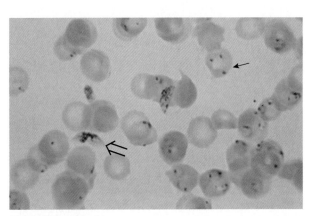

FIGURE 79-1 Thin smear of peripheral blood (Giemsa stain). Note two or more ring forms in a single RBC (*arrow*) and macrogametocyte (crescent; *open arrow*). Note also heavy parasitemia. (*Courtesy of Dr. Paul Southern, Departments of Pathology and Medicine, University of Texas Southwestern Medical Center, Dallas, TX.*)

puncture was performed, which was normal. Blood cultures were drawn and were negative for blood-borne pathogens. Based on the given history of travel, thick and thin blood smears were performed and yielded a diagnosis.

ETIOLOGY

Plasmodium falciparum (malaria)

MICROBIOLOGIC PROPERTIES

Plasmodia are coccidian protozoal agents. The four human-infective species of *Plasmodium* are *P. falciparum*, *P. vivax*, *P. malariae*, and *P. ovale*. Only *P. falciparum* causes life-threatening infection.

The most salient morphologic characteristics of plasmodia belong to the following four developmental stages:

1. **Ring.** Early developmental stage of the asexual erythrocytic parasite, often arranged in a **ring shape around a central vacuole** (Fig. 79-1).
2. **Trophozoite.** In the next developmental stage, the parasite has lost its "ring" appearance and has begun to accumulate pigment. The trophozoite of *P. vivax* is **ameboid** in shape, and the **enlarged infected erythrocyte contains** numerous "**Schüffner dots**" (Fig. 79-2).

3. **Schizont.** In the late developmental stage, the parasite has begun its division into merozoites and thus is characterized by the presence of multiple contiguous chromatin dots.
4. **Gametocyte.** Sexual erythrocytic stages (female is the macrogametocyte [see Fig. 79-1] and male is the microgametocyte).

Laboratory diagnosis is made by demonstration of the parasite within red blood cells. The types of smears obtained are

1. **Thick film.** RBCs are lysed, and WBC, platelets, and parasites are visible. This screening method does not differentiate *Plasmodium* from *Babesia*.
2. **Thin film.** With this method, morphologic features are visible for differentiation of *Plasmodium* from *Babesia* and for definitive species identification.

The following factors are helpful in determining parasite species:

1. Number of intraerythrocytic parasites (multiplicity/cell, see Fig. 79-1).
2. Morphologic characteristics of the parasites (e.g., **crescent-shaped gametocyte in *P. falciparum***; see Fig. 79-1).
3. Degree of parasitemia (number of infected erythrocytes in a blood film): heavy is considered to be greater than or equal to **10% (with *P. falciparum*** [usually seen in severe malaria]; see Fig. 79-1).

Microscopists may, however, occasionally fail to differentiate between species in cases in which morphologic characteristics overlap (especially *P. vivax* and *P. ovale*), as well as in cases in which parasite morphology has been altered by drug treatment or improper storage of the sample. In such cases, the *Plasmodium* species can be determined by using confirmatory molecular diagnostic tests (e.g., PCR).

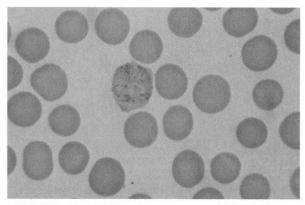

FIGURE 79-2 Important diagnostic features of *Plasmodium vivax*. Note an enlarged RBC (compared with the uninfected RBCs). Merozoites (hyperchromic elements) and Schüffner dots (less dense smaller granules) in a schizont are also seen in the infected RBC. (*Courtesy of Dr. Paul Southern, Departments of Pathology and Medicine, University of Texas Southwestern Medical Center, Dallas, TX.*)

EPIDEMIOLOGY

Malaria remains a devastating global problem, with an estimated 300 to 500 million cases occurring annually. Forty-one percent of the world's population lives in areas where malaria is transmitted. Geographic boundaries are dictated by the presence of the *Anopheles* mosquito vector. Approximately 3 million persons die of malaria each year. *P. vivax* and *P. falciparum* account for more than 95% of all reported cases. *P. malariae* accounts for 4% of cases, and *P. ovale* is extremely rare. *P. falciparum* infection is more severe and life-threatening and requires more aggressive therapy.

Children and the elderly are at particular risk. Blacks are less susceptible than are whites. The absence of the Duffy (a red cell antigen) gene in the West African population is responsible for low incidence of vivax malaria. Sickle-cell trait also provides protection from malaria because RBCs in sickle-cell trait are not able to support growth of the organism. A higher incidence of sickle-cell trait is found in populations where malaria is endemic.

PATHOGENESIS

During a blood meal, a *Plasmodium*-infected female *Anopheles* mosquito inoculates sporozoites into the human host, beginning a complex cycle of replication. Sporozoites infect liver cells (asexual stage) and mature into schizonts, which transform into mero-zoites that are released from the liver to invade red blood cells (RBCs) and attach themselves to specific binding sites on RBCs (Duffy blood group antigen for *P. vivax* entry and a glycophorin antigen for *P. falciparum*). Within erythrocytes, merozoites feed on hemoglobin and other proteins, maturing into trophozoites. Trophozoites undergo nuclear division (without cell division) to form 16 to 32 parasite nuclei (schizonts) within a single RBC. Eventually, the RBC ruptures, and schizonts pinch off to form new mero-zoites, which infect new RBCs, and the erythrocytic cycle repeats. At some point, the parasites (micro- and macrogametocytes) may be taken up by the vector mosquito, where they undergo sexual stages in the midgut of the mosquito to form sporozoites that can be released once more into a human host.

The clinical picture of malaria depends on the age and immune status of the patient and also on the species of parasite (e.g., *P. falciparum* is the most virulent). One brood of parasites becomes dominant and is responsible for the **synchronous nature of the clinical symptoms of malaria (cyclic paroxysms)**. This protozoan brood replication inside the cell induces RBC cytolysis, causing the release of **toxic metabolic byproducts into the bloodstream** as many RBCs rupture at the same time (the host experiences **flu-like symptoms**).

Anemia results from the lysis of infected RBCs, the suppression of hematopoiesis, and the **increased clearance of RBCs by the spleen**. Over time, malaria infection causes thrombocytopenia. **Hepatospleno-megaly** is due to an **influx of host cells**.

Plasmodium falciparum infects all RBCs (young, middle-aged, or old), thus the level of parasitemia is higher than with other *Plasmodium* species. Because more merozoites are produced, more RBCs are destroyed, and a lack of O_2 for body tissues becomes a problem. The parasites derive their energy solely from glucose, and they metabolize it 70 times faster than do the RBCs they inhabit, thereby causing **hypo-glycemia** and **lactic acidosis**.

The dominant (persistent) hepatic stage (hypno-zoites) of *P. vivax* and *P. ovale* may persist in the liver of infected individuals, emerging at a later time and requiring additional drug treatment to prevent malaria relapse. **Immunologic response** to malaria is hard to assess and may involve both antibody-mediated and cell-mediated immunity. A period of time away from exposure is sufficient for the immunity to wane in an exposed person from a highly endemic area.

TREATMENT

Treatment varies according to the infecting species, the geographic area where the infection was acquired, and the severity of the disease. There are three "r" problems associated with management of malaria

1. In **recrudescence**, a controllable number of parasites remains in the blood stream (latent in RBC) because of inadequate immune response or anti-malarial therapy; parasites reactivate on physical trauma or immunosuppression.
2. In **relapse**, sporozoites are dormant in liver and reactivate; dormant sporozoites are referred to as hypnozoites (common with vivax malaria)
3. In **resistance**, antimalarial drugs become ineffective.

Chloroquine remains the drug of choice if the patient is infected with a susceptible strain of *Plasmodium* species. Oral administration of chloroquine for sensitive *P. falciparum*, *P. vivax*, *P. malariae*, and *P. ovale* strains is adequate in drug-sensitive geographic locations. Intrahepatic organisms, such as the hypnozoites of *vivax/ovale*, are not killed by chloroquine; they are killed by **primaquine** (which must be used to prevent relapses). For chloroquine-resistant strains (usually *P. falciparum*), drugs such as **mefloquine**, Malarone (a combination of atovaquone and proguanil), quinine (or intravenous quinidine), or an artemisinin derivative, can be used for successful treatment. Doxycycline or clindamycin may be used in combination with the drugs for *P. falciparum*. In severe cases of falciparum malaria, exchange transfusion is recommended to rapidly reduce the level of parasitemia.

NOTE The morbidity and mortality caused by *P. falciparum* are greater than those caused by other *Plasmodium* species because of the increased parasitemia of *P. falciparum* and its ability to cyto-adhere. When an RBC becomes infected with *P. falciparum*, proteinaceous knobs are produced on the surface of the erythrocytes. The sticky RBCs bind to the endothelial cell lining of the microvasculature of the brain, kidney, lung, and other organs. This leads to aggregation, or rosetting, of uninfected RBCs, a process known as sequestration and, consequently, to clogging of the capillaries in the microvasculature of the CNS and kidneys by the infected RBCs—resulting in cerebral malaria (a fatal consequence) and kidney failure, respectively.

OUTCOME

The patient was treated with intravenous quinidine and exchange transfusion because his level of parasitemia was high (>10%). He gradually recovered, after a prolonged hospital stay and rehabilitation.

PREVENTION

The key to prevention is avoidance of mosquito exposure in endemic areas. Use of netting at night and repellent spray with DEET is helpful. However, chemoprophylaxis is usually recommended for travelers to further reduce risk. Chloroquine is effective, although in many areas of the world endemic for malaria, *P. falciparum* is resistant to this drug. In these areas, mefloquine is usually recommended, although doxycycline and atovaquone/proguanil are also effective. Unfortunately, a vaccine is not available, although much research work is being done in this area.

FURTHER READING

Chen Q, Schlichtherle M, Wahlgren M: Molecular aspects of severe malaria. Clin Microbiol Rev 13:439, 2000.
Kain KC, Harrington MA, Tennyson S: Imported malaria: Prospective analysis of problems in diagnosis and management. Clin Infect Dis 27:142, 1998.
Kain KC, Keystone JS: Malaria in travelers. Epidemiology, disease, and prevention. Infect Dis Clin North Am 12:267, 1998.
Layton M, Parise ME, Campbell CC: Mosquito-transmitted malaria in New York City, 1993. Lancet 346:729, 1995.

On a hot **summer day, a 23-year-old man** was brought to the emergency department 3 days after the onset of **fever, severe headache,** and muscle pain. He was also experiencing nausea, vomiting, and abdominal pain. A fine, **spotted rash** was seen **on his extremities and trunk;** the patient said the rash had appeared earlier that day.

He lived in **North Carolina** and had no prior history of illness. He had received all appropriate childhood immunizations.

PHYSICAL EXAMINATION

VS: T 39.6°C, P 118/min, R 20/min, BP 102/50 mmHg

PE: The patient appeared ill and had an **erythematous maculopapular and petechial rash on his extremities** (Fig. 80-1), including on his palms and soles.

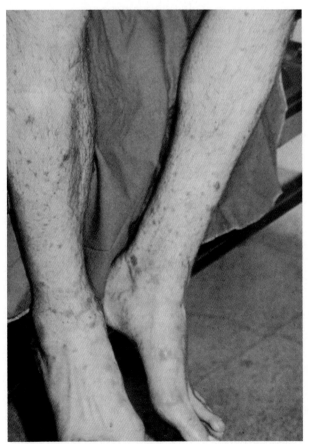

FIGURE 80-1 Spotted petechial rash on both legs of this patient. *(Courtesy of Dr. Paul Southern, Departments of Pathology and Medicine, University of Texas Southwestern Medical Center, Dallas, TX.)*

LABORATORY STUDIES

Blood

ESR: 58 mm/h

Hematocrit: 42%

WBC: 3100/μL

Differential: 40% PMNs, 24% bands, 20% lymphs

Platelets: 58,000/μL

Blood gases: pH 7.34, pCO_2 40 mmHg, pO_2 84 mmHg

Blood electrolytes: BUN 34 mg/dL, creatinine 1.8 mg/dL

Serum chemistries: AST 280 U/L, ALT 230 U/L

Imaging

No imaging studies were done.

Diagnostic Work-Up

Table 80-1 lists the likely causes of illness (differential diagnosis). A presumptive diagnosis based on clinical findings, physical examination, and a reliable patient

---

TABLE 80–1 Differential Diagnosis and Rationale for Inclusion (consideration)

Aseptic meningitis
Dengue hemorrhagic fever
Meningococcemia
Rocky Mountain spotted fever
Scarlet fever
Typhoid fever
Viral exanthems (e.g., measles)

Rationale: Fever and a rash may be signs of serious illness, particularly in a young adult. In the United States, meningococcemia and RMSF should be considered because they are life threatening and prompt therapy must be instituted. Typical viral exanthems are not common in adults, and aseptic meningitis usually is not associated with a rash. Dengue fever and typhoid fever are associated with a specific travel history, although both can be seen in the United States. Scarlet fever does not usually have a petechial rash.

history is important for aggressive management of a patient with these symptoms. Lumbar puncture can be performed if the patient does not have papilledema or lateralizing neurologic findings. Investigational approach may include

- **Gram stain and cultures** of CSF
- **Blood cultures**
- **Biopsy of skin lesion** with immunofluorescence or immunohistologic stain
- In failed diagnosis
  - **Direct antigens** in blood and CSF
  - **Serology**

COURSE

The patient was admitted to the hospital. Within 24 hours of onset, the rash had spread to 75% of his body surface, and many of the lesions had become maculopapular with petechiae. He became hypotensive and was transferred to the ICU. Blood and CSF were collected, and empirical treatment was begun. Blood and CSF on direct antigen testing and on cultures were negative for common bacterial pathogens.

ETIOLOGY

Rickettsia rickettsii (Rocky Mountain spotted fever [RMSF])

MICROBIOLOGIC PROPERTIES

Rickettsia rickettsii belongs to the family of Rickettsiaceae. This genus includes many other species of bacteria associated with systemic illnesses, including those in the spotted fever group and in the typhus group (Table 80-2). Rickettsiae are **small, obligate intracellular bacteria**, structurally similar to Gram-negative rods (LPS is present in the cell wall, but it is not sufficiently endotoxic to cause the sepsis seen in this disease). They are energy dependent, as are chlamydiae (using preformed metabolites from host cells). Rickettsiae can be grown in cell culture, embryonated eggs, or experimental animals. Capsule and cell-wall components are antigenic.

Rickettsiae are difficult to see in tissues by using routine histologic stains; they generally require the use of special staining methods. Culture is hazardous (Biosafety Level 3 facility is required) and is not

TABLE 80–2 Rickettsial Agents and Epidemiologic Features

Disease	Agent	Vector (Mode of Transmission)	Animal Hosts	Geographic Distribution
Spotted Fever Group				
Rocky Mountain spotted fever	*Rickettsia rickettsii*	*Dermacentor variabilis* (dog tick) *Amblyomma americanum* (lone star tick)	Rodents, dogs, and other wild and domestic animals	Western hemisphere
Rickettsialpox	*Rickettsia akari*	*Allodermanyssus sanguineus* (mouse mite)	Common house mouse and other wild rodents; man	Worldwide
Typhus Group				
Endemic (murine)	*Rickettsia typhi*	*Xenopsylla cheopis* (rat flea) *Rickettsia spinulosus* (rat louse)	Rats and other rodents; man	Worldwide (in the U.S., seen primarily in the Southeast and Gulf Coast region)
Epidemic (louse-borne)	*Rickettsia prowazekii*	*Pediculus humanus corporis* (body louse)	Man; flying squirrels and other wild rodents	Worldwide; endemic focus in northeastern U.S.
Scrub typhus	*Orientia tsutsugamushi*	Trombiculid mites	Small mammals; man	Australia, Japan, Korea, India, Vietnam
Other Rickettsiae				
Q fever	*Coxiella burnetii*	Transmission to man generally aerosol	Cows, sheep, goats, and wild animals; man	Worldwide

recommended. Laboratory diagnosis is confirmed by serologic response to specific antigens, but a delay in seropositivity occurs until at least the second week of illness. A greater than fourfold increase in IgG antibody to *R. rickettsii* between acute and convalescent phases is retrospectively diagnostic. Indirect immunofluorescence can also be done on skin biopsies.

EPIDEMIOLOGY

Rocky Mountain spotted fever is the **most severe** and **most frequently reported rickettsial illness in the United States**. RMSF is more commonly found in the eastern and southeastern United States, primarily from April through September. The infectious agent is maintained in nature in hard ticks. *Rickettsia rickettsii* is spread to humans by ticks (Fig. 80-2) found on dogs. At least 4 to 6 hours of attachment and feeding on blood by the tick are required before the rickettsiae become activated and infectious for people. Most cases occur in children and adolescents, reflecting close contact with family pets (usually dogs).

PATHOGENESIS

Rickettsiae are inoculated into the skin by tick bite. Following inoculation in the dermis, rickettsiae attach to the endothelial cell lining via a major surface-protruding rickettsial outer membrane protein (rOmpA) and subsequently spread to the blood stream. Rickettsiae live and multiply primarily within cells that line small- to medium-sized blood vessels; they escape rapidly from the phagosome and replicate in cytoplasm of the host cells (Fig. 80-3). The rickettsiae spread from cell to cell by the polarized actin polymerization of host cells, and the extruding bacteria coordinately infect multiple endothelial cells. Cellular burst after a cycle of **growth leads to rupture of vascular endothelial lining**, causing generalized **vascular damage**. No cytotoxin or exotoxin is known to contribute to pathology. Damage to the vessels and leakage of blood into the skin results in the characteristic spotted (petechial) rash. Endothelial damage triggers homeostatic response, causing activation of platelets and the clotting cascade. Hypotension and hypoproteinemia caused by loss of plasma into the tissues can lead to reduced perfusion of various organs and subsequent organ failure.

TREATMENT

Therapy should not be delayed for diagnostic confirmation because delays in treatment lead to increased mortality. **Doxycycline** is the drug of choice for adults. Chloramphenicol is used for children younger than 8 years of age and for pregnant women. Despite the availability of effective treatment and advances in medical care, approximately 3% to 5% of individuals who become ill with RMSF die from the infection.

FIGURE 80-2 A female "lone star tick," *Amblyomma americanum*, found in the southeastern and mid-Atlantic United States, is a vector of rickettsial diseases. *(Courtesy of Centers for Disease Control and Prevention, Atlanta, GA.)*

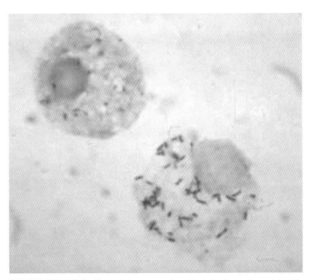

FIGURE 80-3 Gimenez stain of tick hemolymph cells infected with *Rickettsia rickettsii*. Note the intracellular bacteria. *(Courtesy of Centers for Disease Control and Prevention, Atlanta, GA.)*

OUTCOME

Within 48 hours of treatment with intravenous doxycycline, the patient was afebrile and more alert. He completed a course of doxycycline and was discharged home, where he gradually recovered. His rash disappeared on full recovery from the disease. A serologic investigation of paired sera retrospectively confirmed the clinical suspicion of RMSF due to *R. rickettsii*.

PREVENTION

No vaccines are available for any rickettsial disease. Preventive measures are based on avoidance of tick bites and care of household dogs (tick removal). Antibiotic prophylaxis after tick bites is not effective.

FURTHER READING

Raoult D, Roux V: Rickettsioses as paradigms of new or emerging infectious diseases. Clin Micro Rev 10:694,1997.

Thorner AR, Walker DH, Petri WA: Rocky Mountain spotted fever. Clin Infect Dis 27:1353, 1998.

Walker DH: Rocky Mountain spotted fever: A seasonal alert. Clin Infect Dis 20:1111, 1995.

Weber DJ, Walker DH: Rocky Mountain spotted fever. Infect Dis Clin North Am 5:19, 1991.

A **35-year-old** white man went to an urgent care clinic twice over the course of 3 days before being admitted in the month of **August** to the hospital with **high fever, myalgias,** and **severe headache.** Additional symptoms included photophobia, nausea, and anorexia.

The social and medical histories were unremarkable. Ten days before his first visit to the clinic, **he had returned from a boating trip.**

PHYSICAL EXAMINATION

VS: T 38.8°C, P 116/min, R 16/min, BP 110/64 mmHg

PE: An ill-appearing patient with **erythematous and swollen conjunctiva.**

LABORATORY STUDIES

Blood

Hematocrit: 46%

WBC: 7800/μL

Differential: 55% PMNs, 18% bands, 22% lymphs

Platelets: 180,000/μL

Serum chemistries: AST 180 U/L, ALT 240 U/L, total bilirubin 12 mg/dL, creatinine 2.1 mg/dL

Imaging

No imaging studies were done.

Diagnostic Work-Up

Table 81-1 lists the likely causes of illness (differential diagnosis). Investigational approach may include

- **Cultures** of blood, CSF, urine
- **Serology.** Detection of IgM by enzyme-linked immunosorbent assay with confirmatory testing by a single micro-agglutination test

COURSE

The results of examination of the cerebrospinal fluid were normal. Based on the clinical presentation, the patient was started on empirical antibiotics while cultures and serologic tests were pending.

TABLE 81–1   Differential Diagnosis and Rationale for Inclusion (consideration)

Adenovirus

Aseptic meningitis

Brucellosis

Cytomegalovirus (CMV)

Hepatitis A, B, C

Infectious mononucleosis (EBV)

Leptospirosis

Rationale: Flu-like symptoms can be difficult to distinguish from routine viral illnesses. A variety of other blood stream infections may present with similar symptoms, and other factors, such as exposure or travel history, should be explored. The presence of serum markers for hepatitis is seen with many infections (e.g., EBV, CMV) and is not helpful, although it can narrow the differential diagnosis. Leptospirosis follows water exposure (swimming, boating), and brucellosis may be associated with exposure to animals or animal products. Adenovirus may be associated with conjunctivitis.

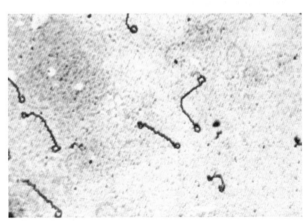

FIGURE 81-1   Silver stain of the pathogen. *(Courtesy of Dr. Paul Southern, Departments of Pathology and Medicine, University of Texas Southwestern Medical Center, Dallas, TX.)*

ETIOLOGY

Leptospira interrogans (a presumptive diagnosis of leptospirosis)

MICROBIOLOGIC PROPERTIES

Leptospires are **long, thin, motile spirochetes with characteristic hooked ends** (Fig. 81-1). The organisms stain poorly but can be examined by darkfield microscopy. The bacteria have a **Gram-negative cell wall**. Leptospires are antigenically complex, with more than 200 known pathogenic serologic variants. All pathogenic leptospires are serovars of *L. interrogans*.

Organisms can be cultured from blood during the first 7 to 10 days; thereafter, they can be cultured only from urine. Laboratory confirmation of a case of presumptive leptospirosis is based on positive result of a screening IgM test (ELISA; done in the state laboratories) with confirmatory testing by a single high titer on microagglutination test (MAT; done at the CDC).

EPIDEMIOLOGY

Leptospirosis is a common global **zoonotic disease**, especially in subtropical regions of the world. It is estimated that 100 to 200 cases are identified annually in the United States. The disease occurs in occupational, military, and recreational settings. **Recreational exposures** can include rafting, kayaking, and swimming, in tropical and temperate climates. Wild and occasionally domestic mammals are reservoirs.

PATHOGENESIS

Leptospirosis occurs mainly through **indirect contact with the urine of infected animals** (e.g., **rats** or domesticated livestock) often through water exposure. The most important features in pathogenesis are the hooked ends (the adhesion factor), two periplasmic flagella that permit burrowing into tissues, and cellular toxicity. The organisms multiply in the blood stream, causing bacteremia, characterized by flu-like symptoms (fever, intense headache, myalgias, and diarrhea).

This acute septicemic phase is followed very quickly by a secondary, "immune," phase involving immune-complex-mediated vasculitis (skin rashes and conjunctivitis). The direct cellular cytotoxicity also contributes to endothelial damage, causing hemorrhages. A systemic inflammatory response can eliminate the organisms from the circulation, but an uncontrolled immune response also contributes to the severity of symptoms. The second phase may progress to an icteric phase, in severe disease, characterized by jaundice. The disease may also progress to aseptic meningitis (antibody associated), renal failure (acute tubular necrosis), severe hemorrhages and hypotension due to vascular collapse (Weil syndrome), and myocarditis.

TREATMENT

Mild infections can be treated with oral doxycycline; patients requiring hospitalization should be treated with IV penicillin. A Jarisch-Herxheimer reaction may also be seen after initial therapy.

OUTCOME

By the third day of hospitalization, after IV rehydration and penicillin, the patient's fever subsided, and his myalgias and headache improved. He was discharged and given doxycycline for 10 days. He had no clinical sequelae.

Serologic tests for infectious hepatitis A, B, and C and for brucella, as well as a Monospot test, were negative. MAT and IgM-ELISA for leptospirosis titers from serum drawn 15 days after the initial exposure were negative; however, convalescent-phase serum drawn 17 days later tested positive for leptospira by IgM-ELISA, and MAT showed the strongest reaction with *L. interrogans*.

PREVENTION

Vaccination of livestock is effective in preventing disease, but not 100%. Avoiding exposure to water sources that may be contaminated (run-off from livestock farms) is important, as is control of vectors such as rats.

FURTHER READING

CDC: Outbreak of leptospirosis among white-water rafters—Costa Rica, 1996. MMWR 46:577, 1997.
Farr RW: Leptospirosis. Clin Infect Dis 21:1, 1995.

Levett PN: Leptospirosis. Clin Microbiol Rev 14:296, 2001.
Sejvar J, Bancroft E, Winthrop K: Leptospirosis in "Eco-Challenge" athletes, Malaysian Borneo, 2000. Emerg Infect Dis 9:702, 2003.

A 55-year-old woman presented with **pain** and **swelling on her right hand.** She had a high **fever, chills,** and **pain in the axilla.**

The patient had been **bitten on her right hand two days before by her cat.** She had otherwise been in good health.

PHYSICAL EXAMINATION

VS: T 39°C, P 94/min, R 16/min, BP 140/80 mmHg

PE: The patient's **right hand was erythematous,** swollen, and tender. A small wound was noted on the dorsum of her first finger. **Axillary tenderness** and **lymphadenopathy** were also noted.

LABORATORY STUDIES

Blood

Hematocrit: 36%
WBC: 16,300/μL
Differential: 54% PMNs, 24% bands, 20% lymphs
Serum chemistries: Normal

Imaging

No imaging studies were done.

TABLE 82–1 Differential Diagnosis and Rationale for Inclusion (consideration)

Capnocytophaga canimorsus
Pasteurella multocida
Staphylococcus aureus
Streptococcus pyogenes

Rationale: Cellulitis is typically caused by *S. aureus* or *S. pyogenes*, both of which can cause severe soft tissue infections. In relation to animal bites, other specific organisms should be considered. *Pasteurella* are the most common causes of bite infections, although *Capnocytophaga* are also seen.

Diagnostic Work-Up

Table 82-1 lists the likely causes of illness (differential diagnosis). Investigational approach may include

- **Blood cultures**
- **Gram stain and cultures** of a biopsy, pus, or exudates

COURSE

The patient was admitted to the hospital. Blood cultures and aspiration of an abscess on the patient's finger were sent for culture, and the patient was taken to the operating room for drainage of the abscess. No joint involvement was noted. The hospital laboratory reported a significant isolate from the abscess the following day.

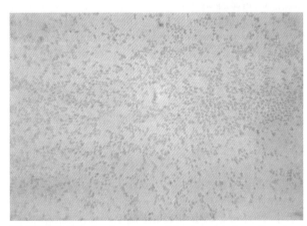

FIGURE 82-1 Gram stain of the clinical isolate. (*Courtesy of Centers for Disease Control and Prevention, Atlanta, GA.*)

ETIOLOGY

Pasteurella multocida (pasteurellosis)

MICROBIOLOGIC PROPERTIES

Pasteurelleae are **short Gram-negative rods** (cocco-bacilli; Fig. 82-1); these bacteria exhibit bipolar staining when examined after staining of tissue specimens. The **fastidious** Pasteurelleae require chocolate agar for growth. The isolates are **oxidase positive** and **encapsulated**.

EPIDEMIOLOGY

Most cats and dogs, even healthy ones, naturally carry this organism in their mouths. Human infection is associated with **cat bites** and, less commonly, with dog bites. (Women receive cat bites more frequently than do men.)

PATHOGENESIS

When an animal bites a person, Pasteurelleae can enter the wound and start an infection. Most bite infections are **polymicrobic**, with a variety of facultative anaerobes and anaerobic organisms in addition to *P. multocida*. Capsule and endotoxin trigger severe **inflammation accompanied by purulent drainage**. Bites to the hand, whether from cats or other animals, are potentially dangerous because of the structure of the hand. There are many bones, tendons, and joints in the hand, and there is less blood circulation in these areas than in other parts of the body. The first signs of pasteurellosis usually occur within 2 to 12 hours of the bite and include pain, reddening, and swelling of the area around the site of the bite.

TREATMENT

Penicillin G is the treatment of choice, although broad-spectrum agents are often used if other organisms are suspected. Amoxicillin/clavulanate and second or third generation cephalosporins are also effective.

NOTE Pasteurellosis can progress quickly, spreading toward the body from the bitten area. Serious infections more frequently follow cat rather than dog bites, and, if untreated, usually occur at the bite sites close to bone or joints and include septic arthritis, tendonitis, and osteomyelitis. Cat tooth, which is long and thin, causes wounds that more readily puncture the tendon sheath (tenosynovitis) or periosteum (osteomyelitis). Unusual complications include bacteremia with septic shock, meningitis, brain abscess, and peritonitis. These complicated infections (e.g., septic arthritis) require operative débridement and prolonged antibiotic treatment.

OUTCOME

The patient received IV penicillin G until she became afebrile, then she completed a total of 10 days of oral therapy at home.

PREVENTION

Individuals who experience cat or dog bites should be given amoxicillin/clavulanate to prevent *P. multocida* infection.

FURTHER READING

Capitini CM, Herrero IA, Patel R, et al: Wound infection with a novel subspecies of *Pasteurella multocida* in a child who sustained a tiger bite. Clin Infect Dis 34:e74, 2002.

Talan DA, Citron DM, Abrahamian FM, et al: Bacteriologic analysis of infected dog and cat bites. N Engl J Med 340:85, 1999.

SECTION VI

Bioterrorism Agents and Emerging Infections

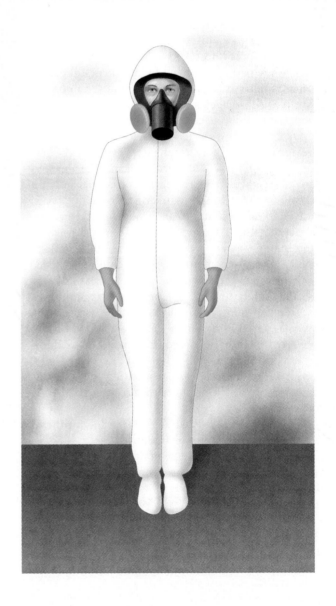

INTRODUCTION TO BIOTERRORISM AGENTS AND EMERGING INFECTIONS

Bioterrorism Agents

On Thursday, October 4, 2001, just 24 days after the tragic events of September 11, the Florida Department of Health and the Centers for Disease Control and Prevention (CDC) confirmed the first case of inhalational anthrax in the United States in more than 25 years. An infectious disease clinician was quick to suspect anthrax. Capable clinical and public health laboratory staff in Florida pursued the evidence, confirming that *Bacillus anthracis* spores had been intentionally distributed through the postal system. This incident led to 22 cases of anthrax, including 5 deaths, forever changing the realm of public health. Although only a few clinical cases were seen in the ensuing weeks, widespread fear and anxiety engendered by such acts are part of the goal of bioterrorists. Biologic warfare is not a new phenomenon, and the development of biologic agents, including the means to disseminate them, has been ongoing for several decades. However, the use of such agents for the purpose of pure terrorism is more recent and has brought the issue to the forefront of modern medicine. Fortunately, there are relatively few pathogens considered suitable for use as biologic weapons, in terms of lethality, availability, and ease of large-scale production. However, for many of these agents, highly effective and easily administered therapy is not available. In addition, these are often uncommonly seen in clinical practice, which may delay recognition. For this reason, it is critically important to be able to recognize and diagnose these infections as rapidly as possible in case of an outbreak caused by bioterrorism. The CDC has compiled a **list of infectious agents** likely to be used for purposes of bioterrorism. They are categorized into levels based on the ease of weaponization and dissemination, risk posed to national security, and lethality (Table VI-1).

Of all the agents listed in Table VI-1, **anthrax** (*Bacillus anthracis*) is one of the most feared. This is due to the hardiness and ease of dissemination of the spores, the nonspecific prodrome of inhalational infection, which can often be mistaken for the flu, and the extremely high mortality (>90%), despite appropriate therapy, once clinical disease has manifested. The outbreak of anthrax cases described earlier illustrates that it can be used as a devastating biologic weapon—should large numbers of people become exposed. The only mitigating factor in its effect is that it is not contagious.

Another, even more frightening, scenario would be the reintroduction of **smallpox** virus, one of the great scourges of mankind, into the world. There have been no cases of smallpox since 1977, when it was eradicated by a concerted effort led by the World Health Organization. As a result, most human populations are considered nonimmune and would be severely affected by an outbreak, especially because smallpox is highly contagious and may be difficult to contain depending on how it was released. The other more practical aspect of its threat is that there is no therapy, mortality is approximately 30%, and stockpiles of vaccine are old and may not be effective. However, there is ongoing work to determine the usefulness of dilute vaccine, to treat large numbers of individuals, and on the development of newer vaccines. Although the smallpox virus is known to exist in only two labs in the world, the possibility that it may fall (or may already have fallen) into the wrong hands is real.

Botulinum toxin, produced by the bacterium *Clostridium botulinum*, is the most potent toxin known to man. Microgram amounts are lethal and there is no therapy or antidote once symptoms have begun, except supportive care—which can be required for weeks. It is not very stable, however, and dissemination may be difficult.

Plague, another disease that has ravaged human populations, is caused by *Yersinia pestis* and is naturally spread by rodent fleas. It has a high mortality if untreated, especially pneumonic plague, which is also contagious (spreading person to person). Sporadic disease of natural origin and outbreaks still occur in certain areas.

Tularemia is caused by *Francisella tularensis* and also has a high mortality. It is unique in that infection often leads to debilitating systemic symptoms that can last for months, even with therapy.

Viral hemorrhagic fevers have caused rare outbreaks in specific geographic regions, but these are not seen in the United States. There is no therapy currently, and they all have a high mortality, particularly Ebola.

Numerous other infections are also possible bioterrorism agents (see Table VI-1). Many of these other pathogens are considered to be somewhat less virulent, and they have lower mortality rates than those just described. Nonetheless, they could still cause large outbreaks with significant morbidity if disseminated widely. All physicians should be familiar with the manifestations of the most likely and deadly bio-

TABLE VI–1	Classification of Bioterrorism Agents and Diseases*
Bioterrorism Categories	**Diseases (Agents)**
Category A	High-priority agents include organisms that pose a risk to national security because they • Can be easily disseminated or transmitted from person to person • Result in high mortality rates and have the potential for major public health impact • Might cause public panic and social disruption • Require special action for public health preparedness Anthrax (*Bacillus anthracis*) Botulism (*Clostridium botulinum* toxin) Plague (*Yersinia pestis*) Smallpox (Variola major) Tularemia (*Francisella tularensis*) Viral hemorrhagic fevers (filoviruses [e.g., Ebola, Marburg] and arenaviruses [e.g., Lassa, Machupo])
Category B	Second highest priority agents include those that • Are moderately easy to disseminate • Result in moderate morbidity rates and low mortality rates • Require enhanced disease surveillance Brucellosis (*Brucella* species) Food poisoning (Epsilon toxin of *Clostridium perfringens*; staphylococcal enterotoxin B) Food safety threats (*Salmonella* species, *Escherichia coli* O157:H7, *Shigella*) Glanders (*Burkholderia mallei*) Melioidosis (*Burkholderia pseudomallei*) Psittacosis (*Chlamydia psittaci*) Q fever (*Coxiella burnetii*) Pulmonary edema (Ricin toxin from *Ricinus communis* [castor beans]) Typhus fever (*Rickettsia prowazekii*) Viral encephalitis (Alphaviruses [e.g., Venezuelan equine encephalitis, eastern equine encephalitis, western equine encephalitis]) Water safety threats (*Vibrio cholerae, Cryptosporidium parvum*)
Category C	Third highest priority agents include emerging pathogens that could be engineered for mass dissemination in the future because of • Availability • Ease of production and dissemination • Potential for high morbidity and mortality rates and major health impact Pulmonary disease (Hantavirus) Encephalitis (Nipah virus)

*Adapted from CDC.

terrorism agents. The dangerous organisms, and those most likely to be used in a bioterrorist event, will be presented in the subsequent cases. It is our intention that, at the end of this important and timely section, students will know (1) how one acquires the disease of concern under normal circumstances; (2) the key symptoms; (3) the incubation period; (4) what to do next when confronted with a bioterrorism-related disease; and (5) what to do in times of uncertainty and questionable diagnosis.

Note that some of the threat agents (e.g., anthrax, plague, tularemia) have normal risks of animal-to-person transmission and risks of vector-borne transmission. The key to dealing with a bioterrorism event is to remain knowledgeable about the epidemiologic and clinical aspects of threat agents and to keep a high index of suspicion when seeing patients who present with unusual symptoms, or clusters of patients appearing in a short time frame. Once suspicion has been confirmed, the next step after starting appropriate medical management is to immediately notify the local health authorities, as shown in an algorithm from the CDC (Fig. VI-1). On the local level, clinicians and laboratory workers play a key role in this process. When not sure, it is strongly recommended that one knows "when to hold and when to fold" and one does not waste time if a bioterrorism agent is suspected, but refers it to an appropriate authority. Knowledge of infection control issues and use of appropriate barrier precautions and biosafety techniques are essential when evaluating a suspected case or handling the clinical specimens.

The CDC has established rapid response teams of individuals with expertise in field operations, epidemiology, and microbiology (in the Laboratory Response Network [LRN] for Bioterrorism). In each state in the United States, there is at least one specialty laboratory participating in this network, enabling widespread testing for microbes that might be used in a terrorist attack to cause illnesses such as anthrax, tularemia, plague, and botulism. Suspicious specimens should be referred to one of these laboratories. Multiple systems are now in place, both in the United States and internationally, to detect initial cases.

EMERGING INFECTIOUS DISEASES

Although our understanding of infectious diseases and microbes has grown tremendously in the past several decades, there will always be new and emerging infectious agents (Table VI-2) to contend with. These may be pathogens that previously have not been seen in a particular geographic region (i.e., the introduction of

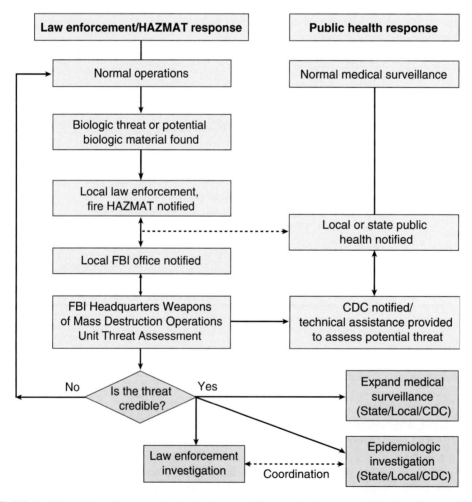

FIGURE VI-1 Likely flow of communication during overt bioterrorism in most (*solid line*) and some (*dashed line*) jurisdictions. HAZMAT, hazardous materials management personnel. (*Courtesy of Centers for Disease Control and Prevention, Atlanta, GA.*)

TABLE VI–2 Examples of Pathogenic Microbes and Infectious Diseases Recognized Since 1973

Year	Microbe	Disease
1973	Rotavirus	Infantile diarrhea
1975	Parvovirus B19	Aplastic crisis in chronic hemolytic anemia
1976	*Cryptosporidium parvum*	Waterborne diarrhea
1977	*Legionella pneumophila*	Legionnaires disease
	Ebola virus	Hemorrhagic fever
	Hantaan virus	Hemorrhagic fever
	Campylobacter jejuni	Diarrhea/dysentery
1980	T-cell leukemia/lymphoma virus 1 (HTLV-1)	Lymphoma/leukemia
1981	Toxic producing strains of *Staphylococcus aureus*	Toxic shock syndrome (associated with tampon use)
1982	*Escherichia coli* O157:H7	Hemorrhagic colitis; hemolytic-uremic syndrome
	HTLV-2	Hairy cell leukemia
	Borrelia burgdorferi	Lyme disease
1983	*Helicobacter pylori*	Gastric ulcers and adenocarcinoma
	HIV	AIDS
1985	*Enterocytozoon bieneusi*	Protracted diarrhea in AIDS
1986	*Cyclospora cayetanensis*	Persistent diarrhea
1987	*Ehrlichia chaffeensis*	Human monocytic ehrlichiosis
1988	Hepatitis E virus	Enterically transmitted non-A hepatitis
	Human herpesvirus 6	Roseola (infantum) subitum—sixth viral (febrile) exanthem illness
1989	Hepatitis C virus	Parenterally transmitted non-A, non-B hepatitis; hepatocellular carcinoma
1991	Guanarito virus	Venezuelan hemorrhagic fever
1992	*Vibrio cholerae* O139 Bengal	New epidemic cholera (aborted)
	Bartonella henselae	Bacillary angiomatosis (in AIDS), cat-scratch disease
1993	Sin Nombre (hanta) virus	Hantavirus pulmonary syndrome
1994	*Ehrlichia phagocytophila*	Human granulocytic ehrlichiosis
	Sabia virus	Brazilian hemorrhagic fever
1995	Human herpesvirus 8	Kaposi sarcoma
1997	Prions	New variant Creutzfeldt-Jakob disease (spongiform encephalopathy)
1999	West Nile virus in North America	Aseptic meningitis and encephalitis
2003	Coronavirus type 4	Severe acute respiratory syndrome (SARS)

West Nile virus into the United States in 1999) or entirely new pathogens not previously seen or known to infect humans (e.g., severe acute respiratory syndrome [SARS] virus).

Hantavirus is an example of a virus that had not been described before, although it may have caused human disease without being recognized in the past. It was brought to nationwide attention by the deaths of several young, otherwise healthy, individuals in the Four Corners area of New Mexico in 1993. This cluster of unusual deaths due to a respiratory illness prompted an investigation by the CDC that led to the identification of a newly described virus, Sin Nombre virus (literally 'without a name'). It was later found that it belonged to a group of closely related viruses that caused similar diseases. Transmission was found to be from deer mice. It is likely that prior, sporadic cases occurred but were not recognized.

New-variant **Creutzfeldt-Jakob disease** (nvCJD) is a disease caused by **prions**—apparently from an animal source—likely cattle afflicted with bovine spongiform encephalopathy (BSE). Although this link has not been established with certainty, it is clear that this prion disease manifests quite differently from known prion diseases, such as sporadic and familial CJD. It was first recognized in the United Kingdom, where an epidemic of BSE cases had been ongoing. Once the possible connection was made, measures were taken to reduce the incidence of BSE in their cattle. This appears to have been effective in decreasing the number of new cases in recent years.

West Nile virus encephalitis is a disease that crossed the Atlantic from West Africa in 1999 to enter the United States. It remains unknown how the infection arrived in New York, the first city with human cases. However, it has spread widely in a very short time, with almost all of the continental United States reporting cases, and thousands of documented cases. Although West Nile virus is a well-described pathogen, because of its recent arrival in the United States, many physicians are unfamiliar with the signs and symptoms suggestive of infection. The most severe complication is encephalitis, although this is uncommon, with the vast majority of individuals having asymptomatic or minimally symptomatic infection. The elderly are particularly susceptible to severe infection, including encephalitis. This disease is in all likelihood here to stay in the United States—an example of how quickly a pathogen can establish itself in entirely new areas when provided the opportunity.

SARS is another example of a new and emerging disease, caused by a novel coronavirus that was previously not seen in humans. The first cases were seen in China, and then the virus spread rapidly to nearby countries and across the globe with travelers. Coronaviruses are known to cause usually mild upper respiratory infections—generally in children. The strain associated with SARS is much more likely to cause life-threatening disease, even in otherwise normal individuals, which led to widespread fear and concern in countries that had cases, such as China, countries in Southeast Asia, and Canada. It was thought that transmission from animal sources (namely the civet cat) in large market places in China may have been responsible. Effective public health measures and communication led to the end of the outbreak after several months—although more cases may be likely in the future because contact between humans and animals will undoubtedly continue.

The preceding examples illustrate how new and emerging pathogens continue to be a problem, necessitating active surveillance, a solid public health infrastructure, and physicians who are current on the latest developments in these areas. Effective **preparedness** to confront future emerging infectious diseases and other threats to public health will pay off in terms of reducing morbidity and mortality and improving the health and well-being of our world.

FURTHER READING

CDC: Web-based Script: "Response to Bioterrorism: Agents of Bioterrorism." Web access: *http://www.bt.cdc.gov/training/btresponse/btagentsscript99.asp#5*

CDC: Biological and chemical terrorism: Strategic plan for preparedness and response. Recommendations of the CDC Strategic Planning Workgroup. MMWR 49(RR-4), 2000.

Klietmann WF, Ruoff KL: Bioterrorism: Implications for the clinical microbiologist. Clin Microbiol Rev 14:364, 2001.

Lederberg J: Emerging infections: An evolutionary perspective. Web access: *http://www.cdc.gov/ncidod/eid/vol4no3/lederberg.htm*

Questions on the case (problem) topics discussed in this section can be found in Appendix F. Practice question numbers are listed in the following table for students' convenience.

Self-Assessment Subject Areas (book section)	Question Numbers
Bioterrorism-related infections and emerging infectious diseases (Section VI)	18-20, 62, 64, 100-101, 217-219, 247, 249-253

A 63-year-old man awoke early with nausea, vomiting, and confusion and was taken to a local emergency department for evaluation. His illness, which began 5 days prior to arrival at the ED, during a trip to North Carolina, was characterized by **fever, chills, sweats, malaise, anorexia, and headache.** He had a dry **cough for the past 2 days.**

No history of chest pain, myalgias, dyspnea, abdominal pain, diarrhea, or skin lesions was reported. Past medical history included hypertension and cardiovascular disease. He did not smoke.

PHYSICAL EXAMINATION

VS: T 39.2°C, P 109/min, R 24/min, BP 110/64 mmHg

PE: On admission, the patient was alert and interactive but spoke nonsensically. Initial pulmonary, cardiac, and abdominal examinations were reported as normal. No nuchal rigidity was observed. He was **not oriented to person, place,** or **time.**

LABORATORY STUDIES

Blood

WBC: 8200/μL
Differential: 40% PMNs, 32% bands, 22% lymphs

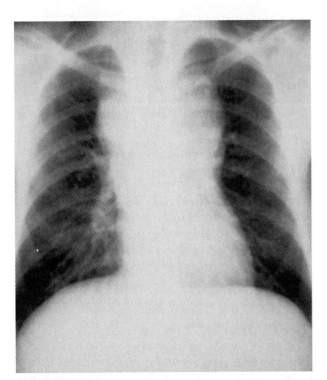

FIGURE 83-1 An admission chest x-ray revealed a widened mediastinum. *(Courtesy of Dr. Paul Southern, Departments of Pathology and Medicine, University of Texas Southwestern Medical Center, Dallas, TX.)*

Platelets: 80,000/μL
Blood gases: pH 7.36, pCO_2 34, pO_2 68
Serum chemistries: Sodium 133 mmol/L

Imaging

A chest X-ray showed a **widened mediastinum** (Fig. 83-1).

Diagnostic Work-Up

Table 83-1 lists the likely causes of illness (differential diagnosis). Investigational approach may include

- **Gram stain and cultures** of sputum, pleural fluid, and cerebrospinal fluid (CSF)
- **Blood cultures.** Gram stain of an unspun peripheral blood smear
- **Direct antigen detection** in problem scenarios

TABLE 83–1 Differential Diagnosis and Rationale for Inclusion (consideration)

Histoplasmosis
Inhalation anthrax
Meningitis (bacterial or viral)
Plague
Pneumonia
   Atypical (e.g., *Mycoplasma, Chlamydophila*)
   Typical (e.g., *Streptococcus pneumoniae*)
   Viral (e.g., influenza, adenovirus [ARD], coronavirus [SARS], hantavirus pulmonary syndrome)
Tularemia
Viral encephalitis (HSV-1)

Rationale: The presentation is nonspecific and may be due to a variety of organisms. The most common are usually viral, such as influenza, although influenza is associated with a cough. Widened mediastinum in this type of clinical setting is often seen with histoplasmosis and inhalational anthrax. Mental status changes are distinctly uncommon for most respiratory infections and should prompt evaluation for CNS infection, including meningitis or encephalitis.

● In failed investigation of pneumonia, **serology** for atypical pneumonic pathogens and endemic fungal pathogens

COURSE

The patient was admitted to the hospital with a diagnosis of meningitis. CSF analysis showed WBC count 4750/μL (81% neutrophils), red blood cell count 1375/μL, glucose 57 mg/dL (serum glucose 174 mg/dL), and protein 666 mg/dL. Microscopic examination of the CSF showed many Gram-positive bacilli. The same Gram-positive bacillus was isolated from CSF after 7 hours of incubation and from blood cultures within 24 hours of incubation. After a single dose of cefotaxime, the patient was started on multiple antibiotics to cover potential pathogens. Shortly after admission, he had generalized seizures and was intubated for airway protection.

ETIOLOGY

Bacillus anthracis (anthrax)

MICROBIOLOGIC PROPERTIES

B. anthracis organisms are **large, Gram-positive, nonmotile, spore-forming rods** that are found in chains (Fig. 83-2A). Virulent strains of *B. anthracis* are pathogenic for animals and are encapsulated, with a capsule composed of poly-D-glutamic acid. **Aerobic anthrax bacteria grow well on blood agar** media after overnight incubation at 35°C without CO_2. Characteristic colonies (2 to 5 mm) have a ground-glass appearance and are nonhemolytic, nonpigmented, edge-irregular with comma projections ("Medusa head" colonies [see Fig. 83-2B]).

A clue to the diagnosis of bioterrorism-related anthrax is the presence of large Gram-positive bacilli in specimens taken from direct blood smears or skin lesions. A *Bacillus* sp recovered from a set of blood culture bottles after 10 hours of incubation indicates a high initial inoculum, not consistent with a "venipuncture contaminant," and should motivate the laboratory to check on the clinical condition of the patient. A large number of Gram-positive bacilli cultured from swabs of the nares might also support such a suspicion. The recovery rate from sputum is very low because

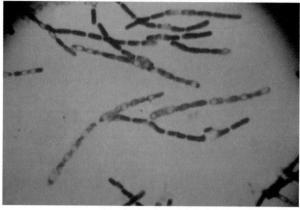

A

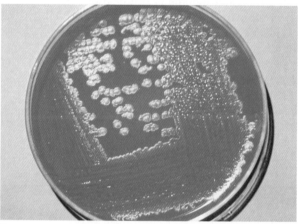

B

FIGURE 83-2 *A,* Gram stain of *Bacillus anthracis.* Note the endospores encased in the box-shaped rods that are arranged in chains. *B,* Blood agar culture of *Bacillus anthracis.* (*A, Courtesy of Dr. Paul Southern, Departments of Pathology and Medicine, University of Texas Southwestern Medical School, Dallas, TX; B, Courtesy of Centers for Disease Control and Prevention, Atlanta, GA.*)

inhalational anthrax is a mediastinitis and not usually a true pneumonia. A *Bacillus* sp that is **nonmotile** and shows **no hemolysis on sheep blood agar** should be considered presumptive *anthracis* and should be referred immediately to a Level B laboratory for final identification.

EPIDEMIOLOGY

Bioterrorism attacks using hardy fine spores of *B. anthracis* sent through the mail in 2001 resulted in 15 anthrax cases (including this case). These initial anthrax cases occurred among persons with known or suspected contact with opened letters contaminated with *B. anthracis* spores. No sources of exposure were identified for two women who were presumably exposed to secondarily contaminated mail.

Natural anthrax occurs worldwide; the number of human cases (all of agricultural origin) may be as high as 100,000 per year. In the United States, the incidence of naturally acquired anthrax is extremely low. For humans, the source of natural infection is infected livestock and wild animals or contaminated animal products. Contact with infected animals and bites of infective flies (after taking blood from the infected animals) can cause **cutaneous anthrax,** which is by far the most common form of anthrax. **Inhalation anthrax** occurs in persons working in certain occupations where spores may be forced into the air from contaminated animal products, such as animal hair processing (wool sorter disease) or textile milling. Occupational risk groups include those coming into contact with livestock or products from livestock (e.g., veterinarians, animal handlers, and laboratorians). Person-to-person transmission of inhalation anthrax does not occur. Consumption of contaminated meat can cause **gastrointestinal anthrax.** Inhalational anthrax is extremely rare in the United States; less than 20 cases were reported from 1900 to 2000.

PATHOGENESIS

The risk for airborne infection with *B. anthracis* (in a suspected bioterrorism event) is determined not only by the virulence of the organism but also by the balance between infectious aerosol production and removal, pulmonary ventilation rate, duration of exposure, and host susceptibility factors. **Anthrax spores, which measure between 1.5 and 3 μm, lend themselves very well to aerosolization**. Inhalational anthrax generally has an incubation period of 1 to 6 days. Macrophages carry the spores to tracheobronchial or mediastinal lymph nodes, where *B. anthracis* finds a favorable milieu for growth and germination.

The vegetative bacteria have three virulence factors: **an antiphagocytic capsule** and **two toxins**. A plasmid codes for the peptide **capsule consisting of poly-D-glutamic acid**, which plays a major role in virulence (evasion of phagocytosis). Three proteins, known as **lethal factor** (LF, a zinc metalloprotease), **edema factor** (EF, an adenylate cyclase), and **protective antigen** (PA), found on a second plasmid,

make up the two functional toxins. Following the A-B toxin model of toxicity (see Appendix B), PA serves as a necessary carrier molecule for EF and LF and permits penetration into cells.

Edema toxin, EF, is a calmodulin-dependent **adenylate cyclase. Lethal toxin, LF,** is a specific endopeptidase that cleaves the kinase family of proteins and interferes with intracellular signaling. LF targets the macrophages, causing hemorrhagic necrosis in the lymph nodes (Fig. 83-3), resulting in the release of large numbers of *B. anthracis*. The organisms gain access to the vascular circulation and multiply, causing bacteremia. A nonspecific flu-like illness develops, characterized by fever, muscle aches, headache, and a nonproductive cough with mild chest discomfort. An overwhelming mediastinal edema is caused by the intracellular cyclic AMP that is produced by the enzymatic action of EF. **A chest x-ray is often pathognomonic, revealing a widening of the mediastinum with pleural effusions without infiltrates**. The nonspecific flu-like illness is followed by an overwhelming secondary bacteremia (a direct Gram stain of peripheral blood may reveal broad Gram-positive rods), sepsis, meningitis, and death.

NOTE The same toxins (LF and EF) cause edematous cutaneous anthrax lesions, characterized by necrosis (black eschar) and edema, at the site of inoculation of organisms. Bacteremia occurs in all types of anthrax, and **uncontrolled bacteremia** subsequently leads to **meningitis**.

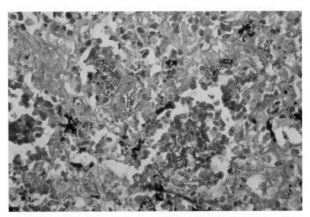

FIGURE 83-3 Histopathology of mediastinal lymph node in human anthrax. Note necrosis of lymph node due to anthrax. *(Courtesy of Dr. Paul Southern, Departments of Pathology and Medicine, University of Texas Southwestern Medical Center, Dallas, TX.)*

TREATMENT

Ciprofloxacin or **doxycycline** should be used for initial intravenous therapy until antimicrobial susceptibility results are known. Because of the extremely high mortality associated with inhalational anthrax, two or more antimicrobial agents predicted to be effective are recommended. Treatment of systemic *B. anthracis* infection with penicillins, cephalosporins, or trimethoprim/sulfamethoxazole is not recommended because bioterrorism-related strains may be resistant to these drugs. Supportive care includes controlling pleural effusions.

NOTE Direct contact with spores or bacilli may cause **cutaneous anthrax** (usually an immediate response within 1 day). Localized itching is followed by papular lesions that become vesicular and by the subsequent development of black eschar within 7 to 10 days of the initial lesion. Diagnosis is the same as for inhalational anthrax (culture of skin lesion is highly recommended). For cutaneous anthrax, ciprofloxacin is also a first-line therapy.

OUTCOME

On hospital day 2, penicillin G, levofloxacin, and clindamycin were begun; ampicillin, ceftazidime, and trimethoprim-sulfamethoxazole were discontinued when the specific diagnosis was obtained. The patient remained febrile and became unresponsive to deep stimuli. His condition progressively deteriorated, with hypotension and worsening renal insufficiency. The patient died 3 days after he was brought to the hospital.

PREVENTION

Clinical suspicion should be high for bioterrorism-related inhalational anthrax, although practically no initial cases will be diagnosed early because symptoms are nonspecific. In the event of a presumptive **bioterrorism** event, **medical personnel must be alert to coordinate** testing, packaging, and transporting with the public health laboratory. The **Laboratory Response Network** will need to be contacted and appropriate specimens (e.g., blood [essential], pleural fluid, CSF, and skin lesion specimens) should be obtained. Suspected or confirmed anthrax **cases must be reported immediately to local or state departments of health.** Standard contact precautions are adequate. Direct contact with wound or wound drainage should be avoided.

FURTHER READING

Bush LM, Abrams BH, Beall A, Johnson CC: Brief report: Index case of fatal inhalational anthrax due to bioterrorism in the United States. N Engl J Med 345:1607, 2001.

Inglesby TV, O'Toole T, Donald A, et al: Anthrax as a biological weapon, 2002: Updated recommendations for management. (Consensus Statement) JAMA 287:2236, 2002.

Kyriacou DN, Stein AC, Yarnold PR, et al: Clinical predictors of bioterrorism-related inhalational anthrax. Lancet 364:449, 2004.

Lawrence D: Genome provides clues to anthrax virulence. Lancet 361:1529, 2003.

Mayer TA, Bersoff-Matcha S, Murphy C, et al: Clinical presentation of inhalational anthrax following bioterrorism exposure: Report of two surviving patients. JAMA 286:2549, 2001.

Mogridge J: Anthrax and bioterrorism: Are we prepared? Lancet 364:393, 2004.

Swartz MN: Recognition and management of anthrax—An update (Current Concepts). N Engl J Med 345:1621, 2001.

A 24-year-old man presented to the emergency department with a complaint of a **chickenpox-like rash** that he noticed the day before. Four days earlier, he had developed the sudden onset of fever, severe headache, and back pain, which were resolving when the rash began.

He thought the rash was unusual because he was certain **he had had chickenpox as a child.** He was otherwise healthy and had no medical problems.

PHYSICAL EXAMINATION

VS: T 38°C, P 84/min, R 14/min, BP 124/80 mmHg

PE: Papulovesicular lesions were noted mostly **on the face and extremities** (including the palms and soles), with a few on the chest and abdomen. The **bumpy lesions**, filled with a thick, opaque fluid, **appeared to be at a similar stage of development** (Fig. 84-1).

LABORATORY STUDIES

Blood

Hematocrit: 46%
WBC: Normal
Differential: Normal

Blood gases: Normal
Serum chemistries: Normal

Imaging

No imaging studies were done.

Diagnostic Work-Up

Table 84-1 lists the likely causes of illness (differential diagnosis). An illness with acute onset of fever higher than 38.3°C followed by a rash characterized by firm, deep-seated vesicles or pustules in the same stage of development without other apparent cause requires immediate investigation. Investigational approach may include

- **EM examination** of vesicular fluid (variola virus)
- Cytology (**Tzanck preparation**) on vesicular fluid (VZV)

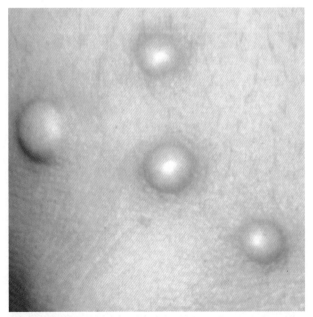

FIGURE 84-1 The bumpy lesions filled with a thick, opaque fluid were noted to be at a similar stage of development in a patient. *(Courtesy of Centers for Disease Control and Prevention, Atlanta, GA.)*

TABLE 84–1 Differential Diagnosis and Rationale for Inclusion (consideration)
Chickenpox (varicella-zoster virus [VZV])
Drug reactions
Herpes lesions
Molluscum contagiosum
Rickettsialpox
Smallpox (variola virus)

Rationale: There are many causes of vesicular and pustular rash diseases (e.g., varicella, herpes simplex, drug reactions, rickettsialpox). The likelihood of reintroduction of smallpox is extremely low. In the unlikely event of an outbreak, it is an extremely important task to first rule out chickenpox. Chickenpox is more prominent on the trunk and does not usually involve the palms and soles; lesions may appear in crops. Smallpox is more prominent on the face and extremities and often involves the palms and soles. Chickenpox lesions reveal different stages of development. In contrast, smallpox lesions are all at the same stage of development. The other rash diseases, notably rickettsialpox and molluscum, can have vesicular lesions and should also be considered. Herpes does not usually disseminate in a normally healthy individual. Drug reactions may be vesicular.

- **Polymerase chain reaction** (PCR) identification of viral DNA in a clinical specimen
- **Isolation** of virus from a clinical specimen (WHO Reference laboratory or laboratory with appropriate reference capabilities)

COURSE

The patient was admitted and placed in isolation. Several lesions were unroofed and scraped for microscopic and immunologic tests, as well as PCR testing. Later that day several other patients presented to the ED with similar illnesses.

ETIOLOGY

Variola major (smallpox)

MICROBIOLOGIC PROPERTIES

Variola virus is **a large, complex DNA virus** (300 nm in length) and is easily distinguished from herpes simplex virus and varicella-zoster by electron microscopy. It has a **dumbbell-shaped core** (Fig. 84-2) and a **complex membrane system**, containing many

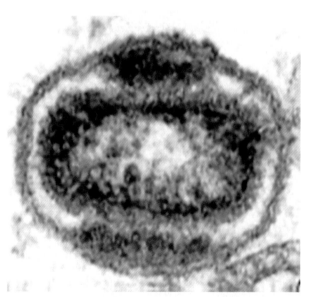

FIGURE 84-2   A transmission electron micrograph of a large variola virion with dumbbell-shaped core. *(Courtesy of Centers for Disease Control and Prevention, Atlanta, GA.)*

proteins (more than 100). The viral genome has a **linear double-stranded DNA** of 130 to 300 kb. Variola virus exists (in the specialized laboratories) as two strains: (1) variola major—a virulent strain causing the severe and most common form of smallpox—with an extensive rash and higher fever; and (2) variola minor (mild form).

Vaccinia virus, a related virus used for the smallpox vaccine, has similar antigenicity, inducing both specific and cross-reacting antibodies—hence, the ability to vaccinate against variola virus.

In the quest for the cause of a presumptive bioterrorism-related smallpox outbreak, the key responsibility for a laboratory is to **rule out chickenpox** and then to **obtain instructions for properly collecting vesicular fluid**. Laboratory diagnostic testing for variola virus should be conducted in a CDC Laboratory Response Network (LRN) using LRN-approved PCR tests and protocols for variola virus. In case of vaccine-associated smallpox, virologic and immunologic laboratory testing can help confirm the diagnosis of progressive vaccinia. This should be done after consultation with state or local health authorities, experienced infectious disease clinicians, or immunology experts, and/or the CDC.

EPIDEMIOLOGY

This is a fictitious bioterrorism scenario. Because smallpox was eradicated many years ago, a case of smallpox today would be the result of an intentional act. A single confirmed case of smallpox would be considered an emergency. Detailed information and instructions are available on the CDC web site and through other media, such as radio and television.

In 1966, the WHO announced its goal to eradicate smallpox in the world, and when this goal was thought to have been achieved in 1971, routine vaccination in the United States was discontinued. The last case of endemic smallpox was reported in 1977 from Somalia. In 1980, the WHO officially declared that smallpox had been eliminated worldwide.

Humans are the only natural hosts of variola. Smallpox is not a hardy virus; it is killed by sunlight and heat. Variola is acquired by inhalation of respiratory droplets from an infected person and direct contact with infected bodily fluids. **In the event of an intentional act**, a group of persons can become infected via **exposure to an aerosol release of smallpox** (the virus is put in the air), and secondary spread can occur via **prolonged face-to-face contact with someone who has smallpox** or by **direct contact with infected bodily fluids** or an object such as bedding or clothing that has the virus on it.

PATHOGENESIS

After an incubation period of 7 to 17 days, the individual becomes febrile and develops flu-like symptoms, including backache, headache, and vomiting. Two or 3 days later a maculopapular eruption appears, prominently on the oropharyngeal mucosa and face, and the patient becomes infectious to other persons. A person with smallpox is sometimes contagious before the onset of fever (prodrome phase), but the person becomes most contagious with the onset of rash. The characteristic vesicles may not become apparent for a few

days, but they then appear all over the body, including the palms and soles. The infected person is contagious until the last smallpox scab falls off (Table 84-2).

TREATMENT

There is no proven treatment for smallpox. Patients with smallpox may be helped by intravenous fluids, medicine to control fever or pain, and antibiotics for any secondary bacterial infections that may occur.

OUTCOME

The patient was kept in isolation until all his lesions had completely healed, scabs had fallen off, and he could be discharged. However, many other patients

TABLE 84-2 Important Facts About Smallpox Disease*		
Stages (duration)	**Contagiousness**	**Manifestations**
Prodrome (2 to 4 days)	Sometimes contagious	The first symptoms of smallpox include fever, malaise, head and body aches, and sometimes vomiting. The fever is usually high, in the range of 101 to 104°F.
Early Rash (4 days)	Highly contagious	A rash emerges first as small red spots on the tongue and in the mouth.
		These spots develop into sores that break open and spread large amounts of the virus into the mouth and throat.
		Around the time the sores in the mouth break down, a rash appears on the skin, starting on the face and spreading to the arms and legs and then to the hands and feet.
		Usually the rash spreads to all parts of the body within 24 hours. By the third day of the rash, the rash becomes raised bumps. By the fourth day, the **bumps fill with a thick, opaque fluid and often have a depression** (a major distinguishing characteristic of smallpox [Fig. 84-1]).
Pustular rash (5 days)	Contagious	The bumps become sharply raised pustules, usually round and firm to the touch.
Pustules and scabs (5 days)	Contagious	The pustules begin to form a crust and then scab. By the end of the second week after the rash appears, most of the sores have scabbed over.
Resolving scabs (about 6 days)	Contagious	The scabs begin to fall off, leaving marks on the skin that eventually become pitted scars.
Scabs resolved	Not contagious	Scabs have fallen off. Individual is no longer contagious.

*Adapted from CDC.

were admitted with smallpox, and a massive investigation was launched by the CDC and law enforcement agencies to determine the source of the outbreak and to vaccinate possible contacts.

PREVENTION

Although smallpox has been eradicated, there are concerns about the potential use of variola virus as a weapon of terror. One of the best ways to prevent smallpox is through vaccination. If given to a person before exposure to smallpox, the vaccine offers complete protection. Currently, the smallpox vaccine is not widely available to the general public, although certain "first responders" have been vaccinated for the remote possibility of an outbreak. Research is underway to determine if use of stored vaccine in a diluted form is effective. New vaccines are also being developed.

For vaccination to be considered successful, a painful pustule must develop (the "Jennerian vesicle"), indicating that the virus has multiplied in the recipient and elicited an inflammatory response. Approximately 1 in 1000 vaccinees develop one of the following untoward reactions: (1) autoinoculation; (2) disseminated vaccinia; (3) eczema vaccinatum; (4) progressive vaccinia (vaccinia necrosum), a potentially fatal illness occurring in patients with immunodeficiency; or (5) postvaccinia encephalitis. Contraindications to vaccination are eczema, immunocompromise, or close, unavoidable contact with someone who has eczema or is immunocompromised or is pregnant. It is thought that vaccination is effective even if it is administered a few days after exposure.

To prevent smallpox from spreading, anyone who has been in contact with a person with smallpox but who decides not to get the vaccine may need to be isolated for at least 18 days. During this time, they will be checked for symptoms of smallpox.

FURTHER READING

Blendon RJ, DesRoches CM, Benson JM, et al: The public and the smallpox threat. N Engl J Med 348:426, 2003.

Breman JG, Henderson DA: Diagnosis and management of smallpox (current concepts). N Engl J Med 346:1300, 2002.

Drazen JM: Smallpox and bioterrorism. N Engl J Med 346:1262, 2002.

Grabenstein JD, Winkenwerder W, Jr: U.S. military smallpox vaccination program experience. JAMA 289:3278, 2003.

Henderson DA, Inglesby TV, Bartlett JG, et al: Smallpox as a biological weapon: Medical and public health management (consensus statement). JAMA 281:2127, 1999.

Madeley CR: Diagnosing smallpox in possible bioterrorist attack. Lancet 361:97, 2003.

On a summer day in August, an 18-year-old man was taken to the emergency department of a local hospital because of a 2-day history of **fever, weakness, pain in his left groin,** and diarrhea. The groin pain was so severe that he walked with a limp, with his left leg abducted. He also had small rashes on his leg.

The patient lived in Flagstaff, Arizona. He had maintained good health before the current event.

PHYSICAL EXAMINATION

VS: T 39.3°C, P 126/minute, R 28/min, BP 90/52 mmHg

PE: The patient was alert but restless and was in moderate distress due to pain. A **left groin mass** (~6 cm), which was firm and **exquisitely tender,** was noted, with mild erythema, and there were **small hemorrhages on the skin** of his right leg (Fig. 85-1).

LABORATORY STUDIES

Blood

ESR: 55 mm/h
Hematocrit: 44%
WBC: 20,500/µL
Differential: 46% PMNs, 25% bands, 16% lymphs
Blood gases: Normal
Serum chemistries: AST 210 U/L, ALT 190 U/L, bilirubin 2.1 mg/dL

Imaging

No imaging was done.

Diagnostic Work-Up

Table 85-1 lists the likely causes of illness (differential diagnosis). The patient was considered to have lymphadenitis and sepsis. Investigational approach may include

- **Blood cultures**
- **Routine bacterial cultures** of lymph node aspirates

COURSE

The patient was admitted to the hospital, and blood and lymph node aspirate cultures were obtained. He was started on empirical antibiotics because of his serious condition. The following day, cultures of blood samples obtained in the ED were presumptively positive.

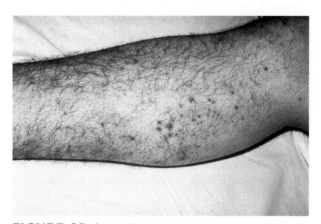

FIGURE 85-1 Small hemorrhages on the skin of leg in this patient. *(Courtesy of Centers for Disease Control and Prevention, Atlanta, GA.)*

TABLE 85–1 Differential Diagnosis and Rationale for Inclusion (consideration)
Abscess
Cat-scratch disease (*Bartonella*)
Plague
Sexually transmitted diseases (e.g., chancroid, lymphogranuloma venereum)
Rationale: Lymphadenopathy can be caused by many infectious and noninfectious processes. A careful history will usually elicit some risk factor, such as cat exposure for cat-scratch disease, or recent sexual activity for chancroid or other STD. An abscess would be fluctuant. Except for plague, the other diseases listed generally do not produce severe systemic illness.

ETIOLOGY

Yersinia pestis (plague, a presumptive diagnosis)

MICROBIOLOGIC PROPERTIES

Yersinia pestis is a **nonmotile, nonspore-forming, Gram-negative, bipolar, ovoid, "safety-pin-shaped"** bacterium (Fig. 85-2). These bacteria grow well on most standard laboratory media. After 48 to 72 hours, colonies appear gray-white to slightly yellow opaque, with raised, irregular "fried egg" morphology. Like all other members of *Enterobacteriaceae*, colonies of *Y. pestis* are **oxidase negative**; they **ferment glucose, and reduce nitrates to nitrites** but are otherwise biochemically nonreactive. The positive identification of an isolate is achieved by a direct fluorescent antibody (DFA) stain specific for F1 antigen, a plasmid-encoded envelope glycoprotein.

EPIDEMIOLOGY

The World Health Organization reports 1000 to 3000 cases of plague every year, globally. Plague is endemic in China, Indonesia, Mongolia, Burma, and India. *Yersinia pestis*, a **zoonotic pathogen**, is maintained in the reservoir, which includes wild rodents, carnivores, and domestic cats and dogs. **Natural transmission of *Y. pestis* is through the bite of infected fleas.** On average, in the United States, there are 10 to 15 cases of plague in humans per year, primarily in Native Americans in the Southwest.

PATHOGENESIS

There are three clinical forms of plague: (1) **bubonic** (infection of lymph nodes); (2) **septicemic** (bloodstream infection, the deadliest form); and (3) **pneumonic**.

Once inside human tissue, F1 glycoprotein allows the organisms to survive phagocytosis and to multiply within macrophages. The organisms produce several virulence factors (coded by *Y. pestis* plasmids) in addition to antiphagocytic F1 glycoprotein:

- a protease (activates plasminogen and degrades serum complement)
- a coagulase
- an exotoxin (adrenergic antagonist)

The organisms multiply explosively and **spread to regional lymph nodes** via the lymphatics. The nodes become enlarged in **bubonic plague**, leading to characteristic lesions called **"buboes."** Dissemination takes place via the blood stream, where the organisms multiply. The monocytes, which can phagocytize *Y. pestis* organisms without destroying them, carry the organisms in the blood stream. An endotoxin associated with Gram-negative lipopolysaccharide (LPS) of *Y. pestis* triggers a systemic inflammatory response. Elevated levels of proinflammatory cytokines, such as IL-1, TNF, and IL-6, are detectable in blood. Diffuse hemorrhages or widespread petechial lesions (thus the term "black death") are a result of disseminated intravascular coagulation and vascular necrosis in the event of systemic inflammatory response syndrome (**septicemic plague**). The monocytes also facilitate hematogenous dissemination of the organisms to the lungs, causing **interstitial pneumonic plague**.

> **NOTE** Pneumonic plague patients may aerosolize the organism, causing rapid airborne spread of bacteria (secondary infections) to household and other contacts (outbreak or epidemic of **pneumonic plague**). The best specimens for recovering *Y. pestis* in the pneumonic form of the disease are sputum and bronchial washings. Secondary cases lead to lobar pneumonia and can occur by direct inhalation of **airborne droplets** from symptomatic patients with primary disseminated illness. The secondary transmissions have caused outbreaks of pneumonic plague in the past. Intentional

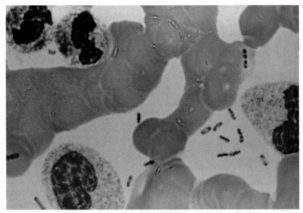

FIGURE 85-2 Wright stain of *Yersinia pestis* from blood of an infected patient. Note bipolar, ovoid, "safety-pin-shaped" bacteria. *(Courtesy of Centers for Disease Control and Prevention, Atlanta, GA.)*

aerosolization of this highly pathogenic agent has the **potential for bioterrorism,** and the CDC has classified *Y. pestis* as one of the category A bioterrorism agents along with anthrax and smallpox.

TREATMENT

The **drug of choice for all clinical forms of plague is streptomycin** (not easily available in the United States). Alternative drugs include gentamicin, tetracycline (doxycycline; also a choice for chemoprophylaxis), and chloramphenicol. Given that untreated plague has a very high mortality, appropriate antibiotics should be given as soon as possible once cultures are obtained from a likely case.

OUTCOME

After the culture results, the patient was given streptomycin. He completed a 10-day course of treatment, and his symptoms improved. An epidemiologic investigation by the public health officials indicated that the case patient most likely became infected 5 days before the onset of symptoms, as the result of bites by *Y. pestis*-infected fleas while trapping prairie dogs in Navajo country. High antibody titers against an envelope glycoprotein F1 antigen of *Y. pestis* were detected in the blood of the patient and in two of four pet dogs living in houses near the prairie dog colony.

PREVENTION

Patients with suspected *Y. pestis* infections should be reported immediately to local or state health departments to enable prompt initiation of appropriate public health control and prevention activities. These may include rodent control, avoidance of contact with infected rodents, and insecticide in areas where the disease is prevalent or outbreaks occur. During epidemics or for sporadic cases, strict isolation with precautions against airborne spread is necessary. All contacts should be provided with **chemoprophylaxis with tetracycline** for 1 week. Vaccine is no longer available because the only manufacturer in the United States has discontinued production.

FURTHER READING

Chanteau S, Rahalison L, Ralafiarisoa L, et al: Development and testing of a rapid diagnostic test for bubonic and pneumonic plague. Lancet 361:211, 2003.
Inglesby TY, Dennis DT, Henderson DA, et al: Plague as a biological weapon: Medical and public health management (consensus statement). JAMA 283:2281, 2000.
Perry RD, Fetherston JD: *Yersinia pestis*-etiologic agent of plague. Clin Microbiol Rev 10:35, 1997.

A 38-year-old man developed a **flu-like syndrome** with fever, headache, and anorexia. During the ensuing 8 weeks, he noted **continuing fevers, sweats, 30-lb weight loss, and depression.** He saw his family physician, complaining of acute **pain in joints, especially his lower back.**

He was a computer programmer and had recently traveled to Crete to visit his grandmother. During his stay on the Mediterranean island he enjoyed drinking **unpasteurized goat milk.** Six weeks after his return to the United States he developed the current illness.

PHYSICAL EXAMINATION

VS: T 38.2°C, P 84/min, R 14/min, BP 124/72 mmHg

PE: Generalized **lymphadenopathy** and mild **splenomegaly** were noted. **Tenderness in his sacroiliac joint** was also noted.

LABORATORY STUDIES

Blood

Hematocrit: 34%
Hemoglobin: 11 g/dL
WBC: 4500/μL
Differential: 62% PMNs, 21% lymphs, 12% monos
Blood gases: Normal
Serum chemistries: AST 102 U/L, ALT 94 U/L

Imaging

Plain x-rays of the lumbosacral spine were unrevealing.

Diagnostic Work-Up

Table 86-1 lists the likely causes of illness (differential diagnosis). Investigational approach may include

- **Skin tests** (e.g., tuberculin, histoplasmin)
- **Gram and acid-fast stain** of respiratory specimens, if possible
- **Cultures** of blood, bone marrow, other tissues (biopsy)
- **Serology** to measure IgG antibody

TABLE 86-1   Differential Diagnosis and Rationale for Inclusion (consideration)

Brucellosis (*Brucella* spp)
Dengue fever
Hepatitis A, B, or C
Histoplasmosis (*Histoplasma capsulatum*)
Leptospirosis (*Leptospira interrogans*)
Lymphoma
Subacute bacterial endocarditis
Tuberculosis (*Mycobacterium tuberculosis*)
Tularemia (*Francisella tularensis*)
Typhoid fever

Rationale: Chronic fevers and weight loss may be due to a variety of factors. Noninfectious etiologies (such as lymphoma) should be considered as well as infectious etiologies. Hepatitis may be considered owing to the elevated transaminases, but hepatitis would not usually cause such a prolonged illness. Dengue fever causes musculoskeletal pains, but it is not a chronic disease. The other etiologies (e.g., histoplasmosis, leptospirosis, tuberculosis, and tularemia) certainly may cause long illnesses, but they often have specific epidemiologic features or history of exposure. *Brucella* and *Francisella* are potential agents of bioterrorism, and if several cases occur without the usual risks present, that possibility should be considered. Endocarditis would be unusual without a history of IV drug use in a patient of this age.

COURSE

The patient was admitted to the hospital. Bone scan revealed uptake in the sacroiliac region. Blood culture after overnight incubation did not yield any significant pathogen. A liver biopsy contained granulomas.

ETIOLOGY

Brucella melitensis (presumptive diagnosis of brucellosis)

MICROBIOLOGIC PROPERTIES

Brucellosis is a zoonotic disease caused by four species of *Brucella*: *B. abortus*, *B. melitensis* (the most virulent and common species), *B. suis*, and *B. canis*. Brucellae are **poorly staining**, **small**, **Gram-negative coccobacilli**, **seen mostly as single cells and appearing like "fine sand"** (Fig. 86-1*A*). These hardy organisms can remain viable for more than 40 days in moist soil.

All species of *Brucella* are biosafety level (BSL)-3 agents when growth on laboratory media is attempted or being manipulated. Brucellae are **recovered from the blood** or bone marrow of an infected patient, generally **after prolonged incubation** (at least 21 days). **On blood agar, colonies are small, convex, nonhemolytic,** and **translucent** (see Fig. 86-1*B*). Oxidase and urease tests can help differentiate *Brucella* species.

EPIDEMIOLOGY

Brucellosis is a **zoonosis** with a worldwide distribution. In the United States, the current incidence rate is less than 120 cases per year (most in Texas and California). Brucellae are maintained in cattle (*B. abortus*), swine (*B. suis*), dogs (*B. canis*), and goats and sheep (*B. melitensis*). The organisms are transmitted by ingestion of raw milk and dairy products (unpasteurized milk and cheese) from infected animals and by contact with tissues, blood, urine, and aborted fetuses from infected animals. Aerosols of the most virulent species, *B. melitensis*, are highly infectious, making it a potential agent for weaponization and airborne dissemination in **bioterrorism**.

PATHOGENESIS

The infective dose of *Brucella* is fewer than 100 organisms, and the incubation period ranges from 5 days to 6 months. On entry into the body, the bacteria are opsonized by serum factors and are engulfed by PMNs. The facultative intracellular brucellae survive phagocytic killing by virtue of their suppression of the myeloperoxide/hydrogen peroxide-halide system and the production of superoxide dismutase. They replicate in phagocytes in the regional lymph nodes, and **disseminate throughout the body via the blood stream**, initiating **multisystem infections** involving the liver, spleen, bones, kidneys, lymph nodes, heart valves, nervous system, and testes. A systemic inflammatory response is triggered by the lipopolysaccharide (LPS) endotoxin of the Gram-negative bacteria. Cytokines (e.g., interleukin [IL]-1 and tumor

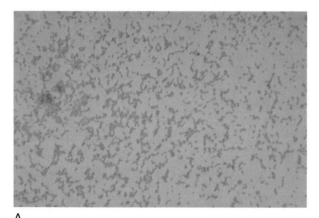
A

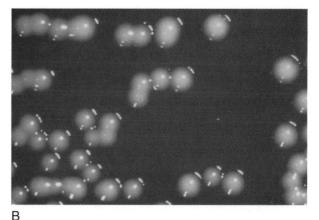
B

FIGURE 86-1 *A*, Gram stain of *Brucella* spp. *B*, Culture of *Brucella melitensis* on sheep blood agar, showing the smooth, translucent, nonhemolytic colonies. (*Courtesy of Dr. Paul Southern, Departments of Pathology and Medicine, University of Texas Southwestern Medical Center, Dallas, TX.*)

necrosis factor) cause irregular fever of variable duration and weight loss (wasting), the key manifestations of brucellosis. The systemic cytokine response also contributes to profuse sweating, chills, arthralgias, and generalized aching.

NOTE The disease can linger for months if not adequately treated. Osteoarticular complications (particularly of the sacroiliac joints) are seen in 20% to 60% of cases. Genito-urinary involvement is reported in up to 20% of cases, with orchitis and epididymitis the most common. Endocarditis, which occurs in fewer than 2% of cases, accounts for the majority of brucellosis-related deaths.

TREATMENT

A combination of doxycycline and either streptomycin or rifampin for at least 6 weeks is clinically effective. Quinolones also have activity. In patients with neurologic disease, corticosteroids may be helpful.

OUTCOME

Skin tests were negative for the systemic fungal pathogens and tuberculin. Blood culture on extended incubation grew *Brucella melitensis*. The patient received a course of doxycycline and streptomycin for 6 weeks and gradually recovered over the course of 2 months.

PREVENTION

No vaccine is currently available to protect against brucellosis in humans. There are vaccines for *B. abortus* and *B. melitensis* for use in animals; there are no vaccines for *B. suis* or *B. canis*. It is important to pasteurize milk and dairy products from cows, sheep, and goats. Care should be exercised in handling and disposing of placentas, discharges, and fetuses from aborted animals.

FURTHER READING

Brouqui P, Raoult D: Endocarditis due to rare and fastidious bacteria. Clin Microbiol Rev 14:177, 2001.
CDC: Suspected brucellosis case prompts investigation of possible bioterrorism-related activity—New Hampshire and Massachusetts, 1999. MMWR 49:509, 2000.

Chomel BB, DeBess EE, Mangiamele DM, et al: Changing trends in the epidemiology of human brucellosis in California from 1973 to 1992: A shift toward food-borne transmission. J Infect Dis 170:1216, 1994.

In **late September,** a 51-year-old man with a **one-week history of fever, headache,** and **muscle aches** was taken to a clinic. He noted **an ulcer on his right hand that had not healed and pain in his right axilla.**

The patient **lived in Oklahoma** and had a hobby of sewing together tanned rabbit hides to make blankets. In the week before the onset of his illness, he had **skinned and tanned a rabbit** killed by the family dog.

PHYSICAL EXAMINATION

VS: T 39.4°C, P 112/min, R 16/min, BP 142/70 mmHg

PE: Remarkable physical findings included an **indurated erythematous rash with an ulcerated lesion** on the dorsal skin of the right hand and **painful axillary adenopathy** (Fig. 87-1).

LABORATORY STUDIES

Blood

Hematocrit: 38%
WBC: 14,200/μL
Differential: 72% PMNs, 18% lymphs
Blood gases: Normal
Serum chemistries: Normal

Imaging

No imaging studies were done.

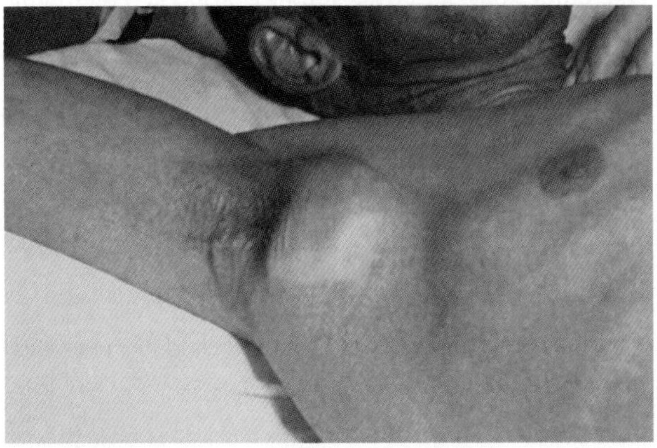

Diagnostic Work-Up

Table 87-1 lists the likely causes of the man's illness (differential diagnosis). Investigational approach may include

- **Gram stain** of aspirates or biopsies from the ulcerative lesion and the suppurated lymph nodes
- **Cultures** of blood, pus, or exudates
- **Serology.** Acute- and convalescent-phase sera are collected for investigation of species-specific IgG levels.

COURSE

Cultures of blood and from the ulcerative lesion and the suppurated lymph nodes were obtained. The patient was treated for cellulitis of undetermined etiology with an oral cephalosporin, but without success. Serologic tests were ordered.

FIGURE 87-1 A painful, massively enlarged axillary node (glandular adenopathy). *(Courtesy of Dr. Paul Southern, Departments of Pathology and Medicine, University of Texas Southwestern Medical Center, Dallas, TX.)*

TABLE 87-1 Differential Diagnosis and Rationale for Inclusion (consideration)
Brucellosis (*Brucella* spp) Bubonic plague (*Yersinia pestis*) Cat-scratch fever (*Bartonella henselae*) Pasteurellosis (*Pasteurella multocida*) Sporotrichosis (*Sporothrix schenckii*) Staphylococcal and streptococcal infections Tularemia (*Francisella tularensis*)
Rationale: Fevers and local soft tissue infection can be associated with several organisms. *Staphylococcus* and *Streptococcus* are the most frequent causes. Often, specific exposures will lead toward a specific diagnosis, such as cats or dogs for *Pasteurella*, ticks or wild animals for tularemia, animals for plague, and cats for cat-scratch fever. Sporotrichosis is often related to minor trauma involving plants, and brucellosis is associated with unpasteurized dairy products.

ETIOLOGY

Francisella tularensis (presumptive diagnosis of tularemia)

MICROBIOLOGIC PROPERTIES

F. tularensis is a poorly staining, **very tiny Gram-negative, nonmotile coccobacillus**, seen mostly as single cells. The organisms have a thin layer of polysaccharide capsule. Bacterial culture is difficult; diagnosis is most commonly made clinically. This **fastidious pathogen**, when grown aerobically on chocolate medium, yields gray-white colonies (1 to 3 mm) at 48 to 72 hours (biosafety level 3 is required for culturing). The **clinical diagnosis is confirmed by a rise in IgG-class antibodies** (which usually appear in the second week of the disease).

EPIDEMIOLOGY

In the United States, the disease is endemic in Missouri, Arkansas, and Oklahoma. *F. tularensis* is maintained in nature in wild animals (e.g., rabbits, hares, muskrats, and deer) and some domestic animals. The common mode of acquisition of *F. tularensis* is accidental, via the **bite of infective ticks** and by direct contact with infected animal tissues through skin abrasions. The incidence rate is 150 to 300 cases per year in the United States. The bacterium *F. tularensis* is considered to be a dangerous **potential bioterrorism agent** because of its **extreme infectivity, ease of dissemination** (inhalation of a small dose of aerosolized bacteria can cause disease), and substantial **capacity for causing pneumonitis** and **death**.

PATHOGENESIS

Tularemia has an **extremely low infective dose** of 10 to 50 organisms. The infection usually starts when the organism penetrates the skin (e.g., from an infective tick bite). Following an incubation period of 3 to 5 days, an erythematous papule appears at the site of introduction of the bacteria, ulcerating 2 to 4 days later. The ulcer is erythematous, indurated, and non-healing and has a "punched out" appearance at 1 to 3 weeks. The bacteria are transported via the lymphatics to the regional lymph nodes, where they grow in phagocytic cells. The lymph nodes enlarge and frequently become necrotic (**ulceroglandular type**; seen in the case patient), characterized by pyogenic-granulomatous pathology. Other clinical forms of infection include oculoglandular, pharyngeal, typhoidal, and pneumonic, depending on the route of transmission.

NOTE | In untreated patients, the organisms enter the blood from the lymphatics. Blood stream invasion and dissemination of bacteria-carrying monocytes to various organs are similar to those seen in plague. Lipopolysaccharide-associated endotoxin is the major virulence factor in the development of systemic toxicity. Hematogenous spread of organisms can result in pneumonia (as in pneumonic plague), with diffuse interstitial infiltration visible in the chest x-ray. Clinically, because of bubo (**suppurative lymphadenopathy**) and following hematogenous dissemination, a subsequent pneumonia in complicated cases, tularemia may be confused with plague. Unlike plague, secondary cases or person-to-person transmission of pneumonic tularemia rarely occurs.

TREATMENT

Streptomycin given for 7 to 14 days is the drug of choice. Gentamicin or doxycycline are alternative choices.

OUTCOME

When the patient did not improve after 3 days, he was taken to his regular physician, who suspected tularemia. After taking doxycycline for 10 days, he recovered. Paired serum titers for *F. tularensis* collected at the first clinic visit and 15 days later on a follow-up visit were 1:40 and 1:1024, retrospectively confirming the diagnosis of tularemia.

PREVENTION

No vaccines are available. Care should be exercised when handling animal carcasses, and tick repellent should be used in infested areas.

FURTHER READING

Dennis DT, Inglesby TV, Henderson DA, et al: Tularemia as a biological weapon: Medical and public health management. JAMA 285:2763, 2001.

Ellis J, Oyston PCF, Green M, Titball RW: Tularemia. Clin Microbiol Rev 15:631, 2002.

Feldman KA, Enscore R, Lathrop S, et al: Outbreak of primary pneumonic tularemia on Martha's Vineyard. N Engl J Med 345:1601, 2001.

Haristoy X, Lozniewski A, Tram C, et al: *Francisella tularensis* bacteremia. J Clin Microbiol 41:2774, 2003.

In early summer, a **previously healthy 42-year-old** man was admitted to his local hospital with a 1-week history of **fever, muscle aches,** and **malaise.** For 2 days before the admission, he noted shortness of breath, and on the day of admission he felt extremely weak.

In the 3 weeks before becoming unwell, the patient had been stationed at a **rural campsite in New Mexico.** He had reported that in and around his tent, there had been many **deer mice,** although he had not been bitten.

PHYSICAL EXAMINATION

VS: T 38.0°C, P 124/min, R 36/min, **BP 76/40** mmHg
PE: The patient appeared ill and crackles were heard on lung examination.

LABORATORY STUDIES

Blood

ESR: Normal
Hematocrit: 56%
WBC: 22,000/µL
Differential: 54% PMNs, **22% bands**, 16% lymphs
Platelets: **80,000/µL**

Blood gases: pH 7.28, pCO$_2$ 22 mmHg, **pO$_2$ 40 mmHg**
Serum chemistries: normal

Imaging

Chest x-ray revealed bilateral interstitial infiltrates with hilar indistinctness (Fig. 88-1).

Diagnostic Work-Up

Table 88-1 lists the likely causes of this man's illness (differential diagnosis).

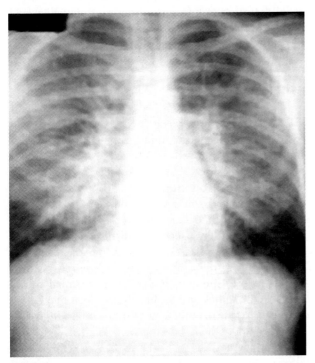

FIGURE 88-1 Chest film showing marked interstitial edema with hilar indistinctness in this patient. *(Courtesy of Centers for Disease Control and Prevention, Atlanta, GA.)*

TABLE 88–1 Differential Diagnosis and Rationale for Inclusion (consideration)

Atypical bacterial pneumonia (e.g., *Mycoplasma*, *Chlamydia*, Q fever)
Dengue hemorrhagic shock syndrome
Meningococcemia (sepsis)
Plague
Respiratory viral infections
    Adenovirus (acute respiratory distress syndrome [ARDS])
    Influenza
    Severe acute respiratory syndrome: coronavirus (SARS-CoV)
Rickettsial infection (e.g., Rocky Mountain spotted fever [RMSF])
Sin Nombre virus
Tularemia
Typical bacterial pneumonia (e.g., *Streptococcus pneumoniae*)

Rationale: Although the manifestation of severe pneumonia and hypotension is not common, it can be due to typical bacterial pneumonia. However, atypical bacterial pneumonia and other unusual zoonotic infections (e.g., tularemia, plague, RMSF) should also be considered. Meningococcemia may manifest with hypotension, but lobar pneumonia is uncommon. *Mycoplasma*, *Chlamydia*, and Q fever do not generally cause hypotension. Although adenovirus may cause severe pneumonia, other viruses, such as the SARS-CoV, hantavirus, or dengue hemorrhagic viruses, would be more likely. These would be expected to be seen in certain geographic areas or with specific exposures.

Investigational approach may include

- **Rapid antigen** detection or **direct immuno-fluorescence antibody staining** of nasopharyngeal secretions
- **Gram** and **acid-fast stain** of respiratory secretions or lung biopsy
- **Sputum cultures** for acid-fast bacilli, fungi, or bacteria
- In failed tests
    - **Serology** to demonstrate virus-specific IgG
    - **PCR** of target DNA from blood clots and lung biopsy specimens

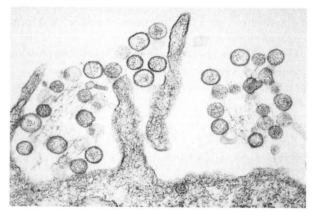

FIGURE 88-2 Electron micrograph of hantaviruses. Note that the enveloped viruses are spherical. *(Courtesy of Centers for Disease Control and Prevention, Atlanta, GA.)*

COURSE

Following admission to the hospital, the patient became progressively hypoxemic. He was treated with intravenous erythromycin and ceftriaxone. Nevertheless, he developed progressive respiratory failure, necessitating mechanical ventilation. His condition steadily worsened for his first 10 days in the hospital. On the 12th day of hospitalization, an open-lung biopsy was performed, but tests did not show acid-fast bacilli, fungi, or bacteria. At that point, serology was performed by enzyme immunoassay for a virus, which was diagnostic.

ETIOLOGY

Sin Nombre virus (hantavirus pulmonary syndrome [HPS])

MICROBIOLOGIC PROPERTIES

Hantaviruses belong to the bunyavirus family of viruses. There are five genera within the family: bunyavirus, phlebovirus, nairovirus, tospovirus, and hantavirus. All of these genera include viruses that are arthropod borne, with the exception of hantavirus, which is rodent borne. Like arthropod-borne members of the bunyavirus family, hantaviruses are **enveloped viruses** (**spherical** [Fig. 88-2]; 80 to 120 nm virions) with a **genome that consists of a triple-segmented, circular, single-stranded, negative-sense RNA** (11 to 21 kb in overall size). Virion particles contain three circular, **helically symmetrical nucleocapsids**. Hantaviruses replicate exclusively in the host cell cytoplasm. Several members of the hantavirus genus cause different forms of hemorrhagic fever with renal syndrome (HFRS). Hantaan virus from Korea is associated with a severe form of HFRS characterized by renal failure that can precede pulmonary edema and disseminated intravascular coagulation.

Laboratory diagnosis of HPS includes (1) detection of hantavirus-specific IgM (even in prodrome); or (2) rising titers of hantavirus-specific immunoglobulin G; or (3) detection of hantavirus-specific ribonucleic acid sequence by PCR in clinical specimens (e.g., blood clots and lung biopsy specimens).

EPIDEMIOLOGY

HPS, which is now recognized as a pan-American zoonosis, exhibits an expanding clinical spectrum. It is caused by many novel viruses, each of which has a distinct rodent host. **Human infection is strongly associated with rodent urine** or droppings in a tight environment (e.g., under a tent in a remote campsite). Transmission can occur when dried materials contaminated by rodent excreta are disturbed and inhaled. HPS has not been associated with person-to-person transmission in the United States. However, person-to-person transmission, including nosocomial transmission of Andes virus, has been documented in southern Argentina and Chile. Adult men are at risk for HPS; individuals bearing one particular B-locus allele, B*35, appear to be at higher risk for severe disease. **Aerosols of the virus** could be a **potential agent for weaponization** and airborne dissemination in **bioterrorism**.

PATHOGENESIS

The incubation period after exposure to infective rodent droppings is 1 to 5 weeks. The b3 integrins are the cellular receptors for pathogenic hantaviruses. Tumor necrosis factor and interleukin-2 are involved in the pathogenesis of HPS. Activated CD8+ T cells produce holes in infected pneumocytes. The non-specific prodrome consists of fever, chills, myalgia, headache, and gastrointestinal symptoms. Prodrome is followed by bilateral interstitial pulmonary infiltrates and respiratory compromise, clinically resembling ARDS. Thrombocytopenia (significantly reduced platelet level) and a left shift (>15% bands) are almost always evident; thrombocytopenia is a key early clinical marker. An increase in nuclear permeability causes plasma leakage into alveolae, producing **pulmonary edema** and **hemoconcentration**. In the most severe cases, disseminated intravenous coagulation develops.

> **NOTE** An intentional spread of the aerosolized virus strain has the potential for an outbreak of a febrile illness associated with clinical manifestations, including rash and shock, 3 to 21 days later.

TREATMENT

The primary mode of therapy is supportive, with aggressive ventilator management and use of systemic vasopressors. Nevertheless, patients should be placed on broad-spectrum antibiotics until the diagnosis of HPS is well established, because septic shock is far more common than is shock due to hantavirus. The antiviral agent ribavirin has not been shown to be effective for pulmonary syndrome caused by Sin Nombre virus.

OUTCOME

The patient was placed on mechanical ventilation and required high doses of vasopressors for several days, but he slowly recovered over the next few weeks in the hospital. He was discharged from the hospital without any complications.

PREVENTION

Since rodent exposure in endemic areas is critical to the development of HPS, limiting such exposure should be the primary focus of efforts, including rodent control, preventing entry of rodents into indoor areas, and proper clean-up of rodent-infested areas. No vaccines are available.

FURTHER READING

Borio L, Inglesby T, Peters CJ, et al: Hemorrhagic fever viruses as biological weapons: Medical and public health management (consensus statements). JAMA 287:2391, 2002.

CDC: Hantavirus pulmonary syndrome-Vermont, 2000. JAMA 286:912 , 2001.

CDC: Hantavirus pulmonary syndrome-Panama, 1999-2000. JAMA 283:2232, 2000.

Duchin JS, Koster FT, Peters CJ, et al: Hantavirus pulmonary syndrome: A clinical description of 17 patients with a newly recognized disease. N Engl J Med 330:949, 1994.

In **August**, a friend brought a **69-year-old woman** with a history of diabetes mellitus to the emergency department following a 2-day history of **fever, headache, vomiting, weakness, and confusion.**

The woman had been in good health previously, and her diabetes was well controlled. She commonly **spent much time outdoors** in her yard and garden.

PHYSICAL EXAMINATION

VS: T 38.5°C, P 102/min, R 20/min, BP 118/60 mmHg

PE: On examination, the patient was difficult to arouse and did not respond appropriately to questioning. A **coarse tremor was present in the chin and upper and lower extremities**.

LABORATORY STUDIES

Blood

Hematocrit: 37%
WBC: 11,600/μL

TABLE 89–1   Differential Diagnosis and Rationale for Inclusion (consideration)

Arthropod-borne encephalitides
   California/LaCrosse virus
   Eastern equine virus
   St. Louis virus
   West Nile virus
   Western equine virus
Aseptic (enteroviral) meningitis
Aseptic (nonviral) meningitides (e.g., tuberculous, parasitic, or syphilitic)
Bacterial meningitis
Brain abscess
Cryptococcal meningoencephalitis
Herpes simplex virus (HSV) encephalitis

Rationale: The presence of fever and altered mental status suggests encephalitis, which has many causes that are difficult to distinguish clinically. Viral causes include arboviral or mosquito-transmitted disease and sporadic HSV encephalitis. Bacterial meningitis should also be considered, but bacteria often cause a more acute illness, with mental status changes later in the course. *Cryptococcus* can also cause a similar syndrome, although often more indolent. Enteroviruses usually cause meningitis, with occasional overlapping symptoms suggestive of encephalitis. Other aseptic causes (e.g., tuberculosis, neurosyphilis) should also be considered. Brain abscess may cause mental status changes, but it more often leads to focal neurologic deficits or seizures.

Differential: 72% PMNs, 8% lymphs, 15% monos
Blood gases: Normal
Serum chemistries: AST 58 U/L

Imaging

Head CT was unremarkable, and **brain MRI** showed increased signal in the basal ganglia.

Diagnostic Work-Up

Table 89-1 lists the likely causes of illness (differential diagnosis). Diagnosis of encephalitis on clinical grounds alone is difficult. There are not enough specific findings on physical examination. A high index of suspicion in the summer months in the appropriate geographic areas may support a presumptive diagnosis of arboviral encephalitis. Lumbar puncture is performed, and CSF is examined to rule out bacterial and viral meningitis. The CSF of patients with arboviral encephalitis may reveal lymphocytic leukocytosis, with a normal protein and a normal glucose level. Further investigational approach may include

- **Cultivation** of neurotropic virus in CSF
- **Direct cryptococcal antigen** in CSF

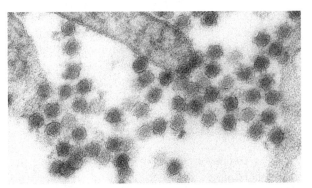

FIGURE 89-1   An electron micrograph of the viral pathogen. *(Courtesy of Centers for Disease Control and Prevention, Atlanta, GA.)*

● **Viral serology**

  ● Presence of specific **IgM** in acute-phase serum or CSF
  ● Elevated **arboviral IgG** between early and late specimens of serum
  ● **PCR amplification** of DNA from CSF to rule out a neurotropic virus (e.g., HSV-1)

COURSE

Lumbar puncture was performed. Analysis of the patient's spinal fluid demonstrated 42 cells/µL (82% lymphs), protein concentration of 70 mg/dL, and glucose concentration of 80 mg/dL. IgM and neutralizing antibody tests confirmed an acute infection due to an emerging arthropod-borne virus.

ETIOLOGY

West Nile virus (encephalitis)

MICROBIOLOGIC PROPERTIES

West Nile virus (WNV) is a flavivirus in the Flaviviridae family (Japanese encephalitis antigenic complex). This complex includes several viruses associated with human encephalitis: St. Louis encephalitis virus in the Americas, Japanese encephalitis virus in East Asia, and Murray Valley encephalitis virus and Kunjin virus (a subtype of WNV) in Australia. The viruses in the Flaviviridae family on appearance in the electron microscope (Fig. 89-1) share a common size (40 to 60 nm), symmetry **(enveloped, icosahedral nucleocapsid), and nucleic acid configuration (positive-sense, single-stranded RNA)**.

Laboratory confirmation of a clinical diagnosis of WNV requires the measurement of IgM antibodies in blood or CSF. The test is positive in most infected people within 8 days of onset of symptoms. Reference laboratories with biosafety level 3 facilities should have flavivirus isolation and identification capabilities.

EPIDEMIOLOGY

WNV is commonly found in Africa, West Asia, and the Middle East and is **transmitted among wild birds by mosquitoes of the genus *Culex*** and also,

rarely, of the genus *Anopheles*. It has **emerged in recent years in North America**, presenting a threat to public health. It is not known exactly when and how WNV was introduced into North America. International travel may have played a role, or the virus may have been transported by an infected migratory or imported bird. In 1999, necropsy pathology confirmed deaths from WNV among bird populations in the northeastern United States. The first human cases of aseptic meningitis and encephalitis were primarily among the elderly. Since then, human cases of WNV have been on the rise in the United States.

PATHOGENESIS

WNV gains access to humans via the bite of an infective mosquito. The virus localizes in the vascular endothelium and the lymphatic cell of the reticuloendothelial system, where replication occurs, causing primary viremia. In immunocompetent healthy individuals, primary viremia presents with a febrile illness without involvement of the CNS. In the elderly and immunocompromised individuals, another event of viral replication in the blood stream (secondary viremia) leads to **hematogenous dissemination of the virus, which localizes primarily in the CNS**. The virus causes inflammation of the brain (cerebellum) and vascular tissues; characteristic **brain pathology reveals scattered microglial nodules and perivascular inflammatory infiltrates** of lymphocytes. Head CT and brain MRI may show chronic microvascular ischemic changes. Acute WNV infection also has been associated with acute flaccid paralysis attributed to a peripheral demyelinating process, or to an anterior myelitis; a severe, asymmetrical process affects anterior horn cells and/or their axons.

TREATMENT

No specific antiviral therapy is available; treatment is supportive.

OUTCOME

The patient became comatose the following day and remained unresponsive for several days. After regaining consciousness, she gradually recovered. It took several weeks for her strength to return to normal.

PREVENTION

WNV control and prevention require (1) surveillance to detect the presence of WNV in areas where humans are at risk; (2) sustained and integrated mosquito control; and (3) public education on the use of personal protective behaviors and peri-domestic mosquito control to reduce the risk for mosquito bites. Standing water should be eliminated wherever possible. The use of DEET is highly effective in reducing mosquito bites. Although a vaccine is not available, work is progressing toward one for human use.

FURTHER READING

Lanciotti RS, Roehrig JT, Deubel V, et al: Origin of the West Nile virus responsible for an outbreak of encephalitis in the northeastern United States. Science 286:2333, 1999.

Morse DL: West Nile Virus—Not a passing phenomenon (perspective). N Engl J Med 348:2173, 2003.

Nash D, Mostashari F, Fine A, et al: The outbreak of West Nile virus infection in the New York City area in 1999. N Engl J Med 344:1807, 2001.

Petersen LR, Marfin AA, Gubler DJ: West Nile virus. JAMA 290:524, 2003.

Sejvar JJ, Haddad MB, Tierney BC, et al: Neurologic manifestations and outcome of West Nile virus infection. JAMA 290:511, 2003.

A **46-year-old man** presented with a 6-month history of **increasing forgetfulness, depression, and personality changes.** Electroencephalographic (EEG) examination and head CT scan were unremarkable, and no specific diagnosis was made. A few months later, he was hospitalized for **increasing confusion, ataxia,** and **movement tremors of his extremities.**

He had lived in the United Kingdom for several years, and he returned to the United States 2 years ago.

PHYSICAL EXAMINATION

VS: T 37°C, P 76/min, R 12/min, BP 124/82 mmHg

PE: The patient was disoriented and confused and had **resting tremors of all extremities. Myoclonus** was noted when he was startled.

LABORATORY STUDIES

Blood

Hematocrit: Normal
WBC: Normal
Differential: Normal
Blood gases: Normal
Serum chemistries: Normal

Imaging

Chest x-ray was normal. A magnetic resonance image (MRI) of the brain demonstrated mild, nonspecific enhancement along the inferior parasagittal occipital lobe. A repeat **EEG showed normal waves.**

Diagnostic Work-Up

Table 90-1 lists the likely causes of the patient's illness (differential diagnosis). Investigational approach may include

- **Blood cultures**
- **CSF examination and cultures**
- **Histopathology of brain biopsy or imuno-histochemical tests**
- **PCR** analysis

COURSE

Lumbar puncture was performed; blood culture and all analyses on CSF, including cultures, were negative for any infectious etiology. The patient's clinical con-

---

TABLE 90–1   Differential Diagnosis and Rationale for Inclusion (consideration)

Alzheimer disease
Prion disease (e.g., Creutzfeldt-Jakob disease)
Progressive multifocal encephalopathy (JC virus)
Subacute sclerosing panencephalitis (SSPE)
Toxic and metabolic encephalopathies
Tumors and other space-occupying lesions

Rationale: Dementia may be due to many etiologies, which are often difficult to distinguish clinically. Gradual onset over a period of years usually suggests Alzheimer disease. Metabolic encephalopathies should always be considered because they are frequently reversible; tumors may be treatable. SSPE and prion-associated diseases are not treatable, are more difficult to diagnose, and have important prognostic implications.

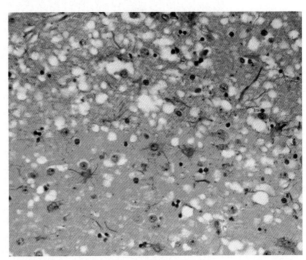

FIGURE 90-1 Histopathologic changes in brain biopsy. Note the presence of vacuoles (microscopic "holes" in the gray matter), which give the brain a sponge-like appearance. *(Courtesy of Centers for Disease Control and Prevention, Atlanta, GA.)*

dition progressively worsened; a panic attack was diagnosed, and the patient was treated with an antianxiety medication.

ETIOLOGY

Prion agent (new-variant Creutzfeldt-Jakob disease [nvCJD]; a presumptive diagnosis)

MICROBIOLOGIC PROPERTIES

Prions are mainly composed of an abnormal, protease-resistant conformer (PrP^{sc}) of a normal cellular protein (PrP^c). PrP^c is rich in α-helix polypeptide, while PrPs^{sc} has a high content of β-sheet polypeptide.

EPIDEMIOLOGY

Prions are believed to cause **transmissible spongiform encephalopathies** (TSEs) that include scrapie in sheep, bovine spongiform encephalopathy (BSE) in cattle, chronic wasting disease (CWD) in deer and elk, and **CJD in humans** (Table 90-2). Two major forms of CJD have been recognized—**classic CJD** and **new-variant form (nvCJD)**. The classic CJD, which was recognized in the early 1920s, occurs in the United States at a rate of approximately one case per 1 million population per year.

Classic CJD is believed to occur in three forms: **sporadic, familial (genetically determined), and acquired (by infection).** Sporadic CJD is the most common form. It is of unknown source, and risk increases with age (>50 years). Familial CJD is less common, but its study has greatly aided our understanding of the disease. Iatrogenic CJD is extremely rare and has been associated with transplantation of infected corneas or dural grafts, contaminated neurosurgical instruments, and administration of contaminated growth hormone or pituitary gonadotropin. The incubation period for iatrogenic CJD is long—12 months to possibly more than 30 years.

A new-variant form of CJD (nvCJD), with unique clinical and pathologic features, was first described in 1996 in the United Kingdom. The median age at death for nvCJD patients is 28 years, as compared with 68 years for patients with classic CJD. nvCJD is believed to be associated with eating cattle products infected with the prion that causes BSE.

PATHOGENESIS

Conversion of α-helix (in PrP^c) to β-sheet (PrP^{sc}) and its deposition in brain tissue is the fundamental concept in the pathogenesis of prion diseases.

TABLE 90-2   Prion Diseases: Method of Acquisition and Year of Finding

Disease	Presumptive Method of Acquisition	Year of Finding
Animal Diseases		
Scrapie	Infection in genetically susceptible sheep	1730
Transmissible mink encephalopathy	Infection with prions from sheep (species jump)	1947
Chronic wasting syndrome (in mule, deer, and elk)	Unknown	1980
Bovine spongiform encephalopathy (BSE)	Infection with scrapie sheep-rendered meal (species jump)	1987
Feline spongiform encephalopathy	Infection with prion-contaminated feed	1990
Human Diseases		
Creutzfeldt-Jakob disease (CJD)	Infection through contaminated hormone extracts and neurosurgical implants	1920
Gerstmann-Straussler syndrome	Germline mutation in *PrP* gene	1928
Kuru	Infection through cannibalism	1957
Fatal familial insomnia	A mutation in codon 178 of the *PrP* gene	1991
New-variant Creutzfeldt-Jakob disease (nvCJD)	Possibly by ingestion of tainted meat from BSE outbreak	1996

Although tissue damage occurs in only the central nervous system, no alteration in disease manifestations by immunosuppression or immunopotentiation is evident; no interferon production is known. Typical neuropathologic findings in CJD include (1) **diffuse loss of neurons**; (2) intense proliferation with fibrous gliosis; (3) **intracytoplasmic vacuolation (diffuse spongiform degeneration**; Fig. 90-1); and (4) swelling of both neuronal and astroglial processes. Deposition of PrP^{sc} in the brain tissues was found to correlate with neuronal vacuolation.

> NOTE Neuropathologic evaluation, particularly by immuno-histochemistry or Western blot, is the most definitive method to (1) diagnose human prion diseases; (2) monitor for nvCJD and various subtypes of CJD; and (3) detect the possible **emergence of new prion diseases** in the United States. In most CJD patients, the presence of a 14-3-3 protein (designated by electrophoretic mobility in Western blot) in the CSF is diagnostic for CJD. This protein also appears in CSF from patients with nvCJD. The nvCJD has specific neuropathologic and molecular characteristics that distinguish it from sporadic CJD. Therefore, a confirmatory diagnosis of CJD requires neuropathologic and/or immunodiagnostic testing of brain tissue obtained either at biopsy or at autopsy.

TREATMENT

None is available.

OUTCOME

The patient received experimental treatment with quinacrine for 3 months. Several months later, he had become bedridden, experienced considerable weight loss requiring surgical insertion of a feeding tube, and was no longer communicating with family members. The man died later that month; neuropathologic examination of brain tissue during autopsy indicated subacute spongiform encephalopathy, compatible with CJD.

PREVENTION

Standard precautions in hospitals to eliminate iatrogenic transfer of prions should be followed. Prions are not destroyed by standard decontamination methods, and optimal procedures are not known. Suggested sterilization procedures for CJD tissues and contaminated materials include steam autoclaving for 4.5 hr at 121°C (15 psi) or immersion in 1N NaOH (30 min ×3) at room temperature. General precautions should be followed when caring for patients with CJD, and special precautions are needed during autopsy. Severe restrictions are in place on the importation from countries where BSE is known to exist in live ruminants, such as cattle, sheep, and goats, and certain ruminant products.

FURTHER READING

Belay E, Schonberger L: Variant Creutzfeldt-Jakob disease and bovine spongiform encephalopathy. Clin Lab Med 22:849, 2002.

Brown P, Will RG, Bradley R, et al: Bovine spongiform encephalopathy and variant Creutzfeldt-Jakob disease: Background, evolution, and current concerns. Emerg Infect Dis 7:6, 2001.

Hampton T: What now, mad cow?: Experts put risk to U.S. public in perspective. JAMA 291:543, 2004.

Harrington MG, Merril CR, Asher DM, Gajdusek DC: Abnormal proteins in the cerebrospinal fluid of patients with Creutzfeldt-Jakob disease. N Engl J Med 315:279, 1986.

Harris DA: Cellular biology of prion diseases. Clin Microbiol Rev 12:429, 1999.

Johnson RT, Gibbs CJ: Creutzfeldt-Jakob disease and related transmissible spongiform encephalopathies: Medical progress. N Engl J Med 339:1994, 1998.

A 44-year-old female presented to the emergency department with complaints of **fever, cough, myalgias,** and **mild shortness of breath** for 2 days. She also had a moderate headache and had experienced several episodes of **diarrhea** in the last 24 hours.

She was a nurse's aide and had been working in a busy medical unit of a hospital. She **had taken care of a patient with severe respiratory illness** 4 days before feeling ill. She did not have underlying disease.

PHYSICAL EXAMINATION

VS: T 38.6°C, P 110/min, R 22/min, BP 104/50 mmHg

PE: The patient appeared anxious and in mild respiratory distress, but she was able to speak in complete sentences. Exam revealed inspiratory crackles bilaterally. Pulse oximetry was 94% on room air.

LABORATORY STUDIES

Blood

Hematocrit: 38%

WBC: 5400/μL

Differential: 68% PMNs, 16% lymphs, 18% monos

TABLE 91-1   Differential Diagnosis and Rationale for Inclusion (consideration)
Adenovirus (acute respiratory distress syndrome [ARDS])
Atypical bacterial pneumonia
Chlamydia pneumoniae
Chlamydia psittaci
Coxiella burnetii (Q fever)
Legionella pneumophila
Mycoplasma pneumoniae
Atypical viral pneumonia
Influenza (types A, B, or C)
Respiratory syncytial virus (RSV)
Severe acute respiratory syndrome-associated coronavirus (SARS-CoV)
Typical bacterial pneumonia (e.g., *Streptococcus pneumoniae*)

Rationale: Community-acquired pneumonia should be considered, with *S. pneumoniae* being the most common, although the lack of a productive cough makes it less likely. The remainder of the pathogens generally cause atypical pneumonia and are often similar to each other in presentation. *L. pneumophila* and influenza often present with severe headache in addition to respiratory symptoms. RSV and adenoviruses do not usually cause severe disease in adults; neither do *Chlamydia* nor *Mycoplasma*. The presence of significant hypoxia is worrisome, and contact with a possible communicable lower respiratory disease while taking care of a symptomatic patient makes the similar diagnosis likely.

Blood gases: pH 7.42, pCO_2 36 mmHg, pO_2 70 mmHg (on room air)

Serum chemistries: AST 84 U/L, LDH 254 U/L, creatinine kinase 230 U/L

Imaging

Chest x-ray showed bilateral lower lobe interstitial infiltrates.

Diagnostic Work-Up

Table 91-1 lists the likely causes of her illness (differential diagnosis). A clinical diagnosis of pneumonia was considered. Investigational approach may include

- **Blood cultures**
- **Sputum Gram stain** and **culture**
- **Urinary antigen testing** for *Legionella*
- **Virus cultures**
- **DFA** of nasopharyngeal secretions
- **Serology**
- **RT-PCR** of hard-to-cultivate pathogens

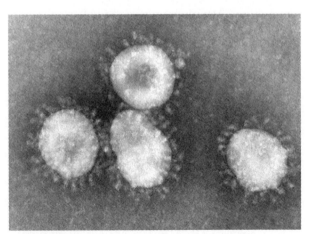

FIGURE 91-1 Electron micrograph of the viral pathogen. *(Courtesy of Centers for Disease Control and Prevention, Atlanta, GA.)*

COURSE

The patient was admitted to the hospital with a diagnosis of pneumonia. She was placed in respiratory droplet isolation, and blood and sputum cultures were obtained, as well as serology for several respiratory viruses. Ribavirin and steroids were started empirically in addition to therapy for community-acquired pneumonia. On the second day of hospitalization, she developed respiratory failure and was intubated.

ETIOLOGY

Presumptive SARS-associated coronavirus (SARS-CoV)

MICROBIOLOGIC PROPERTIES

Severe acute respiratory syndrome (SARS) is a viral respiratory illness caused by a coronavirus, called SARS-associated coronavirus. The coronaviruses (family Coronaviridae, genus Coronavirus) are **large, enveloped, positive-stranded RNA viruses**. Coronavirus particles are irregularly shaped (60 to 220 nm in diameter). They have an **outer envelope** bearing distinctive **club-shaped peplomers, providing a crown-like** (i.e., "corona") **appearance** (Fig. 91-1). This nonsegmented, single-stranded, positive-sense RNA (27 to 31 kb) is the longest of any RNA virus. Three groups of coronaviruses have historically been known to cause upper respiratory and enteric diseases in humans and other animals. Groups 1 and 2 include mammalian viruses, whereas group 3 includes only avian viruses. Most human coronaviruses do not grow in cultured cells. The sequence of the entire genome of SARS-CoV is known, and it demonstrates that although this virus has features typical of a coronavirus, it also has features that distinguish it from other coronaviruses (groups 1 to 3). SARS-CoV is now called a group 4 coronavirus.

The virus can be found in nasopharyngeal aspirate, urine, and stools of SARS patients. Laboratory confirmation of SARS-CoV infection is based on (1) isolation in cell culture of SARS-CoV from a clinical specimen, or (2) detection of serum antibodies to SARS-CoV in a single serum specimen, or (3) detection of SARS-CoV RNA by RT-PCR validated by the CDC, with confirmation in a reference laboratory.

EPIDEMIOLOGY

According to the World Health Organization (WHO), a total of 8098 people worldwide became sick with SARS during the 2003 outbreak. Of these, 774 died. In the United States, only eight people had laboratory evidence of SARS-CoV infection. All of these people had traveled to parts of the world where SARS was in evidence. SARS did not spread more widely in the United States. Probable major modes of transmission of the virus are large droplet aerosolization and contact (direct and fomite). Transmission efficiency may vary among individuals. SARS-CoV is transmitted most readily by respiratory droplets produced when an infected person coughs or sneezes. In addition, it is possible that owing to its very small nuclei (<5 μm) the SARS virus might spread easily through the air (airborne spread) or by other ways that are not known.

PATHOGENESIS

Coronaviruses are transmitted by aerosols of respiratory secretions, by the fecal-oral route, and by mechanical transmission. Incubation period ranges from 2 to 10 days. Whereas in common cold-type respiratory infections, growth of classic coronaviruses appears to be localized to the epithelium of the upper respiratory tract, SARS is a form of viral pneumonia in which infection encompasses the lower respiratory tract. The genome of SARS-CoV has several unique features that are of biologic significance and that may contribute to virulence of this virus. Pathologic cytoarchitectural changes in the infected lungs reveal **diffuse alveolar damage** (DAD), with a mixture of inflammatory infiltrate, as well as **multinucleated giant cells with no conspicuous viral inclusions** (Fig. 91-2). Desquamation of pneumocytes is prominent and consistent. The role of cytokines in the pathogenesis of SARS is still unclear. With the progression of illness, the pathology changes to more intense DAD, with increased fibrosis, squamous metaplasia, and multinucleated giant cells.

TREATMENT

Several therapeutic agents, including steroids, have been tried, but none have been found to be effective. **Supportive care** needs to be optimized. Other potential causes of community-acquired pneumonia of unknown etiology should be treated until SARS-CoV

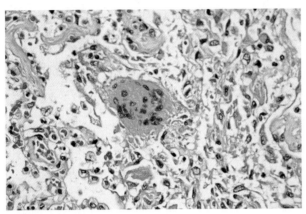

FIGURE 91-2 Pathology of lung tissue in SARS. Note pathologic cytoarchitectural changes with a mixture of inflammatory infiltrate and multinucleated giant cells with no conspicuous viral inclusions. *(Courtesy of Centers for Disease Control and Prevention, Atlanta, GA.)*

positivity is determined and verified by a reference laboratory (e.g., CDC).

> **NOTE** Older age and underlying illness are markers for poor prognosis and severe disease. Estimated mortality overall is 10.5%.

OUTCOME

Sputum culture and Gram stain were both negative. Urine tests were also negative for pneumococcal and *Legionella* antigens. Other serologic tests, including those for chlamydiae, mycoplasmas, rickettsiae, influenza virus, parainfluenza virus, adenovirus, RSV, and Coxsackie virus remained negative. Results of all cultures and serology were negative except for SARS virus. The patient slowly recovered over the course of 3 weeks and was extubated. After rehabilitation, she was discharged home. Several other patients and health-care workers were diagnosed as probable SARS cases; one of these died from respiratory failure.

PREVENTION

Infection control measures are of utmost importance in the wake of an index SARS case. Early recognition and isolation are key. Transmission may occur during the early symptomatic phase, potentially before either fever or respiratory symptoms develop. Heightened suspicion and aggressive triage procedures may aid in effective control and prevention of secondary cases. Infection control measures must include airborne precautions, hand washing, contact precautions (gloves and gown), and environmental cleaning—because there is some evidence that transmission may occur through contaminated stool. Much more remains to be learned regarding transmission and effective preventive measures.

FURTHER READING

Drosten C, Gunther S, Preiser W, et al: Identification of a novel coronavirus in patients with severe acute respiratory syndrome. N Engl J Med 348:1967, 2003.

Ksiazek TG, Erdman D, Goldsmith CS, et al: A novel coronavirus associated with severe acute respiratory syndrome. N Engl J Med 348:1953, 2003.

Lee N, Hui D, Wu A, et al: A major outbreak of severe acute respiratory syndrome in Hong Kong. N Engl J Med 348:1986, 2003.

Peiris JSM, Lai ST, Poon LLM, et al: Coronavirus as a possible cause of severe acute respiratory syndrome. Lancet 361:1319, 2003.

Poutanen SM, Low DE, Henry B, et al: Identification of severe acute respiratory syndrome in Canada. N Engl J Med 348:1995, 2003.

Rubenfeld GD: Is SARS just ARDS? JAMA 290:397, 2003.

Seto WH, Tsang D, Yung RWH, et al: Effectiveness of precautions against droplets and contact in prevention of nosocomial transmission of severe acute respiratory syndrome (SARS). Lancet 361:1519, 2003.

APPENDIX A

Pathogenic Viruses: Concepts

Michael Gale, Jr, PhD

VIRUS STRUCTURE

Viruses are entities whose genomes are elements of nucleic acid that replicate inside living (host) cells. Viruses use the host cellular synthetic machinery to synthesize specialized elements that can transfer the viral genome to other cells. A virus, then, is an **obligate intracellular parasite** having a strict dependence on the host cell for the metabolic components required for virus replication. The nucleic acid core of a virus is surrounded by a complex assembly of proteins (capsid), which in turn can be surrounded by a lipid membrane (envelope). A completely assembled nucleoprotein particle (membrane bound or not) that is competent for infection is also called a **virion**. Viruses are very diverse and range in size, overall morphology, genome composition, and structure.

Virion Size and Design

Clinically relevant viruses range in size from 20 to 30 nm (picornaviruses) to 300 nm (poxviruses). In **naked viruses**, the outermost layer surrounding the viral genome is called the **capsid** (Fig. A-1*A*). **Enveloped viruses** differ from naked viruses in that the capsid is covered by a lipid bilayer envelope (see Fig. A-1*B*). All enveloped viruses acquire their envelopes from the host cell internal membrane or plasma membrane when they exit from the cell.

The **nucleocapsid** is the capsid-covered viral genome. The protein subunits that are assembled to form the capsid are called **capsomeres**. Capsomeres assume a particular architecture, either **icosahedral** or **helical**. An **icosahedron** comprises 20 equilateral triangles with characteristics of rotational symmetry. Each triangular face of the icosahedron is composed of

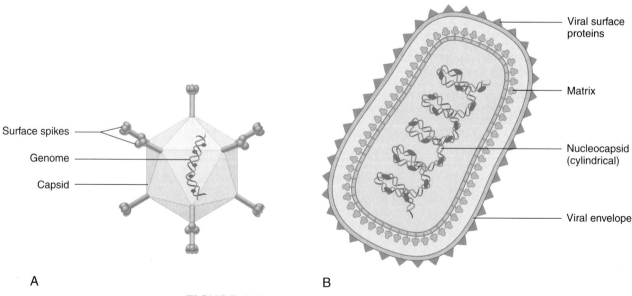

FIGURE A-1 *A*, Naked virus. *B*, Enveloped virus.

repetitive subunits, or "protomers." Such repetitive structures give the icosahedron a high level of structural integrity. Many icosahedral viruses are resistant to extremes in pH, temperature, and solvents, and they can persist outside the body's environment. A capsid that demonstrates a helical architecture is composed of identical protein subunits (protomers) self-assembled into a **helical array** surrounding the nucleic acid. The self-assembly process results in flexible filaments.

Essential enzymes are proteins that are carried in the virion and/or encoded within the viral genome and play an essential role in catalyzing viral replication. The **viral genome** is typically composed of DNA or RNA. The nucleic acid of the viral genome exhibits diverse structural characteristics and may be single stranded or double stranded, linear or circular, and segmented or unsegmented. Single-stranded (ss) RNA may have a positive or negative sense. The naked genome of positive-sense ss-RNA viruses (e.g., flaviviruses, picornaviruses) is **directly infectious** because it functions directly as an mRNA. This type of virus genome can generate viral progeny when introduced into permissive cells. The naked genome of viruses with a DNA, negative-strand RNA, or ds-RNA genome is not directly infectious because viral mRNA must first be produced to support infection and virus replication. **Virus ploidy** refers to the number of copies of the virus genome present in a single virion or virus particle. Most viruses are monoploid and carry a single genome copy. Retroviruses (e.g., HIV) are diploid (two genome copies).

Virus Classification

More than 30,000 different viruses have been described, including 21 virus families that infect humans. The classification (Table A-1) of viruses can be based on

1. Physical/chemical parameters
2. Enveloped vs. naked virion
3. Capsid symmetry (icosahedral vs. helical)
4. Structural features of the viral genome
   a. DNA vs. RNA
   b. Nucleotide sequence homology
5. Clinical syndromes of infection (criteria commonly used in clinical settings)

VIRAL MULTIPLICATION AND LIFE CYCLE

The replicative cycle, or "life cycle," of viruses follows a well-defined series of events (Fig. A-2). These events, and their order within the overall life cycle, are (1) attachment; (2) penetration; (3) uncoating; (4) synthetic events (protein production and genome replication); (5) assembly of virions; and (6) release of virions.

Virion proteins (**viral adhesins**) **direct the attachment** (binding) of the virus to target host cell-surface molecules (receptors). Viral adhesins include hemagglutinin (e.g., influenza virus), gp120 (e.g., HIV-1), and multiple glycoprotein molecules (e.g., HSV). **Cellular receptors** (cell surface moieties) include proteins, glycoproteins, fatty acid side chains, and polyions. Examples of cellular receptors include chemokine receptors and CD4 for HIV-1, complement receptor (CD21) for EBV, acetylcholine receptor for rabies, CD81 for HCV, and sialic acid for many viruses. Cells lacking a specific receptor are not susceptible to infection, even though the cell and tissue type may support viral replication. The cell-expressed viral receptor is, therefore, the primary important determinant of viral tropism. Viral attachment is a dynamic process, sometimes leading to irreversible changes in the structure of the virion and/or the cell surface. Sometimes attachment is not followed by virus penetration into the cell, and the virus can detach itself and readsorb to a different cell. Attachment is dependent on the affinity of the viral adhesin for its cognate cell-surface receptor ligand.

Penetration into the cell involves one of three mechanisms: (1) translocation of the entire virion across the plasma membrane, (2) endocytosis of the virus particle, resulting in accumulation inside cytoplasmic vacuoles, or (3) fusion of the cellular membrane with the virion envelope.

Uncoating is a process through which the viral genome is separated from its envelope or capsid and prepares to express its functions. Many different mechanisms are involved in the uncoating process. Uncoating marks the beginning of the **eclipse phase**, a period during which no infectious virions can be recovered from the cell.

Synthetic events (Fig. A-3) are a series of highly ordered events that involve viral protein synthesis, mRNA production, and genome replication. These events require the function of one or more specific viral enzymes. **Immediate early** events include the synthesis of viral proteins (transactivators) required for transcription and translation in the next event. **Early** events include the synthesis of viral proteins required for nucleic acid replication. **Late** events include the production of the viral genome and proteins necessary for the formation of mature viral particles (viruses, or virions). **Replication** of the viral

TABLE A–1   Major Families of Medically Important Viruses

DNA Viruses[1]

Virus Families (in order of increasing size)	Envelope	Capsid Symmetry	Genome Structure	Important Members
Parvoviridae	No	Icosahedral (smallest DNA virus)	ss, linear DNA	B19 virus; adeno-associated viruses
Papovaviridae	No	Icosahedral	ds, circular DNA (super-coiled)	Papillomaviruses; polyoma virus (monkey); JC, BK viruses
Hepadnaviridae	Yes	Icosahedral	ds, incomplete circular DNA	Hepatitis B virus
Adenoviridae	No	Icosahedral	ds, linear, DNA	Adenoviruses
Herpesviridae	Yes	Icosahedral	ds, linear DNA	Herpes simplex viruses types 1,2; varicella-zoster virus; cytomegalovirus; Epstein-Barr virus; human herpesvirus (HHV)-6, HHV-7 (no disease association), HHV-8 (KSHV)
Poxviridae	Yes	Helical/amorphous (largest DNA virus)	ds, linear DNA	Smallpox; vaccinia virus; molluscum contagiosum

[1]All **DNA viruses** have an icosahedral capsid except poxviruses, which have a complex capsid. Poxviruses, herpesviruses, and hepadnaviruses have a lipid envelope.

ds, double stranded; KSHV, Kaposi sarcoma-related herpes virus; SARS, severe acute respiratory syndrome; ss, single stranded.

TABLE A–1 Major Families of Medically Important Viruses—cont'd

RNA Viruses[2]

Virus Families (in order of increasing size)	Envelope	Capsid Symmetry	Genome Structure	Important Members
Picornaviridae	No	Icosahedral	ss, linear (+)[3] RNA	Polioviruses; rhinoviruses; echoviruses; coxsackieviruses; enteroviruses; hepatitis A virus
Caliciviridae	No	Icosahedral ("Star of David")	ss, linear (+) RNA	Noroviruses (formerly Norwalk viruses)
Flaviviridae	Yes	Icosahedral	ss, linear (+) RNA	Hepatitis C virus; dengue virus; yellow fever virus; West Nile virus; Japanese encephalitis virus
Togaviridae	Yes	Icosahedral	ss, linear (+) RNA	Rubella virus; Eastern equine encephalitis virus; Western equine encephalitis virus
Reoviridae	No ("Rota" or wheel-shaped, virus)	Icosahedral	ds, linear RNA (10 segments)	Rotaviruses; Colorado tick fever; orbiviruses
Orthomyxoviridae	Yes	Helical	ss, linear (−) RNA (8 segments)	Human influenza viruses types A, B, and C; swine and avian influenza viruses
Retroviridae	Yes	Icosahedral	ss, linear (+) RNA (diploid)	Human immunodeficiency virus (HIV)-1 and -2; human T-cell lymphotropic virus (HTLV)-1 and 2
Bunyaviridae	Yes	Helical	ss, linear (−) RNA (3 segments)	California (LaCrosse) encephalitis virus; Hantaan (Sin Nombre) viruses; Rift valley viruses Respiratory syncytial virus
Arenaviridae	Yes	Helical	ss, circular,[4] (−) RNA (2 segments)	Lymphocytic choriomeningitis virus; Lassa virus

[2]All **RNA viruses** are enveloped (exceptions include caliciviruses, picornaviruses, and reoviruses). All members of Caliciviridae, Picornaviridae (e.g., enteroviruses), and Reoviridae families are naked (without envelope), are stable to gastric acid and enzymes, and pass through the gut to cause infection.

[3]Positive (+)-sense strand of ss-RNA genome functions as mRNA.

[4]The circular genome serves as template for both mRNA and genome replication.

Continued

TABLE A–1 Major Families of Medically Important Viruses—cont'd

RNA Viruses

Virus Families (in order of increasing size)	Envelope	Capsid Symmetry	Genome Structure	Important Members
Coronaviridae	Yes	Helical (crown-shaped virus)	ss, linear (+) RNA	Coronavirus (CoV) types 1-3; CoV type 4 (SARS) virus
Rhabdoviridae	Yes	Helical (bullet-shaped virus)	ss, linear (−) RNA	Rabies virus
Filoviridae	Yes	Complex/amorphous	ss, linear (−) RNA	Marburg and Ebola viruses
Paramyxoviridae	Yes	Helical (largest RNA viruses)	ss, linear (−) RNA	Mumps Measles Parainfluenza viruses Respiratory syncytial virus

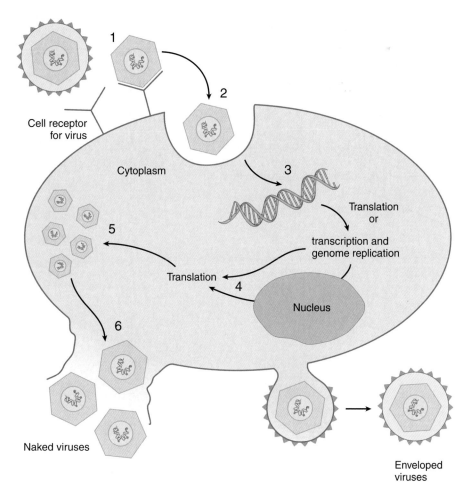

FIGURE A-2 Viral life cycle. The life cycle is carried out by a virus through a series of events: (1) attachment; (2) penetration; (3) uncoating; (4) synthetic events (translocation and transcription and genome replication, translation); (5) assembly of virions; and (6) release of virions. Antiviral drugs can potentially target each of these events.

genome is dependent on specific enzymatic activities (Table A-2) encoded by the virus. Such enzymatic activities also represent attractive therapeutic targets for antiviral compounds for treating viral infection.

DNA virus gene expression is often regulated through interaction with cellular (host) transcription factors. The specific replicative strategy of a given virus is dictated by the nature of the viral genome. All DNA viruses except poxviruses replicate in the nucleus of the host cell. DNA viruses except poxviruses use host cell RNA Pol II to transcribe their genes to mRNA. A hallmark of reproduction of DNA viruses is that viral gene expression and viral DNA synthesis typically occur in a strictly defined, orderly pattern of immediate-early, early, and late gene expression (see Fig. A-3). **Immediate-early gene expression** is required for producing viral non-structural proteins and transcription factors that have regulatory functions. These proteins are produced to

induce and regulate the expression of other viral proteins and specific host functions.

Early gene expression involves the expression of viral proteins (e.g., viral DNA polymerase, helicase) that participate in viral genome replication (see Table A-2 for a list of viral polymerases required for genome replication). Viral **DNA replication** is performed by the host cell DNA-dependent DNA polymerase (e.g., papilloma, parvoviruses). This strategy requires that the host cell is in S phase for efficient viral replication to occur. In **viral polymerase-mediated replication**, the viruses (e.g., adenovirus, herpes, pox viruses) encode their own DNA-dependent DNA polymerase. **Late-gene expression** allows production of viral structural proteins (protein components that build the virion).

The **replicative strategies for RNA viruses** are directed by the nature of the viral genome. All RNA viruses except influenza and retroviruses replicate in

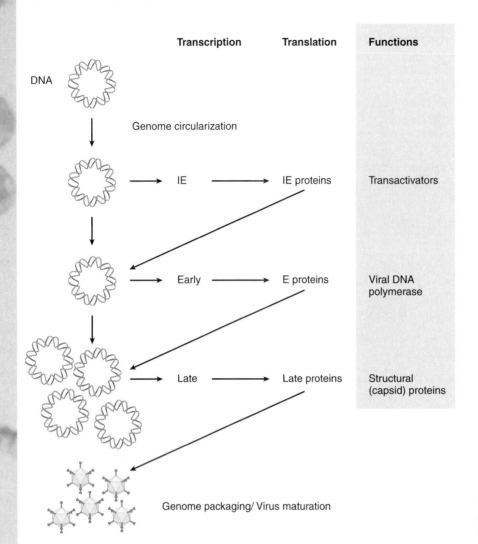

FIGURE A-3 Cascade of expression of viral (CMV) genes during the synthetic events.

the cytoplasm. In viruses with ss-RNA genomes, the positive-sense genome directly codes for proteins; thus translation is the first step, followed by viral RNA replication. Viruses containing a negative-sense genome or a ds-RNA genome must first use the genomic RNA as a template for transcription of mRNA. This requires the virus to carry an RNA-dependent RNA polymerase in the virion. Viral mRNA production is followed by protein synthesis and viral genome replication. Retroviruses must first reverse-transcribe the viral RNA genome into a DNA template, which is then integrated into the host cell chromosome. The DNA copy of the viral genome then replicates as the host cell DNA replicates. Viral mRNA is directly transcribed from the integrated viral DNA genome template.

For the **assembly of infectious virions**, each viral subunit (nucleic acid, structural proteins, and enzymes) has to fit into a compact arrangement. The program for this assembly is carried within the physical properties of each subunit. Assembly of enveloped viruses requires association of the nucleocapsid with the infected host cell membranes, which have been modified by viral proteins.

Release of virions is the final stage of the infection cycle. Enveloped viruses are typically released gradually by budding from the plasma membrane or through the process of exocytosis. Naked viruses

TABLE A–2   Viral Polymerase Activities Required for Genome Replication

Enzyme	Virus	Product or Function
dsRNA polymerase	ds-RNA viruses	mRNA production and genome replication
ssRNA polymerase	Negative strand ss-RNA viruses	mRNA production and genome replication
RNA-dependent DNA-polymerase (reverse transcriptases)	Retroviruses	Reverse-transcription of the viral RNA genome to the DNA genome template
DNA-dependent DNA polymerase	DNA viruses	Viral genome replication

ds, double stranded; ss, single stranded.

typically accumulate within the cytoplasm of the host cell and are released during cell lysis.

VIRAL GENETICS

Viruses are constantly changing. A **mutation** is an error that is incorporated into the viral genome. Mutations can be deleterious, neutral, or favorable to the virus. Only those mutations not interfering with essential functions persist in a virus population.

DNA viruses have a low rate of mutation; they possess proofreading polymerases and they exhibit a low error rate of nucleotide incorporation during replication (about 10^{-11} errors per nucleotide per round of replication). These viruses derive about one mutant per several hundred to many thousand genome copies. The **viral polymerase in RNA viruses** lacks proofreading ability, and rates of 10^{-3} to 10^{-4} errors per nucleotide per round of replication can occur. Mutants may arise as often as once per genome copy. High mutation rates lead to the generation of viral "quasispecies," making the development of antiviral therapies and vaccines problematic. Mutations can produce viruses with altered antigenic determinants, allowing the virus to escape host immune defenses or to replicate in a previously resistant host. This is referred to as "**antigenic drift,**" or "quasispeciation."

Recombination is a crossing over mechanism by which two sequence-related viral nucleic acids recombine in regions of homology, generating a distinct genetic motif that encodes a gene with potentially altered or novel function(s). Major genetic changes occur through **recombination**.

Coinfecting viruses may exchange genetic information, creating a novel virus by **independent reassortment**. This process, in which genes residing in different pieces of nucleic acid are randomly sorted, occurs among viruses with segmented genomes and results in the generation of a virus with new antigenic determinants and new host ranges (classically called "**antigenic shift**"). This can cause major epidemics (e.g., the influenza virus epidemic of 1918).

Complementation is a process by which a defective or unfit virus is complemented by novel genetic determinants directly through actions presented by a second virus during a dual infection.

Phenotypic mixing is typically related to a superinfection with one or more viruses of dominant phenotype over distinct traits. The dominant virus determines infectivity but the underlying genome of the weaker virus replicates to produce progeny viruses.

Mechanisms of Viral Pathogenesis

The **pathogenesis** of viral infection involves (1) **implantation** of the virus at a portal of entry; (2) local **replication** and local **spread**; (3) systemic **dissemination** of some viruses through the circulation; (4) arrival into target organs and **secondary multiplication** there; and (5) **negative effects** on end organs. Viruses most commonly **implant** within cells via respiratory, gastrointestinal, skin-penetrating, and genital routes (Fig. A-4). This process requires that the tissues at the portals of entry express the viral receptor and support at least limited viral **replication**. Most viruses **spread** extracellularly, although some also spread intracellularly, via cell fusion. The viruses establish local infection and cause local disease, which is followed by localized shedding of virus. Subsequent viremia through the circulation allows systemic

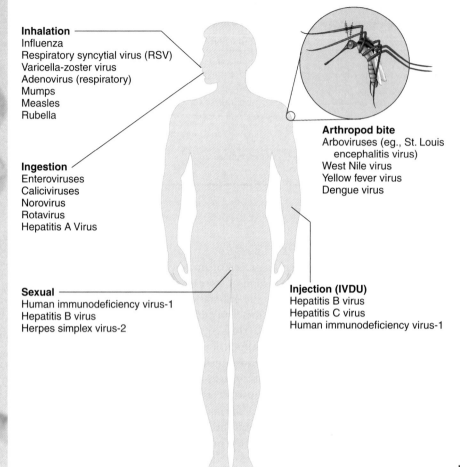

Inhalation
Influenza
Respiratory syncytial virus (RSV)
Varicella-zoster virus
Adenovirus (respiratory)
Mumps
Measles
Rubella

Ingestion
Enteroviruses
Caliciviruses
Norovirus
Rotavirus
Hepatitis A Virus

Sexual
Human immunodeficiency virus-1
Hepatitis B virus
Herpes simplex virus-2

Arthropod bite
Arboviruses (eg., St. Louis
  encephalitis virus)
West Nile virus
Yellow fever virus
Dengue virus

Injection (IVDU)
Hepatitis B virus
Hepatitis C virus
Human immunodeficiency virus-1

FIGURE A-4 Sites of entry of various pathogenic viruses.

dissemination of the virus. Virus enters **target organs** from capillaries by **multiplying** in endothelial cells or fixed macrophages, diffusing through gaps or being carried by migrating leukocytes. Neurotropic viruses (e.g., rabies virus, herpesvirus, or poliovirus) disseminate along nerve axons. Release (shedding) of virus from the infected organism commonly occurs through body fluids/secretions, causing various effects on end organs including respiratory, gastrointestinal, urogenital, and blood routes. Viruses are very diverse and can be shed at virtually any site.

End-organ (tissue) **tropism** (affinity) is defined by the cell or tissue type that supports the replication of a given virus. Tropism is dependent on (1) cell receptors for the virus, (2) proper expression of cellular transcription factors and/or replication cofactors, and (3) ability of cells to support viral protein synthesis.

Balance between virus multiplication and host defenses will determine organ pathology or dysfunction. The **incubation period** (time between exposure to the virus and onset of disease) extends from implantation until the virus replicates sufficiently in the target organ to cause symptoms. The length of incubation varies from very short (polio virus) to long (HIV) to very long (human papillomavirus; hepatitis B and C). **Effects on end organs** may occur by a variety of mechanisms and include

- **Toxic effects** of viral structures (e.g., adenovirus—fibers)
- **Host response** to infected cells expressing viral proteins (e.g., liver injury during hepatitis C virus infection).
- **Lytic destruction** of host cells (acute viral diseases)

Host cells are completely permissive for viral replication; infected cells round up and detach from each other (**cytopathic effect** [CPE]) and ultimately die owing to lysis. Viruses can disrupt normal cellular processes by reprogramming host cell metabolic and/or immune defense functions to favor virus infection. This reprogramming can result in tissue destruction and disease. **Cytopathogenesis** results when host cell processes are disrupted. For example, expression of viral proteins within the infected cell can lead to reprogramming of host cell protein synthesis to favor the production of viral proteins. This can result in a "host shutoff" of cellular protein synthesis and can result in cell death. In addition, viral protein and/or viral genome accumulation can cause the formation of large **nuclear** or **cytoplasmic inclusion bodies**. This generally defines the site of genome synthesis and/or viral genome packaging into new virions. During this process the expression of viral proteins on the cell surface can lead to fusion of the infected cell with uninfected cells, resulting in the formation of large **multinucleated giant cells**, which lose all cellular functions.

Certain mutated genes, called oncogenes, produce oncoproteins, which can alter the growth regulation of infected cells to induce their uncontrolled proliferation and malignant progression, leading to tumor formation. **Oncogenesis** is a multistep process that results in the growth of successive populations of cells in which mutations have accumulated. These mutations affect various steps in the pathways that regulate (1) cell communication; (2) cell growth; and (3) cell division. As a result, there are (a) uncontrolled growth; (b) increasing cellular disorganization; and, ultimately, (c) cancer. In most cases these mutations are inherited or arise as a consequence of exposure to environmental carcinogens. However, in a definite subset of cancers, a relationship with particular viruses has been established. Transducing retroviruses (e.g., HTLV-1 and -2) carry a cellular oncogene within the retrovirus genome. Nontransducing retroviruses transform via integration of the viral genome in the vicinity of the cellular oncogene. Cis-acting retroviral genes close to the insertion site become deregulated. Oncoproteins encoded by **oncogenic DNA viruses** (HBV, HCV, HPV-16 and 18, EBV, and HHV-8) are required for productive viral replication. Transformation by DNA viruses is extremely inefficient (<1 in 10,000 cells infected will become transformed). The oncogenic events mediated by DNA tumor viruses reflect their ability to stimulate a quiescent cell to enter the cell cycle.

Pathogenesis and Immunity

The natural barriers of the body include (1) physical barriers (skin and mucous membranes); (2) secretory IgA that may cross-react with virus; (3) natural antibodies of the IgM class that react with pathogens.

The upper respiratory tract has a lower temperature than the rest of the body, which is not favorable for the growth of many microorganisms. Nonspecific inflammatory responses (**innate immune defenses**) are induced that can many times limit the ability of a virus to infect. This can result from the many cytokines released almost immediately after infection, including interleukin (IL)-1, IL-6, tumor necrosis factor (TNF), and interferons (IFN). **Proinflammatory cytokines** (e.g., IL-1, IL-6) induce the expression of genes whose products direct intracellular antiviral actions. These and other cytokines also serve to activate cells, including macrophages and natural killer (NK) cells, which can ingest or kill virus-infected cells. This allows a response to be made against low doses of pathogens and prevents colonization or infection. Recognition of invading viruses is partly mediated by "pattern recognition receptors" including Toll-like receptors. **Toll-like receptor** signaling leads to activation of transcription factors, resulting in transcription of proinflammatory cytokines as well as expression of other membrane molecules to assist in immune response.

Interferons (IFNs)-α, -β, -γ are proteins that cause uninfected cells to progress to an antiviral state. IFNs induce the production of proteins that inhibit viral protein synthesis by degrading viral mRNAs but not host mRNAs. IFNs bind IFN receptors on cells, causing activation of interferon-responsive genes and resulting in (1) upregulation of MHC molecules; (2) activation of NK cells; and (3) production of translational inhibitors. Other antiviral proteins are also made.

NK cells are nonspecific host cells that are important in the early phase of infection and are part of the innate response. NK cells show activity against virus-infected cells that is not MHC restricted or virus specific. They kill target cells using cytolytic molecules in their granules, as well as secrete interferons. NK cell killing by an antibody-dependent mechanism is referred to as antibody-dependent cell-mediated cytotoxicity (ADCC).

Cell-mediated immune mechanisms are critical to **recovery** from many viral infections, especially CMV, EBV, and chickenpox—as well as herpesviruses and measles. The major T cell in this process is the

cytotoxic T lymphocyte (CTL). This cell recognizes viral peptides presented by self-HLA class I molecules. Because class I molecules are expressed on all cells, the CTL can recognize and destroy infected cells before viral progeny are released. CTLs kill by at least two major mechanisms

- **Release of granules** containing perforin and granzyme B
- **Fas signaling**

Perforin is similar to complement component C9 and polymerizes in the presence of Ca^{++} to form pores that allow granzyme B to enter target cells. This molecule activates the **apoptotic pathway** in the target cell, leading to lysis. The CTL, which expresses FasL, binds Fas on target cells, which activates the apoptotic pathway. In addition, CTLs can also secrete both IFN and TNF, which bind receptors on target cells, resulting in antiviral activity.

Phagocytosis of viruses by **macrophages** can lead to several outcomes, including disruption of virus (influenza, herpes), harboring of virus (measles), or propagation of virus (HIV). IFN action on macrophages increases the ability of these cells to mediate an antiviral effect (**activation of macrophages**). Macrophages can also show cytotoxic activity against neighboring virally infected cells.

Mechanisms of evasion of the innate immune response by viruses include

- Antigen variation (influenza A or HIV)
- Inhibition of antigen processing (CMV shunts class I molecules back to the cytosol for degradation)
- Production of cytokine-receptor homologs (pox virus-infected cells encode and secrete receptors for TNF, IL-1, and IFN)
- Production of immunosuppressive cytokines (EBV secretes an IL-10 homolog)
- Escape from IFN by
  - Prevention of its secretion (e.g., EBV secretes IL-10, which prevents CD4+ cells from secreting IFN-γ)
  - Encoding of secreted receptors (e.g., pox viruses)
  - Inhibition of cellular IFN-stimulated gene expression and/or function

Following viral infections, another sequence of events (**humoral immunity**) is observed in response to antigenic stimulus. IgM antibody is detected early, followed by IgG. Antibody can act to **neutralize virus**, **promote phagocytosis**, cause lysis of virus in the presence of complement, and allow NK cells to kill infected targets via ADCC or lyse infected cells in the presence of complement. **Serum antibodies** (IgM and IgG) are important in protection against polio, type A hepatitis, and measles. **Secretory antibodies** (sIgA) are of great importance in infections of mucous surfaces in viral infections. Examples include RSV, influenza virus, and enteroviruses. Persistence of viral antibodies after natural infection or after immunization with live attenuated vaccines is extremely long lasting. In most viral infections, reinfection is common, but after initial viral replication, the infection is often aborted, resulting in a booster response to the immunologic apparatus.

Viral persistence may occur in many viral diseases (e.g., HSV-1, CMV) and the constant antigenic challenge causes the serum antibody level to remain elevated. Viral neutralization can occur at (1) attachment of the virus to the cell surface; (2) penetration of the virus into the cell; or (3) uncoating of the virus within the cell. For those infections in which viremia is an essential link in the pathogenesis of the disease, such as measles and viral meningitis, the degree of protection following the infection is directly related to the level of neutralizing antibodies sustained in the serum.

VACCINES

Live, attenuated vaccines induce humoral and cell-mediated immunity but can revert to virulence in a very few individuals. Successful live, attenuated vaccines have been developed against adenovirus, measles, mumps, rubella, polio (Sabin), yellow fever virus, and varicella-zoster virus (VZV). Live, attenuated vaccines are under development for cytomegalovirus and parainfluenza virus.

Inactivated (killed) virus vaccines induce only humoral immunity and are stable. Current successful killed virus vaccines include polio (Salk), rabies, influenza, hepatitis B, and hepatitis A virus. Inactivated vaccines under development are cytomegalovirus, RSV, parainfluenza virus, and HIV.

Antibodies are sometimes used to provide **passive immunity** against viruses such as VZV, RSV (in development), HAV, and HBV.

Pathogenic Bacteria: Concepts

Kevin McIver, PhD

BACTERIAL STRUCTURE

Bacteria are **prokaryotes**—simple, single-cell organisms. They lack a distinct nucleus (i.e., no nucleolus, no nuclear membrane), and they often possess **a single, circular chromosome**. Transcription (mRNA synthesis) and translation (protein synthesis) occur simultaneously in the cytoplasm. In comparison, higher organisms (eukaryotes) have a true nucleus with a nuclear membrane containing multiple chromosomes; transcription of mRNA occurs in the nucleus and must be translocated to the cytoplasm to be translated by ribosomes.

Bacterial Morphology

Bacteria are microscopic in size, ranging from about 0.2 to 2.0 μm in diameter. Bacteria have distinct shapes (Table B-1; Fig. B-1*A-F*). For cocci (spheres), cell division can occur in various planes, resulting in several arrangements: **diplococci**, **chains**, and **clusters**. The bacilli (rods) are most frequently found among the medically important bacteria, and they vary in length and detailed shapes: **short rods**, **coccobacilli**, **fusiform**, and **curved** (comma-shaped, spiral) **rods**.

Gram staining is used to differentiate bacteria based on their cell-wall characteristics. The process is based on the ability of acetone or alcohol to extract an iodine-crystal violet (a basic purple dye) complex through the cell wall (see Appendix D for detailed methodology). The washed bacteria are examined after counterstaining with safranin (red dye), and the bacteria are either Gram positive or Gram negative. **Gram-positive** bacteria have a single membrane surrounded by a thick cell wall that retains crystal violet (and thus stain **purple**). **Gram-negative** bacteria have an inner and an outer membrane with a thin cell envelope between the membranes. Gram-negative bacteria cannot retain crystal violet, but they can retain safranin counterstain (and thus stain **red**). In older cultures of Gram-positive bacteria, some of the bacteria may appear Gram-negative or Gram-variable (and appear to be a mixture of purple and pink) because dye-retaining properties are lost over time.

Bacterial Ultrastructures

Bacteria do not have a clearly defined nucleus; rather most have a **single chromosome** attached to an invagination of the cytoplasmic membrane, which plays a key role in segregation of the two daughter chromosomes following chromosomal replication. Some bacteria have circular DNA molecules, called **plasmids**, which replicate autonomously from the chromosome and usually **contain antibiotic-resistance and/or virulence genes**. Plasmids are often exchanged between different bacteria via conjugation, transformation, or even transduction. **Ribosomes** are found in the cytoplasm or associated with the cell membrane. **Storage granules** contain reserve polymers (e.g., glycogen) synthesized in the presence of an excess carbon source for utilization later but do not increase the osmotic pressure of the cell. Polymerized metaphosphate $[(PO_3)]_n$, also called "**metachromatic granules**" (e.g., *Corynebacterium diphtheriae*), can be stained and examined by using methylene blue.

The bacterial **cytoplasmic membrane** is a permeability barrier and the site of active transport of small organic compounds in all bacteria (this is comparable to the eukaryotic plasma membrane). The cytoplasmic membrane is a typical lipid bilayer composed of phospholipids and proteins. **Bacterial membranes contain no cholesterol or sterols**, with the exception of wall-less bacteria, *Mycoplasma*. Cytochromes and enzymes at the cytoplasmic membrane are involved in electron transport and oxidative phosphorylation (analogous to mitochondria).

TABLE B–1 Bacterial Morphologies

Shapes/Arrangement	Mechanism of Division or Detailed Shapes	Examples
Cocci (Spheres)		
Diplococci	Pairs dividing in one plane (Fig. B-1*A*)	*Streptococcus pneumoniae, Neisseria gonorrhoeae*
Chains	Divide in one plane, no separation (Fig. B-1*B*)	*Streptococcus pyogenes*
Clusters	Divide in three planes (Fig. B-1*C*)	*Staphylococcus aureus*
Bacilli (Rods)		
Short rods	Rounded ends (Fig. B-1*D*)	*Pseudomonas aeruginosa*
Coccobacilli	Short, rounded	*Francisella tularensis*
Fusiform	Tapered ends	*Fusobacterium necrophorum*
Curved Rods		
Vibrios	Hook or comma shaped (Fig. B-1*E*)	*Vibrio cholerae*
Filamentous forms	Beaded	*Nocardia asteroides, Mycobacterium tuberculosis*

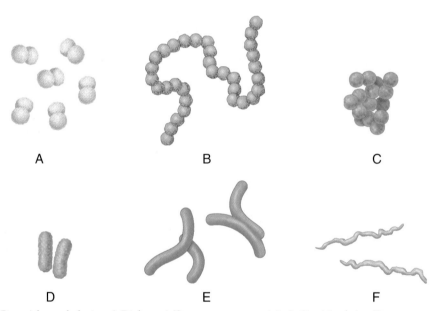

A   B   C

D   E   F

FIGURE B-1 Bacterial morphologies. *A,* Diplococci (*Streptococcus pneumoniae*). *B,* Cocci in chains (*Streptococcus pyogenes*). *C,* Cocci in clusters (*Staphylococcus aureus*). *D,* Rods (*Pseudomonas aeruginosa*) *E,* Curved rods-vibrios (*Vibrio cholerae*). *F,* Spiral-shaped bacilli (*Treponema pallidum*).

Gram-positive and Gram-negative bacteria differ in their structures, their chemical compositions, and the thickness of their cell walls (Fig. B-2). The **cell wall** (cell envelope) is composed of **peptidoglycan** in all bacteria; it is responsible for **structural rigidity** and **resistance to the internal osmotic pressure** of the bacterial cell.

Gram-positive bacteria have no outer membrane, but they do have a **thick cell wall**, with as many as **40 sheets of peptidoglycan**, composing up to 50% of the cell-wall material. Some Gram-positive bacteria carry an outer structure, called **teichoic acids**, which are repeating units of ribitol phosphate or glycerol phosphate joined through phosphodiester linkages to *N*-acetylglucosamine or through covalent linkages to a glycolipid (lipoteichoic acids) in the inner membrane. Teichoic acids in these Gram-positive bacteria are virulence factors, which are capable of eliciting

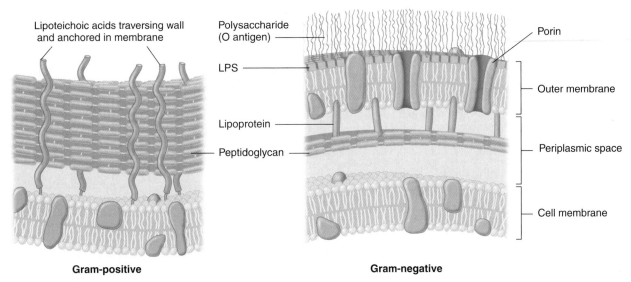

Lipoteichoic acids traversing wall and anchored in membrane

Polysaccharide (O antigen)

LPS

Lipoprotein

Peptidoglycan

Porin

Outer membrane

Periplasmic space

Cell membrane

Gram-positive

Gram-negative

FIGURE B-2 Structures of Gram-positive cell wall and Gram-negative bacterial envelope.

immune responses in humans, and may act as receptors for bacterial viruses.

Gram-negative bacteria have a thin peptidoglycan layer compared to Gram-positive organisms. There appear to be **only one to two peptidoglycan sheets**, composing 5% to 10% of the wall material. Gram-negative bacteria have an **outer membrane** outside the peptidoglycan. The outer membrane is a bilayer composed of phospholipid and protein and is the site of lipopolysaccharide antigens and endotoxin (described later). The outer membrane is impermeable to many small and large organic molecules; however, impermeability of the outer membrane to small molecules is overcome by active transport and porins. The space between the outer membrane and cytoplasmic membrane in Gram-negative bacteria is called the **periplasmic space**, where **a thin layer of peptidoglycan floats in a gel-like material**. A number of hydrolytic and cell-wall biosynthetic enzymes are also found in the periplasmic space.

The **peptidoglycan backbone** of the bacterial cell wall is composed of alternating *N*-acetylglucosamine **(NAG)** and *N*-acetylmuramic acid **(NAM)**, a set of identical **tetrapeptide side chains** (attached to NAM), and a set of **peptide cross-links** (Fig. B-3). The backbone is the same in all bacterial species; however, the tetrapeptide side chains and the peptide cross-links may vary from species to species. In all species, the tetrapeptide side chains have L-alanine at position 1 (attached to NAM); D-glutamate at position 2; and D-alanine at position 4 (see Fig. B-3). Position 3 is the most variable amino acid; most

Gram-negative bacteria have diaminopimelic acid (DAP) at this position, to which is linked the lipoprotein cell-wall component (i.e., outer membrane).

Gram-positive bacteria may have DAP, L-lysine, or any of several other L-amino acids at position 3. In many Gram-negative cell walls, the **cross-links** (see Fig. B-3) consist of a direct peptide linkage between the DAP amino group of one side chain and the carboxyl group of the terminal D-alanine of a second side chain. **Cross-linking** between sheets of peptidoglycan allows sac-like scaffolding around the cell and can vary among species. The biosynthesis of peptidoglycan begins with stepwise synthesis in the cytoplasm of UDP-*N*-acetylmuramic acid-pentapeptide. The UDP-*N*-acetylmuramic acid-pentapeptide is attached to **bactroprenol** (a lipid of the cell membrane) and receives a molecule of *N*-acetylglucosamine from UDP. The bridge peptide (e.g., pentaglycine in *Staphylococcus aureus*) is next formed in a series of reactions. The completed disaccharide-peptide bridge is transferred to the growing end of a glycopeptide polymer in the cell wall (in the Gram-positive bacteria) and in the periplasm (in the Gram-negative bacteria). Final cross-linking is accomplished by a **transpeptidation reaction** in which the free amino group of a pentaglycine residue displaces the terminal D-alanine residue of a neighboring pentapeptide. Transpeptidation is catalyzed by one of a set of enzymes called **penicillin-binding proteins (PBPs)**. During growth of bacteria, bacterial enzymes insert new murein chains. Therefore, **cell-wall biosynthesis**, especially **cross-linking**, provides **targets for antimicrobial**

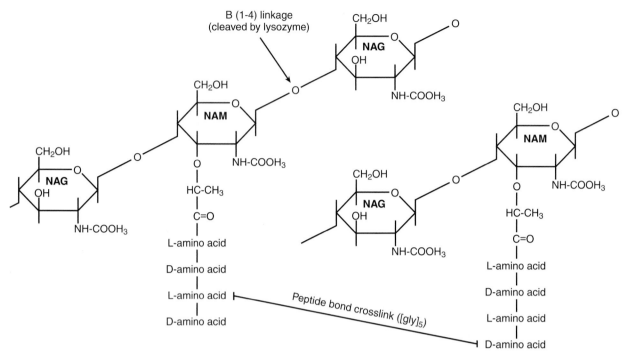

FIGURE B-3 Peptidoglycan structure in bacterial cell wall. Note the structure is composed of alternating *N*-acetylglucosamine (NAG) and *N*-acetylmuramic acid (NAM) sugars, tetrapeptide side chains (attached to NAM), and a peptide cross-link.

agents such as **penicillins, cephalosporins,** and **glycopeptide antibiotics.**

In Gram-negative bacteria, **lipopolysaccharide (LPS),** anchored to the outer portion of the outer membrane, contains three components (Fig. B-4): (1) a disaccharide-diphosphate core; (2) polysaccharide O-antigens; and (3) lipid A (branched chain fatty acids). The somatic O-polysaccharides are **major antigenic determinants** in Gram-negative bacteria.

The **bacterial capsule** is found outside the cell wall and envelope of some Gram-positive and Gram-negative bacteria, respectively. The capsule is very important in virulence because it usually **protects pathogenic bacteria from phagocytosis.** Most capsules are composed of polysaccharides. However, *Bacillus anthracis*, which causes anthrax, contains a capsule made of polyglutamic acid. Capsules are also important **antigenic determinants (K antigens)** and can elicit vigorous immune responses in the host. The capsular polysaccharides for different bacteria have different sugars and linkages, which can be used for the classification (serotyping) of encapsulated bacteria.

Some bacteria (e.g., *Staphylococcus epidermidis*, *Streptococcus mutans*, *Pseudomonas aeruginosa*) produce a **slime** layer (rather than a capsule), also composed of polysaccharide. This outermost layer, also known as **glyco-**calyx, mediates adherence to mucosal surfaces, central venous lines, and foreign devices. The glycocalyx is refractory to phagocytosis and penetration by antibiotics.

Pathogenic bacteria exhibit a variety of **appendages** on the cell surface. **Motile bacteria** have projections called **flagella** that **function in locomotion** (Fig. B-5*A*). Flagella are about 3 to 12 μm long (about 125 to 250 angstroms thick) and require special stains to be seen. They can be localized at one end or at both ends (polar flagella), or they can be placed all around the bacterium (peritrichous). They are composed of a large number of identical subunits of a single protein called flagellin. Flagella carry antigenic determinants known as the **H-antigens,** which can also be used to **serotype bacteria.**

Pili (also called **fimbriae**) are short, hairlike projections. They are protein in nature, but much thinner and shorter than flagella (see Fig. B-5*B*). In many bacteria, they are involved in **bacterial adhesion to host tissue** or to other bacteria (via sex pili). The sex pilus, also known as the fertility (F)-pilus, is involved in conjugation, a mechanism of gene transfer between donor and recipient bacteria (described subsequently).

Spores are found in two genera of bacteria of medical significance, ***Bacillus*** and ***Clostridium***, both

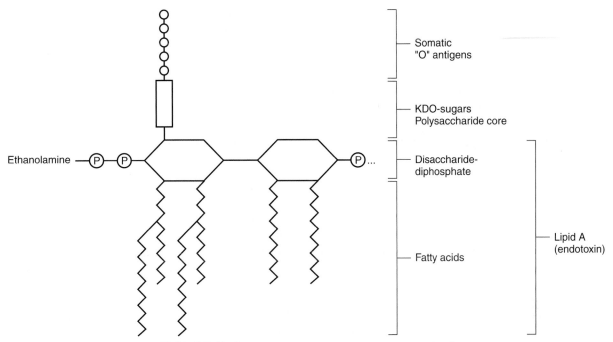

FIGURE B-4 Lipopolysaccharide of Gram-negative cell wall.

FIGURE B-5 *A,* Flagella, polar. *B,* Pili of *Neisseria.*

of which are Gram-positive rod-shaped organisms. These spores are dormant stages of the bacteria and are **highly resistant to desiccation**, **heat**, or **chemical (bactericidal) agents**. When conditions are favorable, spores germinate and produce vegetative cells that, in most cases, release toxins.

BACTERIAL GROWTH

Bacterial growth factors include oxygen, temperature, pH, and osmotic pressure. However, many species of bacteria are deficient in one or more biosynthetic pathways and require exogenous sources of nutritional factors, amino acids, B vitamins, and/or nucleic acid constituents, which they cannot synthesize themselves. **Autotrophic** bacteria derive energy from the oxidation of inorganic substrates, using this energy to fix and convert CO_2 into bacterial mass. **Heterotrophic** bacteria require one or more organic carbon components as energy sources and for biosynthetic precursors. Bacteria of medical significance are typically heterotrophic.

Bacteria require a **source of energy** for the many biosynthetic pathways involved in growth. During bacterial growth and metabolism in the presence of molecular oxygen two toxic substances are produced, **hydrogen peroxide** (H_2O_2) and the considerably more toxic **superoxide ion** (O_2^-). These may be detoxified in the bacteria by **catalase** and **superoxide dismutase** (SOD), respectively. Bacteria exhibit an almost continuous spectrum of oxygen requirements and tolerances, usually classified into four main groups: (1) obligate aerobes; (2) obligate (and aerotolerant) anaerobes; (3) facultative anaerobes; and (4) microaerophilic bacteria.

Obligate (or strict) **aerobes** (e.g., *Bacillus, Corynebacterium, Mycobacterium, Nocardia, Pseudomonas*) require oxygen for growth; they almost invariably possess catalase and SOD and carry out the full spectrum of catabolic (oxidative) processes and ATP generation by using an electron transport chain and oxidative phosphorylation.

Obligate anaerobes (e.g., *Fusobacterium, Prevotella, Propionibacterium*) grow in the absence of oxygen; they usually possess neither catalase nor SOD, and are fermentative only. These anaerobes require low reducing conditions (low redox potential) and protection from oxygen. **Aerotolerant anaerobes** (e.g., *Clostridium, Bacteroides, Peptostreptococcus*) can survive in the presence of oxygen but grow best in the absence of oxygen and are more virulent than their counterpart strict anaerobes.

Facultative anaerobes (e.g., *Escherichia coli, Salmonella, Shigella, Staphylococcus, Vibrio*) can grow either in the presence or in the absence of oxygen; they usually possess both catalase and SOD, and can switch between oxidative and fermentative metabolic pathways. The clinical isolates from patients with bacterial infections are frequently facultative anaerobes.

Microaerophilics (e.g., *Campylobacter, Helicobacter*) grow best at reduced oxygen tension but can tolerate normal atmospheric levels; they often possess SOD but not catalase, and they are usually fermentative.

In the **presence of oxygen,** strict aerobes and facultative anaerobes will use **oxygen as a terminal electron acceptor**; the products of this respiratory metabolism are primarily carbon dioxide and water plus relatively high amounts of energy per molecule metabolized. CO_2 is an atmospheric component required by most bacteria. Although endogenously produced CO_2 or the normal atmospheric level is usually sufficient, there are a few species that do require elevated levels of CO_2 in the gaseous atmosphere for growth. In the **absence of oxygen,** strict anaerobes and facultative anaerobes and microaerophilics (while growing in the presence of CO_2) use **metabolic intermediates as electron acceptors**; the products of this anaerobic metabolism are large amounts of organic compounds with or without concomitant gas production and relatively low levels of energy per molecule of substrate used. The range of substrates that a given species of bacteria can use and the end products of substrate use are of considerable importance in identifying bacteria in the clinical laboratory.

Bacterial species vary considerably in their ability to grow at different **temperatures**, with each species characterized by a temperature range within which growth can occur, and an optimum at which most rapid growth occurs. **Most pathogenic bacteria** (e.g., *E. coli, Staphylococcus, Streptococcus*) found in clinical specimens grow well at 37°C (body temperature); however, **selected bacteria** (e.g., *Campylobacter, P. aeruginosa*) grow at 42°C. Most bacteria grow best near physiologic ranges of **pH** (6.0 to 7.4). Bacteriologic media are usually adjusted or buffered to provide an initial pH in this range. Medically important

bacteria usually grow best at the **osmotic pressure** equivalent to physiologic saline (0.9%). However, the ability to grow at elevated pressures (such as in 6% NaCl or higher) is a useful property for isolation and differentiation applications.

Growth Curve

Bacteria **multiply by binary fission**, and each parental cell produces two daughter cells on division. When small numbers of bacteria are inoculated into fresh, sterile medium, which is then incubated under appropriate conditions, the increase in bacterial numbers (determined by viable counts of sequential samples) follows a sequence of four distinct phases of growth (Fig. B-6). Practically all species of bacteria exhibit these same growth phases, differing in the duration of the phases, maximum growth rates achieved, and maximum numbers attained. During the **lag phase,** the inoculum adapts to the new medium and incubation conditions. There is little or no increase in numbers, but the cells become metabolically more active, and cell mass increases. When cells start to divide at the end of the lag phase, they rapidly increase in numbers in the **logarithmic phase,** until a maximum logarithmic rate is achieved. Cell numbers double at a constant maximum rate (called generation time), and all growth parameters (mass, protein,

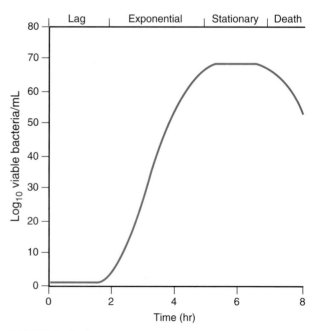

FIGURE B-6 Phases of bacterial growth in pure culture (in liquid medium).

nucleic acid, and so on) increase at the same rate. Cells in the logarithmic growth phase usually are large and rich in ribosomes. They are most active metabolically, and are most sensitive to antibiotics (e.g., cell-wall synthesis inhibitors that affect growth).

In the **stationary phase**, nutrients are depleted rapidly, and significant amounts of metabolic products accumulate when relatively high cell concentrations are approached. This causes the growth rate to slow until the viable count becomes constant. Cells in stationary phase are usually smaller, less active metabolically, and more resistant to toxic agents than are those in the logarithmic phase. At the end of the stationary phase, viable numbers start to decrease at a rapid rate (**death or decline phase**). This continues until only a few viable cells remain or until no viable cells can be detected.

Pathogenic bacteria that have penetrated a host's defenses must multiply to produce disease. To elicit overt illness, numbers of bacteria must reach a level at which toxins are elaborated and normal host functions are disrupted. If this critical level is not reached before growth ceases, an **abortive, subclinical,** or **inapparent infection** results, which may be evidenced only by the presence of circulating antibodies (**seroconversion**). The ability of a pathogen to reach this critical level depends on the **infective dose** and **multiplication within the host,** two parameters that are difficult to estimate in humans because of individual differences in resistance, as well as the fluctuation of resistance in a given individual with time. The apparent generation time of a pathogen growing in a host is usually considerably longer than that which the same organism can achieve in pure culture on laboratory medium. Unlike the stable, optimal environment of the laboratory, the pathogen in an infected host is often in a dynamic, changing environment brought about by the interplay of various host factors and responses to the growth of the pathogen.

STERILIZATION AND DISINFECTION

Sterilization implies complete killing of all living organisms, including pathogenic organisms. **Moist heat** is more effective than dry heat for sterilization. Autoclaving (using steam under pressure) is probably the most widely employed sterilization method and involves treating materials in a pressure chamber with steam at 15 lbs. pressure (equivalent to 121°C) for at least 15 minutes. Sterilization by **dry heat** can be performed in an oven; 2 hours at 160 to 180°C is adequate.

Ultraviolet radiation or germicidal lamps are effective for sterilization, owing to the sensitivity of DNA to UV radiation. UV radiation has little penetrating power and is used primarily for the disinfection of surfaces and air masses. Ionizing radiation is an effective but expensive sterilization method and is not used routinely. A variety of **filtration methods** can selectively remove bacteria. Filtration is the usual method for sterilization of biologicals and other substances that are prone to denaturation in heat-based sterilization methods.

Disinfection implies the **killing of potentially pathogenic microorganisms** that contaminate skin, surfaces, and instruments without removal of all vegetative and dormant bacteria (i.e., spores). A variety of chemicals are available as disinfectants, germicides, antiseptics, and preservatives, with various modes of action and efficacy, ranging from **bactericidal** to **bacteriostatic**. Some important groups of chemical agents are

- **Alcohols** (e.g., 70% ethanol). Primary action is disruption of cell membranes.
- **Detergents** (e.g., benzalkonium chloride). Surface-active agents disrupt cell membranes.
- **Phenols** (e.g., cresol [Lysol], hexachlorophene). Act as cell membrane and protein denaturant
- **Halogens** (e.g., chlorine, hypochlorite [Clorox], tincture of iodine, iodophors such as Betadine). These oxidizing agents are particularly effective against sulfhydryl groups of proteins.
- **Heavy metals** (e.g., Merthiolate, Mercurochrome, silver nitrate, silver sulfadiazine). These agents bind to sulfhydryl groups of bacterial proteins, blocking enzyme activity.
- **Hydrogen peroxide**. Potent oxidizing agent, but less effective against bacteria rich in catalase
- **Formaldehyde** and **glutaraldehyde**. Potent alkylating agents, particularly of $-NH_2$ and $-OH$ groups on proteins and nucleic acids
- **Ethylene oxide**. Also a potent alkylating agent, but in gaseous form; probably second to autoclaving in popularity but used for heat-sensitive materials such as plastics and tubing

BACTERIAL GENETICS

Rapid generation times (often minutes) and **multiple mechanisms of exchanging DNA** enable rapid transfer of certain traits among bacteria. The best clinical example of the ability of bacteria to adapt through mutation and genetic transfer is the **development** and spread of antibiotic resistance over time. The ability to mutate and adapt is also an important **virulence mechanism** for pathogenic bacteria. A **mutation** is simply any heritable change or alteration in the genetic material (e.g., DNA). Mutations occur randomly at some specific rate (mutation rate) that correlates to DNA replication errors and other forms of mutagenesis. A **mutant** is any direct offspring of the **wild type** or parent strain of bacteria. The **genotype** is the actual DNA and any mutations it may contain, whereas **phenotype** refers to the observable properties of an organism, such as a "mutant phenotype." **Alleles** are different forms of the same gene, such as mutant and wild type. Bacteria are **haploid** (a single chromosome per cell); a mutation in bacteria often results in a direct phenotype. Because there is only one allele for each gene, dominant and recessive alleles do not mask mutations in bacteria.

Single base pair mutations, or **point mutations** (Fig. B-7*A*), can have quite varied effects depending on the resulting codon. Some have little or no implication for the phenotype—**silent** mutations code for the same amino acid, and **missense** mutations code for a different amino acid. **Nonsense** mutations insert a stop codon in place of the expected residue and can result in dramatic changes in phenotype. A second spontaneous mutation occurring at the same rate as the first can rescue the original mutation, either back to the initial codon (**reversion**) or to a redundant codon (**suppression**).

The **genetic code** is nonoverlapping; therefore mutations that result in a net gain or loss of a base pair affect the **reading frame** of codons located downstream or 3′ to the mutation, resulting in **frame-shift** mutations (see Fig. B-7*B, C*). It does not change the codons located prior to the mutation.

Alterations in DNA can occur either naturally or via an external process (e.g., UV irradiation). **Spontaneous mutations** result from imperfect replication of DNA or by the insertion of moveable genetic elements called transposons (see subsequently).

Recombination is the breakage and rejoining of DNA into new combinations (Fig. B-8*A, B*) for survival of a strain and is a basic mechanism found in all organisms. Like mutagenesis, it is used to adapt and speed up evolution, change genome order, and aid in DNA repair of large chromosomal breaks. Recombination (either **homologous** [see Fig. B-8*A*] or **site-specific** [see Fig. B-8*B*]) is critical for genetic transfer in bacteria.

Transposons are genetic elements that can "hop" from one location in DNA to another (Fig. B-9). A transposon can duplicate and appear in different locations of the same or different genetic elements. The movement of these elements is called *transposition* and requires a site-specific recombinase enzyme called a *transposase*. There are many types of transposons, ranging in size and complexity from small insertion sequence (IS) elements to large composite transposons. Because **transposons often carry antibiotic resistance** and can move between different bacteria through genetic transfer, they can manifest with a major clinical problem.

A **plasmid** (a **replicon**) is any DNA molecule that replicates independently of the chromosome. Plasmids are primarily circular, but some bacteria (e.g., *Borrelia*) have linear versions. Resistance (**R**) **plasmids** are medically relevant because they can carry multiple drug resistance genes and are capable of conjugation. Other types of R plasmids produce small proteins that kill other bacteria and have an immunity system to prevent self-killing. Many bacterial pathogens, such as *Bacillus anthracis*, *Yersinia* spp, and pathogenic *E. coli*, require **large virulence plasmids** that possess genes encoding toxins, invasins, and other factors important for causing disease. These virulence plasmids can convert a nonpathogenic bacterium (normal flora) to a pathogenic one.

On the genome in pathogenic bacteria reside large islands of genes (called **pathogenicity islands** [PI]) that originated from a different organism. These PIs confer on normal bacterial flora the ability to cause disease. These regions of DNA, which are introduced into the chromosome via site-specific recombination, are often unstable. The majority of genes encoding virulence factors tend to be clustered together in PIs, which have a higher G + C content than other regions of the chromosome. PIs are usually inserted within or close to tRNA genes. They are **found** only **in pathogenic bacteria** and **encode virulence factors**, **adhesins**, **specialized secretory systems**, and **toxins**. PIs are **acquired by bacteria through horizontal gene transfer**, via either bacteriophages or transposons (discussed subsequently).

Genetic Exchange

The ability of bacteria to grow and mutate rapidly allows them to readily adapt to challenges and environments. Recombination provides a mechanism for incorporating genetic material from other sources. However, it is the ability of bacteria to **exchange useful mutations via genetic transfer** to other bacteria that allows an adaptation to spread among a

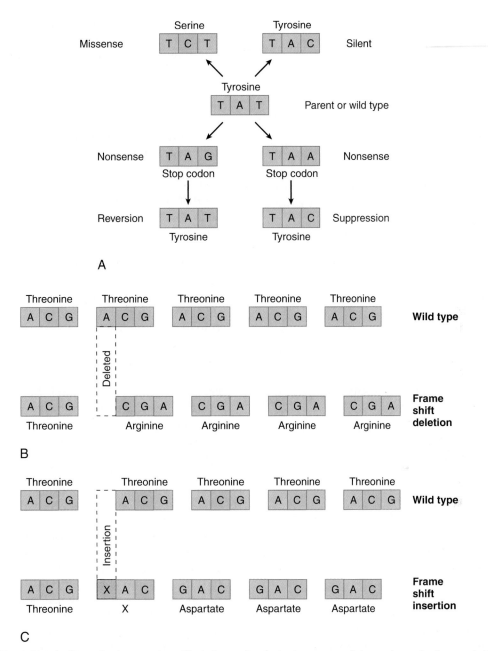

FIGURE B-7 *A,* Varied effects of point mutations. Single base pair substitutions can result in no change in the encoded amino acid (silent), new amino acid (missense), or a stop codon (nonsense). A nonsense mutation can be overcome by a second point mutation to a new amino acid (suppression) or the wild type (reversion). *B,* Nucleotide deletion resulting in frame-shift mutation. *C,* Nucleotide insertion resulting in frame-shift mutation.

population of bacteria in the human body. Even more impressive is the ability to exchange genetic material between different genera of bacteria. The three main types of genetic exchange in bacteria are transformation, conjugation, and transduction.

Certain bacteria possess the ability to uptake DNA in nature (e.g., *Streptococcus pneumoniae, Neisseria* spp),

whereby naked DNA released from lysed donor bacteria is taken up by nearby bacteria that are competent for transformation. **Transformation** is a mechanism by which **naked or purified DNA** is **introduced into a competent bacterial cell,** followed by its insertion into the chromosome or its existence as a plasmid (Fig. B-10*A*). Because transformation involves naked DNA

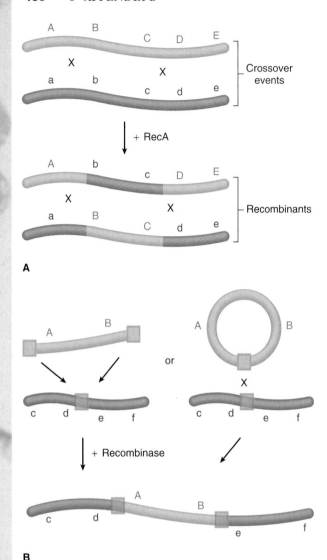

A

B

FIGURE B-8 *A,* Generalized recombination. *B,* Site-specific or specialized recombination.

molecules, the process is sensitive to enzymes that degrade DNA (DNAases). **Homologous recombination** between resident and incoming DNA molecules is required for the formation of a stable recombinant genome.

Conjugation involves **direct contact between the donor and the recipient bacterial cell** via an appendage (called a pilus) on the donor bacterium (see Fig. B-10*B*). Conjugation is almost exclusively mediated by plasmids, which are protected during the process and are therefore resistant to DNAse. **Conjugative plasmids** are low in copy number in each bacterial cell (one to three copies) and possess

genes that allow their transfer to other bacteria. Nonconjugative plasmids cannot transfer themselves alone and are found in high copy number per bacterial cell (>10 copies). A well-studied conjugative plasmid is the so-called fertility (F) factor. Male (F+) bacteria are donor cells and possess the F plasmid, which encodes the F pilus and all the proteins involved in transferring the plasmid to a recipient. Female (F−) bacteria lack the F plasmid, do not produce the F pilus, and act as recipient cells. Conjugation involves direct interaction between the F+ donor and an F− recipient followed by transfer of a copy of the F plasmid into the recipient cell in sequential manner from an origin of transfer. The process converts the F− cell to an F+ cell that expresses the F pilus on its surface. **Conjugative transposons** are mobile genetic elements that, in addition to hopping into new DNA, contain the ability to transfer themselves to a recipient strain via conjugation.

Transduction is the third mechanism of **genetic transfer between bacteria that requires a specific bacteriophage, or bacterial virus, to mediate exchange** (see Fig. B-10*C*). Bacteria, like higher organisms, can be infected by viruses, which can replicate and then move on to infect another cell. Phage, meaning "to eat," describes the process by which most bacterial viruses destroy a cell, releasing viral progeny. **Generalized transduction** involves nonspecific packaging and transfer of genetic material (chromosome or plasmid) in a phage particle to the recipient bacteria. Some rare phages are produced by aberrant packaging that contain large fragments of donor chromosomal DNA instead of the phage genome. Because these transducing particles do not possess their own genetic material, they have no lytic properties. Therefore, they can be used to inject recombinant DNA into a recipient host bacterium without its eventual lysis. The injected donor DNA can be incorporated into the recipient genome via homologous recombination. Certain phages, called temperate phages, can enter a lysogenic cycle involving integration of the phage genome (prophage) into a bacterial chromosome to form a bacterial strain called a lysogen. The prophage remains in the chromosome of the lysogen quietly until the cell is stressed, leading to a lytic cycle and the death of the cell.

Specialized transduction allows transmission of a few specific donor genes to a recipient bacterium. These rare recombinants are formed during aberrant excision of the prophage upon stress; they contain a small region of chromosome adjacent to the prophage attachment site. This can then be transduced to a

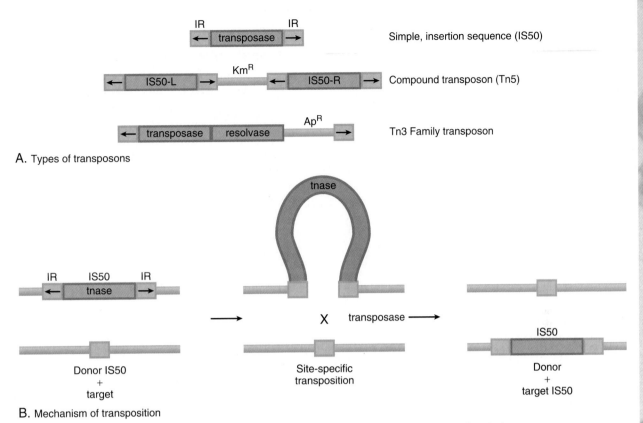

A. Types of transposons

B. Mechanism of transposition

FIGURE B-9 Transposon structure and mechanism of transposition (hopping).

recipient through **lysogeny**. If a temperate phage carries genes encoding virulence factors in addition to normal phage genes, formation of a lysogen may convert a nonpathogenic bacterium to one that can cause disease **(phage conversion)**. This is **a common method for transmitting virulence genes in nature**. Several examples of pathogenic bacteria and phage-encoded virulence factors are: toxin coregulated pilus and cholera toxin in *Vibrio cholerae*, botulinum toxin in *Clostridium botulinum*, pyrogenic exotoxins or superantigens in *Streptococcus pyogenes*, and Shiga toxin in enterohemorrhagic *E. coli*.

BACTERIAL VIRULENCE

Normal microbial flora or microbiota (endogenous organisms) do not cause disease and live in symbiosis with the host. A diverse microbial flora colonizes certain sites of the human body (e.g., colon, skin, and mouth). The normal flora aids in the digestion of food and the production of some vitamins and protects the host from colonization by pathogenic microbes. **Common normal flora** include

- Anaerobes (e.g., *Clostridium, Bacteroides, Actinomyces*)
- Enterobacteriaceae (e.g., *E. coli, Klebsiella*)
- *Staphylococcus* and *Streptococcus viridans*
- *Enterococcus*
- *Gardnerella* and *Mobiluncus*

These **normal bacterial flora** do not cause disease in healthy individuals. They can, however, cause disease in individuals with underlying medical risks (e.g., immunosuppression or HIV infection) or when they leave their normal niche and make their way to a sterile site, where they are not welcome; therefore, these bacteria are known as **opportunistic** pathogens.

Primary pathogens gain entry into the human body through various portals of entry (Fig. B-11). The human body has several **natural defense barriers** against microorganisms: skin, ciliated epithelium, antibacterial secretions (e.g., lysozyme, bile), and mucous layer. There are **no microbes** (with the exception of

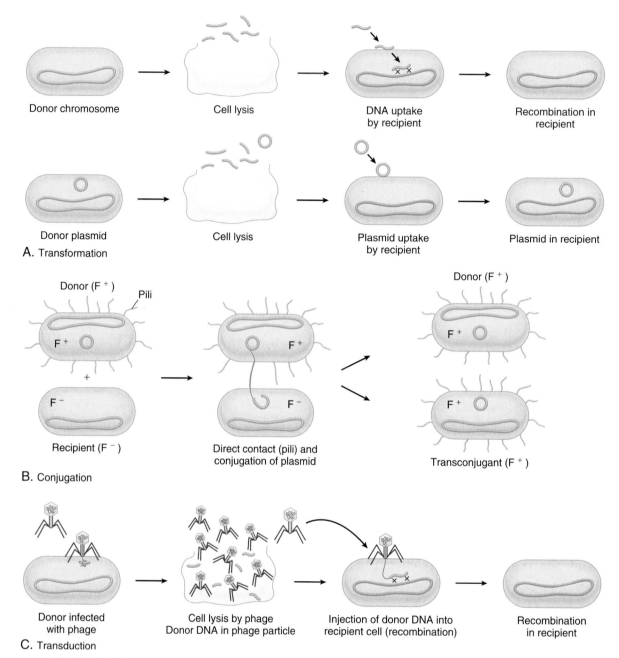

FIGURE B-10 Mechanisms of gene transfer between bacteria: *A*, transformation. *B*, conjugation. *C*, transduction.

some nematodes) **known to penetrate the skin**. Usually bacteria penetrate broken skin (e.g., following surgery, catheterization, or wounds) or skin trauma created by biting arthropods. **Body cavities** (gastrointestinal and respiratory tracts, eyes, urogenital tract) are protected by mucus, ciliated epithelium, and antibacterial secretions. Yet, bacteria have mechanisms to circumvent these defenses. Bacteria may **lack mucin receptors**, thereby avoiding being trapped in the mucin layer. Some pathogenic bacteria produce **mucin-degrading enzymes**. Because of **flagella**, motile bacteria can move through viscous material, such as mucin, to reach epithelial cells. In gastrointestinal mucosa, the mucin layer is thinner and the role of M cells is to sample material passing through the intestine and to deliver it to the immune system.

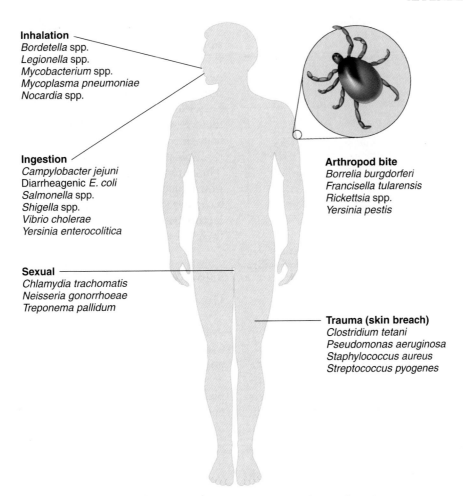

Inhalation
Bordetella spp.
Legionella spp.
Mycobacterium spp.
Mycoplasma pneumoniae
Nocardia spp.

Ingestion
Campylobacter jejuni
Diarrheagenic *E. coli*
Salmonella spp.
Shigella spp.
Vibrio cholerae
Yersinia enterocolitica

Sexual
Chlamydia trachomatis
Neisseria gonorrhoeae
Treponema pallidum

Arthropod bite
Borrelia burgdorferi
Francisella tularensis
Rickettsia spp.
Yersinia pestis

Trauma (skin breach)
Clostridium tetani
Pseudomonas aeruginosa
Staphylococcus aureus
Streptococcus pyogenes

FIGURE B-11 Sites of entry of various pathogenic bacteria.

Colonization (adherence) is crucial for bacterial adherence to host cells because otherwise fluids might wash the bacteria away from the site. **Pili** (fimbriae) are rod-shaped protein surface structures, differing in thickness and length that allow the initial contact of bacteria and host cell. Pili are formed by ordered array of single subunits. Their tip structure specifically attaches to receptors on the host cell. These receptors are usually carbohydrate residues of glycoproteins or glycolipids. The connection with the host cell is often loose (Fig. B-12*A*). Pili are constantly lost and re-formed, partially due to their fragility. Pili-mediated **adherence facilitates mechanisms of tighter attachment** and **signal transduction** in both the bacterial and human cells. Numerous bacterial gene products (e.g., type I fimbriae, colonization factor 1 [CFA1], P fimbriae in *E. coli*; fimbriae in *Neisseria gonorrhoeae*; type IV pili in *Vibrio cholerae*) are expressed or turned on and off in response to

communication with host cells. **Afimbrial adhesins** are bacterial cell surface proteins that do not form pili. They mediate tighter binding of bacteria to the host cells and may bind proteins rather than carbohydrates (see Fig. B-12*B*; e.g., AIDA and Intimin in *E. coli*).

Bacteria bind to each other on surfaces within polysaccharide slime, also called **biofilm**. Biofilms are ordered three-dimensional structures composed of pillars of bacteria surrounded by water channels that allow nutrients to reach the bacteria and allow toxic metabolites to diffuse out. The complex structure of a biofilm is refractory to disinfectants and antibiotics. Biofilms are important because they can be the source of aerosolized bacteria (e.g., *Legionella pneumophila*); and can form on body surfaces (e.g., *Streptococcus mutans* in dental plaque; mucoid strains of *P. aeruginosa* in lungs of cystic fibrosis patients) and on plastic tubing (e.g., *S. epidermidis* in catheters and central

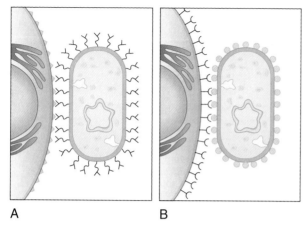

A   B

FIGURE B-12 *A,* Fimbriae or pili: hair-like structures used for adherence to the host cells. *B,* Afimbrial adhesins: cell surface proteins used for adherence that do not form pili.

venous lines) or plastic implants (e.g., *S. aureus* in heart valves and prosthetic joints).

Some bacteria (e.g., *Shigella* spp., *Listeria monocytogenes, Salmonella typhi, Brucella* spp) have evolved mechanisms to **invade** host cells that are normally nonphagocytic. These bacteria attach to the host cells and provoke a rearrangement of the host cytoskeleton that results in the engulfment of the bacteria. Bacteria (e.g., *Mycobacterium tuberculosis, S. typhi*) can also induce their own uptake by phagocytes. Microbes establish intracellular residence in several ways. Some (*M. tuberculosis, Salmonella* spp) **prevent phagosome-lysosome fusion**. Some (e.g., *Listeria, Shigella, Rickettsia*) **escape the phagocytic vacuole** by disrupting its membrane through degrading the membrane lipids or by forming pores in the membrane.

Bacteria that grow within the cytoplasm (e.g. *L. monocytogenes, Shigella* spp) interact with actin, promoting actin condensation on one end of the bacteria to propel the bacteria through the host cell cytoplasm and into adjacent cells. There are many advantages to growing in the cytoplasm: abundance of nutrients, protection from antibodies and complement, and partial protection from some antibiotics that cannot penetrate mammalian cells. The effective **host defenses** to bacteria (e.g., *Shigella* spp) that evade phagocytosis are **natural killer** (NK) cells and the **cytotoxic T-cell response**. Some intracellular bacteria induce the formation of specialized phagosomes (*Legionella, Borrelia*); these bacteria adapt to live within the phagolysosome; these bacteria produce enzymes that detoxify reactive forms of oxygen or prevent oxidative burst.

Iron is essential for bacterial growth. Concentration of free iron is very low in the human body, where it is bound to lactoferrin, transferrin, ferritin, and hemoglobin. Therefore, bacteria have developed different mechanisms to sequester iron. For instance, bacteria secrete **siderophores,** which are low-molecular-weight compounds that chelate iron. The iron-siderophore complex is taken up by bacteria through special siderophore receptors on the surface. The internalized complex is cleaved to release the iron molecule inside the bacterial cell. Some bacteria have surface receptors that bind to host **transferrin, lactoferrin, and ferritin** and sequester iron from these host proteins.

Evasion of Host Defenses

Capsules are composed of a network of polymers covering the bacterial surface; these polymers are usually polysaccharides, but rarely they are proteins. **Capsules protect against complement activation and phagocyte-mediated killing**. Some capsules prevent the formation of C3 convertase in the complement pathways. An effective host response against encapsulated bacteria is to produce antibodies that bind to the capsule. Some bacteria subvert this type of protective response by having capsules that resemble host polysaccharides. The **common encapsulated bacterial pathogens** that are associated with invasive infection in healthy individuals are

- *S. pneumoniae* (**serotype III**; pneumonia and bacteremia)
- *N. meningitidis* (serotypes **A, B, C, Y, W 135**; meningitis)
- *E. coli* (serotype **K1**; sepsis and meningitis in pediatric patients)
- *Klebsiella pneumoniae* (undetermined serotypes; pneumonia and bacteremia)
- *Streptococcus agalactiae* (serotype **IIIa**; sepsis and meningitis in pediatric patients)
- *S. typhi* (**Vi antigen**; sepsis in typhoid fever)
- *Haemophilus influenzae* (**serotype b**; this disease is rare because of a successful immunization program)

The following additional **evasion mechanisms** are of importance to bacterial pathogens and in the causation of infectious diseases:

- **Inactivation of antibody.** Cleavage of IgA by IgA protease; binding to Fc receptors of Ig prevents interaction with phagocyte Fc receptors.

- **Inhibition of complement action**. Blocking deposition of complement on bacterial surface, cleavage of chemotactic complement components, inactivation of the membrane attack complex
- **Antigenic (phase) variation.** Alteration of epitopes on major bacterial surface molecules, allowing evasion of antibody-mediated immune response.
- **Molecular mimicry**. Pathogen expression of products that resemble host factors and thereby allow evasion of the adaptive immune response. This can sometimes lead to **autoimmune sequelae**.
- **Lipopolysaccharide (LPS) modification**. Targeting of LPS by complement in Gram-negative bacteria. Therefore, modification in the LPS can prevent phagocytosis and subsequent formation of membrane attack complex (MAC). Two types of LPS modification affect the interaction of LPS and complement components:

  - Attachment of sialic acid to LPS "O" antigen prevents formation of C3 convertase.
  - Changes in length of LPS "O" antigen prevent MAC killing, probably because MAC forms farther from the membrane.

Toxins

The LPS-associated toxin from Gram-negative bacteria is highly toxic and is named **endotoxin**, given that it is embedded in the membrane (see Fig. B-4). **Lipid A**, with β-hydroxymyristic acid as the prime toxic moiety, is the endotoxin of Gram-negative organisms. This endotoxin, which has a vast array of biologic properties, is often a very important contributing factor to the severity of multiorgan system disease. Endotoxic shock or septic shock occurs on the release of even nanogram amounts of lipid A into the blood stream after attacks on the bacteria by the serum factors and immune cells, and results in the release of proinflammatory cytokines (TNF, IL-1). An overreactive systemic inflammatory response, mediated by the proinflammatory cytokines, can damage target tissue(s) and organ(s), causing manifestations of multiorgan system failure, and irreversible hypotension and intravascular coagulation. Teichoic acids and other peptidoglycan fragments from the cell wall of Gram-positive pathogens (which lack LPS) can also induce toxicity, but at a much lower level than does endotoxin.

Exotoxins comprise bacterial proteins that are toxic to host cells. Table B-2 lists the exotoxins and the associated bacterial species. They are found in both Gram-positive and Gram-negative bacteria. The majority of bacterial toxins are encoded on mobile genetic elements such as bacteriophages (e.g., diphtheria toxin, cholera toxin, Shiga toxin) and plasmids (e.g., heat stable toxin and heat labile toxin from enterotoxigenic *E. coli*).

Type I toxins are **superantigens** that bind to the MHC class II of macrophages and to the receptors on T cells that interact with MHC. In the normal scenario, macrophages process protein antigens by cleaving them into peptides and displaying one of the resulting peptides complexed with an MHC class II molecule on the macrophage surface. This peptide-MHC class II complex is recognized by a few helper T cells. Superantigens are not processed, and because

TABLE B–2	**Exotoxin-Producing Bacteria**
Organisms	**Exotoxins**
Bacillus anthracis	Edema factor and lethal factor carried by protective antigen
Bordetella pertussis	A-B subunit toxin (plasmid encoded): ADP ribosylating activity
Clostridium botulinum	Botulinum toxin (prophage)
Clostridium difficile	Enterotoxin A, cytotoxin B
Clostridium perfringens	Phospholipase C; enterotoxin
Clostridium tetani	Tetanospasmin (plasmid)
Corynebacterium diphtheriae	A-B subunit toxin (prophage)
Escherichia coli	LT (plasmid), ST (plasmid), Shiga toxin
Pseudomonas aeruginosa	Exotoxin A
Shigella dysenteriae	Shiga toxin; enterotoxin
Staphylococcus aureus	TSST-1, exfoliating toxin (plasmid), alpha toxin
Streptococcus pyogenes	Pyrogenic exotoxin SpeA, SpeC (prophage), streptolysins O & S

they bind indiscriminately to MHC class II and T-cell receptors, many more macrophage-T helper cell pairs are formed. This results in the release of **excessive levels of cytokines** (especially IL-2), giving rise to a variety of symptoms (nausea, malaise, vomiting, and fever) that culminate in **toxic shock syndrome,** caused by *S. aureus* or *S. pyogenes.*

Type II (pore forming) toxins (e.g., alpha-toxin from *S. aureus* and listeriolysin O from *L. monocytogenes*) destroy the integrity of eukaryotic cell membranes. There are two types of membrane-disrupting toxins: a protein that forms channels in the membrane (because of differences in osmotic strength between the host cell cytoplasm and the environment, these holes in the membrane trigger a rush of water into the cell, rupturing the cell); and an enzyme (phospholipase) that degrades phospholipids in the membrane, removing the polar head group of the phospholipid.

Type III toxins, or **A-B toxins**, are composed of two types of subunits: a **"B" subunit**, which recognizes and binds to the host cell receptor (usually a carbohydrate moiety), and an **"A" subunit**, which has enzymatic activity. A-B toxins can be simple, with only one A and one B subunit, or they can be composed of several B subunits and only one A subunit. The A and B subunits are usually separated by a proteolytic cleaving event, and they remain connected through a disulfide bond. Both the simple and the compound toxins

bind to and enter the host cell. The B subunit binds to the host receptor (which is what determines the host cell specificity of the toxin). Following this binding, the toxin is endocytosed by the cell and the A subunit (active enzymatic subunit) is translocated to the cytoplasm (Fig. B-13). The A subunits of different toxins enter different cell types according to the distribution of the receptors for the B subunits (which confer host-cell specificity). However, most A subunits catalyze the same reaction: they remove the ADP-ribosyl group from NAD and attach it covalently to a host-cell protein.

The effects of this reaction vary according to the host cell protein that has been ADP ribosylated. The A subunit of **diphtheria toxin** ADP-ribosylates elongation factor-2 (EF-2), a protein that plays an essential role in host cell protein synthesis, thereby killing the host cell by stopping protein synthesis. The A subunit of **cholera toxin** ADP-ribosylates a regulatory enzyme that controls cyclic AMP levels in the host cell; this prevents the enzyme from being turned off, causing the host cell to lose control of ion flow and resulting in a massive loss of host cell water, which is seen as diarrhea. Other A subunits have different activities. **Shiga-toxin A** subunit cleaves a host cell rRNA molecule, preventing protein synthesis. Table B-3 depicts a summary of the comparable properties of exotoxins and endotoxins.

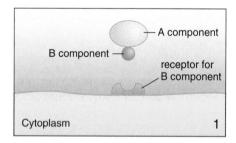

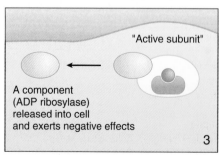

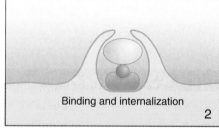

FIGURE B-13 Mechanism of entry and enzymatic action of A-B toxins. See text for the description of the entry and function of toxin. Enzymatic activity of the A component of the toxin is responsible for (a) inhibition of protein synthesis and death of target cells (e.g., posterior pharynx in diphtheria) or (b) increased cAMP production and release of fluid (e.g., in the small intestine in cholera).

TABLE B–3 Major Differences Between Exotoxins and Endotoxins

Property	Exotoxin	Endotoxin
Source	Certain species of Gram-positive and Gram-negative bacteria	LPS (lipid A) of most Gram-negative bacteria
Secreted from cell	Yes	No
Chemistry	Polypeptide	Lipopolysaccharide (LPS)
Location of genes	Plasmid or bacteriophage	Bacterial chromosome
Toxicity	High (fatal dose is 1 μg)	Low (fatal dose is hundreds of μg)
Clinical effects	Various effects (see text)	Fever, DIC, shock
Mode of action	Various modes (see text)	Includes interleukin (IL)-1 and tumor necrosis factor via the mediation of IL-2
Antigenicity	Induces high-titer antibodies or antitoxins	Poorly antigenic
Vaccines	Toxoids used as vaccines	No toxoids formed; no vaccine available
Heat stability	Destroyed at 60°C (except *Staphylococcus* enterotoxin)	Stable at 100°C for 1 hr
Typical diseases	Tetanus, botulism, diphtheria, toxic shock	Meningococcemia; sepsis by Gram-negative rods

APPENDIX C

Pathogenic Fungi and Parasites: Concepts

Introduction to Pathogenic Fungi

Fungi are nonmotile eukaryotic organisms that have cell walls and produce filamentous structures and spores but do not contain chlorophyll. They are considered saprophytes, and they primarily decompose organic matter. The distinction between fungi and bacteria is based on the fundamental difference in nuclear structure of eukaryotes and prokaryotes. Because of this difference, antibacterial agents, which are uniquely targeted at the prokaryotic structures and functions, do not generally affect fungal cells.

Structures, Physiology, and Reproduction

The fungal cells are typically surrounded by cell walls that contain complex polysaccharides (e.g., **chitin** and **glucans**). A cross-section of a fungal cell is shown in Figure C-1. The chitin layer is composed of a polymer of *N*-acetylglucosamine. Layered on the chitin are glucans (β[1,3]-D and β[1,6]-D glucans) and other complex polysaccharides (e.g., mannans) in association with polypeptides. The plasma membrane of a fungus contains sterols, primarily **ergosterol**. Ergosterol is the site of action of most antifungal drugs (e.g., amphotericin B, azoles). Changes in ergosterol synthesis or its composition in the cell membrane can lead to drug resistance.

Fungal respiration is almost exclusively oxidative, and metabolism is exclusively heterotrophic (i.e., dependent on organic nutrients). Generation time (doubling time) is long (hours) compared with most bacteria (minutes). Fungi can be divided into two groups on the basis of their growth pattern: yeasts and molds. **Yeasts** are unicellular fungi and predominantly reproduce by budding (Fig. C-2*A*); in culture, they produce colonies that are soft in consistency and

similar in form to those of bacteria. **Molds** are multicellular fungi with elongated hyphal structures and attached conidia (analogous to bacterial spores; see Fig. C-2*B*); in culture at room temperature they produce cotton-like colonies with aerial growth.

Medically important molds reproduce asexually via conidia (spore) formation. The growth cycle begins with the germination of a conidium (a reproductive structure), a germ tube that grows from the tip of a germinating spore by an apical extension. The continued extension of the germ tube with side branching results in a network of filaments called **hyphae** (see Fig. C-2*B*). The collection of hyphae so generated gives rise to a cotton-like colony called a **mycelium**. The cytoplasm of one type of hyphae is regularly interrupted by septa, and such hyphae are accordingly called **septate hyphae**. The cytoplasm of another type of hyphae is uninterrupted by any cross walls or septa throughout the entire mycelium; such hyphae are termed **aseptate**. The conidia have various morphologies.

- **Arthroconidia** are produced by the conversion of a preexisting hyphal element to conidia that break loose from each other (thalic process) and initiate another cycle of growth by germination and apical extension of growth (Fig. C-3*A*).
- **Blastoconidia** (yeasts) are produced by a blastic (budding) process. The yeast cell itself is referred to as a **blastoconidium**. Sometimes the blastoconidia do not separate at maturity. They continue to grow, elongating into sausage-shaped filaments termed **pseudohyphae** (see Fig. C-3*B*).
- **Macroconidia/microconidia** are produced by some fungi and may be produced either blastically or thalically (see Fig. C-3*C*).
- **Sporangiospores** are produced by cytoplasmic cleavage within a structure called a **sporangium**. The hyphae of sporangiospore-producing fungi are nonseptate (see Fig. C-3*D*).

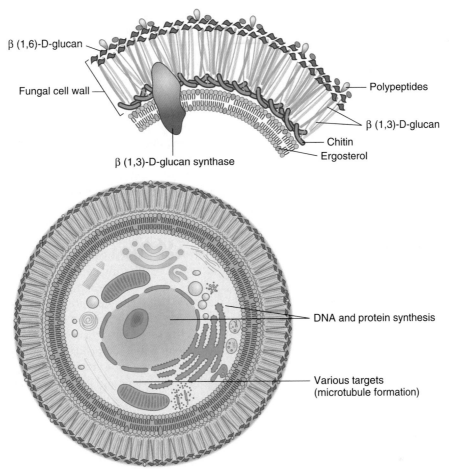

β (1,6)-D-glucan

Fungal cell wall

β (1,3)-D-glucan synthase

Polypeptides

β (1,3)-D-glucan

Chitin
Ergosterol

DNA and protein synthesis

Various targets
(microtubule formation)

FIGURE C-1 Fungus architecture. Note the layers of structures and internal organelles that are targets for antifungal drugs.

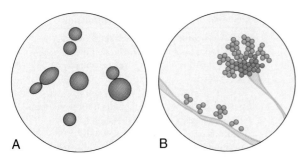

A B

FIGURE C-2 Structures of yeasts and molds. *A,* Unicellular budding yeast. *B,* Multicellular mold (true, septate hyphae) and conidial (spore) structure.

Fungal Diseases and Classification

There are four types of fungal (mycotic) diseases: (1) hypersensitivity (an allergic reaction to molds and spores); (2) mycotoxicoses (poisoning of humans and animals by toxins produced by fungi usually growing on improperly stored grains); (3) mycetismus (mushroom poisoning); and (4) mycosis (infection caused by medically important fungi). We shall be concerned here only with mycoses.

The most practical method of classification of fungi is the clinical taxonomy (Table C-1), which divides the fungi into those causing

- Cutaneous mycoses
- Subcutaneous mycoses
- Systemic mycoses
- Opportunistic mycoses

The **cutaneous mycoses** (e.g., ringworm lesions) are fungal diseases that are confined to the outer layers of the skin (keratinized layers), nail, or hair. Rarely, they invade the deeper tissue or viscera. The fungi involved are called **dermatophytes**.

The **subcutaneous mycoses** (e.g., **sporotrichosis**, mycetoma, chromomycosis) are confined to the subcutaneous tissue and only rarely spread systemically. They usually form deep, ulcerated skin lesions or

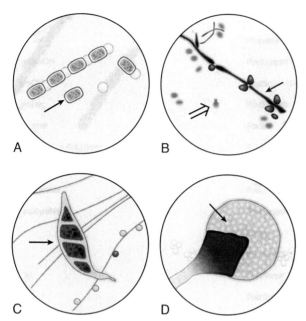

A

B

C

D

FIGURE C-3 Processes of conidiogenesis and structures of conidia. *A,* Arthroconidia (*arrowhead*); *B,* Blastoconidia (*open arrow*) and pseudohyphae (*arrow*). *C,* Macroconidia (*arrowhead*) and microconidia. *D,* Sporangium (with attached aseptate hypha) containing columella and sporangiospores (*arrow*).

fungating masses, most commonly involving the lower extremities. The causative organisms are soil saprophytes, which are introduced through trauma to the feet or legs.

The **systemic mycoses** (e.g., histoplasmosis, blastomycosis) usually involve the lungs. They may also involve other deep organs and become widely disseminated. Systemic mycoses are usually caused by **primary, endemic fungal pathogens** that are capable of causing illness in healthy individuals. In high-risk individuals, systemic mycoses are more severe and are more likely to involve more than one organ system.

Whereas **opportunistic mycoses** are caused by fungi with low inherent virulence that may cause disease when host defensive mechanisms are deficient, they usually do not cause disease in otherwise healthy individuals. The incidence of opportunistic mycoses has significantly increased, owing to a growing population of susceptible individuals. **Risk factors** for opportunistic mycoses (e.g., aspergillosis, systemic candidiasis, zygomycosis) include AIDS, diabetes, lymphomas, broad-spectrum antibiotic therapy, and immunosuppressive therapy (e.g., corticosteroids, radiation therapy, and cytotoxic drugs).

TABLE C-1   Classification of Medically Important Fungi

Fungal Diseases	Major Agents	Anatomic Location	Disease	Infective Form (Mode of Acquisition)
Cutaneous mycoses	*Malassezia*	Dead layer of skin	Tinea versicolor	Yeast (endogenous)
	Dermatophytes:	Skin, hair, nail	Dermatophytosis	Arthroconidia, hyphae
	Trichophyton		(ringworm)	(close contact)
	Epidermophyton			
	Microsporum			
Subcutaneous mycoses	*Sporothrix*	Subcutis, lymphatics	Sporotrichosis	Conidia (skin penetration)
Systemic mycoses	Primary (systemic) pathogens:	Deep organs (e.g., lungs, blood stream, kidney, CNS) and skin	Histoplasmosis	Microconidia
	Histoplasma		Blastomycosis	Microconidia
	Blastomyces		Coccidioidomycosis	Arthroconidia (inhalation in all)
	Coccidioides			
Opportunistic mycoses	Opportunistic pathogens:	Deep organs (e.g., lungs, blood stream, CNS) (dissemination)		
	Cryptococcus		Cryptococcosis	Yeast (inhalation)
	Candida		Candidiasis	Yeast (endogenous; overgrowth)
	Aspergillus		Aspergillosis	Conidia (inhalation)
	Zygomycetes:		Zygomycosis	
	Rhizopus			Conidia (inhalation, penetration)
	Absidia			
	Mucor			

Transmission and Pathogenesis

Fungi are soil **saprophytes,** and most people are exposed to them. With the exception of *Candida* and some dermatophytes, fungi do not generally cause disease that is communicable from person to person.

Many fungi grow morphologically in two distinct forms, a feature that is termed **dimorphism.** Dimorphic fungi exist in nature as molds (infective form) and grow in the target tissue (e.g., lungs) of an infected host as yeast. The mold form also appears in cultures in the laboratory. Most dimorphic fungi that cause systemic infections have a characteristic endemicity in the geographic regions of the world. This habitat is the primary source of molds, where a human, as an incidental host, is exposed to conidia (spores) and fungal elements. The endemic fungal pathogens (e.g., *Histoplasma capsulatum*, *Blastomyces dermatitidis*, *Coccidioides immitis*, and *Paracoccidioides braziliensis*) that are capable of causing systemic mycoses are all dimorphic. *Sporothrix schenckii* is also a dimorphic fungal pathogen, but it causes localized subcutaneous mycosis rather than a systemic disease.

The establishment of a mycosis usually depends on the **size of the inoculum** and on the host's level of **natural resistance** to infection. **Inhalation of airborne conidia** (spores) is the common mode of transmission of mycoses, except in candidiasis, which is caused by endogenous mucosal colonization or overgrowth of *Candida albicans*, and in subcutaneous mycoses, which are caused by penetration of fungal conidia through a breach in the skin. In general, slow-growing fungi usually cause chronic infections in exposed at-risk individuals. Little is known about virulence factors and interactions between the host and the fungi. Adherence and invasion are dependent on the infecting species. **Adhesins** on fungi and lectin-like receptors on macrophages are involved in the colonization of lower respiratory airways. Pathogenic fungi do not produce toxins, but they do show physiologic modifications during the **invasion** process. Most pathogenic fungi are also thermotolerant. The temperature-dependent transition from the hyphal to the yeast form in an appropriate anatomic location of the host is one of the most critical determinants for establishing infection by dimorphic fungi.

Neutrophils are actively recruited in response to fungal infections, and fungal hyphae are destroyed by neutrophils in normal hosts. Invasive fungal diseases develop in patients with severe neutropenia. **Alveolar macrophages** are the major cells involved in the **phagocytosis** of airborne fungal pathogens that gain entry in the lungs. Some fungi are able to evade phagocytosis or phagocytic killing, and immune deficiencies can compromise host ability to control fungal invasion.

Pneumocystis jiroveci and yeast forms of *H. capsulatum* have the ability to grow intracellularly in unactivated alveolar macrophages. HIV infection alters the mannose receptor-mediated binding and phagocytosis of *P. jiroveci* by alveolar macrophages. Unless activated by cytokines, macrophages are unable to kill the intracellular yeasts of *H. capsulatum*. Yet again, *Histoplasma* and other dimorphic fungi can also resist the effects of the active oxygen radicals released during the respiratory burst of phagocytes and are able to withstand phagocytic killing.

These mechanisms of evading phagocytic killing allow many dimorphic fungi to multiply sufficiently to produce an infection that can be controlled only by the walling-off effect of **granulomatous inflammation,** similar to that seen in tuberculosis. The **granulomas** consist of a mixture of mononuclear phagocytes, lymphocytes, and, principally, **cytotoxic T cells.** Patients with T-cell dysfunction are predisposed to serious mycoses. An intense granulomatous reaction produces characteristic pathology in the lungs in systemic mycoses, and the ensuing cytokine response leads to the clinical manifestations (e.g., fever, fatigue, weight loss) of these illnesses. Without control by **cell-mediated immunity**, some fungi may disseminate via the blood stream (fungemia) and cause disease at multiple sites throughout the body, notably the skin, bone, and CNS.

INTRODUCTION TO PARASITES

Parasitism is an ecologic, reciprocal relationship between a large entity (host) and a smaller entity (parasite). Parasitism is usually a temporary association in which the existence of the **parasite** depends on, and takes advantage of, the host in all aspects—regardless of the outcome to the host. Parasitism is different from a **symbiotic** association in which two entities can coexist permanently, depending on each other. In a **mutualistic** association, two entities benefit from each other, whereas in a **commensalistic** association, one entity benefits from the association and the other one does not. An **ectoparasite** lives on the outside of the body of a host, whereas an **endoparasite** lives within the body of the host harboring it. For our discussion in this section and the rest of the book, all endoparasites will simply be called

parasites. A microbial parasite lacks the necessary organs for assimilating nutrients and is, therefore, physiologically or metabolically dependent on its host, which also supports the reproductive activities of the parasite.

MORPHOLOGY, PHYSIOLOGY, AND CLASSIFICATION

Pathogenic Protozoa

Parasitic protozoa are small (**2 to 100 μm**), **single-celled eukaryotes**. The protoplasm of protozoa is enclosed by a cell membrane. The necessary functions for existence within a host are carried out by various intracellular structures called **organelles**. These organelles include food storage granules and contractile and digestive vacuoles. The parasite nucleus is essential for life and contains clumped or dispersed chromatin and a **karyosome**, which plays a role in promitosis. Differentiation of various species of protozoa is based on the structure of the nucleus (particularly the arrangement of the chromatin and karyosome). **Organs of motility** vary from simple cytoplasmic extrusions or pseudopods to more complex structures such as flagella, cilia, or undulating membranes.

The protozoa are most abundant in warmer countries, but the intestinal species are frequently found in the temperate regions. Assimilation of organic nutrients by **pathogenic protozoa** is carried out via pseudopods and by simple diffusion. Respiration in most parasitic protozoa is accomplished by facultative anaerobic processes. Parasitic protozoa have many adaptive capabilities for energy metabolism, including specialized organelles. To ensure survival under harsh or unfavorable environmental conditions, many parasitic protozoa develop into a cyst form. The cyst form is resistant to chlorination and drying (similar to spores in bacteria and conidia in fungi). The cyst form facilitates transmission in the community; it can pass out of the body and subsequently be taken up by another host. Protozoa infect all major tissues and organs of the body. They live as intracellular parasites in a wide variety of cells and as extracellular parasites in the blood, intestine, and urogenital system. Intracellular species obtain nutrients from the host cell, either by direct uptake of small molecules or by ingestion of host cytoplasm. Extracellular species feed by direct nutrient uptake or by ingestion of host cells. Reproduction of protozoa in humans is usually asexual, by binary fission. Sexual reproduction is normally absent or restricted to the insect vector phase (e.g., *Plasmodium* spp). *Cryptosporidium* is exceptional in undergoing sexual reproduction in human hosts.

Protozoa are **classified into four main groups** (see Table C-1). Amebae (rhizoids) and ciliates are free living, and sporozoa are obligatory intracellular. Some flagellates are occasionally (facultatively) intracellular, although most are free living. Each group of protozoa exhibits morphologic characteristics (e.g., outer structures, intracytoplasmic structures) by which it can be differentiated. Each group also exhibits a characteristic means of locomotion and mode of reproduction.

Amebae

The single-celled organisms **amebae** (see figure in Table C-2) are characterized by having **pseudopods** for motility. The motile, reproducible feeding stage (**trophozoite**) lives most commonly in the gut lumen. The nonfeeding, nonmotile stage (**cyst**) of the common gastrointestinal amebae is shed in feces.

Entamoeba histolytica is well recognized as a pathogenic ameba. Each trophozoite of *E. histolytica* has a single nucleus that has a central **karyosome** and uniformly distributed **peripheral chromatin**. The cytoplasm has a granular, or "ground-glass," appearance. Trophozoites of *E. histolytica* measure 15 to 20 μm in size and multiply by binary fission to produce **cysts**, which are passed in feces. The cyst is round (10 to 20 μm size) and contains up to four nuclei; it may contain cigar-shaped **chromatoid bars**. *E. histolytica* should be differentiated from the intestinal commensalistic protozoa, which include nonpathogenic *E. coli*, *Entamoeba hartmanni*, *Entamoeba gingivalis*, *Endolimax nana*, and *Iodamoeba buetschlii* and pathogenic *Entamoeba polecki*. Differentiation between all of these organisms is based on morphologic differences of the cysts and trophozoites.

Acanthamoeba and *Naegleria* are free-living amebae that are capable of surviving without resorting to parasitism. *Naegleria fowleri* has three stages: cysts, trophozoites (ameboid), and flagellates. In amoeboid form, *Naegleria* has sucker-like structures for phagocytosis. *Balamuthia mandrilaris*, another agent of human disease, is morphologically similar to *Acanthamoeba*. Unlike *Naegleria fowleri*, *Acanthamoeba* and *Balamuthia* have only two stages, cysts and trophozoites. The trophozoites replicate by mitosis. *Acanthamoeba* spp and *B. mandrilaris* cysts and trophozoites are found in tissue.

TABLE C–2 Classification of Protozoa

Class of Protozoa	Morphology	Reproduction	Organelles/ Locomotion	Pathogenic Protozoa
Amebae	Unicellular trophozoite and cyst	Binary fission	Pseudopods	*Entamoeba histolytica, E. coli, E. hartmanni, E. gingivalis, Endolimax nana, Iodamoeba buetschlii, Acanthamoeba* spp, *Naegleria* spp
Flagellates	Unicellular trophozoite (*top*) and cyst (*bottom*); cyst may be absent in some flagellates	Binary fission	Flagella	*Giardia lamblia, Dientamoeba fragilis, Trichomonas vaginalis, Trypanosoma brucei, Trypanosoma cruzi, Leishmania* spp (many species in various geographic regions)
Ciliates	Large, unicellular trophozoite with cilia on the surface of the organism and cyst	Binary fission or conjugation	Cilia	*Balantidium coli*
Sporozoa	Unicellular, frequently intracellular with multiple forms including trophozoites, sporozoites, oocysts, and gametes	Schizogony and sporogony	None	*Toxoplasma gondii, Plasmodium* spp, *Cryptosporidium* spp, *Cyclospora cayetanensis, Isospora belli*

Flagellates

The luminal flagellate parasites (see figure in Table C-2) are characterized by the presence of **flagella** and special organs, such as sucking disk, axostyle, and undulating membrane. Luminal flagellate parasites have both trophozoite and cyst stages in the life cycle and in the trophozoite stage; they move by means of flagella. *Giardia lamblia* (also known as *G. intestinalis*) is the most common pathogen in this group. Trophozoites of *G. lamblia* (9 to 21 μm in length) have two nuclei with a large, central karyosome. With four pairs of flagella, these trophozoites resemble an "old man with glasses": this finding in fresh stool is diagnostic. Cysts of *Giardia lamblia* (8 to 12 μm in length) also have two nuclei each (more mature ones have four nuclei). The trophozoites multiply by longitudinal binary fission, remaining in the lumen of the proximal small bowel where they can be free or attached to the mucosa by a ventral sucking disk.

Trichomonas vaginalis, a flagellate, resides in the female lower genital tract and the male urethra and prostate, where it replicates by binary fission. This tissue flagellate is the most common pathogenic protozoan of humans in industrialized countries and is the cause of sexually transmitted diseases.

Trypanosoma spp and *Leishmania* spp are protozoan blood/tissue flagellates. These so-called **hemoflagellates** are pleomorphic, with different morphologies in the human host and arthropods that transmit them. Trypanosomes are very small, motile, fusiform protozoa.

Ciliates

The ciliates are single-celled and are the largest of all protozoa; they move by means of cilia located on the cell surface. The cilia beat in a coordinated rhythmic pattern, causing the trophozoite to move in a spiral path. Ciliates also have cyst stages. *Balantidium*, which means "a little bag," is the only pathogenic ciliate known. It has a large, sac-like structure (65×45 μm) and is found predominantly in developing countries. It is acquired by fecal-oral transmission.

Sporozoa

The sporozoa (see figure in Table C-2) are the **smallest** (1 to 10 μm in diameter) of all groups of protozoa and are also characterized by **intracellular** multiplication. Representative organisms in the **intestinal group of sporozoa** include *Cryptosporidium parvum*, *Cyclospora cayetanensis*, *Isospora belli*, *Sarcocystis* spp, and *Toxoplasma gondii*. A sexual multiplication process (schizogony) in a definitive host (e.g., *T. gondii* in cats) results in an oocyst, which after sporogony contains sporocysts or sporozoites that are shed in the feces and subsequently picked up by an **intermediate host** (e.g., pigs and cows) for **asexual multiplication**. The oocyst is the most common infective stage. The **blood-borne sporozoa** (*Plasmodium* and *Babesia*) undergo exoerythrocytic and pigment-producing erythrocytic schizogony in humans, and a sexual stage is followed by sporogony in mosquitoes.

Microsporidia are a group of intracellular sporozoan parasites that are the smallest (1 to 2 μm) of all sporozoa. The infective form is called the **spore**; each spore contains a polar tubule that is used to penetrate host cells.

Pathogenic Helminths

Helminths (Table C-3) are complex multicellular organisms. Like protozoa, they are most abundant in warmer countries, but the intestinal species are frequently found in the temperate regions. All the major groups of helminths can cause human disease. The helminths are macroscopic (a millimeter to a meter or more). Adult worms may be meters long. Infectious larval stages may measure only 100 to 200 μm. All worms have a complex body organization. The external surface of some is covered with a protective cuticle. Frequently, worms possess elaborate attachment structures such as hooks, suckers, teeth, or plates. Worms have primitive nervous and excretory systems. Some have an alimentary tract; none have a circulatory system.

Helminths are egg-laying parasites. The resulting larvae are always morphologically distinct from the adult parasites. The larvae must undergo several developmental stages before attaining adulthood, either in the host or in a soil environment.

Nutritional requirements are met by active ingestion of host tissue or body fluids. Muscular motility expends energy, and worms metabolize carbohydrates. Nutrients are stored in the form of glycogen. Adult helminth worms catabolize carbohydrates using an adapted anaerobic glycolytic cycle as the source of ATP, and they appear unable to catabolize fatty acids and amino acids as an energy source.

There are three main groups of helminths: **nematodes** (roundworms), **cestodes** (tapeworms), and **trematodes** (flukes).

Nematodes

Nematodes are also known as **roundworms**, because they appear round when examined in cross sections; see figure in Table C-3). A large variety of species exists worldwide and they vary in lengths from a few millimeters to more than a meter in length. They are major parasites of humans, although the vast majority of them are free living. Body organization is fairly complex, with an outer cuticle layer and a pseudocele containing all systems (digestive, excretory, nervous, and reproductive). In spite of their rigid morphologic and developmental organization, nematodes are biochemically and physiologically highly adaptable. Nematodes have **separate sexes**, with the male being smaller than the female. Humans acquire infection by ingestion of larval eggs or skin penetration of larval forms from the soil. Although humans do not support direct multiplication of adult worms, humans support their life process and are considered a definitive host for the intestinal nematodes. Oviparous female worms lay several hundred to several million eggs, depending on the species. Eggs in most species mature in soil

TABLE C–3 Classification of Helminths

Class of Helminths (Common Names)	Morphology	Reproduction	Organelles and Locomotion	Pathogenic Helminths
Nematodes (roundworms)	Multicellular, round, smooth, spindle shaped, tubular alimentary tract, may possess teeth or plates for attachment	Separate sexes	No single organelle; active muscular motility	*Ancylostoma duodenale* (hookworm), *Ascaris lumbricoides, Enterobius vermicularis* (pinworm), *Strongyloides stercoralis, Toxocara canis, Trichinella spiralis, Wuchereria bancrofti, Brugia malayi, Onchocerca volvulus, Loa loa*
Cestodes (tapeworms) Parasitic adult Gravid proglottis	Multicellular, head (scolex, equipped with hook and suckers for attachment) with segmented body (proglottids), no alimentary tract	Hermaphroditic	No single organelle; usually remain attached to mucosa, proglottids may display muscular motility	*Diphyllobothrium latum* (dwarf tapeworm), *Echinococcus granulosus, Taenia saginata, Taenia solium*
Trematodes (flukes) Adult fluke Male and female schistosomes	Multicellular, leaf shaped, with oral and ventral suckers, blind alimentary tract	Hermaphroditic (*Schistosoma* group has separate sexes, also called digenetic blood flukes)	No single organelle, muscular, directed motility	*Clonorchis sinensis* (Chinese or oriental liver fluke), *Paragonimus westermani* (oriental lung fluke), *Schistosoma haematobium, Schistosoma japonicum, Schistosoma mansoni*

and with time become infective to a new host. In a few species, eggs may be immediately infective to a new host.

Cestodes

The cestodes are commonly called **tapeworms**, because of their usual long (up to meters in length), **ribbon-like structure** and **flattened appearance** in cross-sections. The adult tapeworm contains a chain of egg-producing segments called **proglottids**, which develop from the neck region of the attachment organ, the **scolex** (see figure in Table C-3). The scolex is usually equipped with four cup-shaped **suckers**, and some species also have a crown of **hooks** that facilitate attachment of the worms to the intestinal mucosa. Proglottids develop by budding from the posterior end (the location of germinal tissue) of the scolex.

Physiologically, cestodes are an interesting group because they have no mouth and no gut. The outer tegument of the adult tapeworm is a naked cytoplasmic layer covered with microvilli. Tapeworms have no digestive enzymes of their own and instead rely on their hosts' enzymes, acquiring low molecular weight nutrients by absorption through the outer surface of their bodies.

Tapeworms are **hermaphrodites**, so each mature proglottid contains both male and female reproductive organs. Self-fertilization results in embryonated eggs (known as onchospheres); the posterior segments carrying the fertilized eggs break off and are expelled in feces. Each embryonated egg bears six tiny hooklets that facilitate entry of the embryo into the intestinal mucosa of a new host.

Trematodes

Trematodes **(flukes)** are **nonsegmented, leaf-shaped** parasites. Flukes vary in size from a few millimeters to several centimeters in length. Most have cup-shaped (muscular) oral and ventral suckers. Flukes are among the most common and abundant of the helminth parasites, and about a dozen species are important parasites of man. The most important of these flukes are the schistosomes, of which four major species infect some 200 million people in 75 countries.

With the exception of the schistosomes, flukes are **hermaphroditic**, having male and female reproductive organs with complex branched structures. Schistosomes live as unisexual male or female (yet they coexist as pairs; see figure in Table C-3). In both schistosomal and nonschistosomal types of flukes, sexual reproduction occurs in humans, releasing millions of eggs into the circulation during a protracted course of infection.

Transmission (Life Cycle) of Parasites and Pathogenesis

The pathogenesis of parasitic infections depends on several factors: (1) infective dose (number of parasites); (2) modes of acquisition (Figs. C-4 and C-5); (3) passage and target organ; (4) parasite and antigenic load; and (5) host response to various stages of parasite development. An interesting differentiating characteristic of pathogenesis is the fact that **protozoa multiply within the human host, whereas the helminths do not multiply** in the human host. The pathology of helminthic infection may result from any of these events: (1) migration of larvae through various body tissues originating in the intestine; (2) blood-sucking activities of worms; (3) innate immune response; (4) allergic reactions to antigenic products—secreted by adult worms or larvae; or (5) cell-mediated immune responses to coated antigens of parasites, eggs, or larvae.

Infection by *Entamoeba histolytica* (a **single-host amebic parasite**) occurs by ingestion of tough, environment-resistant protozoal cysts in fecally contaminated food or water (Fig. C-6). Excystation leads to the development of trophozoites in the intestine. The motile trophozoites of *E. histolytica* invade the intestinal mucosa. The **cytolytic effect** of amebas after direct contact with target mucosal cells causes "flask-shaped" ulcerations of intestinal wall, resulting in dysentery. In complicated infections, extraintestinal invasion by trophozoites, also involving a cytolytic effect, causes liver abscess.

Naegleria fowleri and *Acanthamoeba* spp are commonly found in lakes, swimming pools, and tap water. The trophozoites of *N. fowleri* gain entry through the nasal mucosa when the host is diving and swimming in a warm body of water. **Invasion of the olfactory neuroepithelium** causes primary amebic meningoencephalitis, manifesting as fever, severe headache, vomiting, and focal neurologic deficits, and progressing rapidly to coma and death. In individuals with compromised immune systems, *Acanthamoeba* spp and *B. mandrilaris* (opportunistic free-living amebas) cause granulomatous amebic encephalitis, with prominence of mononuclear pleocytosis in cerebrospinal fluid. *Acanthamoeba* spp can also cause keratitis and corneal ulcers following corneal trauma or in association with

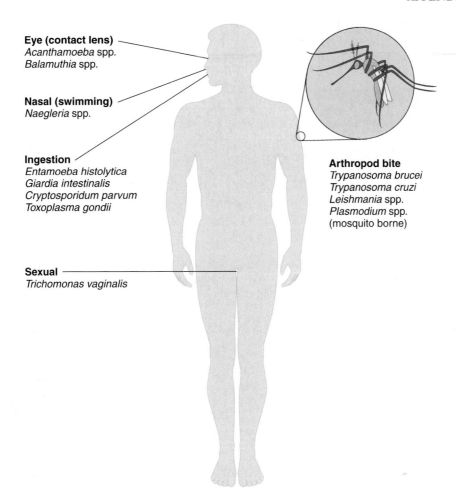

FIGURE C-4 Sites of entry of various pathogenic protozoa.

contact lenses contaminated with tap water containing amebas.

Trypanosoma brucei (blood and tissue flagellate) are characterized by having the motile (trypomastigote) form (found in tsetse [*Glossina*] fly), with the typical undulating membrane and free flagellum at the anterior end. During a blood meal on the human host, an infected tsetse fly injects trypomastigotes into skin tissue. A chancre develops at the site of inoculation. The trypomastigote enters the lymphatic system and passes into the blood stream (hemolymphatic stage), with symptoms that include fever, lymphadenopathy, and pruritus. The blood-borne trypomastigotes cross the blood-brain barrier into the cerebrospinal fluid (meningoencephalitic stage), causing headaches, somnolence, and abnormal behavior and leading to a loss of consciousness and coma (**sleeping sickness**).

Trypanosoma cruzi, a blood/tissue flagellate endemic in poor, rural areas of Central and South America, is transmitted to humans by blood-sucking triatomine bugs during a blood meal. Infection is initiated by wound contamination with bug feces containing the infective flagellates. A local lesion (**chagoma**) appears at the site of inoculation, where the motile **trypomastigotes invade leukocytes and cells of subcutaneous tissues** and **differentiate into amastigotes**, with subsequent development of interstitial edema. The amastigotes then multiply by binary fission, differentiate into trypomastigotes, and are released into the circulation, causing acute febrile illness, a result of parasitemia. Blood-borne trypomastigotes disseminate into the heart and transform into intracellular amastigotes. Manifestations of chronic **Chagas disease** include cardiomyopathy (caused by infiltration of CD8+ T lymphocytes).

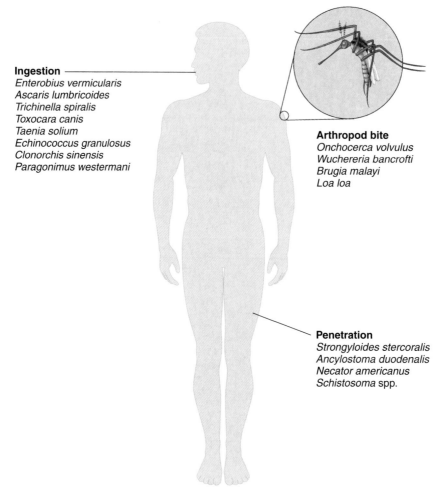

Ingestion
Enterobius vermicularis
Ascaris lumbricoides
Trichinella spiralis
Toxocara canis
Taenia solium
Echinococcus granulosus
Clonorchis sinensis
Paragonimus westermani

Arthropod bite
Onchocerca volvulus
Wuchereria bancrofti
Brugia malayi
Loa loa

Penetration
Strongyloides stercoralis
Ancylostoma duodenalis
Necator americanus
Schistosoma spp.

FIGURE C-5 Sites of entry of various pathogenic helminths.

Leishmania parasites, another group of hemo-flagellate, are transmitted by the bite of female phle-botomine sandflies; multiple species of *Leishmania* are found in many geographic regions of the tropics and subtropics and are morphologically indistinguishable. The sandflies inject the infective stage, flagellate pro-mastigotes, during blood meals. Promastigotes are then phagocytized by macrophages, to transform within phagolysosomes into the nonflagellated intra-cellular amastigotes. Amastigotes multiply by binary fission, causing rupture of unactivated macrophages and leading to **cutaneous leishmaniasis**, charac-terized by one or more sores on the skin where the sandflies have fed. The sores have a raised edge and central crater and can be painless or painful. **Visceral leishmaniasis** manifests as fever (due to parasitemia), also accompanied by leukopenia, thrombocytopenia, and anemia, as well as hepatosplenomegaly. **Killing of**

infected host cells by activated macrophages causes the pathology of leishmaniasis.

Members of the feline family are the only known definitive host for the sexual stages of ***Toxoplasma gondii.*** After the cat ingests tissue cysts or oocysts, viable organisms are released and invade epithelial cells of the small intestine, where they undergo an asexual cycle followed by a sexual cycle and then form oocysts, which are excreted. The unsporulated oocysts sporulate in the soil environment and become infec-tive for pigs and other animals. Human infection occurs by ingestion of meat containing pseudocysts. A pseudocyst represents a large number of *T. gondii* trophozoites enclosed in a macrophage in animal tissue. The parasites form tissue cysts, most common-ly in skeletal muscle, myocardium, and brain; these cysts may remain throughout the life of the human host. In patients with AIDS, toxoplasmic encephalitis

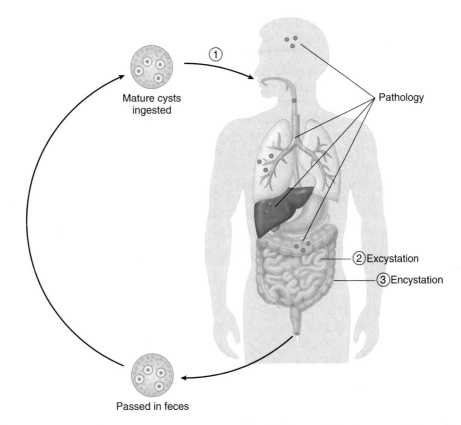

FIGURE C-6 Single-host life cycle of *Entamoeba histolytica*. (1) Infection by *Entamoeba histolytica* occurs by ingestion of mature cysts in fecally contaminated food or water; (2) excystation occurs in the small intestine, and trophozoites are released that migrate to the large intestine; (3) the trophozoites multiply, invade colonic mucosa (causing pathology), and encyst in colon; (4) the cysts are passed in the feces. (Modified from the Centers for Disease Control and Prevention, Atlanta, GA.)

is the most common cause of intracerebral mass lesions and is thought to be caused by reactivation of chronic infection and **progressive focal destruction by the cytolytic tachyzoites** in the absence of an array of immune mechanisms—including cytotoxic antibody, activation of macrophages, and interferon-assisted stimulation of CD8+ T lymphocytes that, in a healthy person, would control the organisms.

Transmission of ***Cryptosporidium parvum*** occurs mainly through ingestion of water contaminated with animal or human excreta. Following ingestion, excystation occurs; the sporozoites are released and parasitize epithelial cells of the gastrointestinal tract. In these cells, the parasites localize themselves in intracellular vacuoles and undergo asexual multiplication (schizogony or merogony) and then sexual multiplication (gametogony), producing microgamonts (male) and macrogamonts (female). On fertilization of the macrogamonts by the microgametes, oocysts develop that sporulate in the infected host. Freshly passed oocysts are infective, allowing **autoinfection** in

a patient with AIDS, with multiple rounds of life cycle in the same host. Protracted diarrhea occurs only in the **absence of cell-mediated immune mechanisms**. The pathogenesis of the diarrhea is unknown; mucosal pathology is unremarkable.

Four species of **Plasmodium**—*P. falciparum, P. vivax, P. ovale,* and *P. malariae,* are known to cause malaria in humans. In a **two-host life cycle**, humans are infected during a blood meal of a *Plasmodium*-infected female *Anopheles* mosquito, which inoculates sporozoites into the human host. Sporozoites infect liver cells and mature into schizonts that rupture and release merozoites. Merozoites infect RBCs. The ring-stage trophozoites mature into schizonts, which rupture, releasing merozoites. The synchronized infection and rupture of RBCs cause heavy parasitemia, manifesting as paroxysmal fever and chills, accompanied by anemia (due to **hemolysis**), with progression to nephrotic dysfunction (immune-complex reaction) or to cerebral dysfunction (owing to **cytoadherence of capillary endothelium by** infected RBCs). The clinical presentation varies depending on the infecting species, the level of parasitemia, and the immune status of the patient. In the RBCs, some *Plasmodium* parasites differentiate into sexual erythrocytic stages (macro- and microgametocytes), which are ingested by an *Anopheles* mosquito during a blood meal. The parasites' sexual multiplication (sporogenic cycle) in the mosquito gut results in release of sporozoites, which make their way to the mosquito's salivary glands. Inoculation of the sporozoites into a new human host perpetuates the life cycle.

Infected rodents are the definitive host for *Trichinella spiralis* tissue parasites, also known as nematodes. *T. spiralis* is found worldwide in many omnivorous and carnivorous animals, which feed on the infected rodents. Humans are accidentally infected when eating improperly processed meat containing cysts (encysted larvae) of *Trichinella*. After exposure to gastric acid and pepsin, the larvae are released from the cysts and invade the small bowel mucosa. Gastrointestinal symptoms develop, **triggered by a large number of parasites**. Adult worms are developed in the small bowel. The female worms release larvae, which migrate into muscle tissues, causing periorbital and facial edema, conjunctivitis, fever (parasitemia), myalgias from **larval encystment in the muscles**, splinter hemorrhages (thrombocytopenia), rashes, and **hypereosinophilia. Local and systemic hypersensitivity reactions** cause much of the pathology.

Hookworms that are infective to humans are two nematode (roundworm) species, *Ancylostoma duodenale* and *Necator americanus*. On contact with the human host, the larvae penetrate the skin, causing local skin manifestations ("ground itch"). The larvae are carried through the veins to the heart and then to the lungs. They penetrate into the pulmonary alveoli, ascend the bronchial tree to the pharynx, and are swallowed. The larvae reach the small intestine, where they reside and mature into adults. Iron deficiency anemia caused by **blood loss at the site of intestinal attachment of the adult worms** is the most common symptom of hookworm infection. Respiratory symptoms (**eosinophilic pneumonitis** [Loeffler syndrome], commonly associated with ascariasis and strongyloidiasis) are associated with migration of the larvae into the lungs.

Filariasis is caused by nematodes (roundworms) that are transmitted by larvae-carrying arthropods during a blood meal. The larvae of *Wuchereria bancrofti* and *Brugia malayi* migrate through lymphatic vessels to lymph nodes and nodules in subcutaneous tissues, where they develop into **microfilariae-producing adult worms**. They cause lymphatic dysfunction (**lymphedema** and elephantiasis). The female worms of *Onchocerca volvulus* migrate in the skin (causing pruritus, dermatitis, and subcutaneous nodules) and invade the eye, causing ocular lesions, sometimes progressing to blindness. **Hypersensitivity reactions to the adult worms** are the most likely cause of the pathology and symptoms of all forms of filariasis.

Adult cestodes (tapeworms) are acquired through ingestion of the larval form, found in poorly cooked or raw meats or fresh water fish. The intermediate host acquires larval forms through ingestion of the adult tapeworm eggs. Humans can serve as both intermediate and definitive hosts. Tissue infection can follow the ingestion of eggs of, or accidental contact with larval forms of, *Taenia solium*, *Echinococcus granulosus*, and several other cestodes. *Diphyllobothrium latum* (fish, or broad, tapeworm) is the largest human tapeworm. After infected fish are ingested, the larvae develop into immature adults and then into mature adult tapeworms that reside in the small intestine. The adult form of *D. latum* attaches to intestinal mucosa by means of the two bilateral grooves of its scolex. Chronic manifestations of diphyllobothriasis include abdominal discomfort, vitamin B_{12} deficiency, and diarrhea, which can be **triggered by large numbers of parasites**. Massive infections may result in **intestinal obstruction,** and **migration of proglottids** can cause cholecystitis or cholangitis.

All **trematode** (fluke) infections, with the exception of schistosomes, are food borne (fresh water fish, mollusks, or plants), and they are emerging as a major public health problem (>40 million people infected with intestinal and liver or lung trematodes). The **development of adult trematodes requires at least two hosts**: a mollusk and a human host (Fig. C-7). The three main species of digenetic blood trematodes (**schistosomes**) infecting humans are *Schistosoma haematobium*, *S. japonicum*, and *S. mansoni*.

Infective cercariae of *Schistosoma* species penetrate the skin of the human host to initiate infection in a two-host life cycle (see Fig. C-7). Schistosomulae-associated **dermatitis** reflects **inflammatory responses** mediated by schistosomulae-specific **IgE. Mast cells** bind to IgE-coated schistosomulae and **degranulate**, releasing cationic basic proteins (e.g., major basic protein), which damage the worms. The schistosomulae then spread to the circulation, and female adult schistosomes deposit eggs in the small venules of the

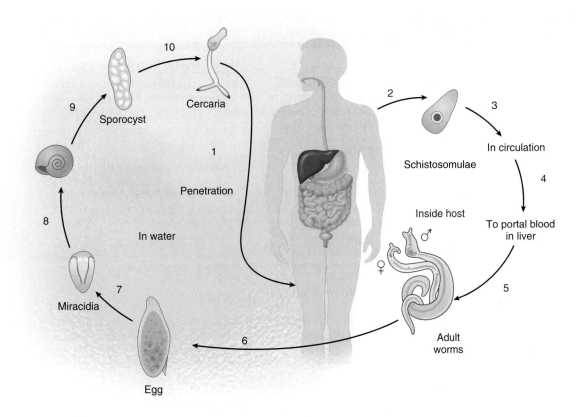

FIGURE C-7 Two-host life cycle of *Schistosoma mansoni*. (1) Free-swimming cercariae penetrate skin; (2) cercariae lose tails and become schistosomulae; (3) schistosomulae in circulation; (4) schistosomulae migrate to portal blood in liver; (5) schistosomulae mature into adult worms; (6) paired adult worms migrate to mesenteric venules of bowel and rectum, lay eggs (passed in feces); (7) eggs (in water) hatch, releasing miracidia; (8) miracidia penetrate snail tissues; (9) miracidia develop into sporocysts in snails (many generations); (10) cercariae are released in water and become free-swimming parasites (infective form). (Modified from the Centers for Disease Control and Prevention, Atlanta, GA.)

portal and perivesical systems. Pathology of infections due to *S. mansoni* and *S. japonicum* includes Katayama fever (**immune complex reaction**) and hepatic dysfunction (caused by a **granulomatous reaction to eggs**). Pathology of *S. haematobium* schistosomiasis includes hematuria, scarring, and calcification in the target organs (**granulomatous inflammation** in response to eggs).

Immunity

Resistance to protozoan parasites involves three interrelated mechanisms: nonspecific factors, cellular immunity, and humoral immunity (Table C-4). However, many worms successfully avoid host defenses in a variety of ways and can survive in the face of otherwise effective host responses. In humans, all immune effector mechanisms to parasitic infections are CD4+ T-cell dependent. CD4+ T cells do not directly kill parasites, however. Intracellular protozoal parasites are contained by macrophages that have been activated by parasite-specific CD4+ T_H1 cells. Helminth infections induce T_H2 responses, characterized by **eosinophilia** and **hypergammaglobulinemia of the IgE isotype** (a hallmark of helminth infections). T_H2 cells induce antiparasite antibodies of the IgE isotype, which exhibit multiple effector functions in the immunity to helminth parasites (see Table C-4). A wide variety of **immunoevasive mechanisms** is patterned to suit the unique features of the parasite and its location within the host. Escape mechanisms (Table C-5) are strategies by which parasites avoid the killing effect of the immune system in an immunocompetent host.

TABLE C–4 Major Mechanisms of Host Immune Response to Control Pathogenic Parasites

Immune Mechanism	Major Targets or Pathogenic Parasites
Antibody-Dependent Cellular Cytotoxicity	All helminthic parasites (e.g., *Ascaris lumbricoides*, *Trichinella spiralis*)
Mediated by **eosinophilic leukocytes** (attracted to the invaders by parasite-specific IgE); eosinophils bind to IgG-coated parasites and degranulate, releasing cationic basic proteins (e.g., major basic protein)	
Mediated by parasite-specific IgE; **mast cells** bind to IgE-coated parasites and degranulate, releasing cationic basic proteins (e.g., major basic protein)	Helminthic parasites: *Schistosoma* spp (skin-embedded schistosomulae are killed) *Trichinella spiralis* (lumen-dwelling adult worms are expelled by peristalsis)
Other Roles of Humoral Immunity	
Opsonization of parasite for phagocytosis by reticuloendothelial system	*Plasmodium* spp
IgG-coated parasites fix complement, leading to the parasite lysis	*Plasmodium* spp
Activation of (parasiticidal) **platelets** by parasite (egg) antigen-specific IgE	*Schistosoma* spp
Activation of macrophages by T_H1 cytokines (e.g., **interferon-γ**) and killing of infected host cells	All protozoa, especially (obligate or facultative) intracellular protozoa (e.g., *Toxoplasma gondii*, *Leishmania* spp)

TABLE C–5 Evasion Mechanisms by Parasites to Avoid Elimination by Host Immune Response

Evasion of Host Defense	Mechanism	Pathogenic Parasites
Antigenic mimicry	Parasites escape immune detection by passively acquiring host blood group agglutinins and MHC antigens	*Schistosoma* spp (adult worms)
Antigenic variation	Parasites change their surface antigens; new antigens "frustrate" and "exhaust" the immune system and escape the immune response to the old antigens	*Trypanosoma* spp *Plasmodium* spp
Blocking of serum factors	Parasites coat themselves with noncytotoxic antibodies that sterically block the binding of cytotoxic IgE, IgG antibodies	Adult worms (e.g., *Schistosoma* spp)
Immunosuppression	Unknown mechanism	Intracellular coccidian protozoa
Intracellular sequestration	Parasites in host cells (e.g., in eye, brain) lacking MHC class I antigens prevent detection by CD8$^+$ cytotoxic T lymphocytes	*Toxoplasma gondii*
Evasion of phagocytic killing	Inhibit macrophage activation by interferon-γ	*Leishmania* spp
	Parasites escape from phagosome	*Trypanosoma* spp
	Prevent phagolysosomal fusion	*Toxoplasma gondii*
	Impervious to lysosomal enzymes	*Leishmania* spp
	Inhibit oxidative burst and production of oxygen radicals	*Leishmania* spp
	Neutralization of superoxide molecules by superoxide dismutase secreted by parasites	*Schistosoma* spp

Diagnostic Methods: Concepts

Dominick Cavuoti, DO

GENERAL PRINCIPLES

The diagnosis of a presumed infectious disease begins with a thorough physical examination and history of the illness. A differential diagnosis is generated, and laboratory (e.g., hematology, serum chemistry) and radiographic studies are performed as necessary to narrow the differential diagnosis. Specimens are collected and submitted to the laboratory, and empirical antimicrobial therapy may be started pending the results of the microbiologic studies. Once an etiologic agent is identified, directed therapy can be instituted (Fig. D-1).

ASSESSING THE PERFORMANCE OF LABORATORY TESTS

No laboratory test is perfect. Therefore, clinicians must have a sense of how reliable the tests are. When a microbiologic test correctly predicts the presence of a pathogen, the result is said to be a **true positive**. Similarly, a negative test obtained in the absence of the pathogen is a **true negative**. A test that is negative in the presence of the pathogen is a **false negative**, and one that is positive in the absence of the pathogen is a **false positive**.

The terms **sensitivity** and **specificity** describe the performance and value of diagnostic tests (Fig. D-2). The sensitivity of a test is the likelihood that it will be positive when the pathogen is present. The specificity of the test measures the likelihood that a test will be negative if the pathogen is not present. In clinical practice, diagnostic tests with 100% sensitivity and

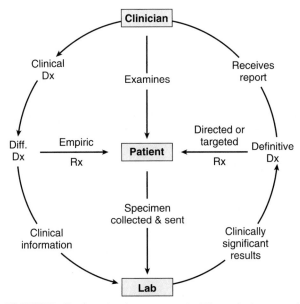

FIGURE D-1 The diagnostic cycle. The cycle is centered around the patient, where the clinical information and all the work-up end with the targeted therapeutic intervention.

Test Results

	Positive	Negative
Organism present	True-positives (TP)	False-negatives (FN)
Organism absent	False-positives (FP)	True-negatives (TN)

$$\text{Sensitivity} = \frac{TP}{TP + FN} \times 100\% \qquad \text{Specificity} = \frac{TN}{TN + FP} \times 100\%$$

$$\text{Positive Predictive Value (PPV)} = \frac{TP}{TP + FP} \times 100\%$$

$$\text{Negative Predictive Value (NPV)} = \frac{TN}{TN + FN} \times 100\%$$

FIGURE D-2 Definitions of terms used in evaluating diagnostic tests.

100% specificity do not exist. **Screening tests,** which rule out a disease from the differential diagnosis, should have very high sensitivity to assure that no patient with the disease is missed. A positive screening test result should be followed by a **confirmatory test** having a very high specificity. Screening tests are usually fairly easy to perform and less technically challenging than are confirmatory tests. The **predictive value** of a diagnostic test is influenced by the frequency of the infection in the population being tested. The interpretation of a test result depends not only on the technical accuracy of the method used but also on the prevalence of the infection in the population.

SPECIMEN COLLECTION

The value of a microbiologic test result is proportional to the quality of specimen. An adequate amount of material, ideally, a few milliliters of fluid or 1 cm³ of tissue, should be submitted. Fluids (collected via needle and syringe aspiration) and tissues are always preferred over submission of specimens on swabs, because only a limited amount of material can be collected on a swab. Swabs should be used only to sample mucosal surfaces (e.g., vaginal or throat swabs) or, for instance, a ruptured eardrum, after the external auditory canal has been thoroughly cleansed. Examples of suggested specimens based on suspected etiology are listed in Table D-1.

Direct specimens are collected from **normally sterile tissues and body fluids** and are used to confirm invasive infections, and sera and sometimes CSF are used directly to detect antibody response.

Meticulous attention to proper skin antisepsis must be maintained when collecting direct specimens.

Indirect specimens are specimens that **have passed through sites colonized with normal flora.** Examples include expectorated sputum or voided urine. Indirect samples are usually easier to collect but more difficult to interpret than are direct specimens. Patients must be given clear and accurate instructions when collecting sputum specimens or clean-catch midstream urine specimens; otherwise, the specimen may become contaminated with the resident microbial flora.

Specimens must be submitted in an appropriate container. In general, specimens should be submitted in sterile, leak-proof containers; stool samples may be submitted in a clean, leak-proof container because sterility is not an issue. Specimens for anaerobes must be submitted in a container devoid of oxygen. When special transport of samples is required, it is wise to contact the laboratory for instructions and supplies. For fluids that are aspirated with a needle and syringe, the needle should be removed, the air expelled, and the syringe capped. This maintains an environment conducive to isolating anaerobes, but it is also acceptable for routine bacterial, fungal, and mycobacterial cultures.

Specimens should be transported to the laboratory as soon as possible after collection (ideally, within 2 hours) to help maintain viability of the organisms and prevent potential pathogens from being overgrown by normal flora. This may not always be possible, however; sterile fluids (e.g., blood, CSF, and joint fluid), surgically collected specimens, and specimens for the isolation of *Neisseria gonorrhoeae* should be considered priority specimens. Refrigeration is

TABLE D–1	Examples of Investigations Using Clinical Specimens to Confirm the Suspected Etiology		
Specimen	**Clinical Diagnosis**	**Methods**	**Suspected Etiology**
Blood (serum)	Syphilis	VDRL, RPR, FTA, MHA-TP serology	*Treponema pallidum*
Throat swab	Pharyngitis	Direct antigen detection or conventional culture	*Streptococcus pyogenes*
Sputum	Pneumonia	Culture	*Streptococcus pneumoniae*
Spinal fluid	Meningitis	Culture	*Neisseria meningitidis*
Stool	Gastroenteritis	Culture	*Salmonella, Shigella, Campylobacter*
Urine midstream urine (MSU)	Urinary tract infection (UTI)	Semiquantitative urine culture	*Escherichia coli*
Genital tract (Cytobrush)	Mucopurulent cervicitis	DFA, cell culture, molecular methods	*Chlamydia trachomatis*
Wound aspirate	Surgical wound infection	Aerobic and anaerobic culture	Mixed infection

recommended for some specimens, especially urine, if transport will be delayed.

The **specimen must be properly labeled** with the appropriate patient information, date and time of collection, initials of collector, ordering physician, and source of the specimen. Additional helpful information would include the suspected diagnosis and current antibiotic therapy, if applicable. The utility of an accurate description of the specimen source cannot be overstated. Knowing the anatomic location can help the laboratory determine whether the culture contains true pathogens or contaminating flora. Whenever an unusual or fastidious pathogen is suspected (e.g., *Corynebacterium diphtheriae* or *Brucella* spp), the laboratory should be notified so that the appropriate transport system and culture media can be used.

It is imperative that the utmost attention to detail be followed when selecting, collecting, and submitting specimens to the microbiology lab. If there is any doubt about how a specimen should be collected or submitted or what is the best specimen for the suspected disease, the laboratory should be contacted before the specimen is collected. This will help avoid having a specimen rejected, prevent the work-up of an inappropriate specimen, or avoid subjecting the patient to another procedure. Maintaining open lines of **communication between the laboratory and clinical staff** benefits everyone involved, especially the patient.

DIAGNOSIS OF BACTERIAL INFECTIONS

The general approaches to laboratory diagnosis vary with different bacteria and anatomic sites investigated. A combination of **direct microscopic examination**, along with **culture**, is generally used. **Serologic studies** to detect antibodies or antigens may be performed in the case of a noncultivable agent or a hard-to-culture agent. **Molecular techniques** may involve the direct, nonamplified detection of an agent through molecular **probes** or amplified techniques (e.g., **PCR**).

Microscopy

Microscopy is used to detect and identify viruses, fungi, and parasites, as well as bacteria. Several methods are available to facilitate the visibility of organisms. **Gram stain**, the most common stain used in the microbiology laboratory, is performed with the following four aims:

1. Separating major groups of bacteria (e.g., **Gram-positive**, **Gram-negative** bacteria [Fig. D-3]).
2. Determining the **quality of the specimen** submitted (important when evaluating potentially contaminated specimens, e.g., sputum)
3. Presumptive diagnosis and **clinical decision making**
4. **Direction** for further investigation

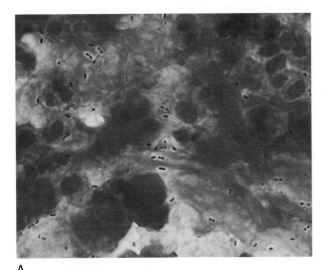
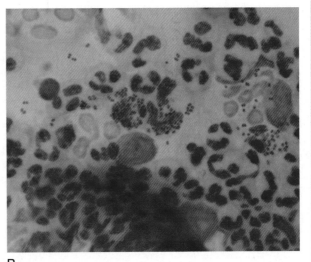

A
B

FIGURE D-3 *A,* Lancet-shaped gram-positive diplococci characteristic of *Streptococcus pneumoniae; B,* Gram-negative diplococci characteristic of *Neisseria* spp.

The **Gram stain procedure** involves a series of steps. Thick specimens (e.g., sputum or purulent exudates) are spread thinly on a glass slide and allowed to air dry. Sterile fluids are centrifuged to concentrate the organisms, which may be present in low numbers. After centrifugation, the sediment is applied to a slide. After drying, the slide is fixed by flooding with 95% methanol and then air dried again. The smear is flooded with crystal violet, rinsed with water, then flooded with an iodine solution, which acts as a mordant (it complexes with the crystal violet). After another rinsing with water, the smear is decolorized with acetone-alcohol. During decolorization, the crystal violet-iodine complex is retained by Gram-positive bacteria (hence, the purple color) and is lost in Gram-negative organisms. The decolorization step is the most critical because over- or under-decolorization can alter the Gram reaction of the organisms. The smear is finally counterstained with safranin, which is retained by Gram-negative organisms (hence, the red color). Because the Gram stain can be done quickly, it can help the physician select an initial antibiotic regimen before the culture investigation is completed. Selected shapes and arrangements can be useful in identifying the genus of bacteria in clinical specimens (Table D-2).

Fluorescence microscopy can be used to detect a variety of organisms. The direct fluorescent antibody (DFA) assay consists of a fluorescent compound bound to an antibody that is specific for a certain organism or group of organisms. It can be used to screen for the presence of an organism in a direct specimen or to confirm the identification of an organism (growing in culture) that is otherwise difficult to detect using any

TABLE D–2 Identification of Genus Based on Gram Stain Characteristics

Gram Characteristics	Examples of Genus
Gram Positive	
Cocci in pairs and chains	*Streptococcus* ("lancet-shaped": *Streptococcus pneumoniae*)
	Enterococcus
	Peptostreptococcus
Cocci in clusters	*Staphylococcus*
	Micrococcus
Rods	*Bacillus* (large; aerobe)
	Clostridium ("box car-shaped": *Clostridium perfringens*)
	Corynebacterium (palisading; aerobe)
	Propionibacterium (pleomorphic; anaerobe)
	Listeria (small)
	Nocardia (branching, filamentous, aerobe)
	Actinomyces (branching, filamentous, anaerobe)
Gram Negative	
Cocci	*Neisseria* (diplococci: "kidney bean-shaped")
	Moraxella catarrhalis (diplococci)
	Veillonella (anaerobe)
Rods	Enterobacteriaceae (*Escherichia coli, Klebsiella, Salmonella*; "safety pin-shaped": *Yersinia pestis*)
	Pseudomonas
	Bacteroides (anaerobe)
	Fusobacterium (anaerobe)
	Haemophilus (pleomorphic)
	Brucella (coccobacillus)
	Vibrio (curved)
	Campylobacter ("seagull" appearance)
	Helicobacter (curved)
Organisms Not Detected on Gram stain include	*Legionella*
	Mycoplasma
	Chlamydophila or *Chlamydia*
	Mycobacterium (may appear as a "ghost" or negative image on Gram stain)
	Coxiella burnetii
	Treponema pallidum

biochemical marker. **Fluorochrome stains** are based not on an antibody antigen reaction but on the binding of the stain to a component of the organism. Examples of fluorochrome stains include auramine-rhodamine (AR) for acid-fast organisms; acridine orange (AO) for bacteria; and Calcofluor white for fungi. These stains offer greater sensitivity in detecting organisms directly from a specimen. A positive AR stain result is usually followed with a conventional acid-fast stain for confirmation, and a positive AO stain is followed by a Gram stain to determine the Gram reaction of the organisms.

Acid fastness is a property of the mycobacteria (Fig. D-4) and related organisms and of the sporozoan parasites (*Cryptosporidium*, *Cyclospora*, and *Isospora*). Acid-fast organisms resist decolorization with an acid alcohol solution. Some organisms (e.g., *Nocardia*) are decolorized by the usual acid alcohol decolorizer but can resist decolorization with a weaker mineral acid and are, therefore, considered partially acid fast. Variants of the acid-fast stain include the Ziehl-Neelsen (hot) stain, the Kinyoun (cold) stain, and the AR stain mentioned in the previous section.

Cultivation of Microorganisms

Most medically important bacteria grow on semisolid nutrient-rich agar media in Petri dishes. Table D-3 lists media that are usually used for routine culture investigation of clinically significant bacteria. Examples of organisms that cannot be cultivated on laboratory media are: *Treponema pallidum*, *Rickettsia*, and *Coxiella*.

TABLE D–3 Routine and Special Bacteriologic Agar Media for Culture Investigation of a Suspected Pathogen

Media	Uses
Enriched media	Media to grow the widest range of bacteria prepared from animal products (e.g., chocolate agar, blood agar)
Selective media	Unwanted organisms can be inhibited (or the culprit organism is selected out) with chemicals or antimicrobial agents (Hektoen or SS agar) for enteric pathogens (*Salmonella*, *Shigella*); Thayer-Martin for *Neisseria. gonorrhoeae*)
Differential media	Characteristic properties of bacteria can be demonstrated by incorporation of a substrate in medium (MacConkey agar)
Anaerobic media	Media for isolation of anaerobic bacteria (e.g., kanamycin, vancomycin laked blood agar, *Bacteroides* bile esculin agar)
Special media for Mycobacteria	Lowenstein-Jensen egg-malachite green agar; Middlebrook 7H10/11 agar

*Incubation temperature and atmospheric conditions can be selective.

A **specimen might be plated on a variety of agar media** and possibly also placed in a broth-type media. A portion of the specimen may be placed on a slide for a Gram stain. Using aseptic technique, a small amount of the specimen is applied to the plate using a sterile loop (Fig. D-5A), swab, or pipet. The specimen is then **streaked** over the agar surface in four quadrants, with the purpose of diluting the specimen to single organisms on the third or fourth quadrant to yield individual colonies. It is important to work with colonies originating from a single organism to produce accurate identification and susceptibility reports. Working with mixed cultures can lead to erroneous results. Colonies from a single organism have consistent and characteristic features, which can aid the technologist in deciding what further testing might be necessary for identification.

The inoculated media are **incubated** at specific temperatures and atmospheres, depending on the specimen source and likely pathogens. One or more microorganisms may grow from a clinical specimen, depending on the source and disease process. The medical technologist **examines the primary culture plates** from overnight incubation and notes the **colony size**, **shape**, and **color**, and the **growth**

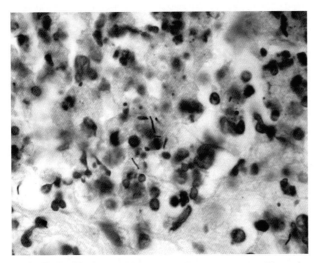

FIGURE D-4 Acid-fast stain of sputum smear. Note the slender red rods of *Mycobacterium tuberculosis*.

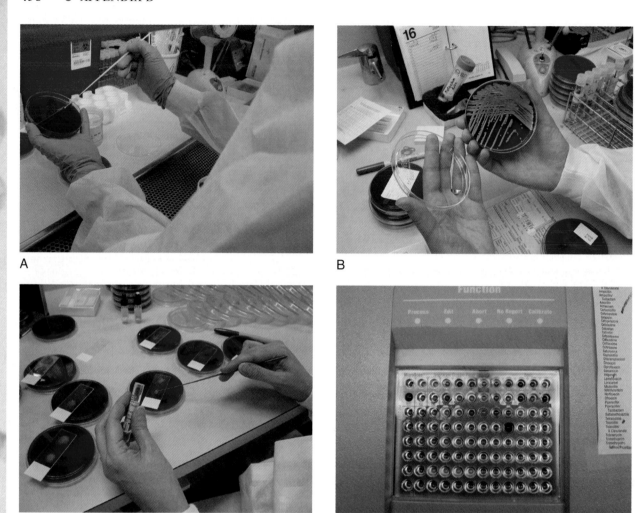

FIGURE D-5 Culture and sensitivity testing in a diagnostic microbiology laboratory. *A*, Planting fresh clinical specimen on standard agar media. *B*, Examination of primary cultures. *C*, Rapid tests for presumptive identification of a significant isolate. *D*, A micro-broth panel with a combination of species identification and antibiotic susceptibility (MIC) testing. *(Courtesy of Diagnostic Microbiology Laboratory, Parkland Memorial Hospital, Dallas, TX.)*

pattern on the various media (see Fig. D-5*B*). The following **primary characteristics** are important for the presumptive identification of the isolate:

- **Gram reaction**, cell morphology (rod or coccus; pairs or chains)
- Cultural characteristics: ability to grow under aerobic or anaerobic conditions, **hemolysis on blood agar** (Table D-4), colony morphology, motility
- Growth requirements: simple or fastidious
- Selective media: growth or no growth
- Presumptive identification of bacterial isolates can be achieved based on **rapid biochemical tests** (see Fig. D-5*C*)

- Ability to produce important enzymes (e.g., catalase, oxidase, and coagulase [Table D-5])
- Motile or nonmotile
- Ability to produce pigments

Final identification of the clinical isolates can be achieved by using a battery of biochemical tests or a commercially available panel (see Fig. D-5*D*). Some common tests for species identification include

- Ability to metabolize sugars (e.g., glucose, lactose, maltose, sucrose, trehalose) fermentatively or oxidatively
- Ability to utilize a range of substrates for growth (e.g., citrate, malonate, acetamide)

TABLE D–4 Microorganisms with Hemolytic Characteristics on Blood Agar Cultures

Type of Hemolysis	Microorganisms
β-**hemolysis** (complete hemolysis of blood with a zone of clearing around the colony)	*Streptococcus pyogenes* (group A)* *Streptococcus agalactiae* (group B) *Staphylococcus aureus* (most) *Listeria monocytogenes* (narrow) *Clostridium perfringens* (double zone, α and β)
α-**hemolysis** (partial hemolysis)	*Streptococcus pneumoniae* viridans streptococci
γ-**hemolysis** (no hemolysis)	*Enterococcus* Group D *Streptococcus* Some viridans streptococci

*Lancefield antigenic classification is based on precipition reactions with homologous antiserum. It distinguishes the streptococci into serogroups A to H and K to V.

- Other enzymatic activities (e.g., tryptophanase, decarboxylases)

Fastidious Gram-negative rods (e.g., *Haemophilus*, *Brucella*, *Francisella*, *Bordetella pertussis*, and *Legionella*) have special nutritional requirements, but their presence should be suspected based on patient history, specimen source, time necessary for detection of visible colonies, Gram stain characteristics, and differential growth characteristics on blood agar, chocolate agar, and specialized media. Most larger labs can presumptively identify them or rule them out with a few tests; however, they are often submitted to reference laboratories for definitive identification.

Antibiotic Susceptibility Tests

After a significant isolate has been identified, the organism may be tested against a battery of antibiotics to help determine appropriate therapy. Testing may include agar dilution, disc diffusion (Fig. D-6*A*), E-test strips (see Fig. D-6*B*), macro-broth dilution (see Fig. D-5*D*), or an automated system using the principles of micro-broth dilution. With the disc diffusion method, the organism is usually first streaked on agar in a Petri dish to confluency. Discs impregnated with various drugs are placed on the agar. The amount of antibiotic in each disc is related to the achievable serum concentration and differs for different antibiotics. After the drugs have had time to work, a clear area appears around the disc of any drug that kills or inhibits the growth of the organism. The area where the drug has eradicated the organism is called the **zone of inhibition** (see Fig. D-6*A*). The zone sizes

TABLE D–5 Presumptive Identification of Microorganisms Based on Rapid Biochemical Tests

Biochemical Properties	Organisms
Catalase (Gram Positive)	
Positive	*Staphylococcus* spp Most aerobic Gram-positive rods (e.g., *Listeria, Bacillus, Corynebacterium*)
Negative	*Streptococcus* spp *Enterococcus* spp
Coagulase (Gram Positive)	
Positive	*Staphylococcus aureus*
Negative	*Staphylococcus epidermidis* *Staphylococcus saprophyticus* There are many other species of coagulase-negative staphylococci.
Oxidase (Gram Negative)	
Positive	*Neisseria* *Haemophilus* *Pseudomonas* *Campylobacter* *Vibrio* *Legionella* (weak)
Negative	Enterobacteriaceae *Stenotrophomonas maltophilia* *Acinetobacter* spp
Motile	*Proteus* (swarming) *Listeria* (tumbling) *Vibrio* (single polar flagellum) *Campylobacter* (polar flagellum) *Pseudomonas* *Legionella* Spirochetes (*Treponema, Borrelia, Leptospira*) Enterobacteriaceae except *Shigella, Klebsiella, Yersinia* (variable depending on species)
Urease Positive	*Proteus* *Helicobacter* *Ureaplasma*
Pigment Producing	*Pseudomonas aeruginosa* (pyocyanin = blue-green; pyoverdin = yellow fluorescent) Some *Mycobacteria* (yellow or orange colonies) *Prevotella* (black pigmented anaerobe) *Serratia* (red colonies)

are compared with those for reference organisms; interpretations of sensitive ("S"), intermediate resistant ("I"), or resistant ("R") are recorded. In the E-test, strips coated with a gradient of antimicrobial on one side and with a minimum inhibitory concentration (MIC) interpretive scale on the other side, are applied to the surface of an agar plate with a lawn of freshly seeded bacteria (similar to the disc diffusion method). Following incubation, MICs are read from the point of intersection of an elliptical zone of inhibition with the interpretive scale (see Fig. D-6*B*).

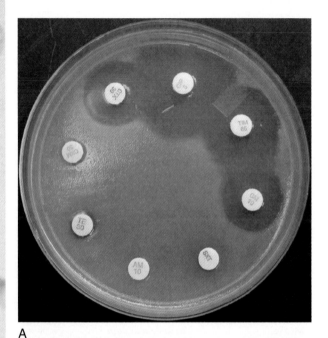

A

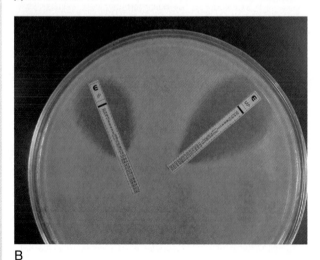

B

FIGURE D-6 Antibiotic susceptibility testing. *A*, Disc diffusion susceptibility of a *Pseudomonas aeruginosa* isolate. Note, at the top of the plate, the isolate is sensitive to four antibiotics (including ciprofloxacin and ticarcillin-clavulanate) and, at the bottom of the plate, it is resistant to four drugs (including ampicillin and tetracycline). *B*, E-test susceptibility of an *E. coli* isolate with imipenem and meropenem strips.

All but the disc diffusion method can give an actual MIC. Most larger laboratories use an automated system of either a microtiter panel (see Fig. D-5*D*) or a multisubstrate card. Although the panels may have a multitude of antimicrobials, the laboratory, in conjunction with the pharmacy and infectious disease departments, decides which antibiotics will be released based on the organism identification, source of the specimen, and age of the patient. Older, less expensive—but equally effective drugs—are reported before newer, more expensive antibiotics, thus allowing for a more prudent usage of antibiotics and slower development of resistance to newer agents. These systems provide in vitro data only. Susceptibility of an organism to a particular agent does not necessarily mean that a particular patient will respond to that agent. Pharmacologic factors, the immunologic status of the host, and the anatomic location of the infection are all important. Laboratories periodically generate antibiograms giving composite susceptibilities of all the significant organisms isolated at the institution. This information is used to trace resistance trends and suggest alternative reporting to slow the development of resistance.

Detection of Microbial Antigens

These methods (usually based on the principles of latex agglutination) suffer from specificity issues and do not usually provide any more information than a well prepared and interpreted Gram stain. Therefore, they are no longer recommended for routine use except in cases of partially treated bacterial meningitis. The agents detected include *Streptococcus pneumoniae* (pneumococcus); *Haemophilus influenzae* type B; *Neisseria meningitidis* Groups A, B, C, Y, and W135 (meningococcus); and Group B streptococcus.

Serologic Investigation

Serologic tests may be used to diagnose recent or chronic infections, determine immune status, or verify a response to vaccination. They can also be helpful in diagnosing some bacterial infections, especially those caused by organisms that are difficult to cultivate (e.g., *Rickettsia*, *Mycoplasma*, *Bartonella*, *Chlamydia/ Chlamydophila*, and *Treponema pallidum*). One of the disadvantages of serology is that the diagnosis may be retrospective, as paired sera (acute and convalescent collected 2 weeks apart) may be necessary. A single serum can be used to demonstrate the presence of IgM, confirming an **acute infection** (current infection; Fig. D-7) in the appropriate clinical setting. Demonstration of an IgG titer elevation (usually a fourfold rise in titer between the acute and convalescent specimens) confirms a **past infection.**

DIAGNOSIS OF FUNGAL INFECTIONS

Diagnosis of fungal infections depends on a combination of clinical observation and laboratory investigation. Superficial or noncomplicated fungal infections often cause characteristic skin or mucosal lesions. With many deep fungal infections, the clinical presentation may be nonspecific, and infections due to bacteria or viruses may be included in the differential diagnosis. Radiologic or other diagnostic imaging methods cannot always distinguish fungal infection from other causes of disease, and laboratory tests (Table D-6) can help establish or confirm a diagnosis, provide objective assessment of treatment response, and monitor resolution of the infection.

Laboratory Methods

To ensure that the most appropriate laboratory tests are performed, it is important for the laboratory to know that a fungal infection is suspected. Background information on underlying illnesses, recent travel or

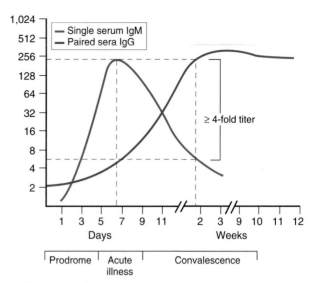

FIGURE D-7 Detection of specific antibody, IgM or IgG. The IgM and IgG titers are measured for nonculturable pathogens that cause invasive infection. These titers are determined to confirm the acute infection and the infection of recent past, respectively.

previous residence in another fungus-endemic area, recent animal contacts, and the patient's occupation are also important.

Direct microscopic examination of clinical specimens is one of the simplest, most rapid, and most helpful tests for laboratory diagnosis of fungal

TABLE D–6	Fungal infections, Causative Agents, and Specimens for Laboratory Diagnosis			
Fungal Diseases	**Fungal Agent**	**Anatomic Location**	**Disease**	**Specimen Collection**
Cutaneous mycoses	Malassezia	Dead layer of skin	Tinea versicolor	Skin scraping
	Dermatophytes: Trichophyton Epidermophyton Microsporum	Skin, hair, nail	Dermatophytosis (ringworm)	Hair, skin or nail scrapings
Subcutaneous mycoses	Sporothrix	Subcutis, lymphatics	Sporotrichosis Mycetoma	Biopsy of lymph node and infected tissue
Systemic mycoses	Primary (systemic) pathogens: Histoplasma Blastomyces Coccidioides	Deep organs (e.g., lungs, blood stream, kidney, CNS) and skin (dissemination)	Histoplasmosis Blastomycosis Coccidioidomycosis	Respiratory secretions, blood, CSF, and tissue biopsy
Opportunistic mycoses	Opportunistic pathogens: Cryptococcus Candida Aspergillus	Deep organs (e.g., lungs, blood stream, CNS) (dissemination)	Cryptococcosis Candidiasis Aspergillosis	CSF, blood, tissue Blood, tissues Respiratory specimens, tissue
	Zygomycetes: Rhizopus Absidia Mucor		Zygomycosis	Biopsy of tissues

infection. **Skin scrapings** or pus from a lesion are mounted in potassium hydroxide (KOH) on a slide. The KOH preparation clears away all organic material except fungi, making the fungi easier to visualize. Unstained wet-mount preparations are examined by lightfield illumination, or dried smears are stained and examined by light microscopy. Because of their unique cell walls and membranes, fungi are not well stained by routine bacterial or tissue stains. Special stains (e.g., Calcofluor white, Gomori methenamine silver [GMS], and periodic acid Schiff [PAS] stains) identify fungi in tissue and even distinguish between various groups of fungi (Table D-7; Fig. D-8). Calcofluor white can also be combined with KOH, and tissue examined using fluorescence microscopy.

A definitive diagnosis requires **culture and identification**. Fungi grow on most routine bacteriologic media, but specialized media augment their growth and improve the rate of isolation and identification. Antibiotics can be added to prevent bacterial contamination and overgrowth. When dimorphic fungi are suspected, two sets of plates can be incubated at different temperatures to grow yeast and mold forms.

Yeasts typically have large, moist, white-to-tan **colonies** on fungal media (Fig. D-9A). A saline wet mount of the growth is examined to determine if the morphology is consistent with *Candida* or *Cryptococcus*, the two most commonly isolated yeasts. Yeasts can be definitively identified using a combination of biochemical testing and macroscopic and microscopic morphology.

Mycelial or mold phases of fungi are best visualized using potato dextrose agar. The **colony color** and **growth** characteristics on the agar surface (see Fig. D-9B) are noted, as is morphology on the underside of the colony. A lactophenol cotton blue preparation is examined (see Fig. D-8B) using light microscopy to determine the genus and, when necessary, the species of the mold. Size, pigmentation, and degree of septation of the hyphae, coupled with the morphology of the fruiting structures, are considered when identifying a fungus. When fungi are isolated from a sterile site, there is usually no question about their significance. However, fungi isolated from nonsterile sites (e.g., sputum) may be difficult to interpret, as they may merely be colonizers.

Respiratory tract fungal infections are best diagnosed with biopsy and histologic examination to determine actual tissue invasion. **Serology** may also be helpful in identifying invasive fungal disease. Latex agglutination may favor detection of IgM antibodies, and double immunodiffusion and complement fixation usually detect IgG antibodies. Some EIA tests are being developed to detect both IgG and IgM antibodies. Other tests can detect circulating fungal antigens (e.g., *Aspergillis*, *Candida*).

DNA probes have shortened the length of time required for identification of some dimorphic fungal pathogens (e.g., *Histoplasma capsulatum*, *Coccidioides*

TABLE D–7	Commonly Used Stains and Applications in Direct Microscopy of Fungal Pathogens	
Stain	Principle	Application
Calcofluor white	Fluorochrome that binds to cellulose in fungal cell walls; fluoresces blue-white under ultraviolet (UV) light	Detection of fungi in clinical specimens
India ink	Used to highlight the capsules of bacteria or yeast cells, which appear as a clear halo around the organism against a black background	Demonstrate bacterial and cryptococcal capsules. Mainly used for CSF.
Gomori methenamine silver (GMS) stain	Used to stain fungi and the cyst stage of *Pneumocystis*	Primarily used to identify: *Pneumocystis* in bronchoalveolar lavage or lung and fungi in any tissue or cytologic specimen
Gram stain	Differential bacterial stain; also stains yeasts	*Candida* and *Cryptococcus* are usually Gram positive
Lactophenol cotton blue (LPCB)	Strongly acidic dye good for fungi	Primarily used to stain filamentous fungi and their characteristic conidial structures in culture

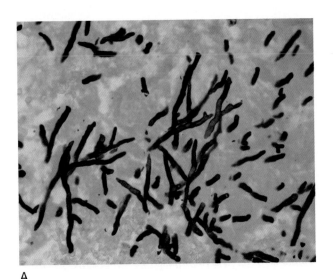
A

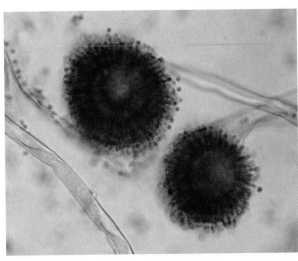
B

FIGURE D-8 Laboratory diagnosis of fungal infection. *A,* Branching septate hyphae suggestive of *Aspergillus.* Methenamine silver stain. *B, Aspergillus flavus.* Characteristic fruiting structures and hyphae. Tease mount of aerial growth stained with lactophenol cotton blue.

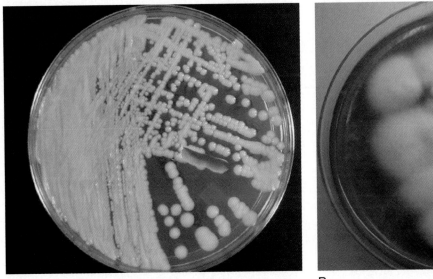
A

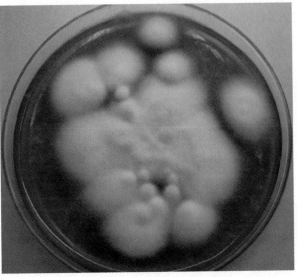
B

FIGURE D-9 Macroscopic features of yeast and mold. *A,* Pasty colonies of yeast. *B,* Aerial (mycelial) growth of mold (*view from the top*).

immitis, and *Blastomyces dermatitidis*) once they have been isolated in culture.

Skin testing (dermal hypersensitivity) was formerly popular as a diagnostic tool but is now discouraged because it may interfere with serologic studies by causing false-positive results. It still may be used to evaluate a patient's immunity, as well as a population exposure index in epidemiologic studies.

DIAGNOSIS OF PARASITIC INFECTIONS

The most commonly submitted specimen for **ova and parasite** examination is stool; however, blood, urine, sputum, duodenal fluid, and tissues are also used. Stool specimens are usually submitted in a two-vial transport system. A vial with 10% formalin preserves protozoan cysts and helminth ova and larvae, and another vial with polyvinyl alcohol (PVA) also preserves protozoan trophozoites. Other systems use a single vial and detect all parasite stages.

Methods for Diagnosis of Protozoan Infections

The concentrated specimen from the formalin vial is viewed microscopically using a wet mount with an iodine or saline solution. The iodine is taken up by the cysts and gives them some contrast against the background of fecal matter that can sometimes obscure the parasites. Trophozoites are usually lysed by the formalin or sufficiently distorted such that they cannot

be accurately identified. A trichrome or iron hematoxylin stain made from the PVA vial is used for identification of protozoan trophozoites and cysts.

Diagnosis of most amebic infections is based on microscopy of stained specimens (e.g., stool). However, direct wet prep of CSF is recommended for identification of free-living amebas (e.g., *Naegleria* spp). Corneal scrapings and skin or brain biopsies are recommended for *Acanthamoeba* species. It is important to measure and note the **internal structure of trophozoite and cyst stages** (Fig. D-10 *A, B*) of pathogenic and commensal protozoa. An understanding of the detailed structures of these stages is important in differentiation of pathogenic *Entamoeba histolytica* from the nonpathogenic amebas (e.g., *Entamoeba coli*, *Entamoeba hartmanni*, *Entamoeba polecki*

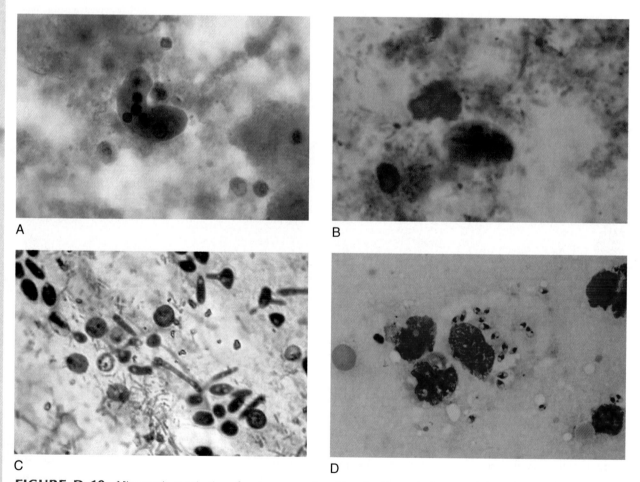

A

B

C

D

FIGURE D-10 Microscopic examination of protozoan parasites. Note the diagnostic features. *A*, A trophozoite from blood stool showing internalized RBCs and characteristic nucleus (diagnostic of *Entamoeba histolytica*). *B*, A characteristic cyst from stool (*Giardia lamblia*). *C*, Oocysts (*red*) in fecal smear after acid-fast staining (*Cryptosporidium parvum*). *D*, Intracellular amastigotes in a histiocyte from bone marrow smear using Giemsa stain technique (*Leishmania donovani*).

[rarely found], *Endolimax nana*, and *Iodamoeba bütschlii*). It is often not possible to differentiate *E. histolytica* from the commensal *E. dispar* by microscopy alone.

Oocysts of the coccidian parasites (*Cryptosporidium, Cyclospora, and Isospora* [see Fig. D-10*C*]) can be detected using an acid-fast stain on stool specimens. Enzyme immunoassays and immunofluorescence assays are available for detecting *Giardia lamblia* (a flagellate protozoan) and *Cryptosporidium.*

Blood or tissue protozoans (e.g., trypanosomes, *Leishmania, Babesia,* malarial parasites) are normally identified to the species level based on geographic exposure history, clinical symptoms, and morphology of the parasite. Both thick and thin blood smears, stained with Wright or Giemsa stains (see Fig. D-10*D*), are examined.

Serologic testing is most commonly used for the diagnosis of extraintestinal *E. histolytica* infections. These most commonly present as liver abscesses. The indirect hemagglutination (IHA) test has been the standard for routine serodiagnosis of amebiasis.

Identification of the cause of a parasite infection is essential for successful treatment, for understanding the epidemiology of the parasite, and for implementing control measures. To date, none of the morphology-based laboratory techniques, which rely on finding the causative agent, is entirely satisfactory; and **molecular approaches** (e.g., PCR) are being developed as alternatives.

Methods for Diagnosis of Helminth Infections

Helminth eggs and larvae are identified using wet mount techniques similar to those used for protozoan parasites. The common **nematodes** (roundworms) reside in the gastrointestinal tract and are thus detected by stool examination. The adult female pinworm (*Enterobius vermicularis*) deposits eggs around the anal opening, so ova should be collected from this site via the "Scotch tape" preparation method, then applied to a glass slide and examined microscopically. Muscle biopsy is recommended for cases of suspected *Trichinella spiralis* infection. *Strongyloides stercoralis* normally inhabits the small bowel but in immunocompromised hosts can cause disseminated infection and may be found in virtually any tissue or fluid. Diagnostic characteristics of rhabditiform-staged *Strongyloides* spp larvae include a length of 200 μm to 250 μm (up to 380 μm), a short buccal cavity, and a prominent genital primordium, i.e., organs that will

become the genitalia of the organism (Fig. D-11*A*). Diagnosis of **filarial infection** is made by identifying larval worms (**microfilariae** [see Fig. D-11*B*]) in the blood, body fluids, or skin, depending on the suspected agent.

The diagnosis of **cestode** (tapeworm) infections is also based on the **identification of eggs** or **proglottids** (see Fig. D-11*C*) in the **ova and parasite examination**. In the deep-seated infections due to *Taenia solium* and *Echinococcus granulosus*, serologic diagnosis is available, and diagnostic imaging provides helpful clues.

The diagnosis of **trematode** (fluke) infections is based on the identification of characteristic eggs in stool specimens; however, sputum is used for identification of *Paragonimus westermani* (oriental lung fluke), and urine for *Schistosoma haematobium* (see Fig. D-11*D*).

DIAGNOSIS OF VIRAL INFECTIONS

A variety of test methodologies are used for diagnosing viral infections, including

- Light microscopy (histopathology or cytopathology)
- Electron microscopy
- Cell culture (traditional tube culture or rapid shell vial)
- Serodiagnosis
- Antigen detection
- Nucleic acid detection (nonamplified or amplified)

Although molecular techniques (e.g., PCR) have revolutionized the diagnosis of viral infections, these techniques are not available in all laboratories, and conventional techniques are still commonly used. Although *Chlamydia* are bacteria, they are similar to viruses in that they are obligate intracellular pathogens. Many of the techniques used in clinical virology are also used in the detection and cultivation of *Chlamydia* in clinical samples.

Light Microscopy

Cytologic preparations used for virus identification include cervical smears stained by the Papanicolaou method (Pap smears) for HPV, and the Tzanck smear (Fig. D-12*A*), which detects members of the herpesvirus family, namely herpes simplex virus (HSV) and varicella zoster virus (VZV). Histopathologic tissue sections stained with H & E are also used for virus detection.

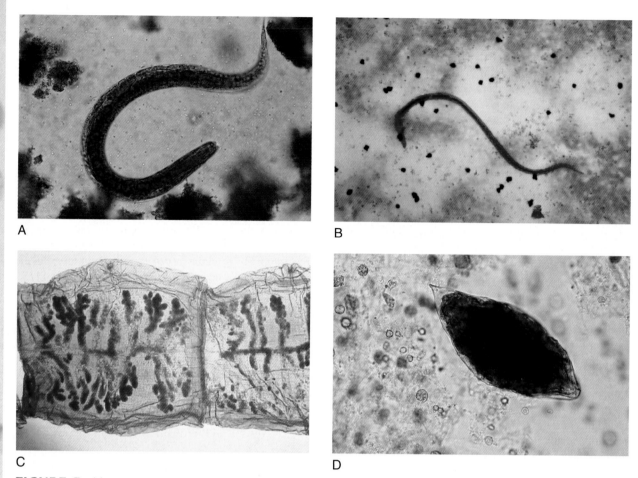

FIGURE D-11 Microscopic examination of helminthic parasites. Note the diagnostic features. *A,* A rhabditiform larva from fecal smear (*Strongyloides stercoralis*). *B,* Microfilaria in a thick blood smear using Giemsa stain technique (*Wuchereria bancrofti*). *C,* Tapeworm proglottids with characteristic uterine branching stained with India ink (*Taenia solium*). *D,* A trematode egg (in urine) with its characteristic vestigial spine (*Schistosoma haematobium*).

Typically, DNA viruses have an intranuclear inclusion (see Fig. D-12*B*), whereas RNA viruses have cytoplasmic inclusions. Not all viruses have distinct inclusions, and other identification techniques may be employed, such as immunohistochemical or immunocytochemical stains or molecular methods (e.g., in-situ hybridization or amplification techniques). Typical inclusions of some commonly encountered viruses are shown in the subsequent figures.

Electron Microscopy

Electron microscopy (EM) may be used to detect viruses associated with gastrointestinal disease, with specimens that have high titers of viruses that are not detectable using standard methods, and for identifying new or unusual viruses. As more sensitive techniques are developed, EM for routine clinical use has declined and is now mainly a research tool.

Cell Culture

As viruses are obligate intracellular pathogens, they require living cells for growth and do not grow on the standard media used for bacteria. There are a number of cell lines available for the cultivation of viruses; however, not all viruses of clinical significance are readily cultivable. Commonly detected viruses in cell cultures are

● Respiratory viruses: respiratory syncytial virus (RSV), adenovirus, parainfluenza, influenza, rhinovirus

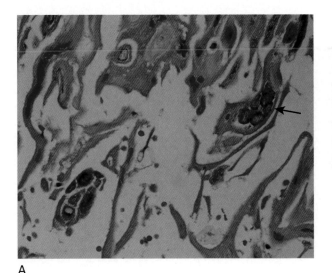

A

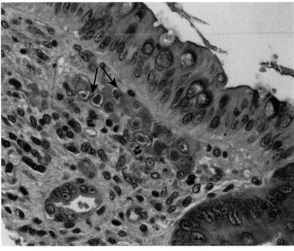

B

FIGURE D-12 Cytologic examination of infected cells. *A*, Nuclear inclusions in squamous epithelial cells (multinucleation [*arrow*]; characteristic of HSV infection). *B*, Typical "owl eye" nuclear inclusions (*arrows*) and granular cytoplasmic inclusions (CMV).

- Herpesviruses: HSV, VZV, cytomegalovirus (CMV)
- Enteroviruses

Specimens are inoculated onto a monolayer of cells in a conventional tube or a shell vial, along with liquid maintenance media and antibiotics (to prevent overgrowth of contaminants). The time to detection of viruses in conventional tube cultures ranges from less than 1 day for HSV to 3 weeks for some strains of CMV. Virus replication can be detected by a number of methods, including

- Development of **cytopathic effects** (CPE), manifesting commonly as rounding and lysis of infected cells (Fig. D-13)
- Adsorption of erythrocytes to virus-infected cells (**hemadsorption**)
- **Interference** with development of cytopathic effect induced by a known virus
- Detection of viral antigens in infected cells (detected by **fluorescent antibody** technique [Fig. D-14])

Serology

Molecular tests and cell culture provide a more direct evaluation of viral infection than the sometimes retrospective diagnosis provided by serology. However, serologic tests are still used to determine immune status or response to immunization and to diagnose

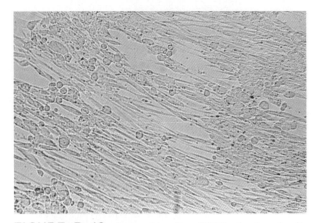

FIGURE D-13 Cytopathic effect (CPE) of cultured cells infected with herpes simplex virus in clinical specimen. Note the rounding and detachment of fibroblastic cells from the monolayer culture. *(Courtesy of Diagnostic Virology Laboratory, Parkland Memorial Hospital, Dallas, TX.)*

infections with viruses that are difficult to cultivate (e.g., EBV, HIV, hepatitis viruses, encephalitis viruses).

Antibody responses can be detected by a variety of methods including: enzyme immunoassay (EIA), immunofluorescence (IFA) assay, complement fixation, and Western blot. IgM antibodies are indicative of acute infection, whereas IgG antibodies are indicative of recent or past infection. EIA and IFA are based on similar principles but differ in the molecules used for detection.

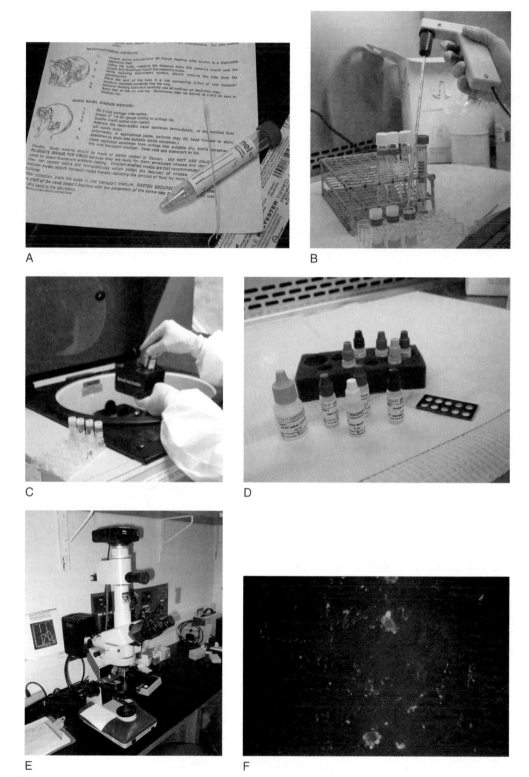

FIGURE D-14 A sequence of steps leading to diagnosis of viral infection (e.g., influenza) by culture and fluorescence microscopy. *A,* Specimen collection. *B,* Preparation of shell-vial culture. *C,* Centrifugation of virus-containing specimen on the cell layer to facilitate infection. *D,* Direct fluorescence antibody staining on a slide of the scraped infected cells from a cell vial culture. *E,* Examination of fluorescence-tagged infected cells by a microscope illuminated with UV light. *F,* Apple-green fluorescence signifying a positive result of a presumed viral infection (influenza). *(Courtesy of Diagnostic Virology Laboratory, Parkland Memorial Hospital, Dallas, TX.)*

Agglutination assays begin with viral antigens bound to erythrocytes or latex particles. Patient serum is mixed with the antigen, and visible clumping of erythrocytes or latex particles indicates the presence of antibodies to the virus.

Complement fixation (CF) is an older method that is less sensitive than other methods. In the CF assay, virus-specific antibodies bind to the added viral antigen, causing complement to bind to the antigen-antibody complexes. Next, antibody-coated erythrocytes are added. Because all of the complement is bound to the antigen antibody complexes, there is no complement-mediated lysis of the erythrocytes. Therefore, the lack of hemolysis indicates a positive test. If no complement-fixing antibodies are present, then complement is free to lyse the erythrocytes, indicating a negative test.

The most common **immunoblot** assay is the Western blot, used to confirm a positive EIA for HIV infection. The HIV virus is lysed, and the viral proteins are separated on a gel by electrophoresis. The gel is blotted on nitrocellulose, and the patient serum is added. Virus-specific antibodies bind to areas on the nitrocellulose corresponding to the separated proteins. Enzyme-linked antihuman antibodies are added, followed by substrate to visualize the various antibodies.

The EIA, IFA, and agglutination techniques used to detect antibody responses can also be used to **detect antigens**. The bound antigen used in an antibody assay is replaced by an antibody. Antigen present in the specimen binds to the antibody and is detected with the corresponding compound (enzyme, fluorescent molecule, latex, and so on). Antigen detection methods can be used to detect viruses directly in clinical samples or to identify viruses growing in culture. IFA techniques used for antigen detection are usually referred to as direct fluorescence assays (DFA). The CMV antigenemia assay is used for the diagnosis and management of systemic CMV disease in immunocompromised patients. Circulating leukocytes (primarily neutrophils) are separated from a blood sample and stained, usually via immunofluorescence, with an antibody to a CMV phosphoprotein, pp65. The presence of pp65 equates with active infection. The results are reported as the number of fluorescing or infected cells per total leukocytes counted. DFA techniques are commonly used to detect respiratory viruses and members of the herpesvirus family, such as HSV and VZV.

Molecular Methods

Molecular diagnostic testing, particularly PCR, has revolutionized the clinical microbiology lab. Once difficult to cultivate or organisms not possible to cultivate can now be detected rapidly and with high sensitivity and specificity. The three main steps in PCR are (1) extraction of the target nucleic acid; (2) amplification of the target (Fig. D-15); and (3) detection of the target.

Extracted target DNA is combined with oligonucleotide primers that are complementary to the target DNA in the presence of DNA polymerase and deoxyribonucleotides. The mixture is heated, causing denaturation of the target DNA into two strands. The temperature is then lowered, allowing the primers to bind to the complementary segments of the target DNA. Then the polymerase and the deoxyribonucleotides make two new segments of double-stranded DNA that are identical to the target DNA. This sequence is repeated multiple times, with the number of target molecules doubling with each cycle. This technology has also been applied to the detection of RNA through the use of a reverse transcriptase that creates a DNA copy of the RNA target. In "real time" PCR, the extraction step is done separately but the amplification and detection steps are done in a closed system. The most common current applications for viral infections include

1. Enteroviral meningitis
2. HSV encephalitis
3. HIV diagnosis and management
4. HCV diagnosis and management
5. CMV in transplant and HIV patients
6. EBV and the post-transplant lymphoproliferative disorder
7. Respiratory viruses
8. Progressive multifocal leukoencephalopathy associated with JC virus in HIV patients
9. BK virus infection in renal transplant patients

PCR can be qualitative (diagnostic) or quantitative (to determine efficacy of therapy). HSV PCR can be performed on CSF for the diagnosis of encephalitis, thus replacing open brain biopsy, and PCR for enterovirus can be performed on CSF in cases of aseptic meningitis. Examples of other molecular methods based on amplification include the **ligase chain reaction** and **branched DNA signal amplification**.

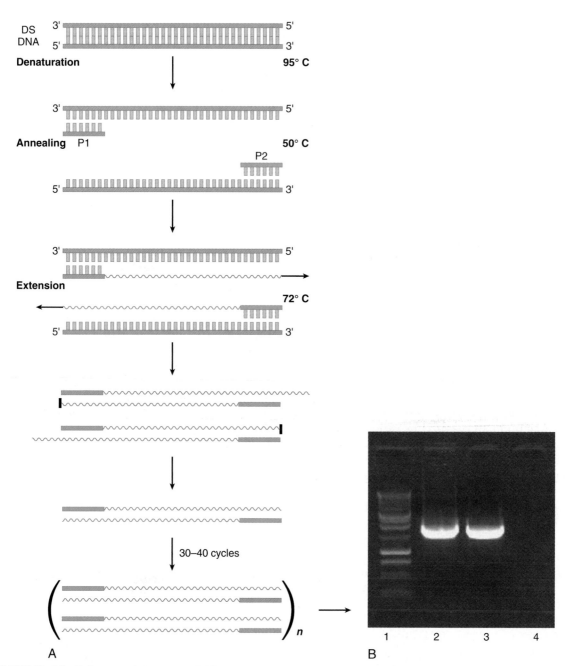

FIGURE D-15 Polymerase chain reaction (PCR). *A,* Denaturation, annealing, and primer extension steps are repeated for many cycles (40 to 50) to produce a large number of copies of primed DNA sequence. *B,* A significant amount of primed and amplified product is visualized in an agarose gel (2%). Lane 1 is the molecular base pair standard, the second and third lanes are amplified product from PCR reactions using the same primer pair on two different strains of bacteria (demonstrating the presence of the common sequence sequestered by the primers), and the fourth lane is a negative control template (*E. coli* genomic DNA), showing no product. The size of the band is 1.5 kb or 1500 base pairs.

Antimicrobial Therapy

APPENDIX E	Antimicrobial Therapy		
Important Drugs	**Mechanism of Action**	**Drug Resistance**	**Clinical Uses**
Antibacterial Drugs			
Natural penicillins (e.g., Penicillin G and V)	Target of action: **final stage of cell-wall synthesis** Mimic D-ala end of peptide; preferentially bind to penicillin binding proteins (PBPs) located inside the bacterial cell wall Irreversibly inhibit transpeptidase activity and cross-linking of cell wall Activate autolysin by interfering with an autolysin inhibitor (bactericidal action)	**Common** occurrence (e.g., *Streptococcus pneumoniae*, *Neisseria gonorrhoeae*) **Mechanism: drug inactivation** (production of a drug-inactivating enzyme); organisms that produce β-lactamase are resistant to the penicillins Penicillin-resistant *Streptococcus pneumoniae* occur by alteration of target site by mutation in the coding gene for a primary PBP.	**Narrow spectrum** and **bactericidal activity** against: Gram-positive cocci (e.g., *Streptococcus pyogenes*, *Enterococcus* spp, *Streptococcus pneumoniae*) Gram-negative cocci (*Neisseria gonorrhoeae*) Spirochetes (e.g., *Treponema pallidum*) Gram-positive rods (e.g., *Corynebacterium diphtheriae*, *Bacillus anthracis*, *Listeria monocytogenes*) Anaerobes (e.g., *Actinomyces israelii*) Pen G is acid sensitive (parenteral use only); pen V is acid-resistant (orally given); long-acting benzathine penicillin G (IM) is also widely used against most infections
Penicillinase-resistant penicillins (e.g., **nafcillin**, oxacillin, cloxacillin)	As for natural penicillins	**Common** occurrence (e.g. staphylococci); staphylococci that are resistant to these drugs are resistant to all β-lactams Mechanism: **production of additional functional target** with poor drug affinity: PBP2a (coded by *mecA*), a new bacterial protein with peptidase activity, but low affinity for these drugs	Narrow spectrum, bactericidal activity against: *Staphylococcus aureus* *S. pyogenes* *Streptococcus viridans*

Continued

APPENDIX E Antimicrobial Therapy—Cont'd

Important Drugs	Mechanism of Action	Drug Resistance	Clinical Uses
Antibacterial Drugs—Cont'd			
Broad spectrum penicillins (e.g., **ampicillin and amoxicillin**)	As for natural penicillins	**Common** occurrence Mechanism: **altered target** or **drug inactivation;** mutations in genes encoding PBPs lead to resistance; some organisms (e.g., *Haemophilus influenzae*) that produce β-lactamase are also resistant to these drugs	**Broad spectrum** and **bactericidal** activity against Gram-positive bacteria and some Gram-negative bacteria, including: *Haemophilus influenzae* *Escherichia coli* *Listeria monocytogenes* *Proteus mirabilis* *Salmonella* Enterococci The drugs are less active against *S. pyogenes, S. pneumoniae,* and *S. agalactiae* than natural penicillins. Both are acid stable (orally given); amoxicillin has greater oral bioavailability than ampicillin May be combined with clavulanic acid (β-lactamase inhibitor) to achieve broader spectrum of coverage and protection from β-lactamase
Extended spectrum penicillins (e.g., piperacillin, ticarcillin, mezlocillin, carbenicillin)	As for natural penicillins	Uncommon; susceptible to β-lactamase produced by some resistant bacteria	**Broad spectrum, bactericidal** activity against: *Pseudomonas aeruginosa* Multidrug-resistant Gram-negative rods (e.g., *Enterobacter* spp, *Klebsiella pneumoniae, Serratia* spp, *Proteus* spp) Use with a β-lactamase inhibitor (e.g., tazobactam, clavulanic acid) has allowed broader coverage
Monobactams (e.g., aztreonam)	As for natural penicillins	Rare; stable against most β-lactamases	**Broad spectrum, bactericidal** activity against aerobic Gram-negative rods only and include *P. aeruginosa* Multidrug-resistant Gram-negative rods (as for broad spectrum bactericidal) No Gram-positive or anaerobic activity
Carbapenems (e.g., imipenem, meropenem, ertapenem)	As for natural penicillins	Rare; stable against most β-lactamases	**Very broad spectrum, bactericidal** activity against most Gram-positive cocci and Gram-negative rods (including resistant bacteria), and anaerobes Imipenem is combined with cilastatin, a reversible, competitive inhibitor of dehydropeptidase-1 (found in the renal tubule that metabolizes imipenem)

APPENDIX E   Antimicrobial Therapy—Cont'd

Important Drugs	Mechanism of Action	Drug Resistance	Clinical Uses
Antibacterial Drugs—Cont'd			
Cephalosporins (1st generation; e.g., **cefazolin**, cephalothin, **cephalexin**, cefadroxil)	Target of action: **final stage of cell-wall synthesis** (same as penicillins) preferentially bind to penicillin binding proteins (PBPs) located inside the bacterial cell wall Irreversibly inhibit transpeptidase activity and cross-linking of cell wall Activate autolysin by interfering with an autolysin inhibitor	**Common** occurrence Mechanism: **altered target** or **drug inactivation**; mutations in genes encoding PBPs lead to resistance; some organisms that produce β-lactamase are also resistant to these drugs	**Narrow spectrum** and **bactericidal activity** against Gram-positive bacteria (e.g., *S. aureus* and *Streptococcus pyogenes*) These parenterally administered drugs are less active against Gram-negative bacteria than are 2nd, 3rd, and 4th generation cephalosporins
Cephalosporins (2nd generation: **cefuroxime**, **cefoxitin**, cefotetan, cefaclor)	As for 1st generation	**Common** occurrence Mechanism: **altered target** (PBPs) or **inactivation of drug** by β-lactamases or **reduced uptake** of drugs into the Gram-negative bacteria	**Broad spectrum** and **bactericidal activity** against Gram-positive and Gram-negative bacteria (similar to broad-spectrum penicillins) These drugs (that are mostly parenterally administered and less often orally given) are less active against Gram-positive bacteria than are 1st generation cephalosporins and more active against Gram-negative bacteria than are the 1st generation cephalosporins Cefoxitin and cefotetan have good activity against anaerobes (e.g., *Bacteroides fragilis*)
Cephalosporins (3rd generation: **cefotaxime**, **ceftriaxome, ceftazidime**)	As for 1st generation	**Uncommon** occurrence Mechanism: **altered target** (PBPs) or **inactivation of drug** by β-lactamases or **reduced uptake** of drugs into the Gram-negative bacteria	**Extended spectrum** and **bactericidal activity** against many Gram-negative bacteria **Cefotaxime** and ceftriaxone have some antimicrobial activity against Gram-negative bacteria (including *Enterobacteriaceae*) and Gram-positive aerobes (particularly *Streptococcus pneumoniae*). Good for empirical and directed therapy of meningitis. Ceftriaxone (long half-life) is DOC for gonorrhea. Ceftazidime and cefoperazone are antipseudomonal cephalosporins. These drugs have less activity against Gram-positive bacteria
Cephalosporins (4th generation: **cefepime**)	As for 1st generation	Uncommon	**Bactericidal** activity against *Staphylococcus, Streptococcus,* and *Pseudomonas* and multiply resistant Gram-negative bacteria

Continued

APPENDIX E Antimicrobial Therapy—Cont'd

Important Drugs	Mechanism of Action	Drug Resistance	Clinical Uses
Antibacterial Drugs—Cont'd			
Glycopeptide inhibitors (e.g., **vancomycin**, teicoplanin)	Target of action: **inhibition of cell-wall synthesis**; inhibits cell-wall synthesis by complexing with D-alanyl-D-alanine, shielding it from transpeptidation reaction; cross-links are not made in the growing cell wall	Uncommon, but seen in *Enterococcus faecium* with *vanA* resistance gene	**Bactericidal activity** against Gram-positive bacteria. No activity against Gram-negative bacteria, because porin channels in the outer membrane of Gram-negative bacterial cell wall are small for the large glycopeptide molecule. Susceptible species include methicillin-resistant *S. aureus*, *Staphylococcus epidermidis*, penicillin-resistant *Enterococcus*. Major indications for use include line-associated infections and other serious infections due to these Gram-positive bacteria
Aminoglycosides (e.g., streptomycin, **gentamicin, tobramycin**, amikacin)	Target of action: **protein synthesis (30S ribosomal subunit)**; inhibit initiation and elongation of translation by "misreading"; the acceptor site on the 30S subunit binds inappropriately activated tRNAs, leading to amino acid substitutions, with resultant nonfunctional proteins. This effect is concentration dependent	**Common** occurrence. Multiple mechanisms: (1) Drug inactivation by enzymatic modification (major), involving transferase enzymes (e.g., acetylase, adenylase, and phosphorylase); or (2) **altered target** (ribosomal binding sites); or (3) **reduced uptake** of drug across outer membrane	**Moderately broad spectrum** and **bactericidal** activity against *S. aureus*, streptococci (except *S. pneumoniae*), and aerobic and facultative anaerobic Gram-negative rods. Antipseudomonal activity: tobramycin >> gentamicin. Antimycobacterial activity: streptomycin. Anaerobic bacteria are not susceptible to aminoglycosides due to lack of active transport system, necessary for uptake of these drugs through the cell membrane. These drugs may be used once daily because of post-antibiotic effect
Tetracyclines (e.g., tetracycline, **doxycycline**, minocycline)	Target of action: **protein synthesis (30S ribosomal subunit)**; reversibly inhibit elongation of protein synthesis by binding to the acceptor site of the amino acyl tRNAs at the 30S ribosome	**Common** occurrence. Mechanism: **decreased drug accessibility** by upregulation of **efflux** pumps	**Bacteriostatic** activity against many bacteria; useful for spirochetal, chlamydial, rickettsial infections
Chloramphenicol	Target of action: **Protein synthesis (50S ribosomal subunit)**; binds close to the peptidyl transferase site and blocks the binding of the aminoacyl tRNA to the enzyme	**Common** occurrence. Mechanism: **drug inactivation** by enzymatic modification of the drug; acetylation of drug molecule by chloramphenicol acetyl transferase causes inactivation	**Broad spectrum, bacteriostatic** activity against bacteria (especially anaerobes), rickettsia, chlamydiae and mycoplasmas. The drug is not a first line agent for any infection due to toxicity

APPENDIX E Antimicrobial Therapy—Cont'd

Antibacterial Drugs—Cont'd

Important Drugs	Mechanism of Action	Drug Resistance	Clinical Uses
Macrolides (e.g., erythromycin, azithromycin, clarithromycin)	Target of action: **protein synthesis (50S ribosomal subunit)**; bind to the 23S rRNA of the 50S ribosomal subunit (proximal to the chloramphenicol-binding site); block translocation and release of tRNA after peptide bond formation	**Common** occurrence Mechanism: **altered target**; 23S rRNA in the 50S ribosomal subunit is altered by methylation of two adenine nucleotides in the RNA; methylase is plasmid mediated Cross-resistance to lincosamides, streptogramins (and chloramphenicol)	**Bacteriostatic** antimicrobial activity against the **intracellular pathogens** and other bacterial agents: *Legionella pneumophila* *Mycoplasma pneumoniae* *Chlamydia pneumoniae* *Chlamydia trachomatis* *Campylobacter jejuni* Hemolytic streptococci in penicillin-allergic patients Empirical coverage against: Upper respiratory infections Pneumonias Sexually transmitted diseases
Lincosamide (e.g., clindamycin)	As for macrolides	Enzymatic modification, altered ribosomal binding sites Cross-resistance to macrolides, streptogramins (and chloramphenicol)	**Bacteriostatic** activity against Gram-positive cocci and anaerobic bacteria
Streptogramins (e.g., quinupristin, dalfopristin)	As for macrolides	Alteration of ribosomal binding sites Cross resistance to macrolides, lincosamides (and also chloramphenicol)	**Bacteriostatic** activity against Gram-positive cocci, including methicillin-resistant *Staphylococcus aureus* (MRSA) and vancomycin-resistant enterococci (VRE)
Ketolides (e.g., telithromycin)	As for macrolides	Unknown Cross-resistance to macrolides, lincosamides, streptogramins (and chloramphenicol) is expected	**Bacteriostatic** activity against *S. pneumoniae, Mycoplasma* spp, *Chlamydia* spp
Oxazolidinones (e.g., linezolid, eperezolid)	Target of action: Protein synthesis (50S ribosomal subunit) These newer agents bind to 50S ribosome at the 30S ribosome interface; inhibit initiation of protein synthesis	Rare In vitro resistance develops with difficulty in enterococci and staphylococci and is associated with mutations in genes encoding the central loop of domain V of 23S rRNA. Clinical resistance against *S. aureus* has recently been documented	**Bacteriostatic** activity against drug-resistant gram-positive bacterial agents: MRSA CoNS VRE *Staphylococcus aureus* with intermediate susceptibility to glycopeptides Penicillin-resistant pneumococci (PRP) Linezolid has demonstrated efficacy in the treatment of: Skin and soft tissue infections CAP requiring hospitalization
Sulfonamides	Target of action: **nucleic acid synthesis** Sulfonamides are analogs of *p*ABA. They are competitive inhibitors of dihydropteroic acid synthase (in folic acid biosynthetic pathway)	**Common** occurrence Mechanism: (1) **increase in concentration of competing substrates** (*p*ABA), or (2) alteration of target, dihydropteroic acid synthase; altered enzyme does not bind as efficiently to sulfonamides	**Bacteriostatic** antimicrobial activity against a broad spectrum of organisms Sulfonamides are usually used in combination with trimethoprim (e.g., sulfamethoxazole-trimethoprim = Bactrim or clotrimoxazole); also used as monotherapy in nocardiosis

Continued

APPENDIX E   Antimicrobial Therapy—Cont'd

Important Drugs	Mechanism of Action	Drug Resistance	Clinical Uses
Antibacterial Drugs—Cont'd			
Trimethoprim	Target of action: **nucleic acid synthesis** Dihydrofolate must be reduced to tetrahydrofolate, the active compound for carbon transfer. This final step in tetrahydrofolate biosynthetic pathway is catalyzed by dihydrofolate reductase (DHFR). Bacterial DHFR is inhibited by trimethoprim	Never used alone; resistance against the combined drug (sulfamethoxazole-trimethoprim) is common	**Bacteriostatic** antimicrobial activity of the combined drugs (sulfamethoxazole-trimethoprim) Active against many Gram-positive (including *S. aureus*) and Gram-negative infections (e.g., UTIs). It is ineffective against anaerobes Therapeutic and prophylactic use against *Pneumocystis jiroveci* Therapeutic use of sulfa and pyrimethamine (another DHFR inhibitor) against coccidian protozoal infections (e.g., malaria, toxoplasmosis)
Fluoroquinolones (e.g., **ciprofloxacin**, **levofloxacin**, norfloxacin, gatifloxacin, moxifloxacin)	Target of action: **nucleic acid synthesis** Drugs bind to DNA gyrase (topoisomerase II); the enzyme facilitates DNA unwinding by making double-strand breaks, passing one strand through another, and then resealing the strands; drugs inhibit DNA replication	Not uncommon Resistance occurs by (1) target alteration (chromosomal mutation, leading to **alteration of DNA gyrase**), and (2) decreased drug accessibility (**efflux pump**); fluoroquinolones do not achieve inhibitory concentration in the cells	**Broad spectrum** and **bactericidal** activity against most Gram-negative bacteria and some Gram-positive bacteria **Good intracellular killing** of *Chlamydia*, *Legionella*, and *Mycobacteria* Indications include: UTIs, bone and soft tissue infections, gonorrhea, bacterial enteritis, and multiply-drug-resistant tuberculosis Ineffective against anaerobes
Metronidazole	Target of action: **DNA** Toxic effects of reduced form of drug intracellularly (DNA damage)	Reduced uptake	**Bactericidal** against trichomonas, anaerobic bacteria, including *Clostridium difficile*. Broadest spectrum anaerobic agent
Anti-TB drugs			
Isoniazid	Target of action: mycobacterial cell wall (**mycolic acid**) Inhibits mycolic acid synthesis	**Common** occurrence Reduced uptake	**Bactericidal** against *M. tuberculosis*
Rifampin	Target of action: **RNA synthesis** Binds to β-subunit of core enzyme, DNA-dependent RNA polymerase	**Common** occurrence Resistance can occur easily via a point mutation on the β-subunit-encoding region	**Bactericidal** Usually used in combination with other drugs (due to resistance problems) Therapeutic use against tuberculosis Drug is concentrated in saliva; prophylactic use against meningococcal meningitis
Other anti-TB drugs: Pyrazinamide	Unknown	Unknown	**Bactericidal** against *M. tuberculosis*
Ethambutol	Unknown	Unknown	**Bacteriostatic** against *M. tuberculosis*

APPENDIX E   Antimicrobial Therapy—Cont'd

Important Drugs	Mechanism of Action	Drug Resistance	Clinical Uses
Antibacterial Drugs—Cont'd			
Cycloserine	Target of action: **inhibition of cell wall synthesis**; the drug, a structural analog of D-alanine, inhibits L-alanine racemase and D-alanine synthetase, required for production of D-alanine from L-alanine, resulting in short supply of this unique component of bacterial cell wall	Rare	**Broad spectrum** activity against Gram-negative bacteria and especially used as an antimycobacterial agent in the combination therapy of tuberculosis due to *Mycobacterium tuberculosis*
Bacitracin	Target of action: **inhibition of cell-wall synthesis**; blocks transfer of mucopeptide subunits from the phospholipid carrier in growing cell wall and interferes with dephosphorylation of phospholipid carrier	Rare	**Broad spectrum,** predominantly **bacteriostatic** activity against Gram-positive bacteria. Topical and parenteral use of the polypeptide drug; often combined with neomycin and polymyxins to broaden the coverage against Gram-negative bacteria
Polymyxins	Target of action: interact with membrane phospholipids to increase permeability	Rare	**Topical, broad spectrum** activity against Gram-negative bacteria
Antiviral Drugs			
Amantadine (and rimantadine)	**Target of action: early events of influenza A virus infection** Bind to viral matrix M2 protein, inhibiting proton channel (acidification) and release viral genome into the nucleus, resulting in abortion of infection	Viral resistance is associated with mutation (changes) in the genetic region for M2 protein	Effective against influenza A only; influenza B and C viruses do not have proton channels and are not affected. Recommended for use: in adults and children in treatment and prophylaxis. Effective when used within 48 hours of illness
Neuraminidase inhibitors (e.g., zanamivir, oseltamivir)	**Target of action: late events of influenza A or B virus infection** Inhibit viral neuraminidase; release of viruses from infected cells and from mucus layer is inhibited and virus spread is decreased	Uncommon	Clinically efficacious antiviral drugs against both influenza A and B viral infections. Effective when used within 2 days of illness: Zanamivir is orally inhaled and was approved for treatment of persons > 7 years of age. Oseltamivir is orally administered and was approved for treatment of persons > 1 year of age. Both are effective for therapy and prophylaxis
Ribavirin	**Target of action: RNA synthesis** Inhibits the synthesis of guanine nucleotides, which are essential for replication of RNA viruses		Ribavirin aerosol is used clinically to treat pneumonia caused by RSV in infants and to treat severe influenza B infections. Oral drug is effective (when given together with interferon-α) against chronic hepatitis due to HCV infection

Continued

Important Drugs	Mechanism of Action	Drug Resistance	Clinical Uses
Antiviral Drugs—Cont'd			
Anti-herpes drugs (e.g., **acyclovir, valacyclovir,** famciclovir)	**Target of action: DNA polymerase (chain termination) of herpes viruses** (except CMV) Prodrugs are phosphorylated by viral thymidine kinase to monophosphate; triphosphates (made by host kinases) inhibit viral DNA polymerase and also cause chain termination once incorporated Only herpes simplex viruses and varicella-zoster virus encode a kinase that efficiently phosphorylates the drug	Acyclovir resistance occurs owing to mutation in thymidine kinase gene	Orally given acyclovir is effective against HSV-1, 2, and VZV (does not eliminate the virus; suppression of viral replication occurs) Intravenous acyclovir is used against HSV-1 encephalitis and serious reactivation in immunocompromised patients Orally given valacyclovir achieves high concentation similar to IV acyclovir; used against serious herpes and VZV infections
Anti-CMV drugs (e.g., ganciclovir, valganciclovir, cidofovir)	**Target of action: DNA polymerase (chain termination) of CMV virus** Same mechanism as for anti-herpes drugs; CMV-encoded kinase mediates the first step of phosphorylation; the enzyme has poor affinity for acyclovir	Resistance due to mutation in CMV-kinase is not uncommon	It is effective in the treatment of retinitis caused by CMV in AIDS patients. The drug may be useful in other disseminated infections caused by CMV Valganciclovir has good oral absorption
Foscarnet	**Target of action: DNA polymerase;** phosphonoformate drug inhibits the DNA polymerases of all herpesviruses, especially CMV Does not require any activation	Rare; alterations in DNA polymerase	Often used in ganciclovir-resistant CMV retinitis
Anti-HIV drugs: nucleoside reverse transcriptase inhibitors (e.g., **zidovudine, lamivudine** [3TC], stavudine [D4T], **tenofovir**)	Target of action: **reverse transcriptase of HIV-1** Triphosphate drugs, activated by host kinase, competitively inhibit reverse transcriptase at the active site. DNA chain elongation is also blocked for the lack of free 3′ hydroxyl group	Resistance due to mutation of the reverse transcriptase is common. Poor fidelity (no proofreading function) of reverse transcriptase leads to errors in genomic replication (mutations)	Must be combined with other anti-HIV drugs for use in highly-active antiretroviral therapy (HAART) to suppress viral replication to undetectable levels Prophylactic use in needlestick injury
Anti-HIV drugs: non-nucleoside reverse transcriptase inhibitors (e.g., **efavirenz, nevirapine,** delavirdine)	Target of action: **reverse transcriptase of HIV-1** Triphosphate drugs bind to the enzyme at an allosteric site. Noncompetitive, steric hindrance leads to inhibition of reverse transcription	Resistance is common	Combined with other anti-HIV drugs for use in HAART Nevirapine can prevent transmission of virus from infected mother to newborn
Anti-HIV drugs: fusion inhibitors (e.g., enfuvertide)	Target of action: **gp41 fusion with cell membrane;** the peptide drug binds to viral protein gp41, preventing it from unfolding (required for fusion)	Unknown	Active in antiretroviral naïve and experienced patients

APPENDIX E Antimicrobial Therapy—Cont'd

Important Drugs	Mechanism of Action	Drug Resistance	Clinical Uses
Antiviral Drugs—Cont'd			
Anti-HIV drugs: protease inhibitors (e.g., ritonavir, indinavir)	**Target of action: viral protease (post-translational processing)** Bind to and block HIV protease activity, necessary for generating functional proteins (e.g., reverse transcriptase)	Resistance is common due to mutations in the protease enzyme	Combined with other reverse transcriptase inhibitors as part of HAART
Interferon-α	**Target of action:** nucleic acid synthesis; activates a ribonuclease that degrades viral mRNA; inhibits elongation of protein synthesis in virus-infected cells	Unknown	Effective against hepatitis B and C infections; broad-spectrum biological may have use against other viral infections Pegylated form is more active
Antifungal Drugs			
Amphotericin B	**Target of action: fungal cell membrane (ergosterol)** Polyene drug preferentially binds to ergosterol in fungal cell membrane; ergosterol molecules can associate with each other in fungal membranes. The polyene side of amphotericin B lines up with the hydrophobic side of ergosterol, whereas the polyhydroxyl side of the drug aligns inward in the membrane, creating aqueous channels and leading to loss of cytoplasmic materials and cell death	Resistance is rare; resistant species include: *Candida lusitaniae* *Fusarium* spp *Malassezia furfur* *Trichosporon beigelii* *Pseudallescheria boydii*	Intravenously administered amphotericin B (ampho B) is rapidly fungicidal The drug has broad spectrum of action. Currently a DOC for life-threatening disseminated mycoses. Indications are: Candidiasis Endemic mycoses Cryptococcal meningitis Invasive aspergillosis (neutropenia) Zygomycosis (only effective therapy) Because of high toxicity new formulations involving drug-lipid combinations are available: liposomal ampho B (L-AmB), ampho B lipid complex (ABLC), and amphotericin B cholesteryl sulfate (ABCD)
Nystatin	Same as for amphotericin B		Used only topically because of toxicity
Imidazoles (e.g., **ketoconazole**, clotrimazole, miconazole)	Target of action: fungal cell membrane (**ergosterol biosynthesis**) These azole drugs inhibit lanosterol 14α-demethylase, leading to depletion of ergosterol and altered membrane permeability	Altered target enzyme, increased efflux of drug	Fungistatic against *Candida*, endemic fungi. Ketoconazole is not well tolerated systemically, so currently a second-line drug. Clotrimazole and miconazole are used topically

Continued

APPENDIX E Antimicrobial Therapy—Cont'd

Important Drugs	Mechanism of Action	Drug Resistance	Clinical Uses
Antifungal Drugs—Cont'd			
Triazoles (e.g., **fluconazole itraconazole**, voriconazole)	Same as for imidazoles	Altered target enzyme, increased efflux of drug	Fluconazole: Active against *Candida* spp, *Cryptococcus neoformans*. Also useful in Coccidioidal meningitis due to excellent CSF penetration Itraconazole: broad spectrum against yeasts, endemic fungi and molds, including *Aspergillus* Voriconazole: broad spectrum against *Candida*, particularly those resistant to fluconazole; *Aspergillus* (DOC), other molds
Flucytosine	Inhibition of DNA and RNA synthesis after intracellular conversion to 5-fluorouridine (5-FU). Mammalian cells cannot convert to active compound	Common on monotherapy Loss of converting enzyme, decreased permeability	Active against *Candida*, *Cryptococcus*; always used in combination with amphotericin B owing to resistance with monotherapy
Echinocandins (e.g., caspofungin, micafungin)	Inhibit synthesis of β-1,3-D-glucan, an important polymer of fungal cell walls	Uncommon	Fungicidal against *Candida* spp, fungistatic against *Aspergillus* spp
Terbinafine	Inhibits ergosterol synthesis at squalene epoxidase	Unknown	Used against dermatophyte infections
Griseofulvin	Inhibits microtubule formation	Decreased uptake	Active only in dermatophyte infections
Antiparasitic Drugs			
Sulfadoxine/pyrimethamine	Synergistic inhibitors (like Bactrim) of tetrahydrofolate synthesis Sulfadoxine inhibits dihydropteroate synthetase; pyrimethamine inhibits dihydrofolate reductase		A combined drug (Fansidar) is effective against *Plasmodium falciparum*
Metronidazole	Reduction of nitroimidazole in a low redox potential (during fermentative metabolism) and auto-oxidation generate free radicals; causes DNA chain breakage		Effective against trichomonas, giardiasis, amebiasis
Chloroquine	Blood schizonticides; inhibit polymerization of heme in the food vacuole of the parasite; heme precursor is toxic to the schizonts		Effective against malaria, although *P. falciparum* is resistant in many parts of the world
Mefloquine	Unknown		Effective against chloroquine-resistant *P. falciparum*
Atovaquone and **proguanil**	Synergistic inhibitors Atovaquone selectively inhibits mitochondrial electron transport, resulting in collapse of membrane potential A metabolite of proguanil inhibits dihydrofolate reductase		Broad spectrum activity Antimalarial prophylaxis in chloroquine-resistant areas

Appendix E Antimicrobial Therapy—Cont'd

Important Drugs	Mechanism of Action	Drug Resistance	Clinical Uses
Primaquine	Hypnozoiticidal		Only drug for treating hypnozoite forms of *Plasmodium ovale* and *Plasmodium vivax* in the liver
Pentamidine	Unknown		Alternative for *Pneumocystis carinii* pneumonia, resistance is uncommon. Also in trypanosomiasis
Nifurtimox	Toxic metabolites from reduction of drug in cell		Used in *Trypanosoma cruzi* infection (Chagas disease)
Suramin	Inhibits enzymes in energy metabolism		Used in African trypanosomiasis, *Onchocerca volvulus*
Pentavalent antimony	Unknown		Used for treating leishmaniasis
Albendazole (mebendazole and thiabendazole)	The benzimidazole antihelmintic drugs inhibit microtubule assembly, causing disruption of absorptive and secretory functions of the cells, which are essential to the worms' survival. Albendazole is also an ovicide and larvicide for hookworms, *Trichuris trichiura*, and *Taenia solium*		Orally given; broad-spectrum drugs; clinically efficacious against *Ascaris lumbricoides* (roundworm) *Ancylostoma/Necator* (hookworms) *Echinococcus granulosus* (dog tapeworm) *Enterobius vermicularis* (pinworm) *Strongyloides stercoralis* (threadworm) *Taenia solium* (pork tapeworm) *Trichinella spiralis* (pork worm) *Trichuris trichiura* (whipworm) Albendazole is DOC for cysticercosis
Pyrantel pamoate	Causes neuromuscular blockade and spastic paralysis of parasites; adult parasites are expelled from the gut		Orally given; clinically efficacious against roundworm infections: *Ascaris* (roundworm) *Necator/Ancylostoma* (hookworms) *Enterobius* (pinworm)
Praziquantel	Causes muscular paralysis and vacuolization in the cells of the parasites		Orally given; broad spectrum drugs; clinically efficacious against cestodes (tapeworms) and trematodes (flukes) Fluid and tissue levels are sufficient to kill larvae
Niclosamide	Inhibits anaerobic fermentation for generating ATP		Orally given; effective against some cestodes (*Taenia solium*/cysticercosis is exception)
Ivermectin	An agonist of glutamate-gated chloride channels causes paralysis of the adult worms		Orally given; effective against onchocerciasis, strongyloidiasis, and scabies

CAP, community-acquired pneumonia; DOC, drug of choice; DHFR, dihydrofolate reductase; IM, intramuscular; Pen, penicillin.

Practice Questions and Answers

PRACTICE QUESTIONS

These practice questions were designed to assess the understanding and application of the medical microbiology concepts illustrated by the cases in this book. The questions evaluate clinical reasoning, recall, and application skills as applied to the practice of medicine, with emphasis on principles and mechanisms of infectious diseases and choice of therapy. Some questions test rote memory per se (necessary for foundational knowledge). Most of the questions, however, are clustered around a simple case (problem) and involve the interpretation of diagnostic test results and scrutiny of microbiologic and pathologic specimens. Each cluster consists of a brief case scenario followed by three to five questions.

The questions call for some or all of the following, in a conceptual framework:

1. Synthesis and application of information in a clinical situation
2. Development of a diagnostic hypothesis
3. Identification of appropriate diagnostic tests and interpretation thereof
4. Understanding of the nature of the etiologic agent, including the following: its mode of acquisition; the mechanism by which it causes disease; specific therapies; and preventive measures.

Each question in a cluster addresses a somewhat different aspect of the associated case, and a limited number of questions about general (core) principles are embedded in the clusters. A review of basic concepts in virology, bacteriology, mycology/parasitology, and principles of diagnostics and therapeutics, presented in appendices A through E, might be in order before beginning self-assessment with these practice questions.

The questions have been scrambled, to encourage a comprehensive understanding of infectious diseases and to aid in long-term retention of the concepts presented in this book. It is possible, however, to work through the questions in a sequence based on the classifications used by the United States Medical Licensing Examination (USMLE) Step 1. The table shows the distribution of questions in each review category.

All questions have five answer choices and, similar to the USMLE, answers are all of the "ONE-BEST ANSWER" type.

DIRECTIONS: Select the ONE letter answer (from the choices: A through E) that is BEST in each question.

Questions 1-3 are linked to the following case:

A 54-year-old man came to the ED with complaints of fever, chills, headache, and cough productive of minimal sputum. He had been attending a conference at a local hotel for the past week. On exam, he appeared ill and was in moderate respiratory distress. Over the course of the next 3 days, several more individuals came in with similar complaints.

1. Laboratory studies are most likely to show
   A. Increased serum pH
   B. Increased serum K^+ concentration
   C. Decreased serum Na^+ concentration
   D. Increased serum Ca^+ concentration
   E. Increased serum HCO_3^- concentration

2. In evaluating the cause of the illness, which of the following is most appropriate?
   A. Direct urine antigen to identify *Legionella pneumophila*
   B. Blood cultures to identify *Streptococcus pneumoniae*
   C. Lung biopsy to identify *Pneumocystis jiroveci*

TABLE F–1 Distribution of Practice Questions Based on USMLE* Step 1-Based Categories and Content Coverage

Review Sections (USMLE Step 1)	Question Numbers
Cardiovascular system	107, 137-139, 149, 152, 181, 226
Respiratory system	1-3, 44-47, 49-57, 59-61, 65-67, 71-74, 81-84, 96-99, 122-124, 126, 133-135, 155-157, 159-161, 167-172, 182-185, 192-194, 196-198, 205-208, 210, 234-246
Gastrointestinal system and liver	7-10, 24-26, 90-93, 111-112, 117-121, 127, 128, 132, 136, 140, 141, 144, 146, 150-151, 153-154, 158, 162, 166, 173-175, 179, 180, 186-189, 191, 195
Urinary system	21-23, 108, 113
Reproductive system	4-6, 35-37, 44-48, 63, 78-80, 89, 94-95, 102-104, 109-110, 116, 125
Central and peripheral nervous system	14-17, 31-34, 38-43, 68-70, 114-115, 145, 199-201, 204, 209, 211-216, 220-224, 228, 250
Skin, connective tissues, and musculoskeletal system	11-13, 27-30, 75-77, 148, 163-165, 176, 178, 202-203, 225, 227, 229-230, 248, 254
Multisystem (including AIDS, bioterrorism, and emerging infectious diseases)	14-20, 31-34, 44-47, 58, 62, 64, 85-88, 100-101, 105-106, 129-131, 142-143, 145, 147, 172, 177, 179, 186, 190, 201, 210, 217-220, 229, 231-233, 247, 249, 251-253

*USMLE, United States Medical Licensing Examination

D. DFA of nasopharyngeal secretion to identify influenza virus

E. Serology using paired sera to identify *Chlamydia pneumoniae*

3. The patient died, despite aggressive management. Autopsy examination to evaluate his infection showed consolidation in both lungs with congestion and bulging of lung tissue (see below).

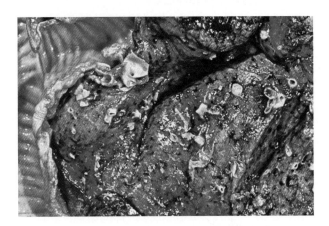

This occurred because of

A. Pulmonary thromboembolus arising from deep venous thrombosis

B. Lobar tissue damage by the reactive mediators from PMNs and T cells

C. Persistent inflammation that led to alveolar fluid and hyaline membranes

D. Acute peribronchial edema caused by chronic inflammation

E. Large caseous lesions caused by granulomatous inflammation

Questions 4-6 are linked to the following case:

A 19-year-old woman presented with a 2-week history of painless ulcers on her labia. She admitted unprotected intercourse with multiple partners since she was 15 years of age. Physical examination was remarkable for labial chancres (see below) and inguinal lymphadenopathy.

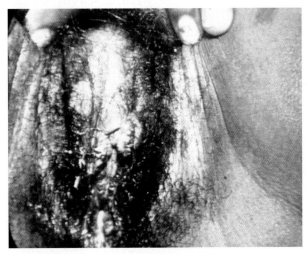

4. Laboratory studies are most likely to show
   A. Neutrophilic leukocytosis
   B. Lymphocytic leukocytosis
   C. Eosinophilic leukocytosis
   D. Normal WBC count and differential
   E. Abnormal WBC count with left shift

5. In evaluating the cause of these genital ulcers, which of the following laboratory tests is most appropriate?
   A. Wet mount examination of biopsy section to identify *Trichomonas vaginalis*
   B. Tzanck smear of biopsy to identify multinucleated giant cells and HSV-2
   C. Direct fluorescence microscopic examination of biopsy to identify *Chlamydia trachomatis*
   D. Gram stain and culture of biopsy to identify *Candida albicans*
   E. Rapid plasma reagin test and serologic titer specific for *Treponema pallidum*

6. A biopsy specimen from the lesion in this case would most likely reveal
   A. Plasma cell-rich mononuclear infiltrate
   B. Granulomas with caseation
   C. Acute inflammation with abscess formation
   D. Epithelioid cells and atypical lymphocytes
   E. A combination of pyogenic and gummatous inflammation

Questions 7-10 are linked to the following case:

A 27-year-old male undergraduate nursing student presented with jaundice and dark yellow urine. He also complained of nausea, low-grade fever, and loss of appetite. He felt pain in the right side of his abdomen and his joints. He had never traveled outside the United States. He admitted to having experienced a needlestick 2 months ago, which he did not report. On physical examination, he was icteric and had tender, firm hepatomegaly.

7. Laboratory studies are most likely to show
   A. Normal CBC and differential; normal ALT, AST, and bilirubin
   B. Eosinophilic leukocytosis; normal ALT, AST, and bilirubin
   C. Lymphocytic leukocytosis; elevated ALT, AST, and bilirubin
   D. Neutrophilic leukocytosis; elevated ALT, AST, and bilirubin
   E. Leukopenia, thrombocytopenia, and reduced alkaline phosphatase

8. The hepatitis serology panel was negative for HAV IgM and HBs IgG and positive for HBsAg. The profile of this man's viral serologic markers was as follows:

   HBsAg positive
   HBeAg positive
   IgM HBc positive
   IgG-HBc positive
   IgG-HBs negative

   Which of the following best describes his illness?
   A. Chronic active hepatitis B
   B. Recent recovery from HBV infection
   C. Hepatitis B infection in the distant past
   D. Acute HBV infection
   E. Persistent (healthy) HBV carrier

9. The patient did not have any remarkable medical history of serious illness or hospitalization. Which of the following modes of transmission is most likely to be implicated in this case?
   A. He recently worked a few shifts in the renal disease and hemodialysis area
   B. He was sexually active with his girlfriend
   C. He played contact sports
   D. He frequently ate imported tropical fruits
   E. He had an accidental needlestick at the work place

10. The patient had a steady girlfriend, who was seronegative. Which of the following preventive measures could protect his girlfriend?
    A. Advise abstinence from sexual intercourse until he becomes asymptomatic
    B. Give both the patient and his girlfriend a course of α-interferon
    C. Administer only hepatitis B immunoglobulin (HBIG)
    D. Administer a single dose of HBIG and the HBV vaccine
    E. Give both the patient and his girlfriend a course of ribavirin

Questions 11-13 are linked to the following case:

A 13-year-old boy had pain in his right arm after a soccer-related injury 2 weeks ago. Physical examination was remarkable for a febrile patient in moderate distress. An x-ray film of his arm showed bony lesions. Laboratory studies were remarkable for elevated ESR and CRP and a neutrophilic leukocytosis with left shift.

11. In evaluating the cause of the illness, which of the following is most appropriate?
    A. Begin empiric antibiotics
    B. Clinical diagnosis with a good immunization history and serum toxin assay
    C. Gram stain and cultures of needle aspirate and blood cultures
    D. Cultures of urine, stool, blood, and joint tap
    E. Echocardiogram and a series of blood cultures

12. Laboratory investigation yielded a significant bacterial agent. Which of the following best describes the most likely etiology?
    A. Gram-positive rods; obligate anaerobic growth with double zone of hemolysis; spore-forming; lecithinase producing
    B. Gram-positive cocci in clusters; β-hemolytic growth, catalase positive, coagulase positive
    C. Gram-positive cocci in chains; α-hemolytic growth, catalase negative, nonencapsulated, Optochin-resistant
    D. Gram-positive cocci in chains; β-hemolytic growth, catalase negative, bacitracin sensitive
    E. Gram-positive cocci in clusters; nonhemolytic colonies, catalase positive, coagulase negative

13. The isolate was resistant to nafcillin. Which of the following genetic loci is responsible for this resistance phenotype?
    A. *mecA*
    B. *revC*
    C. *nafB*
    D. *brmC*
    E. *penB*

Questions 14-17 are linked to the following case:

A 41-year-old man with AIDS presented with a 12-day history of moderate fever, headache, and photophobia. He suffered a seizure 3 hours before his arrival in the ED. Physical examination was remarkable for moderate distress due to the headache and right-side weakness.

14. Head MRI is most likely to show
    A. Normal (no brain lesions) brain scan
    B. Ring-enhancing lesions in one or more lobes
    C. Meningeal enhancement
    D. Multiple nodular (granulomatous) lesions
    E. Subdural hypodense (edematous) lesions

15. In evaluating the cause of this patient's headache and seizure, which of the following is most appropriate?
    A. Examination of CSF with an India ink to identify *Cryptococcus neoformans*
    B. Serology to detect elevation of IgG titer specific for *Toxoplasma gondii*
    C. Histologic examination of a brain biopsy to identify cysticercus of *Taenia solium*
    D. Blood culture, acid-fast stain, and culture of the CSF to identify *Mycobacterium tuberculosis*
    E. A PCR assay to detect herpes simplex virus type 1 DNA in CSF

16. Where in the world is this illness rarely found or not at all found?
    A. Ferret-free islands
    B. Goat-free islands
    C. Sheep-free islands
    D. Feline-free islands
    E. Swine-free islands

17. Which of the following groups is most likely to have serious complications due to infection?
    A. Fetuses
    B. Neonates
    C. Infants
    D. Young children
    E. Older children

Questions 18-20 are linked to the following case:

A 69-year-old man sought treatment for progressive slowing of speech, worsening memory, and personality changes. He was disoriented and confused and had resting tremors of all extremities. Three months later, he was in a coma-like state noted for occasional myoclonus when he was startled.

18. Laboratory studies are most likely to show
    A. Normal CBC and differential; normal MRI head scan; normal EEG
    B. Abnormal CBC and differential; abnormal MRI head scan; normal EEG
    C. Abnormal CBC and differential; normal MRI head scan; abnormal EEG
    D. Normal CBC and differential; abnormal CT head scan; abnormal EEG
    E. Abnormal CBC and differential; abnormal CT head scan; abnormal EEG

19. In evaluating the cause of the illness, which of the following is most appropriate?
    A. Blood cultures to identify *Cryptococcus neoformans*
    B. CSF examination and cultures to identify *Coccidioides immitis*
    C. Serology for *Taenia solium* (neurocysticercosis)
    D. PCR analysis of CSF to identify HSV-1
    E. Histopathology of brain biopsy to demonstrate neuronal vacuolation

20. Further studies showed diffuse spongiform degeneration in the brain. This occurred because of
    A. Deposition of PrP^{sc} in the brain tissues
    B. Deposition of gelatinous capsule in the brain
    C. Cytolytic destruction of brain parenchyma by the virus
    D. Chronic granulomatous inflammation in the brain cortex
    E. Disseminated intravascular coagulation in the brain

Questions 21-23 are linked to the following case:

A 65-year-old man presented with fever, flank pain, frequency in urination, and dysuria. The patient's medical history was remarkable for prostatic hypertrophy. Physical examination was remarkable for fever and right costovertebral angle tenderness.

21. CBC and urinalysis are most likely to demonstrate which one of the following?
    A. Normal white cell counts and proteinuria
    B. Normal white cell counts and hematuria
    C. Lymphocytic leukocytosis and WBC casts
    D. Neutrophilic leukocytosis and WBC casts
    E. Pleocytosis, thrombocytopenia, and normal urinalysis

22. In evaluating the cause of this illness, which one of the following tests is most helpful?
    A. Wet mount, Gram stain, cultures, and serum antigen detection to identify *Candida albicans*
    B. Ova & parasite examination of midstream urine to identify *Schistosoma haematobium*
    C. Enzyme immunoassay of viral antigen in urine sediment to detect adenovirus
    D. Semiquantitative midstream urine and blood cultures to identify *Streptococcus pyogenes*
    E. Semiquantitative midstream urine and blood cultures to identify *Enterobacteriaceae*

23. Further studies to evaluate the patient's low blood pressure (hypotension) showed elevated serum levels of interleukin-1 and TNF. This most likely occurred as a result of which of the following virulence factors?
    A. Endotoxin
    B. Hemolysin
    C. Aerobactin
    D. Capsule
    E. Urease

Questions 24-26 are linked to the following case:

A 59-year-old woman complained of pressure in the upper-abdominal area that radiated to her chest. She stated that it often improved immediately after meals. She also noted occasional heartburn but denied nausea, vomiting, or diarrhea. Abdominal exam revealed mild epigastric tenderness with no rebound.

24. In evaluating the cause of her upper GI pain, which of the following noninvasive tests is most appropriate?
    A. Urea agar test
    B. ^{13}C-urea breath test
    C. Culture of the enteric pathogen from stool
    D. Serology to detect bacteria-specific IgM
    E. Ultrasound

25. Histologic examination of the gastric biopsy (see below).

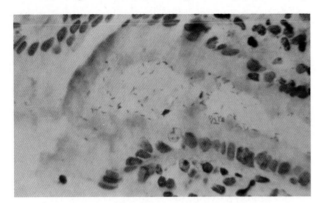

Which of the following microbial features most accurately describes the etiologic agent?

A. Gram-negative rod; facultative growth, oxidase-negative, motile

B. Gram-negative curved rod; microaerophilic slow growth at 42°C, nonmotile

C. Gram-negative curved rod; microaerophilic growth at 37°C, motile

D. Gram-negative short rod; oxidase-negative, motile, lactose fermenting

E. Gram-negative short rod; facultative growth, nonlactose fermenting, H_2S positive

26. Which of the following virulence factors is most likely associated with this organism?

A. Alkaline phosphatase and cytotoxin

B. Acid phosphatase and protease

C. Urease and vacuolating cytotoxin

D. Urealysin and hemolysin

E. Endotoxin and mucinase

Questions 27-30 are linked to the following case:

A 46-year-old hospitalized man had an abrupt-onset, high-grade fever with chills a few hours after he underwent amputation of his left leg due to gangrenous wound infection. Physical examination was remarkable for fever, tachycardia, hypotension, and tachypnea.

27. Laboratory studies would most likely show:

A. Normal erythrocyte sedimentation rate (ESR), normal C-reactive protein (CRP), normal CBC and differential

B. Elevated ESR, elevated CRP, neutrophilic leukocytosis with left shift, thrombocytopenia

C. Elevated ESR, elevated CRP, lymphocytic leukocytosis, thrombocytopenia

D. Lowered ESR, lowered CRP, leukopenia (neutropenia), thrombocytopenia

E. Elevated ESR, reduced CRP, normal CBC and differential, thrombocytopenia

28. In evaluating the cause of acute illness, which of the following is most appropriate?

A. Gram stain and cultures of transtracheal aspirate and blood

B. Detection of proinflammatory cytokines in serum

C. Serologic detection of bacterial antigens

D. Gram stain and cultures of wound and blood

E. Detection of toxin and microbial products in serum

29. Further studies to evaluate his systemic inflammatory response show a dramatic increase in IL-1 and TNF. This occurred because the infectious agent

A. Produced pyrogenic exotoxin (superantigen) that stimulated T lymphocytes and IL-2 production

B. Induced chemotaxis that allowed proliferation of immature granulocytes (bands)

C. Released M proteins that upregulates the proliferation of CD8+ T lymphocytes

D. Released lipoteichoic acid that upregulates the proliferation of CD4+ T lymphocytes

E. Released streptokinase that upregulates the coagulation cascade and platelet destruction

30. Which of the following would be the most appropriate therapeutic agent?

A. Erythromycin   D. Vancomycin

B. Clindamycin   E. Metronidazole

C. Penicillin G

Questions 31-34 are linked to the following case:

A 35-year-old HIV-positive white man presented with a 2-week history of persistent moderate fever and chills and a progressively worsening headache. Physical examination revealed a febrile patient (T: 39°C) with lethargy, mild nuchal rigidity, and diminished reflexes. Chest x-ray was normal.

31. Analysis of cerebrospinal fluid from this patient is most likely to show

A. WBCs <10/µL (all PMNs); protein <40 mg/dL and glucose >50 mg/dL

B. WBCs >1000/µL (>50% PMNs); protein >100 mg/dL and glucose <40 mg/dL

C. WBCs 10-100/µL (>50% lymphs); protein <100 mg/dL and glucose <50 mg/dL

D. WBCs <10/µL (<50% PMNs); protein >100 mg/dL and glucose >40 mg/dL

E. WBCs 200-300/µL (>50% PMNs); protein >100 mg/dL and glucose <40 mg/dL

32. An India ink test of a CSF smear was diagnostic (see below).

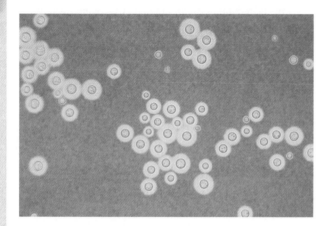

    Which of the following virulence factors is most likely to be involved in the pathogenesis of illness?

A. Soluble β-1,3 glucan polymer

B. Polysaccharide capsule

C. Granuloma-forming soluble antigen

D. Arthroconidial mitogen

E. Atypical cell-wall peptidoglycan

33. The patient's inability to control this infection is most likely due to a deficiency in which of the host responses?

A. PMN response

B. Mucosal IgA response

C. Humoral IgM and IgG response

D. Alternate complement pathway

E. Cell-mediated response

34. Which of the following is the most appropriate therapeutic combination for this disease?

A. Fluconazole and sulfadiazine

B. Penicillin G and rifampin

C. Amphotericin B and 5-fluorocytosine

D. Isoniazid and rifampin

E. Ceftriaxone and dexamethasone

Questions 35-37 are linked to following case:

A 28-year-old male presented with a week of dysuria and urethral discharge. He had had a new sexual partner in the past 2 weeks, but his social and medical histories were otherwise unremarkable. Physical exam was remarkable for a clear urethral discharge.

35. Laboratory studies are most likely to show

A. Normal CBC and differential

B. Abnormal CBC with neutrophilic leukocytosis

C. Abnormal CBC with lymphocytosis and thrombocytopenia

D. Abnormal CBC and decreased CD4$^+$ lymphocytes

E. Abnormal CBC with eosinophilic leukocytosis

36. In evaluating the cause of this patient's urethral discharge, which of the following laboratory tests is most appropriate?

A. Gram stain of urethral discharge to identify *Neisseria gonorrhoeae*

B. Direct mount (microscopy) of urethral discharge to identify *Trichomonas* spp

C. Biphasic culture of urethral discharge to identify *Mycoplasma hominis*

D. Urine antigen or direct fluorescence microscopy of discharge to identify *Chlamydia*

E. Culture of urethral discharge on selective media to identify *Haemophilus ducreyi*

37. A female partner of this patient is at risk for which of the following illnesses?

A. Pelvic inflammatory disease

B. Osteomyelitis

C. Mucopurulent conjunctivitis

D. Trachoma (blindness)

E. Interstitial pneumonia

Questions 38-43 are linked to the following case:

A 21-year-old female college student presented with a 1-day history of fever, chills, severe headache, stiff neck, nausea, and vomiting. She had been healthy with no remarkable medical history. Her vital signs were T: 39.9°C; P: 121/min; BP: 91/61 mmHg. The physical examination revealed an altered sensorium; pallor; moist, cold skin; nuchal rigidity; and a petechial rash in the extremities. CBC was remarkable for 21,000/μL WBCs with 78% PMNs and 18% band forms, and a platelet count of 76,000/μL.

38. Analysis of cerebrospinal fluid in this case would most likely show
    A. WBCs <10/μL (no PMNs); protein <40 mg/dL; glucose >50 mg/dL
    B. WBCs >1000/μL (>50% PMNs); protein >100 mg/dL; glucose <40 mg/dL
    C. WBCs 10-100/μL (>50% PMNs); protein <100 mg/dL; glucose >40 mg/dL
    D. WBCs <10/μL (<50% PMNs); protein >100 mg/dL; glucose >40 mg/dL
    E. WBCs 200-300/μL (<50% PMNs); protein <100 mg/dL; glucose >40 mg/dL

39. In evaluating the cause of this illness, which of the following is most appropriate?
    A. Gram stain of sputum and cultures of blood, sputum, and CSF to identify *Haemophilus influenzae*
    B. Gram stain of CSF, cultures of blood and CSF to identify *Neisseria meningitidis*
    C. Direct antigen detection in CSF to identify *Pseudomonas aeruginosa*
    D. Gram stain and cultures of blood, sputum, and CSF to identify *Streptococcus agalactiae*
    E. Gram stain and cultures of blood and CSF to identify *Listeria monocytogenes*

40. Gram stain of the CSF was significant (see below).

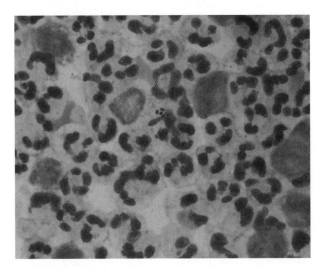

Which of the following pathogenic processes is most likely to have caused the patient's symptoms?
A. Platelet-mediated coagulation cascade
B. Mass lesions in brain parenchyma
C. Subarachnoid and endothelial cellular necrosis
D. Intense pyogenic subarachnoid inflammation
E. Granulomatous inflammation in brain parenchyma

41. Which of the following virulence factors is most likely to be involved in pathogenesis?
    A. Hemolysin
    B. Capsule
    C. Outer membrane
    D. Exotoxin
    E. Pili

42. Which of the following is the most appropriate empirical regimen of drugs based on the Gram smear of CSF in this case?
    A. Vancomycin and cefotaxime
    B. Cefotaxime alone
    C. Ampicillin and gentamicin
    D. Cefotaxime and gentamicin
    E. Vancomycin and rifampin

43. Eleven classmates of this patient were also exposed to the pathogen. Which of the following is the most appropriate prophylactic agent for these individuals?
    A. Hyperimmune serum
    B. Penicillin V
    C. Rifampin
    D. Vaccine (toxoid)
    E. Erythromycin

Questions 44-47 are linked to following case:

A 35-year-old woman presented with 6-week history of progressive dyspnea, cough, and fatigue. She admitted to having unprotected sexual intercourse with a boyfriend 2 years earlier. He told her then that he had tested positive for HIV infection. She had been previously healthy and was taking no medications. She appeared ill and cachectic, and crackles were heard bilaterally on lung exam. Chest x-ray revealed bilateral interstitial infiltrates.

44. Laboratory studies are most likely to show

    A. Leukopenia and CD4$^+$ cell count <200/μL

    B. Normal CBC and differential

    C. Pleocytosis and thrombocytopenia

    D. Neutrophilic leukocytosis and normal platelets

    E. Lymphocytic leukocytosis and normal platelets

45. In evaluating the cause of her illness, which of the following laboratory tests is most appropriate?

    A. Examination of BAL specimen to identify *Pneumocystis jiroveci* and serology (ELISA-Western blot) for HIV

    B. Gram stain of sputum for *Mycobacterium avium-intracellulare* and culture of peripheral blood for HIV

    C. Shell vial culture and direct fluorescence antibody staining to identify CMV and RT-PCR test for HIV

    D. Cell cultures for CMV and HSV and microscopic examination of syncytium of T lymphocytes for HIV

    E. Cultures of sputum and bone marrow for *Histoplasma capsulatum* and direct saliva antigen to identify HIV

46. Further studies showed the ratio of CD4$^+$ T cells and CD8$^+$ T cells to be 0.2. This occurs because the infecting virus

    A. Kills CD8$^+$ T lymphocytes

    B. Kills CD4$^+$ T lymphocytes

    C. Kills macrophages and monocytes

    D. Kills natural killer white cells

    E. Induces proliferation of CD8$^+$ T cells

47. After confirming the clinical diagnosis of the current episode and underlying illness, which of the following regimens would be most appropriate to treat her pneumonia?

    A. TMP-SMX

    B. Cefotaxime

    C. Erythromycin

    D. Ciprofloxacin

    E. Penicillin G

48. A 29-year-old woman complained of itching and abnormal vaginal discharge for 5 days. She denied symptoms of urinary tract infections. Examination revealed a thin discharge (pH is 5.1 [normal <4.5]), which had a distinct "fishy" odor. A wet mount of the discharge was diagnostic (see figure).

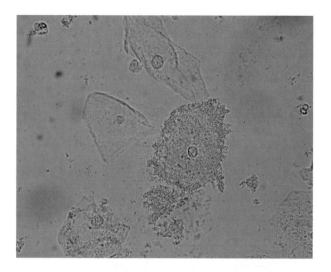

Which of the following pathogens is most likely to have caused her illness?

A. *Chlamydia trachomatis*

B. *Trichomonas vaginalis*

C. *Gardnerella vaginalis*

D. *Candida albicans*

E. *Treponema pallidum*

Questions 49-52 are linked to the following case:

A 61-year-old homeless, alcoholic male was found sleeping on the street on a cold November night. When awakened, he was incoherent and was brought to the emergency department for examination. He had a fever of 40°C and a productive cough, along with the signs of malnutrition. While in the ED, he vomited gastric contents and was noted to have foul-smelling sputum. CBC revealed neutrophilic leukocytosis with left shift.

49. Blood was drawn for culture, and sputum was collected by suction. Which of the following methods should be used to examine his sputum?

    A. Gomori methenamine silver stain

    B. KOH-Calcofluor stain

    C. Gram stain

    D. Acid-fast stain

    E. Metachromatic granule stain

50. Cultures of blood and sputum yielded significant mucoid colonies that were lactose fermentative (see below) and indole negative.

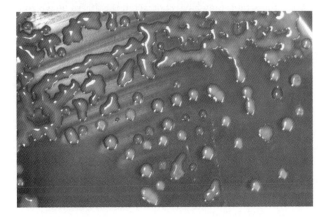

Which of the following pathogens caused his illness?

A. *Klebsiella pneumoniae*

B. *Mycoplasma pneumoniae*

C. *Streptococcus pneumoniae*

D. *Actinomyces israelii*

E. *Pseudomonas aeruginosa*

51. Which of the following is the most likely way he acquired this infection?

A. Contact with another sick individual

B. Intravenous drug use

C. Inhalation of airborne organisms

D. Aspiration

E. Complication following a viral infection

52. If not treated adequately, this patient may go on to develop which of the following conditions?

A. Gas gangrene

B. Cavitary granulomas

C. Bacteremia

D. Bullous emphysema

E. Adenocarcinoma

Questions 53-57 are linked to the following case:

A 65-year-old man presented with a 2-day history of high fever, cough, rusty sputum, and pleuritic left-sided chest pain. His history was remarkable for treatment of congestive heart failure; he was otherwise healthy (no prior respiratory viral infection). Examination revealed fever and tachypnea. His chest radiograph was remarkable for consolidation in the left lower lobe. CBC was remarkable for neutrophilic leukocytosis.

53. Gram stain of sputum smear revealed many PMNs and a significant pathogen. Sputum and blood cultures yielded α-hemolytic colonies that were sensitive to Optochin (see below).

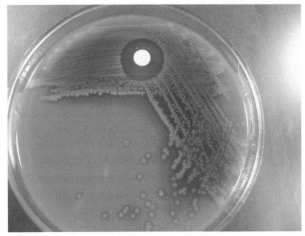

Which of the following pathogens caused his illness?

A. *Streptococcus pneumoniae*

B. *Mycoplasma pneumoniae*

C. *Histoplasma capsulatum*

D. *Staphylococcus aureus*

E. *Streptococcus pyogenes*

54. Which of the following virulence factors is most likely to be involved in the pathogenesis?

A. Polysaccharide capsule

B. Secreted toxin

C. Endospores

D. IgA protease

E. Common pili

55. Which of the following diseases (or complications) is also caused by the etiologic agent?

A. Impetigo

B. Meningitis

C. Infectious diarrhea

D. Glomerulonephritis

E. Myelitis

56. A patient at risk for development of similar infection due to this pathogen should receive which of the following prophylactic drugs or biologicals?

    A. Oral penicillin V
    B. Hyperconcentrated immunoglobulin
    C. 23-valent polysaccharide vaccine
    D. Oral rifampin
    E. Corticosteroids

57. Which of the following populations may also benefit from this vaccine?

    A. Recent immigrants to the United States
    B. Patients with sickle cell disease
    C. Patients with prior respiratory virus infection
    D. Adults in crowded living conditions
    E. Children more than the age of 2 years

58. A 45-year-old traveler presented with a history of cyclical fever to 41.1°C, accompanied by malaise, headache, somnolence, intermittent confusion, jaundice, and anemia. The patient was treated with mefloquine after successful diagnosis. Which of the following findings aided the diagnosis?

    A. Ova in the stool
    B. Elevated antibody titers to sporozoites
    C. Hepatitis A virus-specific IgM in serum
    D. Positive blood cultures
    E. Intraerythrocytic ring stages of parasite in the blood smear

Questions 59-61 are linked to the following case:

A 21-year-old woman presented with a 2-week history of progressively high fever, nonproductive cough, myalgia, and headache. She had a history of sore throat preceding the onset of cough. Despite treatment with cefuroxime (a second generation cephalosporin), the cough did not improve. A chest x-ray revealed diffuse infiltrates in the lungs.

59. Gram stain of sputum and routine cultures of blood and sputum were negative. A cold agglutinin titer was significantly high. Which of the following pathogens is most likely to have caused her illness?

    A. Epstein-Barr virus
    B. *Chlamydia pneumoniae*

    C. Influenza virus
    D. *Legionella pneumophila*
    E. *Mycoplasma pneumoniae*

60. The patient may have contracted the illness by which of the following mechanisms?

    A. Exposure to a pet bird that appeared sick
    B. Inhalation of aerosol from a sick patient with similar illness
    C. Inhalation of air from a dirty air-conditioning unit
    D. Aspiration of endogenous oral flora
    E. Exposure to a dead animal at the roadside

61. Which of the following is the most appropriate therapeutic agent?

    A. Penicillin G
    B. Augmentin
    C. Imipenem
    D. Erythromycin
    E. Vancomycin

62. A 28-year-old female nurse became ill after taking care of a patient with pneumonia 2 days ago and was admitted to the hospital with fever and sore throat. The patient was vomiting but had no diarrhea. CBC was remarkable for a WBC count of 2300/μL and platelets 93,000/μL. Bilateral infiltrates were seen on chest radiograph. She was placed in respiratory droplet isolation and quickly developed respiratory failure, requiring mechanical ventilation. Nine other patients developed pneumonia in the same ward in 10 days. In evaluating the cause of her illness, which of the following is most appropriate?

    A. Blood cultures to identify *Streptococcus pneumoniae*
    B. Urinary antigen testing for *Legionella pneumophila*
    C. Virus cultures to identify influenza A
    D. DFA of nasopharyngeal secretions to identify RSV
    E. Serology to demonstrate antibodies to identify SARS-CoV

63. A 21-year-old woman presented to a clinic with complaints of crampy abdominal pain and vaginal discharge for 3 days. She denied symptoms of urinary tract infections. She admitted to having had three sexual partners in the last month. Examination was remarkable for a vaginal discharge (pH 4.9 [normal <4.5]) that was yellow-green, frothy, and adherent. A Giemsa stain of the discharge was diagnostic (see below).

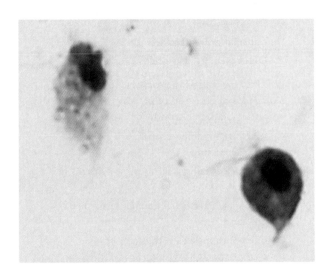

Which of the following pathogens is most likely to have caused her symptoms?

A. *Chlamydia trachomatis*

B. *Trichomonas vaginalis*

C. *Gardnerella vaginalis*

D. *Candida albicans*

E. *Treponema pallidum*

64. A 29-year-old man presented to the ED with a vesicular rash and was very concerned that he might have smallpox. He had had fevers, chills, and malaise for a week, along with a rash that appeared a few days ago. Which of the following characteristics would most likely suggest smallpox?

A. Crops of vesicles appearing in different stages

B. Response to antiviral agents (e.g., ribavirin) and acetaminophen

C. Sore throat and rhinitis before rash

D. Vesicular bumps contain a thick, opaque fluid and have a depression

E. Severe pains of polyneuritis on the face and around the eyes

Questions 65-67 are linked to the following case:

A 55-year-old man presented with a 6-week history of fever, night sweats, and weight loss. He had coughs productive of bloody sputum. He was a recent immigrant from Afghanistan. Examination revealed a malnourished man with abnormal bronchial breath sounds. Chest x-ray was remarkable for a small cavity with streaky infiltrates in the right upper lobe and a calcified lung lesion and lymph node.

65. An acid-fast stain of a section of bronchoscopy-assisted lung biopsy was significant (see below).

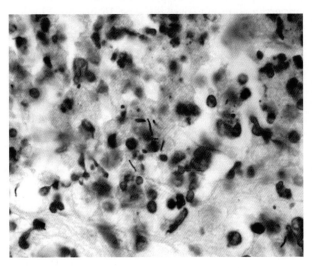

Which of the following pathogens caused his illness?

A. *Klebsiella pneumoniae*

B. *Mycoplasma pneumoniae*

C. *Mycobacterium tuberculosis*

D. *Legionella pneumophila*

E. *Cryptococcus neoformans*

66. The patient most likely acquired this infection by which of the following modes?

A. Intravenous drug use

B. Contact with soiled linens from same household

C. Exposure to airborne organisms from a symptomatic patient

D. Aspiration following a viral infection

E. Inhalation of contaminated dust from soil

67. The patient's 50-year-old wife converted positive on her PPD skin test. A chest x-ray was unremarkable. She was given isoniazid. Three weeks later she had an elevation of liver enzymes. Her elevation of liver enzymes was most likely due to which of the following pathologic processes?

A. Isoniazid-induced liver injury

B. Idiopathic hepatitis

C. Idiopathic pancreatitis

D. Drug-induced reactivation of malaria

E. Drug-induced reactivation of EBV

Questions 68-70 are linked to the following case:

A 3-year-old female presented with one-day history of fever, headache, lethargy, irritability, apnea, and projectile vomiting (over the past 18 hours). Physical examination revealed an irritable, febrile (T: 39.8°C) child with resistance to being touched or moved; no focal neurologic signs; no cranial nerve deficits. The child's immunization history was questionable.

68. CBC revealed neutrophilic leukocytosis. The CSF examination revealed 2500 WBCs/μL with 78% PMNs; glucose 21 mg/dL; protein 220 mg/dL. CSF Gram smear was presumptively diagnostic (see below). Cultures of CSF and blood yielded the same organism (seen in Gram smear) on chocolate agar, but no growth on blood agar.

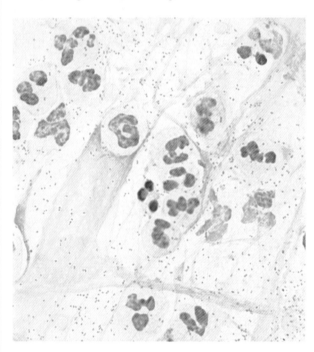

Which of the following pathogens is most likely to have caused this child's illness?

A. *Streptococcus pneumoniae*

B. *Haemophilus influenzae* type b

C. *Escherichia coli* K1

D. *Klebsiella pneumoniae*

E. *Listeria monocytogenes*

69. Which of the following virulence factors is most likely to be involved in pathogenesis?

A. Surface adhesin (lectin)

B. IgA protease

C. Peptidoglycan

D. Capsular polysaccharide

E. Outer membrane proteins

70. Which of the following prophylactic drugs or biologicals can be used to prevent this illness in the pediatric population worldwide?

A. Vaccine containing 23 different capsular polysaccharides

B. Vaccine containing serotype A, C, Y, W135 polysaccharides

C. Conjugate vaccine containing serotype b polysaccharide

D. Chemoprophylaxis with rifampin

E. Passive immunity with IgG from super immune human serum

Questions 71-74 are linked to the following case:

A 19-year-old fully immunized female presented with a 9-day history of fever, sore throat, hoarseness, tiredness, and generalized myalgia. Her vital signs were unremarkable except for a fever of 39.1°C. Physical examination revealed exudative pharyngitis and enlargement of submaxillary and cervical lymph nodes, as well as moderate hepatomegaly, splenomegaly, and mild jaundice.

71. The patient's CBC showed a normal white cell count with 52% monocytes and lymphocytes. A peripheral blood smear was examined by light microscopy (see below).

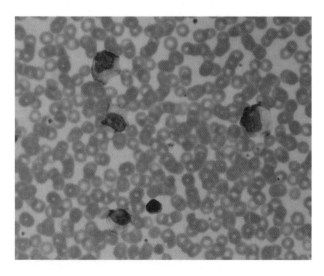

Which of the following findings best describes the microscopy?

A. Neutrophilic leukocytes

B. Eosinophilic leukocytes

C. Basophilic leukocytes

D. Large, atypical T lymphocytes

E. Monocytes

72. Additional biochemical testing revealed elevated liver markers, including aspartate transaminase (AST), alanine transaminase (ALT), alkaline phosphatase, and bilirubin levels. The diagnosis would be confirmed by which of the following tests?

A. Blood culture

B. Testing for a specific toxin

C. Throat culture for bacteria

D. Heterophile antibody in patient's serum

E. Viral antigen detection from nasal wash

73. Her infection was promoted by the binding of the infectious agent to the complement receptor CD21 on the B lymphocytes. Which of the following virologic features best describes the etiologic agent?

A. An enveloped virus with ss-RNA (negative sense) and nonsegmented genome

B. An enveloped virus with ss-RNA (positive sense) and segmented genome

C. An icosahedral virus with capsid-associated fibers and double-stranded, linear DNA genome

D. An enveloped virus with double-stranded, linear DNA genome

E. An enveloped virus with double-stranded, partial circular DNA genome

74. In general, which of the following populations is at the greatest risk for a serious outcome of infection caused by this infectious agent?

A. African children afflicted with malaria

B. Chinese children consuming raw fish

C. Patients with complement deficiencies

D. Patients with phagocyte deficiencies

E. Elderly patients

Questions 75-77 are linked to the following case:

A 21-year-old man presented with a bilateral red pruritic skin lesion in the groin area. He volunteered in a nursing home caring for elderly individuals. Examination was remarkable for circular skin eruption on an erythematous base with an active, advancing periphery over scrotum and perineum.

75. The man's skin lesions are best described as

A. Folliculitis

B. Tinea cruris

C. Type I hypersensitivity

D. Viral exanthem

E. Impetigo

76. In evaluating the cause of the lesions, which of the following would be most appropriate?

A. Testing for drug allergy

B. Microscopy and fungal culture of skin scrapes to detect dermatophytes

C. Direct antigen in serum for *Candida albicans*

D. Serology by agar immunodiffusion to identify *Aspergillus* spp

E. Culture of aspirate to identify *Staphylococcus* or *Streptococcus*

77. Further studies to evaluate his cutaneous lesions yielded the infectious agent (see below).

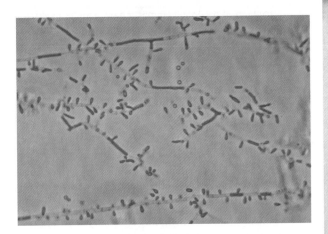

Which of the following is the most appropriate therapeutic agent?

A. Amphotericin B

B. Nafcillin

C. Flucytosine

D. Topical miconazole

E. Penicillin

Questions 78-80 are linked to the following case:

A 26-year-old man presented with headache, low-grade fever, and painful blisters on the penis. He admitted to having had unprotected sex with three female partners in the past month. Examination revealed inguinal lymphadenopathy and blisters on the penis.

78. Tzanck preparation of scrapings from base of vesicles was significant (see below).

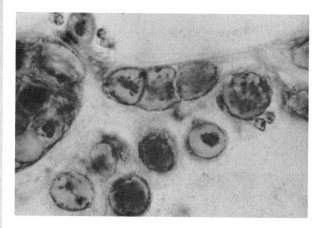

Which of the following pathogens is most likely to have caused his penile blisters?

A. Human papillomavirus
B. *Haemophilus ducreyi*
C. *Treponema pallidum*
D. *Chlamydia trachomatis*
E. Herpes simplex virus

79. The painful blisters result from which of the following processes?

A. Cytopathic effect of the pathogen on cutaneous epithelium and sensory neurons
B. Acute inflammation by proinflammatory mediators in response to OMPs
C. Cytolytic activity of the reticulate bodies of the pathogen on the mucosal cells
D. Desquamation of squamous epithelial cells of the basal layer of the skin due to viral growth
E. Chronic granulomatous inflammation in response to cell wall antigens of the pathogen

80. Which of the following therapeutic agents would be the most appropriate treatment in this case?

A. Acyclovir
B. Ceftriaxone
C. Penicillin
D. Doxycycline
E. Podophyllin

Questions 81-84 are linked to the following case:

A 9-week-old baby boy presented with a 5-day history of choking spells and repetitive coughing, progressing to his turning red and gasping for breath. Two weeks before, a routine vaccination had been postponed because the child had had a common cold illness. Physical examination revealed T: 38.3°C, P: 160/min, R: 32/min. The child's chest radiograph was clear (no evidence of tracheal abnormalities). CBC was 15,000/µL with 70% lymphocytes.

81. Which of the demographic and social/medical history would be most useful in supporting the clinical diagnosis of this case?

A. The patient was born prematurely
B. His parents lived in Chicago, IL, under poor conditions
C. He had four older siblings
D. The delay in immunization of the baby
E. The mother had just recovered from a upper respiratory infection

82. A nasopharyngeal swab was sent for microbiology investigation. Routine microscopy and culture investigations were negative. A selective agar-based culture yielded significant colonies (Gram stain is shown below) that were confirmed by direct fluorescence antibody staining.

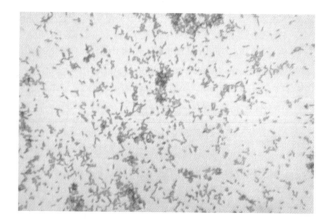

Which of the following pathogens is most likely to have caused his illness?

A. Respiratory syncytial virus

B. *Bordetella pertussis*

C. *Haemophilus influenzae*

D. *Mycoplasma pneumoniae*

E. *Corynebacterium diphtheriae*

83. The etiologic agent was determined to be toxigenic. Which of the following functions of the toxin is most likely to be involved in the pathogenesis?

   A. Inactivation of the Gi-protein complex with a consequent rise in cAMP

   B. Inactivation of the EF-2 with consequent death of the mucosal epithelial cells

   C. Destabilization of membrane integrity of the target cells due to phosphorylation of membrane phospholipid

   D. Creation of pores in target cell membranes by hemolysin

   E. Inactivation of M protein, resulting in the death of mucosal epithelial cells

84. Which of the following treatment plans would most likely lead to the successful recovery of this patient?

   A. Administration of antitoxin concentrated from pooled horse serum

   B. Respiratory care by suctioning to remove mucus; pressurized oxygen

   C. Humidified air produced in the bath from hot water

   D. Administration of penicillin by intramuscular injection

   E. Administration of aerosolized epinephrine via a nebulizer

Questions 85-88 are linked to the following case:

A 23-year-old man returned from a trip to Thailand where he visited refugee camps along the Cambodian border. Five days later, on his return to New York he developed high fevers and disorientation. He was admitted to ICU in a coma. He became progressively somnolent and died a week later. At autopsy, there was marked cerebral edema with areas of cerebral softening.

85. Which of the following pathogens is most likely to have caused this catastrophic illness?

   A. *Trypanosoma brucei rhodesiense*

   B. *Leishmania donovani*

   C. *Plasmodium falciparum*

D. *Toxoplasma gondii*

E. *Entamoeba histolytica*

86. In which of the following organs does this infectious agent initially proliferate after entry in the infected host?

   A. Heart

   B. Liver

   C. Brain

   D. Lymph nodes

   E. Spleen

87. What is the most common mode of transmission of this agent?

   A. Bite of an *Anopheles* mosquito

   B. Blood transfusion

   C. Organ transplantation

   D. Transplacental crossing

   E. Contact with blood parasites

88. Which of the following is the most appropriate first-line treatment when this disease is diagnosed early?

   A. Hepatic merozoiticides

   B. Sporozoiticides

   C. Blood gametocides

   D. Blood schizonticides

   E. Blood trophozoiticides

---

89. A 24-year-old woman complained of itching and burning of vulvar skin and vaginal discharge for 4 days. She had recently had a UTI and recovered after an oral course of trimethoprim/sulfamethoxazole. She had no chills or fever. Examination revealed a white, cottage cheese-like discharge (pH 4.3 [normal <4.5]). Pap smear of the discharge was diagnostic (see below).

Which of the following pathogens is most likely to have caused her itching and vaginal discharge?

A. *Chlamydia trachomatis*

B. *Trichomonas vaginalis*

C. *Gardnerella vaginalis*

D. *Candida albicans*

E. *Treponema pallidum*

Questions 90-93 are linked to the following case:

A 38-year-old male Peace Corps volunteer who recently spent a month in rural Mexico complained of a spiking fever, malaise, headache, and right upper quadrant abdominal pain. He had bloody diarrhea with mucus, and tenesmus that disappeared with some pills given to him by the health facility nurse. Physical examination revealed a fever of 39.6°C, pallor, slight jaundice, and tender hepatomegaly.

90. Which specific information in the case was most helpful in developing the diagnostic hypothesis?

    A. Peace Corps volunteer

    B. Spent a month in rural Mexico

    C. Bloody diarrhea with mucus

    D. Patient has headache

    E. Patient has high fever

91. Routine enteric pathogens cultured by the laboratory were negative. The ova and parasite examination was significant (see below).

A. *Entamoeba histolytica*

B. *Cryptosporidium parvum*

C. *Dientamoeba fragilis*

D. *Giardia lamblia*

E. *Cyclospora cayetanensis*

92. Which of the following toxins is most likely to be involved in pathogenesis of invasive infection with this organism?

    A. Elastin

    B. Leukocidin

    C. Hemolysin

    D. Enterotoxin

    E. Cytotoxin

93. Which of the following is the most appropriate therapeutic combination?

    A. Sulfadiazine and pyrimethamine

    B. Albendazole and diloxanide furoate

    C. Metronidazole and iodoquinol

    D. Praziquantel and ivermectin

    E. Trimethoprim and sulfamethoxazole

Questions 94 and 95 are linked to the following case:

A 31-year-old woman with a history of multiple sexual partners sought medical attention. After pelvic examination, a Pap smear from the cervix was obtained and was found to be abnormal (see below).

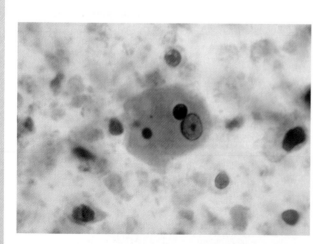

Which of the following pathogens is most likely to have caused his illness?

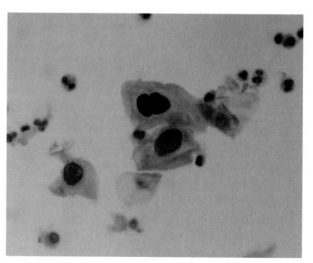

94. Which of the following characteristics best describes the etiologic agent that caused this abnormal Pap smear?

A. An enveloped, double-stranded, linear DNA virus

B. An nonenveloped, double-stranded, circular DNA virus

C. An enveloped, single-stranded RNA (negative-sense) virus

D. An enveloped, single-stranded RNA (positive-sense) virus

E. A nonenveloped (icosahedral), double-stranded, linear DNA virus

95. The patient in this case is predisposed to which of the following complications?

A. Viral encephalitis

B. Ovarian carcinoma

C. Necrotizing gummas

D. Uterine fibroids

E. Cervical carcinoma

Questions 96-99 are linked to the following case:

A 7-year-old child had a sore throat, difficulty swallowing, and fever for 5 days. Examination was remarkable for fever; extensive erythematous rash on neck, groin, axillae; bright red lingual papillae superimposed on a white coat; exudative tonsillitis; and cervical lymphadenopathy.

96. Culture of throat swab on sheep blood agar (SBA) yielded significant colonies that were sensitive to bacitracin (see below).

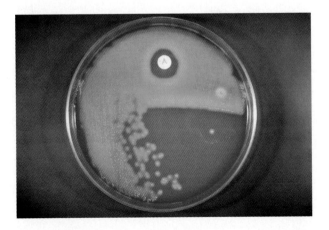

Which of the following microbiologic features best describes the etiologic agent?

A. Gram-positive cocci in clusters; β-hemolytic colonies on SBA, coagulase positive

B. Gram-positive cocci in chains; β-hemolytic growth on SBA, catalase negative

C. Gram-positive diplococci; α-hemolytic growth on SBA, Optochin sensitive

D. Gram-positive diplococci; α-hemolytic growth on SBA, Optochin resistant

E. Gram-positive cocci in chains; γ-hemolytic growth on SBA, catalase negative

97. Which of the following virulence factors is most likely to be involved in the pathogenesis of rash and red papillae on the patient's tongue?

A. Hyaluronate capsule

B. M-protein

C. Erythrogenic toxin

D. Endotoxin

E. Cell wall peptidoglycan

98. Which of the following delayed sequelae of this infection might occur?

A. Endocarditis

B. Pneumonia

C. Hepatitis

D. Acute rheumatic fever

E. Meningitis

99. The patient did not have any allergies. Which of the following is the most appropriate therapeutic agent?

A. Ciprofloxacin

B. Penicillin G

C. Tetracycline

D. Chloramphenicol

E. Gentamicin

Questions 100 and 101 are linked to the following case:

A 21-year-old native American from New Mexico presented with fever, cough, and severe dyspnea. Crackles were heard on lung examination. An arterial pO_2 on room air was 50 mm Hg. The hematocrit was 60%. CBC was remarkable for WBC 40,000/μL, with 8% bands and 30% atypical lymphocytes, and thrombocytopenia. A chest x-ray showed a normal-sized heart and diffuse alveolar infiltrates. Serology performed by enzyme immunoassay for a virus was diagnostic.

100. The man acquired the infection by which of the following modes?

 A. Inhalation of dried materials contaminated by rodent excreta

 B. Inhalation of respiratory droplets from another sick individual

 C. Inhalation of contaminated air from an air-conditioning unit

 D. Forced passage of water through nose during diving in a pond

 E. Aspiration of colonized virus post allergic rhinitis

101. What is the most common complication in severe cases?

 A. Encephalitis

 B. Empyema

 C. Pericarditis

 D. Spastic paralysis

 E. Disseminated intravascular coagulation

Questions 102-104 are linked to the following case:

A 22-year-old woman complained of low-grade fever along with pain and swelling in the left knee. She had purulent vaginal discharge for several weeks. She related that she had been sexually active with multiple partners since she was 14 years of age. Examination was remarkable for swollen, tender, warm left knee and vaginal discharge. Gram stain of pus from cervix was significant (see below).

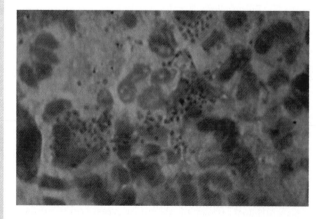

102. Which of the following pathogens is most likely to have caused her illness?

 A. *Chlamydia trachomatis*

 B. *Trichomonas vaginalis*

 C. *Haemophilus ducreyi*

 D. *Neisseria gonorrhoeae*

 E. *Treponema pallidum*

103. Which of the following virulence factors most likely contributed to the presenting symptoms?

 A. A-B subunit toxin

 B. LOS endotoxin

 C. Capsular polysaccharide

 D. Peptidoglycan fragment

 E. Flagella

104. Which of the following therapeutic agents would be the most appropriate in this case?

 A. Penicillin G

 B. Ceftriaxone

 C. Doxycycline

 D. Erythromycin

 E. Clindamycin

Questions 105 and 106 are linked to the following case:

Three weeks after a renal transplantation, a 28-year-old man developed fever and leukopenia, followed by prostration and severe pulmonary and kidney dysfunction. Infected kidney cells showed characteristic inclusions (see below).

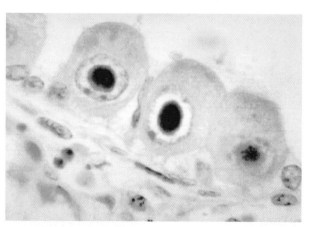

105. What is the most likely etiology?

    A. Cytomegalovirus

    B. Parvovirus B19

    C. Coronavirus

    D. Human herpesvirus type 6

    E. Coxsackievirus

106. Which of the following would be the most appropriate therapeutic agent in this case?

    A. Amantadine

    B. Acyclovir

    C. Ganciclovir

    D. Ribavirin

    E. Valacyclovir

107. A 55-year-old woman came to her internist with complaints of low-grade fevers, chills, and anorexia. She also noted a 10-lb weight loss. Echo showed a mobile vegetation on the mitral valve. Blood cultures revealed significant bacteria (see below) in all bottles.

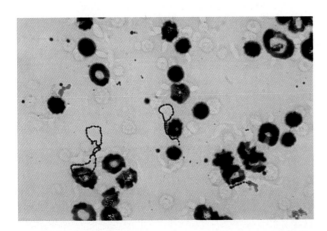

Which of the following is true regarding therapy of this infection?

    A. Bacteriostatic therapy is adequate for most cases.

    B. Quinolones are the drug of choice.

    C. The duration of treatment is usually 4 to 6 weeks.

    D. After initial IV antibiotics, oral antibiotics may be used to complete therapy.

    E. Mortality rates are high even with appropriate antibiotic therapy.

108. A 21-year-old woman presented with dysuria, increased frequency and urgency of urination, and low-grade fever. She was recently married and had no remarkable medical history. Urine examination revealed abundant white cells, low protein, but no casts. Culture of midstream urine on blood agar yielded the growth of a significant uropathogen (see below).

Which of the following organisms is most likely to have caused her illness?

    A. *Staphylococcus saprophyticus*

    B. *Enterococcus faecalis*

    C. *Staphylococcus aureus*

    D. *Proteus vulgaris*

    E. *Candida albicans*

Questions 109 and 110 are linked to the following case:

A 29-year-old man presented with burning urination and urethral pain. Physical examination revealed mucopurulent and slightly blood-tinged urethral discharge and normal testes and epididymis with no urinary retention. He admitted to having had unprotected sexual contact with a prostitute 3 days before.

109. Gram stain of the urethral discharge was significant (see below). Culture of the discharge on a selective agar produced an oxidase-positive isolate that fermented glucose.

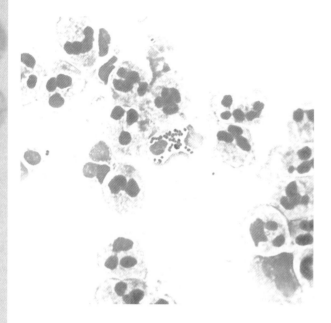

Which of the following pathogenic processes is most likely to have caused his urethral discharge?

A. Intense cell-mediated immune response

B. Mucosal adhesion and toxigenesis

C. Invasive infection with cellular necrosis

D. Pyogenic inflammation

E. Granulomatous inflammation

110. The patient was unsuccessfully treated with penicillin. Isolates pre- and post-treatment were found to differ in their abilities to adhere to mucosal epithelial cells. The change was due to differential gene expression of which of the following virulence factors?

A. Flagella

B. Capsule

C. Lipopolysaccharide

D. Outer membrane

E. Fimbriae

Questions 111 and 112 are linked to the following case:

A 7-month-old male presented with fever and severe vomiting followed by watery diarrhea. Examination was remarkable for fever and tachycardia. Fecal examination was negative for leukocytes. An EIA-based viral stool antigen test was diagnostic, and the patient recovered after fluid replacement therapy.

111. Which of following pathogens is most likely to have caused this infant's diarrhea?

A. Coronavirus

B. Adenovirus

C. Rotavirus

D. Norovirus

E. Calicivirus

112. Which of the following mechanisms is most likely to be involved in the pathogenesis of diarrhea caused by the organism?

A. Damaged villi are replaced by immature crypt cells with low absorptive capacity

B. Increased secretion of sodium and water by epithelial cells

C. Lysis of colonic epithelial cells by the multiplication cycle of the enteric virus

D. Decreased breakdown of lactose to glucose and galactose

E. Stimulation of adenyl cyclase, causing hypersecretion of Cl^- into gut

113. A 22-year-old female presented with pain, burning, and stinging on urination, as well as urgency and frequency. History was unremarkable except for being a newlywed. On examination, the proximal end of the urethra was inflamed. She also had suprapubic pain and tenderness. A semiquantitative urine culture grew significant colonies (Gram-positive oxidase-negative) on MacConkey agar (see below) that were indole positive.

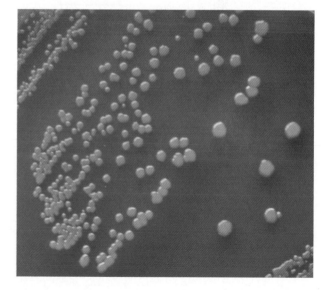

Which of the following therapeutic combinations would be the most appropriate treatment in this case?

A. Piperacillin and tazobactam

B. Ampicillin and gentamicin

C. Erythromycin and chloramphenicol

D. Penicillin and doxycycline

E. Trimethoprim and sulfamethoxazole

114. Meningococcal infections involve both the meninges and the blood stream. Release of endotoxin into the blood stream leads to disseminated intravascular coagulation with skin manifestations. Which of the following is the best description of skin manifestation in this patient with meningococcemia (see below)?

A. Macules

B. Papules

C. Vesicles

D. Purpura

E. Erythema

115. Septicemia can rapidly lead to septic shock and death. Many of the symptoms associated with septicemia and subsequent shock related to Gram-negative infection can be attributed to endotoxin's stimulation of

A. Complement and G-CSF

B. Iron-binding protein

C. Immunoglobulin production

D. IL-1 and TNF

E. IL-2 and IL-4

116. A 22-year-old male college student developed a painless penile ulcer 3 weeks after returning from a spring break. A serologic test was positive. Which of

the following is the most appropriate therapeutic agent for this patient, who has no drug allergies?

A. Penicillin G

B. Acyclovir

C. Ceftriaxone

D. Doxycycline

E. Podophyllin

Questions 117 and 118 are linked to the following case:

A 45-year-old woman with a history of achlorhydria experienced the sudden onset of a severe, voluminous, watery diarrhea immediately after her return from Peru. She became severely dehydrated over the next few days. Microscopic examination of the diarrhea fluid revealed flecks of mucus but no blood and few white blood cells. She recovered in the hospital with intravenous fluid therapy over the next week.

117. Which of the following epidemiologic features best describes this etiologic agent?

A. The agent is transmitted mostly via food to adults and the elderly.

B. The agent originated from a food-borne disease outbreak in Europe.

C. The agent is highly endemic in Peru, where it has caused water-borne epidemics.

D. The agent has a high potential for weaponization and bioterrorism.

E. The agent has endemicity in the areas surrounding the great lakes.

118. Which of the following mechanisms is most likely to be involved in the pathogenesis of such severe diarrhea?

A. The agent induces its effect on the mucosa of small intestine via the mediation of a potent enterotoxin.

B. The agent invades the small intestine and causes cytoskeletal rearrangement of the mucosa via the mediation of cytotoxin.

C. The agent causes intense inflammation at the mucosa of small intestine.

D. The agent causes a penetrating infection in the small intestine leading to diarrhea.

E. The spore-forming agent produces heat-labile enterotoxin during germination in the small intestinal mucosa.

Questions 119-121 are linked to the following case:

A 71-year-old man presented with a 48-hour history of frequent bowel movement while hospitalized post-surgery. One week prior to the current episode, he had undergone surgery for prostate hypertrophy and had received broad-spectrum antibiotics, including clindamycin. His wife, who visited him daily, did not have any diarrheal illness.

119. The patient's diarrhea resolved after his broad-spectrum antibiotics were replaced by oral metronidazole for 7 days. Which of the following pathogens is most likely to have caused his diarrheal illness?

    A. *Giardia intestinalis*
    B. *Shigella sonnei*
    C. *Campylobacter jejuni*
    D. *Clostridium difficile*
    E. *Entamoeba histolytica*

120. Which of the following toxins was most likely involved in pathogenesis of this illness?

    A. Hemolysin
    B. Leukocidin
    C. Enterotoxins
    D. Cytotoxins
    E. Coagulase

121. The infectious agent was cultured from his bedpan. Which of the following is most likely to sterilize the bedpan?

    A. Exposure to glutaraldehyde for 1 hour
    B. Exposure to benzalkonium chloride for 1 hour
    C. Exposure to 0.5% bleach (hypochlorite) for 2 hours
    D. Autoclaving at 15 lbs pressure (121°C) for 15 minutes
    E. Heating in an oven at 150°C for 30 minutes

Questions 122-124 are linked to the following case:

A 59-year-old man with chronic heart disease (CHD) presented with a one-day history of high fever, shortness of breath, left-sided chest pain, and a productive cough. Physical examination revealed T: 39.2°C, R 44/min, BP: 102/60. CBC and differential were remarkable for 27,900 WBCs, 81% PMNs, 12% bands, and 7% lymphocytes. Chest x-ray revealed consolidation of the left lower lobe.

122. Transtracheal aspirate on Gram smear revealed the presumptive etiology (see below). Blood and the respiratory specimens yielded the diagnosis.

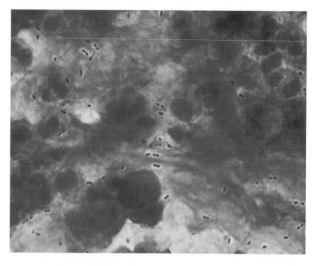

Which of the following pathogens is most likely to have caused his illness?

    A. *Streptococcus viridans*
    B. *Streptococcus pneumoniae*
    C. *Staphylococcus aureus*
    D. *Streptococcus pyogenes*
    E. *Streptococcus agalactiae*

123. A year later the patient was again hospitalized for a similar illness. Which of the following mechanisms explains why the natural immunity from the first infection was not protective against the second infection?

    A. The antibody is against the streptococcal M protein only.
    B. Natural immunity wanes to a nonprotecting level in less than 6 months.
    C. The polysaccharide capsule of the organism is not immunogenic.
    D. The antibodies of more than 80 different capsular types do not cross-protect.
    E. A patient with CHD has a lower level of protective immunity at recovery.

124. The clinical isolate was resistant only to penicillin. Which of the following drugs is the most appropriate therapeutic choice?

    A. Nafcillin
    B. Cefotaxime

C. Sulfamethoxazole

D. Erythromycin

E. Gentamicin

125. A 25-year-old woman, married a month ago, developed a grayish cheesy discharge from her vagina. History showed that she had received broad-spectrum antibiotics recently for her honeymoon cystitis. Which of the following is the most appropriate therapeutic agent to treat this illness?

A. Clotrimazole

B. Penicillin

C. Metronidazole

D. Doxycycline

E. Ceftriaxone

126. A 29-year-old male landscape worker was brought to a hospital with a week-long history of fever, shortness of breath, and dry cough. Examination revealed fever, tachycardia, and mild tachypnea. He had painless "cauliflower" lesions on his left upper arm. A biopsy of the lesions on the arm revealed organisms with characteristic buds (see below).

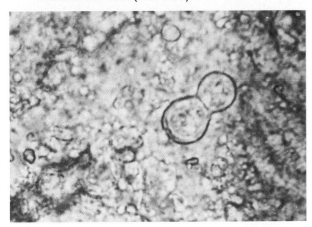

Which of the following pathogens caused his illness?

A. *Cryptococcus neoformans*

B. *Blastomyces dermatitidis*

C. *Mycobacterium tuberculosis*

D. *Candida albicans*

E. *Aspergillus fumigatus*

127. A 31-year-old Hispanic man had a burning feeling in his gut after meals that was relieved with antacids. He responded to treatment with amoxicillin, clarithromycin, and omeprazole for 3 weeks. Which of the following pathogens is most likely to have caused his illness?

A. *Entamoeba histolytica*

B. *Giardia lamblia*

C. *Cryptosporidium parvum*

D. *Helicobacter pylori*

E. *Campylobacter jejuni*

128. Twenty-one of 51 people became ill within 2 hours after eating a potluck dinner at a church gathering. All 21 had nausea, most had vomiting, and several had crampy abdominal pain. All sick patients had eaten the vanilla custard. Which of the following organisms is most likely to have caused this outbreak of food poisoning?

A. Enterotoxigenic *Escherichia coli*

B. *Clostridium perfringens*

C. *Staphylococcus aureus*

D. *Salmonella typhimurium*

E. *Cryptosporidium parvum*

Questions 129-131 are linked to the following case:

A 51-year-old man presented with fever, abdominal pain, and rapidly progressive distention of the abdomen. His appendix had been removed for an abscess a few days before. Examination was remarkable for fever and moderate hypotension.

129. CBC was remarkable for neutrophilic leukocytosis. Ascitic fluid and blood culture yielded a Gram-negative, oxidase-negative, indole-positive bacterium. The anaerobic culture of the needle aspirate of the peritoneal abscess subsequently yielded a significant pathogen (see below).

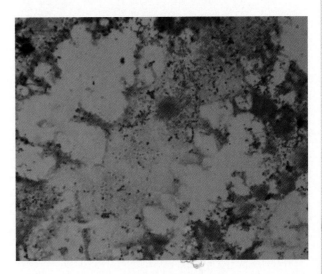

Which of the following major pathogens in synergism with polymicrobic infection is most likely to have caused his abscess?

A. *Bacteroides fragilis*

B. *Clostridium perfringens*

C. *Escherichia coli*

D. *Salmonella typhimurium*

E. *Enterococcus fecalis*

130. Which of the following virulence factors of the anaerobic pathogen is most likely to be involved in pathogenesis of the abscess?

A. Flagella

B. Lipopolysaccharide

C. Peptidoglycan mitogen

D. Polysaccharide capsule

E. Fimbrial adhesin

131. Which of the following would be the most appropriate therapeutic agent?

A. Metronidazole

B. Gentamicin

C. Cefotaxime

D. Trimethoprim/sulfamethoxazole

E. Levofloxacin

132. A 41-year-old female presented with bloody diarrhea after her return from a trip to Mexico. Primary enteric cultures were negative. Histopathology (H and E stain) of the colon showed lesions of nonspecific ulcers surrounded by inflammatory cells (see below).

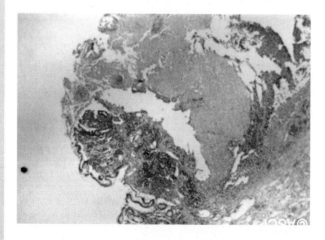

Which of the following pathogens would most likely be the cause of her illness?

A. *Vibrio cholerae*

B. *Salmonella typhimurium*

C. *Campylobacter jejuni*

D. *Entamoeba histolytica*

E. *Giardia lamblia*

Questions 133–135 are linked to the following case:

A 27-year-old man traveled to New Mexico to try out an experimental desert vehicle. Three weeks later, he developed a fever, chest pain, and sore muscles. After 2 days, red tender nodules appeared on the shins, and the right ankle became painful and tender. A chest x-ray film was remarkable for a left pleural effusion.

133. Microscopic examination of an aspirate from a nodule revealed large structures (see below).

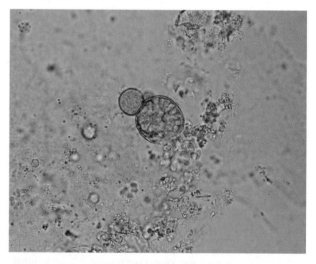

Which of the following pathogens caused his illness?

A. *Blastomyces dermatitidis*

B. *Coccidioides immitis*

C. *Histoplasma capsulatum*

D. Sin Nombre (hanta) virus

E. *Francisella tularensis*

134. Which of the following is the most appropriate therapeutic agent?

A. Isoniazid

B. Fluconazole

C. Tetracycline

D. Albendazole

E. Cefotaxime

135. The structures shown (see below) are inhaled and are deposited in the terminal bronchiole, initiating pathogenesis.

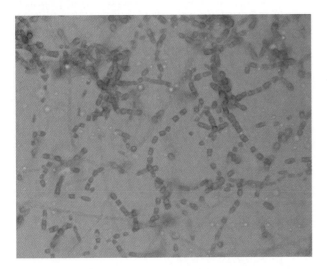

These structures are which of the following?

A. Blastoconidia

B. Macroconidia

C. Microconidia

D. Sporangiospores

E. Arthroconidia

---

136. During a trip to Belize, Central America, a 31-year-old woman developed bloody diarrhea. She did not seek any medical attention, and her diarrhea resolved in 2 weeks. Two months after the initial onset she began to have fever, chills, and abdominal pain, and she came to the ED. Examination was remarkable for right upper quadrant abdominal tenderness. An abdominal CT scan was remarkable (see below).

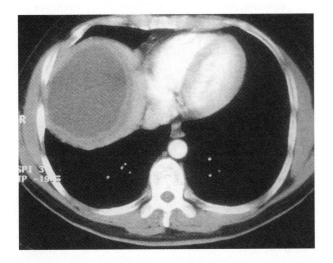

A positive serology subsequently confirmed the diagnosis. Which of the following pathogens is mostly likely to have caused these events?

A. *Giardia lamblia*

B. *Salmonella typhi*

C. *Entamoeba histolytica*

D. *Campylobacter jejuni*

E. *Yersinia enterocolitica*

Questions 137–139 are linked to the following case:

A 36-year-old man presented with a high fever, chills, and multiple skin lesions at injection sites. He had been an IV drug user for the last several months. Examination was remarkable for ejection systolic murmur increasing with inspiration, heard in the tricuspid area. Imaging (Echo) was remarkable for vegetations on the tricupsid valve.

137. Blood cultures were diagnostic (see below).

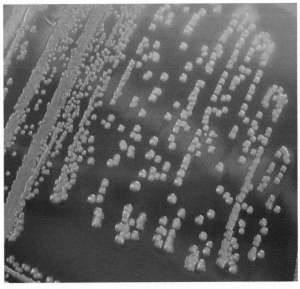

Which of the following pathogens is most likely to have caused his illness?

A. *Staphylococcus epidermidis*

B. *Enterococcus faecalis*

C. *Escherichia coli*

D. *Clostridium perfringens*

E. *Staphylococcus aureus*

138. Which of the following is the most appropriate therapeutic combination?

    A. Nafcillin and gentamicin
    B. Cefotaxime and erythromycin
    C. Vancomycin and rifampin
    D. Levofloxacin and gentamicin
    E. Linezolid and doxycycline

139. Protein A, a major virulence factor of this pathogen, exhibits which of the following functions?

    A. Adhesion to epithelial tissue
    B. Penetration of mucus
    C. Binds to Fc portion of Ig molecules
    D. A cytolytic effect on target cells
    E. A spreading factor causing dissemination

140. A 19-year-old man presented with a week-long history of jaundice and dark yellow urine. He also complained of nausea, vomiting, and malaise, low-grade fever, and abdominal pain. He had recently returned from a vacation in Venezuela, where he said he consumed a lot of shellfish. Which of the following infectious agents is likely to have caused his jaundice?

    A. Hepatitis A virus
    B. Hepatitis B virus
    C. Hepatitis C virus
    D. Hepatitis D virus
    E. Hepatitis E virus

141. A 29-year-old man presented with sudden-onset crampy abdominal pain and diarrhea. The patient complained of low-grade fever with chills, malaise, nausea, and vomiting. He had ingested partially cooked eggs at a poultry farm 24 hours before his symptoms began. Fecal examination was positive for leukocytes. Stool culture on selective agar yielded a significant bacterial pathogen. The patient recovered after fluid replacement therapy without any antibiotics. Which of the following pathogens is most likely to have caused his diarrheal illness?

    A. *Clostridium difficile*
    B. *Escherichia coli* O157:H7
    C. *Salmonella typhimurium*
    D. *Shigella flexneri*
    E. *Campylobacter jejuni*

Questions 142 and 143 are linked to the following case:

A 15-year-old male had a small eschar forming around the site of removal of a tick from his left forearm. A hemorrhagic rash developed over the next few days involving his trunk and extremities and even his palms and soles. Then, small 0.2 to 0.4 cm foci of skin necrosis developed on his fingers and toes. He died of overwhelming septicemia.

142. Examination at autopsy revealed red structures in endothelial cells of a blood vessel indicating immunohistologic staining of organisms. Which of the following organisms is most likely responsible for these findings?

    A. *Brucella abortus*
    B. *Mycobacterium marinum*
    C. *Yersinia pestis*
    D. *Borrelia burgdorferi*
    E. *Rickettsia rickettsii*

143. Which of the following advice is most likely to be effective in preventing this catastrophic illness?

    A. Avoid tick-infested areas and dogs with ticks.
    B. Apply talcum powders to pants and sleeves to deter the ticks.
    C. Search total body every 24 hours when in a tick-infested area.
    D. Seek help while the tick is still attached for removal and identification of tick.
    E. Prophylaxis with penicillin for 7 days is needed after tick removal.

144. A 7-year-old male presented with an extensive purpuric skin rash, oliguria, and marked weakness; he also complained of bloody diarrhea for 1 week. The patient drank apple cider at a roadside store beside a farm 3 days before. He had no fever, and stool examination was negative for leukocytes. Which of the following pathogens is most likely to have caused his illness?

    A. *Escherichia coli* O157:H7
    B. *Salmonella typhimurium*
    C. *Shigella flexnerii*
    D. *Campylobacter jejuni*
    E. *Entamoeba histolytica*

145. A 29-year-old man with a history of HIV-1 infection presented with fever, severe headache, and neck stiffness. He had no papilledema, so a lumbar puncture was performed. A Gram stain of the cerebrospinal fluid revealed many short Gram-positive rods. What is the most likely mode of acquisition of this infection?

    A. Sharing infected needles

    B. Inhalation of respiratory droplets

    C. Inoculation through a cut on the skin

    D. Ingestion of Mexican soft cheese

    E. Sex with a female prostitute

146. A 61-year-old man experienced an acute onset of fever, crampy abdominal pain, and watery diarrhea while visiting his mother in Karachi, Pakistan. Overnight, he noticed mucus and a bloody tinge of the stool. A stool culture became positive for a Gram-negative, nonmotile, facultative anaerobe (biochemically inert), identified by serotyping. If performed, a histologic examination of an endoscopic biopsy of the colon would likely show which of the following findings?

    A. Ulceration of mucosa (endothelium damage) with overlying infiltration of PMNs

    B. Multiple granulomas throughout the wall of the colon with nonspecific infiltrates

    C. Slight increase in the numbers of lymphocytes and plasma cells in the lamina propria

    D. Intranuclear inclusions in the enterocytes and prominence of mucus-secreting cells

    E. Extensive scarring of the lamina propria with stricture formation

147. A 45-year-old man presented with fever, joint pain, and fatigue. The patient had a large annular lesion with an erythematous periphery under his armpit (see below).

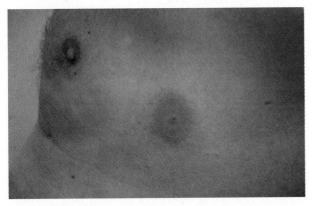

A serologic investigation subsequently confirmed the diagnosis. Which of the following pathogens is most likely to have caused his illness?

A. *Borrelia burgdorferi*

B. *Trichophyton rubrum*

C. *Candida albicans*

D. *Streptococcus pyogenes*

E. *Bacillus anthracis*

148. A 45-year-old woman presented with an ulcer on her right thumb and satellite lesions along the lymphatic drainage. She recalled receiving a thorn prick while she was tending her rose bush 2 weeks before. A Gomori methenamine silver stain of the biopsy revealed organisms (see below).

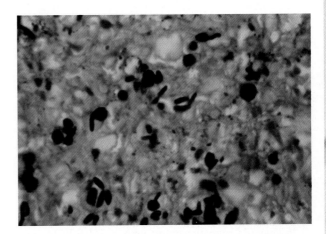

Which of the following pathogens is most likely to have caused her illness?

A. *Staphylococcus aureus*

B. *Bacillus thornii*

C. *Trichophyton rubrum*

D. *Bacillus anthracis*

E. *Sporothrix schenckii*

149. A 41-year-old man presented with rapidly progressive dyspnea and fever. Six weeks before, he had undergone a prosthetic valve replacement for calcific aortic stenosis. Physical examination was remarkable for fever, hypotension, and cardiac auscultation, indicating aortic incompetence. CBC revealed anemia. Echo was remarkable for prosthetic valve dehiscence. Three sets of blood cultures were diagnostic (see figure on page 490).

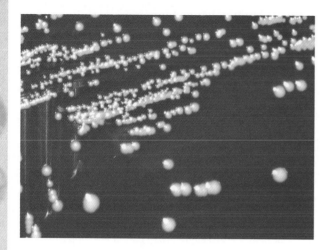

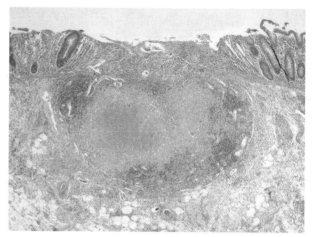

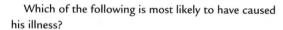

Which of the following is most likely to have caused his illness?

A. *Staphylococcus epidermidis*

B. *Streptococcus mitis*

C. *Streptococcus pyogenes*

D. *Clostridium perfringens*

E. *Pseudomonas aeruginosa*

A. Flask-shaped ulcers, infiltrated by mostly PMNs, with necrosis of the epithelium in the small intestine

B. Focal areas of necrosis and mononuclear inflammatory cells in the base of an ulcer

C. Necrosis of colonic mucosa with fibrosis, mucus, and inflammatory cells

D. Disrupted colonic mucosal epithelium covered by pseudomembrane and interstitial infiltration of PMNs with mucin in glands

E. Inflammatory enteritis involving the entire jejunal mucosa, flattened atropic villi, and necrotic debris in the lumina of crypts

Questions 150 and 151 are linked to the following case:

A 26-year-old man presented with a 5-day history of increasing fever, malaise, headache, and constipation. He had returned from Manila, Philippines, a week before, after a month-long trip for which he did not receive any prior vaccinations. His vital signs revealed bradycardia and fever of 40°C. His physical examination revealed mild hepatosplenomegaly and faint erythematous macules.

150. Which of the following is most likely to have caused this man's illness?

A. *Entamoeba histolytica*

B. *Giardia lamblia*

C. *Cryptosporidium parvum*

D. *Salmonella typhi*

E. *Shigella sonnei*

151. If the following was the finding of a colonic biopsy (see figure above, right), which feature best describes the abnormal histopathology?

152. A 48-year-old male presented to the ED with complaints of fever, chills, and night sweats for a week. Over the last 2 days he had also noted a cough and pleuritic chest pain. He admitted to intravenous drug use recently. Blood cultures were positive for a Gram-positive bacterium. Which of the following is most likely to have caused his illness?

A. *Streptococcus viridans* group

B. *Enterococcus faecalis*

C. *Staphylococcus aureus*

D. *Bacillus cereus*

E. *Propionibacterium acne*

Questions 153 and 154 are linked to the following case:

Within a period of 10 days a series of nine cases of vomiting and diarrhea occurred among young children attending a local day care center. An epidemiologic investigation confirmed a common-source outbreak of diarrhea at the day care center.

153. Duodenal aspirate from two of the nine patients revealed a significant infectious agent (see below).

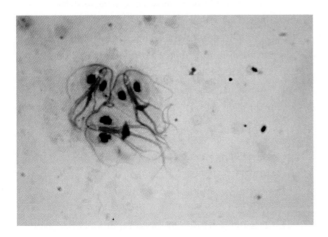

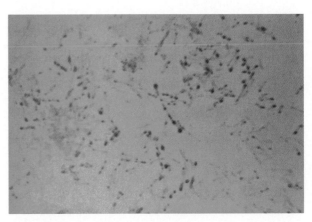

Which of the following pathogens is most likely to have caused this outbreak of diarrhea?

A. *Entamoeba histolytica*

B. *Cryptosporidium parvum*

C. *Giardia lamblia*

D. *Balantidium coli*

E. *Dientamoeba fragilis*

154. Which of the following classes of protozoa best describes the etiologic agent?

A. Ameba

B. Flagellate

C. Ciliate

D. Sporozoa

E. Coccidia

Questions 155-157 are linked to the following case:

A 9-year-old female was admitted to the hospital with a 3-day history of malaise, anorexia, low-grade fever, and increasing delirium. The child was raised in Ghana, Africa, and the family had emigrated to the United States recently. Her immunization status was questionable. On physical examination, she had an acute pharyngitis with an overlying dirty white tough membrane.

155. Microscopic examination of a fibrinopurulent exudate from the pharyngeal membrane revealed numerous small bacteria with metachromatic granules (see figure above, right).

Culture of throat swab on special isolation media yielded the diagnosis. Which of the following best describes the etiology of this illness?

A. Gram-negative rod; facultative anaerobic growth, encapsulated bacteria

B. Gram-negative rod; aerobic growth, lactose nonfermenting, pigment producing

C. Gram-negative coccobacilli; aerobic growth on Bordet-Gengou agar

D. Poor growth on gram stains; growth on buffered charcoal yeast extract agar

E. Gram-positive pleomorphic rods; aerobic growth on tellurite agar

156. The pathology caused by this organism is toxin production. The key characteristic of this toxin is best described as

A. Plasmid-coded edema factor

B. Plasmid-coded exfoliating toxin

C. Bacteriophage-coded A-B toxin

D. Tetanospasmin

E. Pore-forming exotoxin

157. The patient recovered without any complication. The treatment most likely to have effected this outcome is

A. Antitoxin

B. Steroids

C. Humidified air

D. Penicillin

E. Removal of pseudomembrane

158. A 4-year-old girl who recently immigrated with her family to the United States from Venezuela developed abdominal pain, projectile vomiting, and a swollen abdomen. An abdominal x-ray revealed dilated loops of bowel consistent with small bowel obstruction. Ova

and parasite (microscopic) examination of stool revealed a parasite egg (see below).

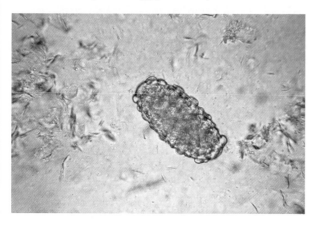

Which of the following parasites is most likely to have caused the child's illness?

A. *Enterobius vermicularis*

B. *Trichuris trichiura*

C. *Ancylostoma duodenale*

D. *Ascaris lumbricoides*

E. *Strongyloides stercoralis*

Questions 159-161 are linked to the following case:

A 30-month-old infant girl had fever, cough, and a respiratory rate of 60/min. Physical exam showed that she was using her accessory muscles of respiration and had diffuse rales and expiratory wheezing.

159. A chest x-ray film is most likely to show which of the following characteristics?

A. Upper airway narrowing

B. Air trapping and infiltrates

C. Normal lung fields

D. Lobar consolidation

E. Infiltrates with lobar cavity

160. In evaluating the cause of the respiratory illness, which of the following is most appropriate?

A. Culture of suctioned sputum to identify *Bordetella pertussis*

B. PCR of nasopharyngeal secretion to identify parainfluenza virus

C. Serology to demonstrate the increase in IgG specific for influenza virus

D. Culture of transtracheal aspirate to identify *Haemophilus influenzae* type b

E. Enzyme immunoassay to detect RSV antigen in nasopharyngeal secretion

161. Wheezing and hyperinflation occur because the pathogen causes

A. Inflammation of the terminal bronchioles, necrosis, and sloughing of the epithelial cells lining the bronchioles

B. Accumulation of dense necrotic coagulum of epithelial cells, fibrin, and leukocytes in the upper airways

C. Loss of cilia, cell damage, and edema, leading to significant airway compromise at the level of the cricoid cartilage

D. Acute exacerbations of bronchitis with excessive tracheobronchial mucus production

E. Dysfunction and degeneration of mucus-secreting, ciliated cells, and other epithelial cells in the large airways

162. A 21-year-old man developed diarrhea that persisted for more than 1 week. He recently returned from a week-long backpacking trip to the backcountry in Colorado. He admitted that he drank untreated water from a mountain stream. The patient had noted flatulence and foul-smelling bulky stools. Which of the following tests is most likely to confirm the clinical diagnosis?

A. Stool antigen test

B. Wet mount of stool

C. Cultures of stool

D. Giemsa stain colonic biopsy

E. Serology to detect IgG

Questions 163-165 are linked to the following case:

A 3-year-old child was brought to the physician's office with symptoms of coryza, conjunctivitis, low-grade fever, and a blotchy reddish-brown rash on his face, trunk, and proximal extremities over 3 days. On physical examination, he had several 0.2 to 0.5 cm ulcerated lesions on the oral cavity mucosa.

163. What is the diagnosis?

A. Mumps

B. Varicella

C. Rubella

D. Mononucleosis

E. Measles

164. The causative agent of this disease belongs to which group of viruses?

    A. Adenovirus

    B. Herpesvirus

    C. Picornavirus

    D. Orthomyxovirus

    E. Paramyxovirus

165. The following year he was again exposed to the same virus, but this time the disease was subclinical. Humoral and cell-mediated immunity after illness is:

    A. Permanent

    B. Short

    C. Nonexistent

    D. Slow to develop

    E. Dependent on booster

166. A 51-year-old man came to the United States from Ghana seeking a lung transplant for idiopathic pulmonary fibrosis. He complained of peptic ulcer-like pain and had an increased number of eosinophils on his blood smear. An ova and parasite examination of stool specimen revealed larvae (see below).

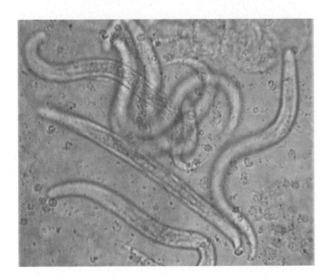

    Before transplantation and treatment with immunosuppressive drugs, which of the following drugs should the patient receive?

    A. Ivermectin

    B. Praziquantel

    C. Pyrantel pamoate

    D. Suramin

    E. Metronidazole

Questions 167 and 168 are linked to the following case:

A 13-year-old girl had a sore throat and fever. Physical exam showed exudative pharyngitis, posterior and anterior cervical lymphadenopathy, and conjunctivitis in the right eye.

167. A virus was recovered from the cell cultures of nasopharyngeal specimen and conjunctival swab and identified by fluorescence antibody staining of virus-infected cells using an antibody reagent. Which of the viruses is the most likely to have caused her illness?

    A. Adenovirus

    B. Coronavirus

    C. Influenza A virus

    D. Epstein-Barr virus

    E. Respiratory syncytial virus

168. Which of the following diseases is also caused by the pathogen of a selected antigenic type?

    A. Allergic rhinitis

    B. Retinitis

    C. Walking pneumonia

    D. Acute exacerbation of chronic bronchitis

    E. Acute respiratory distress syndrome

169. A 14-year-old male was admitted to the local hospital with fever, cough, and enlargement of the lymph nodes. A chest x-ray showed a pattern resembling primary tuberculosis. Cultures of the sputum specimen revealed mycelial growth when incubated at room temperature. Characteristic macroconidia were seen under the microscope (see below).

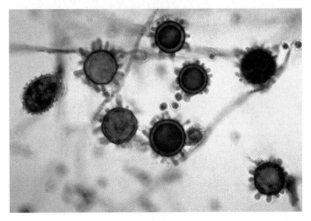

    Which of the following pathogens is most likely to have caused his illness?

A. *Blastomyces dermatitidis*

B. *Candida albicans*

C. *Aspergillus fumigatus*

D. *Histoplasma capsulatum*

E. *Actinomyces israelii*

Questions 170 and 171 are linked to the following case:

A 17-year-old male with cystic fibrosis was seen in the clinic for worsening cough productive of greenish sputum. He had recently been hospitalized for bacterial bronchitis. On exam he was afebrile and in mild respiratory distress.

170. Which of the following organisms is most likely to colonize his lower respiratory tract, resulting in bouts of difficult-to-treat respiratory illnesses?

    A. *Haemophilus influenzae*

    B. *Histoplasma capsulatum*

    C. *Legionella pneumophila*

    D. *Mycobacterium capsulatum*

    E. *Pseudomonas aeruginosa*

171. Mucoid colonies of the pathogen (see below) develop into biofilm within the large airways.

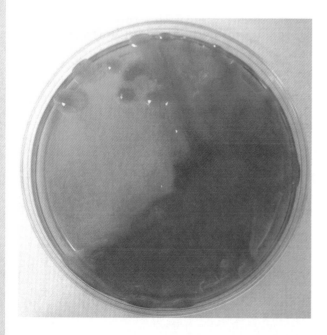

Which of the following virulence factors is most likely to be involved in the development of these colonies?

A. Lipopolysaccharide

B. Elastase

C. Alginate

D. Edema factor

E. Protective antigen

172. A 33-year-old HIV-positive man presented with progressively increasing dyspnea, dry cough, fatigue, and a continuous moderate fever of 3 weeks' duration. Examination was remarkable for fever, tachypnea, and respiratory distress with intercostal retraction. Chest x-ray was remarkable for diffuse, bilaterally symmetrical interstitial and alveolar infiltration. Methenamine silver stain of bronchoalveolar lavage fluid was diagnostic (see below).

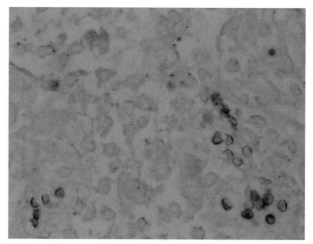

Which of the following organisms is most likely to have caused his current illness?

    A. *Pneumocystis jiroveci*

    B. *Nocardia asteroides*

    C. *Aspergillus fumigatus*

    D. *Histoplasma capsulatum*

    E. *Cryptococcus neoformans*

173. A 23-year-old medical student went to Costa Rica on spring break. He developed diarrhea with abdominal cramps 3 days after arriving in Costa Rica. He did not have any fever. Which of the following is most appropriate for this traveler to a developing county with these symptoms?

    A. Rapid return to the United States for medical attention

    B. Loperamide (an anti-motility drug) and metronidazole

C. Fluid replacement with oral-rehydration solution

D. Rest and relaxation (no specific therapy)

E. Trimethoprim-sulfamethoxazole and loperamide

174. A 21-year-old female college student developed fever, malaise, and frequent headaches after a trip to the Philippines to see her ailing grandmother. The student health service physician found her to be febrile (39.9°C) and to have splenomegaly but no jaundice. She had no diarrhea or grossly bloody stools. Peripheral blood film did not reveal any organisms. Which of the following is most likely to have caused her illness?

A. Hepatitis A

B. *Mycobacterium tuberculosis*

C. *Campylobacter jejuni*

D. *Plasmodium falciparum*

E. *Salmonella typhi*

175. A 31-year-old male returned from Mexico with bloody diarrhea, abdominal pain and cramping, weight loss, and chronic fatigue. The patient should be examined for:

A. Rotavirus

B. *Campylobacter jejuni*

C. *Giardia lamblia*

D. *Entamoeba histolytica*

E. *Shigella sonnei*

176. A 31-year-old breast-feeding woman developed redness and swelling of her right breast. Gram smear of the pus collected by needle aspiration from the abscess revealed the presumptive etiology (see below).

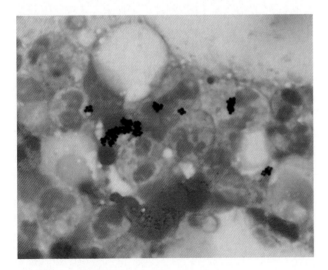

Which of the following pathogens is most likely to have caused this woman's illness?

A. *Staphylococcus aureus*

B. *Staphylococcus epidermidis*

C. *Streptococcus pyogenes*

D. *Clostridium perfringens*

E. *Pseudomonas aeruginosa*

177. A 51-year-old man presented with high fever, chills, severe headache, and confusion. He had recently traveled to Southeast Asia, where he had spent a significant length of time visiting the countryside of Thailand. Examination was remarkable for high fever, tachycardia, profuse sweating, and mild splenomegaly. CBC was remarkable for anemia and thrombocytopenia. Peripheral blood smear was diagnostic (see below).

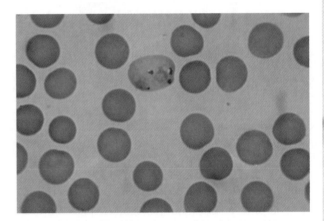

Which of the following is most likely to have caused his illness?

A. West Nile virus

B. Measles

C. *Rickettsia rickettsii*

D. *Plasmodium falciparum*

E. *Plasmodium vivax*

178. A 13-year-old boy complained of severe pain in his right knee. He experienced a soccer injury a week ago, for which he had not sought medical attention. Which of the following is the most appropriate empirical therapeutic agent?

A. Penicillin

B. Nafcillin

C. Ciprofloxacin

D. Doxycycline

E. Linezolid

179. A 31-year-old man with AIDS was hospitalized with fever, pulmonary infiltrates, abdominal pain, and diarrhea. Microscopic examination of the stool revealed larva (see below).

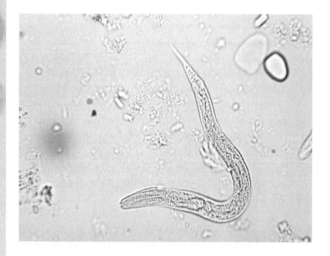

Which of the following should the clinician be concerned about?

A. Osteomyelitis

B. Gram-negative sepsis

C. Pneumonia

D. Hemorrhagic enterocolitis

E. Hepatic abscess

180. Several passengers on a cruise ship presented within 48 hours with a syndrome of severe nausea, vomiting, and watery diarrhea. A low-grade fever was present in some individuals. Over the course of the next 72 hours, more than 90 passengers were afflicted with this illness. Which of the following pathogens is most likely to have caused this outbreak?

A. Norovirus

B. *E. coli* O157:H7

C. *Salmonella typhimurium*

D. Rotavirus

E. Enteric adenovirus

181. A 62-year-old male underwent an aortic valve replacement for severe aortic stenosis. Two weeks following the surgery, he returned for follow-up and complained of several days of worsening fever and chills. His surgical wound looked normal. Blood cultures were positive for a Gram-positive bacterium. Which of the following is true regarding his illness?

A. Transthoracic echo is the most sensitive test for confirming the diagnosis.

B. *Staphylococcus aureus* is the most common etiology.

C. Pneumonia is a rare complication.

D. *Staphylococcus epidermidis* is the most likely etiologic agent.

E. The drug of choice for treatment of this infection is penicillin.

Questions 182-185 are linked to the following case:

A 31-year-old man complained of flu-like symptoms, sustained fever, and increasing fatigue and weight loss for more than 40 days. He had a history of diabetes and was not at risk for HIV infection. He had recently moved from Seattle, Washington to a small town in Tennessee. Physical examination revealed temperature of 38.9°C, pallor, enlarged liver and spleen, crackles throughout his lungs, and lesions in his mouth. A chest x-ray showed scattered, upper lobe 0.4- to 1.5-cm nodules and hilar adenopathy.

182. Gomori methenamine silver stain of a fine-needle aspirate from spleen showed 2- to 5-μm yeast-like organisms (see below).

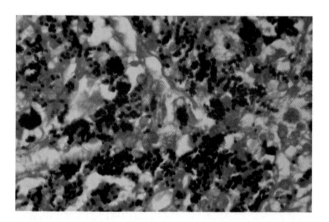

Which of the following is the most likely diagnosis?

A. Coccidioidomycosis

B. Zygomycosis

C. Cryptococcosis

D. Histoplasmosis

E. Blastomycosis

183. Which of the patient's demographics and history was most likely to have been linked to this infection?

   A. The patient is an African American.

   B. The patient is a diabetic.

   C. The patient moved to an endemic area of infection.

   D. The patient has no humoral immunity to the infectious agent.

   E. The patient is skin test negative.

184. How did he most likely acquire this infection?

   A. Mosquito bite on a trip to Africa

   B. Transfusion of packed red blood cells

   C. Injection drug use with shared needles

   D. Ingestion of contaminated milk

   E. Birds roosting on his air conditioner

185. Which of the following is the most appropriate therapeutic agent for the recovery of this patient?

   A. Isoniazid

   B. Itraconazole

   C. Tetracycline

   D. Albendazole

   E. Cefotaxime

186. A 32-year-old man with AIDS presented with chronic, recurrent profuse, nonbloody, watery diarrhea. Examination was remarkable for moderate dehydration and generalized lymphadenopathy. Acid-fast staining of fresh stool revealed oocysts (see below).

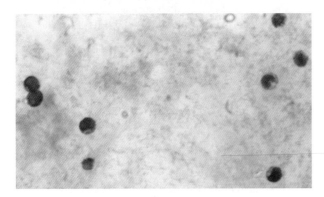

   The patient was treated with paromomycin and fluid replacement therapy. Which of the following pathogens is most likely to have caused his illness?

   A. *Strongyloides stercoralis*

   B. *Entamoeba histolytica*

   C. *Cryptosporidium parvum*

   D. *Isospora belli*

   E. *Cyclospora cayetanensis*

Questions 187 and 188 are linked to the following case:

Two months after returning from a trip to Sudan, a 7-year-old boy developed a febrile illness with symptoms including, headache, abdominal pain, blood in the stool, and granulomatous lesions of the stomach. Ova and parasite examination was positive for large eggs similar to that shown (see below).

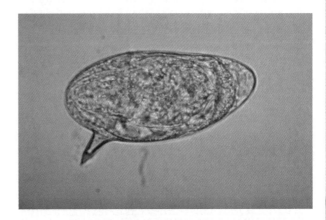

187. Which of the following pathogens is most likely to have caused this boy's illness?

   A. *Strongyloides stercoralis*

   B. *Wuchereria bancrofti*

   C. *Enterobius vermicularis*

   D. *Schistosoma japonicum*

   E. *Schistosoma mansoni*

188. The major cause of granulomatous lesions and fibrosis caused by this agent is a reaction to which of the following parasite stages in the host?

   A. Adult flukes in blood vessels

   B. Blockage of blood vessels by proglottids

   C. Dead eggs in tissues

   D. Calcified cysticerci

   E. Encysted larvae

189. A 23-year old male presented with a 5-day history of fever, nausea, vomiting, anorexia, and jaundice with dark yellow urine. He denied any history of intravenous drug use, sexual contact (in the last 2 months), or tattoo. His travel history was remarkable for a recent visit to Mexico. His physical examination revealed an icteric patient with hepatomegaly but no

evidence of splenomegaly. Laboratory studies were remarkable for liver enzymes and bilirubin. Which of the following serologic markers is most likely to be positive in this case?

A. HAV Ab IgM

B. HBs Ag

C. HCV Ab

D. HBc Ab

E. HBe Ab

190. A 25-year-old man sustained a compound fracture of his right humerus during a motor vehicle accident. Several days later, he experienced marked swelling of his right arm with crepitance. A Gram stain of exudate from the wound site was diagnostic (see below).

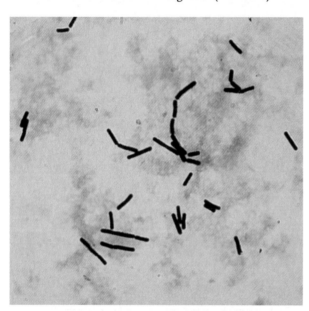

Which of the following pathogens is most likely to have caused his trauma-associated infection?

A. *Clostridium tetani*

B. *Listeria monocytogenes*

C. *Bacillus anthracis*

D. *Clostridium perfringens*

E. *Clostridium botulinum*

191. A 34-year-old white man presented with jaundice, dark yellow urine, and abdominal pain for a week. He had been feeling increasingly weak, nauseated, anorexic, and feverish. History was remarkable for a sexual encounter with a prostitute in Thailand 6 months before. Examination revealed an icteric patient with tender, firm hepatomegaly but no evidence of splenomegaly. CBC showed moderate

leukopenia. Laboratory studies showed hyperbilirubinemia; elevated serum aminotransferases, ALT, AST, and elevated alkaline phosphatase. Which of the following serologic markers would most likely be positive in this case?

A. IgM HAV Ab

B. HCV Ab

C. IgM HBc Ab

D. IgG HBs Ab

E. IgG HBe Ab

Questions 192-194 are linked to the following case:

A 56-year-old white male presented with a large lesion on his right chest wall. The patient was a long-time alcoholic and had a history of severe periodontal disease. He was taken to the surgery department for drainage of the chest lesion. At surgery, 1 L of pus was drained from the chest lesion and the lung.

192. Examination of the pus revealed sulfur granules. Gram stain of these granules was remarkable (see below).

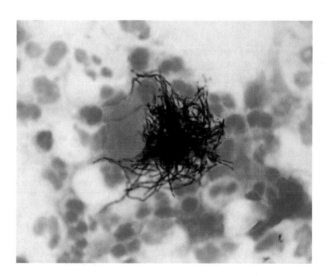

The presence of sulfur granules in the drainage specimen most likely originated from which of the following sources?

A. Dense masses of PMNs

B. Dense masses of hyphal growth

C. Dense masses of filamentous bacteria

D. Pyogenic abscess

E. Granulomatous lesions

193. Which of the following microscopic and growth characteristics best describes the etiologic agent?

    A. Gram-positive cocci in cluster; β-hemolytic growth, catalase-positive

    B. Gram-positive cocci in chains; β-hemolytic growth, catalase negative

    C. Gram nonreactive, acid-fast positive, slender rods; aerobic slow growth

    D. Gram-positive filamentous rods, non-acid-fast; anaerobic growth

    E. Gram-positive filamentous rods, acid-fast; aerobic growth

194. What is the most likely origin of this infectious agent?

    A. Normal oral flora

    B. Another sick individual

    C. Air conditioning unit

    D. Animals

    E. Soil environment

195. A 29-year-old man visited his family physician for a medical evaluation. His social history was unremarkable. He was married with two children and had not traveled outside the United States in 12 years. His medical history was remarkable for a blood transfusion 10 years ago, when he was injured in a motor vehicle accident. Physical examination revealed the patient had firm, tender hepatomegaly. Serum chemistry showed moderate hyperbilirubinemia and elevated serum transaminase levels. If his serum hepatitis markers were to be investigated, which of the following would most likely be positive?

    A. IgM HAV Ab

    B. HBs Ag

    C. HCV Ab

    D. IgM HBc Ab

    E. HBe Ag

Questions 196-198 are linked to the following case:

A 51-year-old man presented with a 6-week history of intermittent fever, chills, chest pain, shortness of breath, and production of foul-smelling, green sputum. The patient had been taking corticosteroids for the treatment of systemic lupus erythematosus. On examination, his vital signs were remarkable for T: 38.9°C; R: 36/minute. A chest x-ray revealed a nodular infiltrate in the right lower lobe.

196. Gram stain (see below) and acid-fast stain of sputum smears were significant.

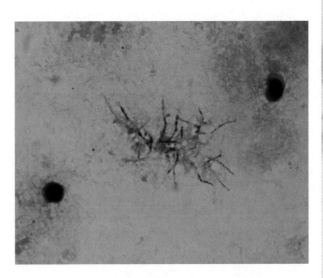

Which of the following microbiologic features best describes the etiologic agent of his illness?

    A. Gram-negative rods; facultative anaerobic growth, lactose fermenting

    B. Gram-positive beaded rods, non-acid-fast bacteria; obligate anaerobic growth

    C. Gram-positive rods, metachromatic granules; aerobic growth on tellurite agar

    D. Acid-fast rods; obligate aerobic growth on Lowenstein-Jensen agar

    E. Gram-positive beaded rods, acid-fast bacteria; obligate aerobic growth

197. This man most likely contracted this infection by which of the following mechanisms?

    A. Inhalation of infective dust from soil

    B. Inhalation of dry arid sand dust

    C. Skin penetration by thorns contaminated with spores

    D. Person to person from another patient

    E. Aspiration of normal flora from the oropharynx

198. In the absence of appropriate therapy, which of the following complications might he most likely develop?

    A. Generalized rash

    B. CNS abscess

    C. Osteomyelitis

    D. Bone abscess

    E. Myocarditis

Questions 199 and 200 are linked to the following case:

A 3-week-old male was brought to the ED because of high fever and convulsions. He had an extensive maculopapular skin rash on his legs and trunk. The patient's mother had an uneventful pregnancy except for an episode of flu-like illness. Physical examination of the patient revealed a febrile (38.6°C) baby with generalized hypotonia and involuntary flexion of hips when flexing neck (Brudzinski sign). Head CT was normal. CSF investigation revealed 1250 WBCs/µL with 85% PMNs; protein 130 mg/dL; glucose 26 mg/dL.

199. Cultures of blood and CSF grew β-hemolytic colonies that have characteristic Gram-stain morphologies (see below) and are motile at 22°C, confirming the clinical diagnosis.

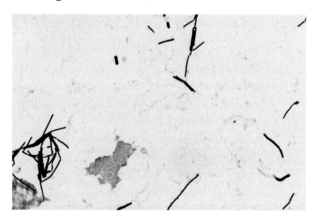

   Which of the following is the most appropriate therapeutic agent to treat this neonate's illness?

   A. Piperacillin and gentamicin
   B. Cefotaxime alone
   C. Cefotaxime and gentamicin
   D. Ampicillin and gentamicin
   E. Cefotaxime and vancomycin

200. What of the following modes of transmission is most likely to have led to the disease of this neonate?

   A. Patient's mother had three cats in her house and he was exposed to cat feces.
   B. Patient's mother had respiratory illness during the first trimester of her pregnancy.
   C. Patient's mother drank unpasteurized milk quite frequently during her pregnancy.
   D. Patient's mother traveled to Mexico and ate sushi at a local wedding reception while pregnant.
   E. Patient's mother ate raw steak with her husband during a candle light dinner during her pregnancy.

201. A 45-year-old patient with AIDS experienced fever and headaches and became lethargic. He died 2 weeks later. At autopsy, a gelatinous meningeal exudate was seen throughout the brain. Which of the following pathogens is most likely to have caused this catastrophic illness?

   A. Histoplasma capsulatum
   B. Taenia solium
   C. Toxoplasma gondii
   D. Streptococcus pneumoniae
   E. Cryptococcus neoformans

202. A 42-year-old woman came to the ED with complaints of severe facial pain and swelling of the eyes. The patient had a high fever for the past 3 days. Physical examination revealed a rash with distinct borders over her face (see below).

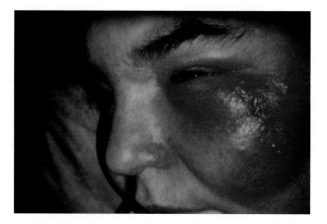

   Which of the following pathogens is most likely to have produced this finding?

   A. Clostridium botulinum
   B. Escherichia coli
   C. Neisseria meningitidis
   D. Pseudomonas aeruginosa
   E. Streptococcus pyogenes

203. A 12-year-old male noticed a slightly itchy eruption on his right arm that eventually developed in a "ring-like," slightly scaly area with central clearing. He went to a physician, who scraped the area with a glass slide and placed the scales on appropriate media for the growth of fungi. Two weeks later an aerial growth on the culture medium (see figure, page 501) was examined microscopically and was diagnostic.

Which of the following is the most likely to have caused his lesions?

A. *Candida albicans*

B. *Malassezia furfur*

C. *Aspergillus fumigatus*

D. *Microsporum canis*

E. *Nocardia asteroides*

204. A 71-year-old woman was brought to the ED complaining of blurred vision and difficulty in swallowing and breathing. She displayed flaccid paralysis in the muscles of her neck and extremities. Her son revealed that the previous day (18 hours earlier), his mother had some home-canned pears. The patient was put on respiratory support. Which of the following is most likely to have caused her illness?

A. Staphylococcal enterotoxin

B. Clostridial lecithinase

C. Gram-negative endotoxin

D. Botulinum toxin

E. Tetanus toxin

Questions 205-208 are linked to the following case:

A 66-year-old female nursing home resident presented with the sudden onset of high fever, chills, diffuse myalgias, and a nonproductive cough, in the month of January. A number of similar cases had been reported from this nursing home in the last month.

205. A nasopharyngeal swab specimen was taken from the patient. A virus strain grew well in monkey kidney cells, inducing hemagglutination (see below).

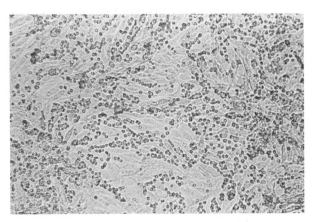

Which of the following pathogens is the most likely cause of her illness?

A. SARS coronavirus

B. Influenza A virus

C. Influenza B virus

D. Enterovirus

E. Respiratory syncytial virus

206. Which of the following preventive measures would be most effective for this disease in a nursing home setting?

A. Immunity in individual residents from prior exposures to the virus over the years

B. Annual immunization of all residents (especially those 65 years of age or older)

C. New neuraminidase inhibitors to be given as soon as an index case arises in the community

D. Ribavirin (a broad-spectrum antiviral drug) and an analgesic to be given as soon as an index case arises in the community

E. Amantadine or rimantadine to be given as soon as an index case arises in the community

207. This virus belongs to a group of viruses that exhibit a range of genetic and biochemical properties. These viruses are distinct from other RNA viruses because they

A. Have a negative-stranded RNA genome that requires a primer DNA for replication

B. Are naked viruses with a genome consisting of nine segments of positive sense RNA

C. Are enveloped viruses with a genome consisting of a single strand of negative-strand RNA

D. Are enveloped viruses with a genome consisting of eight segments of negative-sense RNA

E. Are enveloped viruses whose RNA genome readily undergoes homologous recombination

208. The virulent virus subtype this patient had has caused many epidemics and pandemics in the past. Which of the following properties of this virus is responsible for such devastating spread of infection?

A. Mutations of the hemagglutinins and neuraminidase

B. Increased ability to bind to ICAM-1 receptor

C. Increased ability to bind to CD12 receptor

D. Reassortment of genome segments in a co-infection

E. Increased evasion of mucociliary clearance mechanism

209. In the month of September, 9 older children from a small town developed fever, moderate headache, nausea, and vomiting. All patients recovered spontaneously. How was the infectious agent most likely acquired?

A. Ingestion of contaminated swimming pool water

B. Inhalation of spores from lawn mowing

C. Skin penetration by wild thorns during scavenger hunting

D. Mosquito bites during a camping trip

E. Aspiration by children with poor oral hygiene

210. A 24 year-old man with recently diagnosed HIV infection inquired about opportunistic infections to which he might be susceptible. Which of the following would be very rare as an invasive infection in the early stages of HIV infection?

A. *Aspergillus fumigatus*

B. *Histoplasma capsulatum*

C. *Coccidioides immitis*

D. *Cryptococcus neoformans*

E. *Candida albicans*

211. A 13-year-old girl received a deep laceration when she was water skiing and was taken to the ED. She had had no previous immunization against tetanus. As protection against the possibility of tetanus, the recommended method of treatment would be

A. A mixture of tetanus toxoid and tetanus immune globulin (human) as one injection

B. A tetanus toxoid injection this visit, with an injection of tetanus immune globulin (human) approximately 3 weeks later

C. Separate injections of tetanus toxoid and tetanus immune globulin (human) in different sites on this visit

D. A tetanus immune globulin (human) injection this visit, with an injection of tetanus toxoid about 3 weeks later

E. A tetanus immune globulin (human) injection this visit, with an injection of tetanus toxoid about 5 weeks later

Questions 212 and 213 are linked to the following case:

A group of elderly people in a nursing home developed bacteremia and meningitis after eating a meal that included coleslaw.

212. Which of the following pathogens is most likely to have caused this outbreak?

A. *Rickettsia rickettsii*

B. *Streptococcus agalactiae*

C. *Brucella melitensis*

D. *Listeria monocytogenes*

E. *Clostridium perfringens*

213. Which of the following is the most common clinical manifestation before the onset of CNS symptoms?

A. Diarrhea

B. Dysuria

C. Rash

D. Headache

E. Cough

214. An 11-month-old boy demonstrated evidence of bacterial meningitis on lumbar puncture. His empirical antibiotics should cover which of the following pathogens (differential diagnoses)?

A. *Streptococcus pneumoniae, Streptococcus agalactiae, Neisseria meningitidis, Haemophilus influenzae*

B. *Listeria monocytogenes, Escherichia coli, Pseudomonas aeruginosa, Klebsiella pneumoniae*

C. *Listeria monocytogenes, Escherichia coli, Streptococcus agalactiae, Cryptococcus neoformans*

D. Anaerobes, *Klebsiella pneumoniae, Streptococcus pneumoniae, Streptococcus agalactiae*

E. *Staphylococcus aureus, Escherichia coli, Pseudomonas aeruginosa, Klebsiella pneumoniae*

215. A 9-year-old boy had a fever and headache and vomited while at home. He had received childhood immunizations. On examination, he had a stiff neck and was lethargic. Which of the following pathogens is most likely to have caused his illness?

    A. *Neisseria meningitidis*
    B. *Rickettsia prowazekii*
    C. *Borrelia burgdorferi*
    D. *Treponema denticola*
    E. *Haemophilus influenzae*

216. A 19-year-old woman presented with a high fever, severe headache, and neck stiffness. A lumbar puncture was performed, and the cerebrospinal fluid was sent to the microbiology laboratory with a request for a Gram stain and culture. Gram stain was presumptively diagnostic (see below).

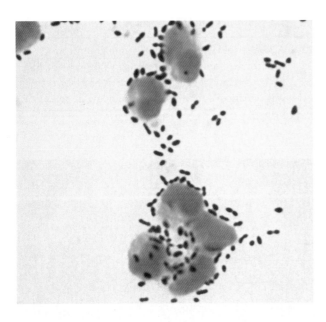

Which of the following is the most appropriate empirical treatment based on Gram stain results?

    A. Vancomycin and cefotaxime
    B. Cefotaxime and ceftazidime
    C. Ampicillin and gentamicin
    D. Cefotaxime and gentamicin
    E. Vancomycin and rifampin

Questions 217-219 are linked to the following case:

A 53-year-old man presented with a 2-day history of fever, fatigue, and painful unilateral inguinal swelling. His history was remarkable for a recent trip to New Mexico. On examination, he appeared ill with diaphoresis, rigors, and lower extremity cyanosis. His vital signs were remarkable for T: 40.2°C; BP: 78/50 mmHg. He had tender left inguinal adenopathy with overlying edema. CBC was remarkable for WBC 24,700/μL and platelet count 72,000/μL. Two of three sets of blood cultures grew a significant pathogen. Gram stain of the isolate revealed the characteristic cellular morphology (see below).

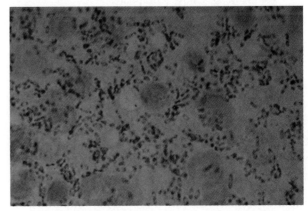

217. The man acquired the infection by which of the following modes?

    A. Inhalation of dried materials contaminated by rodent excreta
    B. Inhalation of respiratory droplets from another sick individual
    C. Inhalation of contaminated air from an air-conditioning unit in a hotel
    D. Exposure to a dead rabbit on the side of a rural road
    E. Bites by infected rodent fleas while on his recent trip

218. Which of the following complications is most likely to develop in severe cases of this infection?

    A. Meningitis
    B. Empyema
    C. Endocarditis
    D. Generalized rash
    E. Septicemia

219. Which of the following characteristics makes it a potential bioterrorist weapon?

    A. High mortality from septic shock

    B. No effective therapy available

    C. Highly infectious (person-to-person transmission)

    D. No prophylaxis available for secondary cases

    E. High mortality from respiratory failure

220. A 51-year-old man with AIDS was hospitalized because of dementia. At admission, he was febrile; nuchal rigidity was present, and Kernig and Brudzinski signs were positive. A lumbar puncture revealed clear CSF with WBC 90/μL; protein 60 mg/dL; glucose 75 mg/dL. In evaluating the cause of his illness, which of the following is most appropriate?

    A. Blood cultures (five sets) using both aerobic and anaerobic bottles to identify *Listeria monocytogenes*

    B. Acid-fast stain of sputum and BACTEC cultures of CSF to identify *Mycobacterium tuberculosis*

    C. Prolonged incubation of blood cultures and serology to identify *Brucella melitensis*

    D. India ink stain of CSF or direct antigen using latex agglutination to identify *Cryptococcus*

    E. Routine bacterial culture from CSF and respiratory specimens

221. A 61-year-old female with a history of SLE presented with symptoms suggestive of infection due to *Mycobacterium tuberculosis*. However, tubercular infection was ruled out by PPD skin test as well as findings of negative chest x-ray and negative acid-fast stain of sputum. Which of the following is most likely to have caused her illness?

    A. *Streptococcus pneumoniae*

    B. *Cryptococcus neoformans*

    C. *Aspergillus fumigatus*

    D. *Taenia solium*

    E. *Candida albicans*

222. A 2-week-old infant girl was brought to the ED with a 24-hour history of fever, poor feeding, irritability, and other symptoms of failure to thrive. She was born preterm with very low birth weight after a normal vaginal delivery. Physical exam was remarkable for an irritable baby with nuchal rigidity. A Gram stain of the CSF showed many PMNs and Gram-positive cocci in chains. The virulence of this organism is related to a bacterial constituent that evades which of the following host phagocyte functions?

    A. Killing by natural killer cells

    B. ADCC-mediated killing

    C. Killing by serum factors

    D. Complement activation and phagocyte-mediated killing

    E. Activation of platelets

223. A 29-year-old IV drug user developed difficulty moving her left arm. Examination showed rigidity of the arm and several needle tracks, plus one small subcutaneous abscess. She was alert, but the next day, she had difficulty opening her jaw. What was the cause of this illness?

    A. *Actinomyces israelii*

    B. *Fusobacterium necrophorum*

    C. *Clostridium tetani*

    D. *Staphylococcus aureus*

    E. *Clostridium botulinum*

224. A 39-year-old woman, a recent immigrant from Mexico, developed a major seizure while at work. She had no history of epileptic disease. A head CT scan was remarkable for a lesion surrounding a scolex (see below).

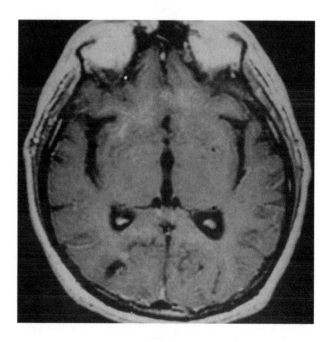

The etiologic agent would most plausibly have been acquired by eating or drinking which of the following food items?

A. Uncooked vegetables

B. Raw pork

C. Raw oysters

D. Unfiltered well water

E. Uncooked fish

225. A 29-year-old woman presented with right upper quadrant pain and fever 6 weeks after surgery for a ruptured appendix. Examination showed a tender, enlarged liver, and a 9-cm fluid-filled abscess in the right lobe was visualized by sonogram. Which of the following is most likely to be correct?

A. *Staphylococcus aureus* is the most likely cause of her problem.

B. The abscess is likely to be caused by a mixed infection by coliforms and anaerobes.

C. This problem could have been avoided if she had seen her doctor for regular check-ups.

D. The development of a liver abscess indicates that she has a previously undiagnosed immunocompromised condition.

E. It is unlikely that the abscess will need to be drained.

226. A 27-year-old male intravenous drug abuser presented to the emergency department with 3 days of fever, pleuritic chest pain, and productive cough. Which of the following pathogens most likely caused his illness?

A. *Streptococcus pyogenes*

B. *Streptococcus mutans*

C. *Staphylococcus aureus*

D. *Haemophilus influenzae*

E. *Staphylococcus epidermidis*

227. A 9-year-old boy stepped on a nail. His mother washed the wound and took him to an emergency department where he was given a tetanus-diphtheria booster. The foot initially felt better, but then swelling and pain set in. Which of the following microbiologic features best describes the mostly likely etiologic agent?

A. S-shaped Gram-negative rod that produces urease

B. Aerobic Gram-negative rod that thrives in water

C. Gram-negative anaerobic rod that thrives in dirt

D. Gram-positive anaerobic rod that contaminates puncture wounds

E. Dermatophyte mold that also causes "athletic foot"

228. Individuals who undergo splenectomy are at risk of developing overwhelming bacteremia due to encapsulated bacteria. These individuals should receive vaccines for protection against such infection. Vaccines to which of the following bacteria should be given?

A. *Neisseria meningitidis*

B. *Streptococcus pyogenes*

C. *Staphylococcus aureus*

D. *Corynebacterium diphtheriae*

E. *Clostridium perfringens*

229. A 59-year-old man complained of worsening intense pain in his right leg 3 days after sustaining an injury while plowing at his farm. The leg gradually became swollen and developed hemorrhagic blisters with crepitance. He underwent extensive surgical débridement of the leg, but he eventually died from septic shock. What caused this man's catastrophic infection?

A. *Clostridium tetani*

B. *Clostridium botulinum*

C. *Clostridium perfringens*

D. *Clostridium difficile*

E. *Clostridium septicum*

230. A 61-year-old female was recovering from surgical removal of her hypernephroma. Chemotherapy for metastases to the lung was underway. She was on hyperalimentation. Persistent fevers developed unresponsive to antibiotics. The patient was at risk for

A. Blastomycosis

B. Cryptococcosis

C. Candidiasis

D. Coccidioidomycosis

E. Aspergillosis

231. A 38-year-old business traveler spent a week in West Africa, during which time he became ill with fever and disoriented. Hemoglobin was markedly decreased. The peripheral blood smear was diagnostic (see figure, page 506).

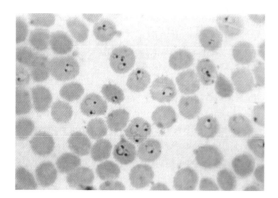

Which of the following pathogens is most likely to have caused this man's illness?

A. *Trypanosoma brucei gambiense*

B. *Trypanosoma cruzi*

C. *Salmonella typhi*

D. *Plasmodium falciparum*

E. *Plasmodium vivax*

232. A 56-year-old woman with end stage renal disease received a renal transplant. Two months after transplant, she had fever to 40°C. Physical exam showed splenomegaly, and laboratory studies showed anemia, leukopenia, and thrombocytopenia, and abnormal liver function tests. A CXR showed interstitial pulmonary infiltrates. This woman's illness could be most easily diagnosed by which of the following tests?

A. HSV-1 IgM antibody

B. CMV antigenemia

C. HHV-6 IgM antibody

D. Culture of urine in erythroblastic cells

E. Monospot test

233. A 24-year-old HIV-positive male had a CD4 count of 50/μL. He presented with increasing respiratory difficulty. A transbronchial biopsy was performed and the histologic finding was diagnostic (see below).

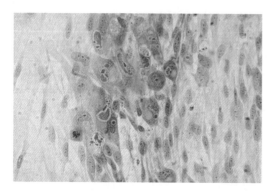

Which of the following organisms is mostly likely to have caused her opportunistic infection?

A. Epstein-Barr virus

B. Cytomegalovirus

C. Respiratory syncytial virus

D. Herpes zoster virus

E. Adenovirus

Questions 234-236 are linked to the following case:

A 67-year-old man presented with a 7-day history of cough and high fever, in the month of January. He had had a flu-like illness 1 week prior to this episode but had otherwise maintained good health. A chest x-ray revealed alveolar infiltrate in the posterior segment of the left lower lobe.

234. The most likely clinical diagnosis is:

A. Primary atypical pneumonia

B. Secondary bacterial pneumonia

C. Cavitary pneumonia

D. Interstitial pneumonia

E. Tuberculosis

235. A Gram smear of sputum was remarkable (see below), and sputum and blood cultures were diagnostic.

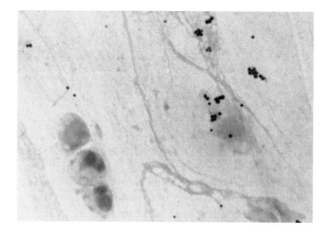

The most likely way the man acquired this infection was which of the following?

A. Inhalation of respiratory droplets

B. Inhalation of airborne aerosol

C. Contact with respiratory discharges

D. Colonization and aspiration

E. Zoonotic exposure

236. Which of the following is the most appropriate therapeutic agent?

A. Gentamicin

B. Nafcillin

C. Penicillin

D. Levofloxacin

E. Doxycycline

237. A 47-year-old man from Alabama who worked for the local electricity company repairing high-tension lines presented to a dermatologist with a granulomatous skin ulcer. Microscopic examination of material from the lesion revealed numerous yeast cells. The dermatologist recommended chest x-ray and hospitalization for which of the following?

A. Aspergillosis

B. Blastomycosis

C. Candidiasis

D. Coccidioidomycosis

E. Tinea corporis

238. A 21-year-old female college student presented with a 5-day history of low-grade fever, myalgias, headache, and nonproductive cough. Gram stain had a few inflammatory cells but did not show any significant pathogen. Which of the following is the most likely cause of her illness?

A. Mixed anaerobic infection

B. *Chlamydia trachomatis*

C. *Streptococcus pneumoniae*

D. *Mycoplasma pneumoniae*

E. *Histoplasma capsulatum*

239. A 35-year-old man with a history of seizure disorder presented with a 4-week history of cough with copious purulent sputum, fever, weight loss, and anemia. Chest x-ray revealed a cavity with an air-fluid level. On physical examination he appeared chronically ill. Which of the following pathogens is most likely to have caused his illness?

A. *Haemophilus influenzae*

B. *Mycoplasma pneumoniae*

C. *Klebsiella pneumoniae*

D. *Mycobacterium tuberculosis*

E. *Histoplasma capsulatum*

240. Sputum was collected from an AIDS patient who had a productive cough. No pathogens were identified in the routine aerobic culture, which was held for 3 days. After 2 weeks, non-branching acid-fast organisms were growing as nonpigmented colonies on blood agar media (see below).

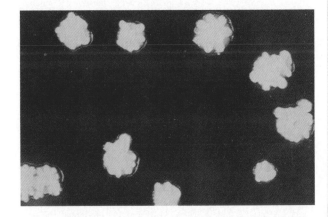

Which of the following organisms was most likely isolated from the patient's sputum?

A. *Candida albicans*

B. *Klebsiella pneumoniae*

C. *Histoplasma capsulatum*

D. *Nocardia asteroides*

E. *Mycobacterium avium* complex

241. A 17-year-old female high school student presented with a 5-day history of low-grade fever, myalgias, headache, and a nonproductive cough. Gram stain did not reveal any significant pathogen. A serologic investigation confirmed the diagnosis retrospectively. Which of the following pathogens is most likely to have caused her illness?

A. *Chlamydia psittaci*

B. *Legionella pneumophila*

C. Mixed anaerobic infection

D. *Streptococcus pneumoniae*

E. *Mycoplasma pneumoniae*

242. A 29-year-old man from the San Joaquin Valley in California presented with a low-grade fever and a nonproductive cough that was aggravated by cigarette smoking. He also had chills, night sweats, anorexia, and headache. He came to the hospital because he began to raise small amounts of mucopurulent sputum that was lightly streaked with blood. Cytologic examination of bronchoscopy-assisted lung biopsy revealed a significant pathogen (see below).

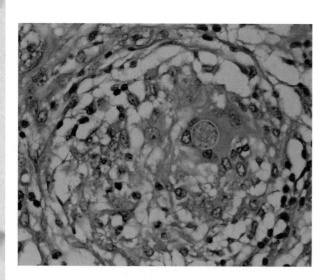

Which of the following pathogens is most likely to have caused his illness?

A. *Histoplasma capsulatum*

B. *Blastomyces dermatitidis*

C. *Coccidioides immitis*

D. *Cryptococcus neoformans*

E. *Nocardia asteroides*

243. A 68-year-old man presented with bloody cough. Three weeks prior to this illness he was working with a group of laborers, demolishing a very old building. Chest x-ray showed a new density inside an old cavity in the right upper lobe. Serum galactomannan antigen test is likely to be positive for which of the following organisms?

A. *Blastomyces dermatitidis*

B. *Nocardia asteroides*

C. *Mycobacterium tuberculosis*

D. *Aspergillus fumigatus*

E. *Coccidioides immitis*

244. A 2-year-old male presented with fever, hoarseness, and respiratory distress because of partial airway obstruction. The patient was from Mexico and had a questionable immunization history. Examination revealed fever, tachypnea, drooling and leaning forward with neck hyperextended, and inspiratory stridor. Throat and blood cultures were diagnostic (significant colonies were seen on lysed blood agar; no growth on nonlysed blood agar; see below).

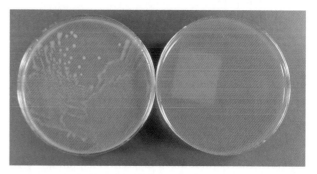

Which of the following pathogens is most likely to have caused this boy's illness?

A. *Haemophilus influenzae*

B. *Legionella pneumophila*

C. *Bordetella pertussis*

D. *Corynebacterium diphtheriae*

E. *Chlamydia pneumoniae*

245. A 31-year-old female with severe neutropenia after induction chemotherapy for acute lymphoblastic leukemia developed fever and difficulty in breathing. A chest x-ray was suggestive of pulmonary infarcts. Microscopic examination of transbronchial biopsy after silver staining revealed that the lung parenchyma and endothelial tissues were occluded by an abundant presence of organisms (see below).

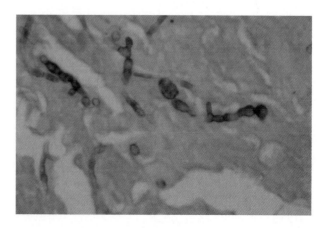

Which of the following is the most appropriate therapeutic agent to treat this serous illness?

A. Voriconazole

B. Amphotericin B

C. Cefotaxime

D. Vancomycin

E. Sulfamethoxazole

246. A 60-year-old woman developed pneumonia while receiving chemotherapy for acute leukemia. A chest x-ray revealed an infiltrate, and WBC was 18,000/mL with 78% PMNs. Cultures of sputum grew a significant bacterial pathogen. The pigment-producing colonies were resistant to several antibiotics (see below).

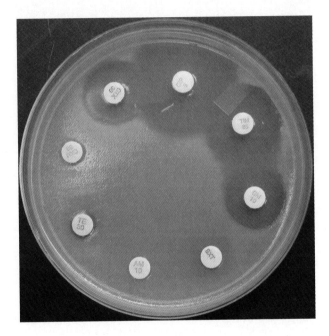

Which of the following is the most significant in the mode of transmission of this pathogen?

A. It may be transmitted to a risk patient in the ICU via mechanical ventilation

B. It is able to grow in diverse environments, including soil and water

C. It is commonly transmitted by ingestion of *Pseudomonas*-laden foods

D. It is borne on the unwashed hands of hospital personnel

E. It is able to cause opportunistic infections in otherwise healthy individuals

247. In July, a 67-year-old man was admitted to a local hospital in Louisiana with a 3-day history of fever, chills, headache, and altered mental status. On physical examination, general weakness and tremors were noted. Brain MRI was normal. A serologic test for IgM, specific for a pathogen, confirmed the clinical diagnosis. Which of the following is the most likely diagnosis?

A. Viral meningitis

B. Herpes simplex encephalitis

C. Rabies

D. Bacterial meningitis

E. West Nile viral encephalitis

248. A 45-year-old woman presented to the ED with fever. She had recently received a round of chemotherapy for acute leukemia. Physical examination was remarkable for fever (T: 39.4°C) and pain at the insertion site of the central venous line (but no erythema). Lab studies were remarkable for neutropenia. An opportunistic bacterial agent, sensitive to vancomycin, was recovered from her blood cultures and the infected central line. Which of the following best describes the culprit agent?

A. Gram-positive rods

B. Gram-positive cocci in chains

C. Gram-negative rods

D. Gram-positive diplococci

E. Gram-positive cocci in clusters

249. A 55-year-old woman presented with fevers and chills for several weeks. She had also noted a 16-lb weight loss. She stated that she ate goat cheese during a recent trip to Macedonia. Blood cultures were positive after prolonged incubation. Which of the following characteristics makes this agent a potential bioterrorist weapon?

A. High mortality from septic shock

B. No effective therapy available

C. Highly infectious if aerosolized

D. Highly contagious from person to person

E. High mortality from respiratory failure

250. A 59-year-old woman was brought to the ED in July for fever and a headache for 3 days and increasing confusion over the past 24 hours. She enjoyed spending time outside in her garden. On exam she was disoriented but there were no focal neurologic deficits or nuchal rigidity. CSF was remarkable for lymphocytic pleocytosis and high protein, but normal

glucose. Which of the following is the most likely diagnosis?

A. Viral meningitis

B. Herpes simplex encephalitis

C. Rabies

D. Bacterial meningitis

E. West Nile viral encephalitis

251. An 18-year-old man was seen in the ED for high fever and chills lasting several days, accompanied by prostration. He had recently skinned several rabbits on a hunting trip. On exam, a deep ulcer with ragged edges was noted on his left arm. The organism responsible for this infection can pose a risk to national security because the agent

A. Is easily cultured on most common laboratory media

B. Is a Gram-positive, box-shaped rod

C. Causes infection with prolonged illness

D. Requires a large inoculum to establish infection

E. Is resistant to most β-lactam antibiotics

252. High-priority bioterrorism agents include organisms that pose a risk to national security because they can be easily disseminated or transmitted from person to person. Which of the following is most likely to be transmitted from person to person in a bioterrorism act?

A. *Bacillus anthracis*

B. *Clostridium botulinum*

C. *Francisella tularensis*

D. *Variola major*

E. *Brucella melitensis*

253. Which of the following characteristics makes *Bacillus anthracis* a potential bioterrorist weapon?

A. High mortality is associated with severe pneumonia

B. No effective therapy is available against this multiply resistant organism

C. Hardy, fine spores can be airborne leading to inhalation

D. Highly contagious from person to person for secondary transmission

E. Monocytes are able to deliver the organism to any organ in an infected host

254. A 6-year-old white male presented with golden-yellow, crusted lesions around his mouth and behind his ears. Examination was remarkable for honey-colored lesions, and for discharge from lesions. Gram stain of the discharge revealed the etiology (see below).

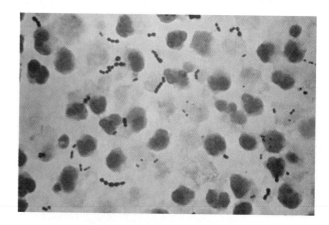

What caused this patient's illness?

A. *Staphylococcus epidermidis*

B. *Trichophyton rubrum*

C. *Candida albicans*

D. *Streptococcus pyogenes*

E. *Clostridium perfringens*

ANSWERS TO PRACTICE QUESTIONS

Question Number	Answer Choice	Page Number*	Question Number	Answer Choice	Page Number
1	C	49	42	B	179, 238
2	A	49	43	C	243
3	B	51	44	A	135, 137
4	D	107	45	A	87, 135
5	E	108	46	B	136
6	A	108	47	A	89, 138
7	C	227	48	C	127
8	D	228	49	C	41
9	E	228	50	A	42
10	D	230	51	D	42
11	C	287	52	C	43
12	B	288	53	A	35
13	A	451	54	A	37
14	B	267	55	B	39
15	B	267	56	C	38
16	D	268	57	B	36, 37
17	A	268	58	E	338
18	A	383	59	E	46
19	E	383	60	B	46
20	A	385	61	D	47
21	D	105	62	E	387
22	E	105	63	B	128
23	A	105	64	D	361
24	B	192	65	C	62
25	C	192	66	C	62
26	C	192	67	A	64
27	B	193, 293	68	B	32, 237, 238
28	D	293	69	D	32, 299
29	A	295	70	C	32
30	C	294	71	D	12
31	C	238, 263	72	D	11, 13
32	B	264	73	D	12
33	E	265	74	A	13
34	C	265	75	B	305
35	A	123	76	B	305
36	D	123	77	D	308
37	A	125	78	E	112
38	B	241	79	A	113
39	B	241	80	A	113
40	D	243	81	D	30
41	B	243	82	B	29

*Explanations in support of the answer choice can be found in these pages.

Question Number	Answer Choice	Page Number	Question Number	Answer Choice	Page Number
83	A	30	126	B	74
84	B	30	127	D	191
85	C	337	128	C	172
86	B	339	129	A	301, 302
87	A	339	130	D	303
88	D	339	131	A	303
89	D	132	132	D	196
90	C	195	133	B	75
91	A	196	134	B	76
92	E	196, 426	135	E	76
93	C	197	136	C	197
94	B	116	137	E	326
95	E	116	138	A	289
96	B	16	139	C	289
97	C	17	140	A	224
98	D	17	141	C	152
99	B	16	142	E	341
100	A	376	143	A	344
101	E	377	144	A	164
102	D	120	145	D	250
103	B	121	146	A	160
104	B	120	147	A	333
105	A	329	148	E	310
106	C	330	149	A	326
107	C	327	150	D	155
108	A	103, 439	151	B	156
109	D	120	152	C	326
110	E	120	153	C	199, 200
111	C	183	154	B	200
112	A	184, 399	155	E	20
113	E	103, 105	156	C	20
114	D	241, 243	157	A	20
115	D	243	158	D	208
116	A	243	159	B	57
117	C	168	160	E	57
118	A	169, 416	161	A	58
119	D	180	162	A	199
120	D	180	163	E	313
121	D	180, 407	164	E	314
122	B	36	165	A	461
123	D	36	166	A	212
124	B	37	167	A	7
125	A	132	168	E	9

Question Number	Answer Choice	Page Number	Question Number	Answer Choice	Page Number
169	D	70	212	D	250
170	E	91	213	A	249
171	C	92	214	A	236
172	A	88	215	A	241
173	C	148	216	A	236, 238
174	E	155	217	E	364
175	D	195	218	E	364
176	A	284, 288	219	C	364
177	E	338	220	D	264
178	B	289	221	B	263, 264
179	B	212	222	D	247, 414
180	A	188	223	C	279
181	D	326	224	B	271, 272
182	D	69	225	B	93. 301
183	C	70	226	C	327
184	E	70	227	B	92
185	B	71	228	A	243
186	C	204	229	C	297, 299
187	E	220	230	C	133
188	C	220	231	D	338
189	A	224	232	B	229, 230
190	D	298	233	B	330
191	C	228	234	B	95
192	C	84	235	D	96
193	D	84	236	B	97
194	A	84	237	B	74
195	C	233	238	D	46
196	E	80	239	C	42
197	A	80	240	D	80
198	B	80	241	E	45
199	D	251	242	C	76
200	C	250	243	D	65
201	E	265	244	A	32
202	E	295	245	A	66
203	D	305, 306	246	A	92, 93
204	D	176	247	E	379, 380
205	B	53	248	E	289, 290
206	B	55	249	C	368
207	D	54	250	A	257
208	D	54	251	D	372
209	A	255	252	D	360
210	A	138	253	C	351,356
211	C	280	254	D	294

APPENDIX G

Microbes/Diseases

APPENDIX G   Microbes/Diseases

Systems/Organisms	Section/Case Numbers	Page Numbers
System Infections and Other Introductory Studies		
Respiratory tract infections: a perspective	Section I	1
Urogenital (reproductive tract) infections: a perspective	Section II	99
Gastrointestinal infections: a perspective	Section III	141
Central nervous system infections: a perspective	Section IV	235
Skin, wound, and multisystem (cardiovascular) infections: a perspective	Section V	283
Bioterrorism and emerging infectious diseases: a perspective	Section VI	349
Medical Virology		
Adenoviruses (pharyngitis, conjunctivitis)	Case 1	7
Arboviruses (vector-borne encephalitis)	Case 60	257
Caliciviruses (epidemic nonbacterial gastroenteritis)	Case 44	187
Coronavirus (severe acute respiratory syndrome; emerging infectious disease)	Case 91	387
Cytomegalovirus (multiorgan diseases in AIDS)	Case 77	329
Enteroviruses (aseptic meningitis)	Case 59	253
Epstein-Barr virus (infectious mononucleosis)	Case 2	11
Hepatitis viruses		
A (enterically transmitted hepatitis)	Case 53	223
B (parenterally and sexually transmitted hepatitis)	Case 54	227
C (parenterally transmitted hepatitis)	Case 55	231
Herpes simplex viruses		
type 1 (temporal lobe encephalitis)	Case 61	261
type 2 (genital ulcers)	Case 27	111
Human immunodeficiency virus (HIV)-1 (AIDS)	Case 33	135
Influenza virus (influenza)	Case 13	53
Measles (febrile rash illness)	Case 73	313
Mumps (febrile parotitis)	Case 5	23
Papillomavirus (genital warts and cancer of cervix)	Case 28	115
Parainfluenza viruses (croup)	Case 6	25
Prions (Creutzfeldt-Jakob disease; emerging infectious diseases)	Case 90	383
Rabies (encephalitis)	Case 65	275
Respiratory syncytial virus (bronchiolitis; pneumonia in infants)	Case 14	57
Rotaviruses (infantile gastroenteritis)	Case 43	183
Rubella (febrile rash illness; congenital rubella syndrome)	Case 74	317
Sin Nombre hantavirus (pulmonary syndrome; emerging infectious disease)	Case 88	375
Varicella-zoster virus (chickenpox; shingles)	Case 75	321
Variola (smallpox; bioterrorism)	Case 84	359
West Nile virus (arboviral encephalitis; emerging infectious disease)	Case 89	379

APPENDIX G Microbes/Diseases—Cont'd

Systems/Organisms	Section/Case Numbers	Page Numbers
Medical Bacteriology		
Actinomyces (thoracic abscess)	Case 21	83
Bacillus (anthrax; bioterrorism)	Case 83	355
Bacteroides (anaerobic, intra-abdominal infections)	Case 70	301
Bordetella (whooping cough)	Case 7	29
Borrelia (vector-borne infections; Lyme disease)	Case 78	333
Brucella (protracted febrile, multisystem disease)	Case 86	367
Campylobacter (gastroenteritis)	Case 34	147
Chlamydia (sexually transmitted diseases)	Case 30	123
Clostridium botulinum (botulism)	Case 41	175
C. difficile (antibiotic-associated colitis)	Case 42	179
C. perfringens (gas gangrene; cellulitis)	Case 69	297
C. tetani (tetanus)	Case 66	279
Corynebacterium (diphtheria)	Case 4	19
Escherichia		
E. coli (urinary tract infections)	Case 25	103
E. coli O157:H7 (hemorrhagic colitis)	Case 38	163
Francisella (tularemia)	Case 87	371
Haemophilus (acute exacerbation of chronic bronchitis)	Case 8	31
Helicobacter (peptic ulcer disease)	Case 45	191
Klebsiella (aspiration pneumonia)	Case 10	41
Legionella (Legionnaires' disease)	Case 12	49
Leptospira (multisystem zoonotic infections)	Case 81	345
Listeria (meningitis)	Case 58	249
Mycobacterium (tuberculosis)	Case 15	61
Mycoplasma (primary atypical pneumonia)	Case 11	45
Neisseria		
N. gonorrhoeae (gonorrhea)	Case 29	119
N. meningitidis (acute bacterial meningitis)	Case 56	241
Nocardia (pneumonia in compromised patients)	Case 20	79
Pasteurella (cat-bite fever)	Case 82	347
Pseudomonas (pneumonia in CF; nosocomial infections)	Case 23	91
Rickettsia (rickettsial infections; Rocky Mountain spotted fever)	Case 80	341
Salmonella		
non typhoidal (infectious diarrhea)	Case 35	151
typhoidal (enteric fever)	Case 36	155
Shigella (dysentery)	Case 37	159
Staphylococcus		
(food poisoning)	Case 40	171
(secondary bacterial pneumonia)	Case 24	95
(skin abscess; osteomyelitis; toxic shock syndrome)	Case 67	287
Streptococcus		
Group A (pharyngitis; rheumatic fever)	Case 3	15
(skin/wound infections; toxic shock)	Case 68	293
Group B (sepsis and meningitis in newborns)	Case 57	245
S. viridans group (infective endocarditis)	Case 76	325
Streptococcus pneumoniae (pneumococcus; community-acquired pneumonia)	Case 9	35
Treponema (genital ulcers and systemic diseases)	Case 26	107
Vibrio (epidemic diarrhea)	Case 39	167
Yersinia (plague)	Case 85	363

APPENDIX G   Microbes/Diseases—Cont'd

Systems/Organisms	Section/Case Numbers	Page Numbers
Medical Mycology		
Aspergillus (aspergillosis; pneumonia in comorbid patients)	Case 16	65
Blastomyces (endemic mycosis: blastomycosis)	Case 18	73
Candida (vaginitis; other opportunistic infections)	Case 32	131
Coccidioides (endemic mycosis: coccidioidomycosis)	Case 19	75
Cryptococcus (meningitis in AIDS)	Case 62	263
Dermatophytes (cutaneous mycoses)	Case 71	305
Histoplasma (endemic mycosis: histoplasmosis)	Case 17	69
Pneumocystis (pneumonia in AIDS)	Case 22	87
Sporothrix (subcutaneous mycosis)	Case 72	309
Medical Parasitology		
Ascaris (large intestinal roundworm infection)	Case 49	207
Cryptosporidium (protracted diarrhea in AIDS)	Case 48	203
Echinococcus (hydatid cyst disease)	Case 51	215
Entamoeba (amebic dysentery)	Case 46	195
Giardia (traveler's diarrhea)	Case 47	199
Plasmodium (malaria)	Case 79	337
Schistosoma (granulomatous liver disease)	Case 52	219
Strongyloides (intestinal helminth infection)	Case 50	211
Taenia (neurocysticercosis)	Case 64	271
Toxoplasma (encephalitis in AIDS)	Case 63	267
Trichomonas (vaginitis)	Case 31	127

Index

Note: Page numbers followed by f indicate figures; those followed by t indicate tables.

A

A-B toxins, 416, 416f
Abscesses, 284
  of brain, 239
  intra-abdominal, *Escherichia coli*, 106t
  respiratory, streptococcal, 17t
  retroperitoneal, 284
  of skin, 290t
Absidia, 420t, 440t
Acanthamoeba, 422, 423t, 426–427, 444
Acid fastness, 437, 437f
Acinetobacter, 439t
Acquired immunodeficiency syndrome
    (AIDS). *See also* Human
    immunodeficiency virus (HIV).
  opportunistic infections in, 331t
Actinomyces, 302t, 411
Actinomyces israelii, 83f, 83t, 83–85, 84f, 451
Acute post-streptococcal
    glomerulonephritis (APSGN), 295t
Acute respiratory distress syndrome
    (ARDS), 4, 9t
Acyclovir, 458
Adenoviruses, 3, 3t, 4, 7f, 7t, 7–9, 8f, 9t,
    392t
Adherence, bacterial, 413, 414f
Adhesins, fungal, 421
Aerobes, 284
  microaerophilic, 406
  obligate, 405
Aeromonas, 284
Aerotolerant anaerobes, 406
Afimbrial adhesins, 413
Agar media, 437, 437t
Agglutination assays, 449
Albendazole, 461
Alleles, 408
Allergic bronchopulmonary aspergillosis
    (ABPA), 67t
Allergic dermatitis, and *Schistosoma mansoni*,
    221t
Allergic disease, respiratory, *Aspergillus
    fumigatus* and, 67t

Alveolar macrophages, 421
Alveolitis. *See* Pneumonia.
Amantadine, 457
Amebae, 422, 423t
Amebic colitis, 144
Amebic dysentery, 195f, 195t, 195–197,
    196f
Amikacin, 454
Aminoglycosides, 454
Amoxicillin, 452
Amphotericin B, 459
Ampicillin, 452
Anaerobes
  aerotolerant, 406
  as normal flora, 411
  facultative, 406
  neurologic infection by, 237t
  obligate, 405–406
  respiratory infection by, 4t
Ancylostoma duodenale, 425t, 430
Anthrax, 350, 355f–357f, 355t, 355–358
Antibacterial drugs, 451–457
Antibodies, 400
Antifungal drugs, 459–460
Antigen detection, 449
Antigenic determinants, 403
Anti-HIV drugs, 458–459
Antimicrobial therapy, 451–461
Antimony, pentavalent, 461
Antiparasitic drugs, 460–461
Antituberculosis drugs, 456
Antiviral drugs, 457–459
Apoptotic pathway, 400
Arenaviridae, 393t
Arthroconidia, 418, 420f
Ascaris, 461
Ascaris lumbricoides, 207f, 207t, 207–208,
    425t, 432t, 461
Aspergillomas, 67t
Aspergillus, 3, 4t, 65f, 65t, 65–67, 66f, 67t,
    420t, 440t, 442, 460
Aspergillus fumigatus, 5, 65f, 65t, 65–67, 66f,
    67t
Atovaquone, 460

Autotrophic bacteria, 405
Azithromycin, 455
Aztreonam, 452

B

Babesia, 424
Bacillary dysentery, 159f, 159t, 159–161,
    160f
Bacilli, 401, 402f, 402t
Bacillus, 404–405, 405, 439t
Bacillus anthracis, 350, 355f–357f, 355t,
    355–358, 404, 415t, 451
Bacillus cereus, 143t
Bacitracin, 457
Bacteremia, 285
  Klebsiella pneumoniae, 43
  puncture-wound, 93t
  Streptococcus pneumoniae, 39t
Bacterial flora, normal, 411
Bacterial infection(s), diagnostic methods
    for
  antibiotic susceptibility tests and,
      439–440, 440f
  cultivation of microorganisms and, 437t,
      437–439, 438f, 439t
  microbial antigen detection and, 440
  microscopy as, 435f, 435–437, 436t, 437f
  serology as, 440–441, 441f
Bacterial pathogens, 401–417
  disinfection and, 407
  genetics of, 407–411, 409f–411f
    genetic exchange and, 408–411, 412f
  growth of, 405–407, 406f
  motile, 404, 405f
  sterilization and, 407
  structure of, 401–405
    morphology and, 401, 402f, 402t
    ultrastructure and, 401–405, 403f–405f
  virulence of, 411–417, 413f, 414f
    evasion of host defenses and, 414–415
    toxins and, 415t, 415–416, 416f, 417t
Bacteriuria, significant, 100

Bacteroides, 406, 411, 437t
Bacteroides fragilis, 453
  cutaneous infection by, 284
  intra-abdominal infection by, 284, 301f, 301t, 301–303, 302t
Bacteroides gingivalis, 302t
Bacteroides melaninogenicus, 302t
Bactroprenol, 403
Balamuthia mandrillaris, 422, 426
Balantidium, 424
Balantidium coli, 423t
Bartonella, 440
Bilharziasis, 219f, 219t, 219–221, 220f, 221t
Biofilms, 413–414
Bioterrorism agents, 350–352, 351t, 352f
Bipolaris, 3
Bladder, schistosomiasis of, chronic, 221t
Blastoconidia, 418, 420f
Blastomyces, 420t, 440t
Blastomyces dermatiditis, 73f, 73t, 73–74, 421, 443
Bordetella, 413f
Bordetella pertussis, 3, 29f, 29t, 29–30, 415t, 439
Borrelia, 439t
Borrelia burgdorferi, 285, 333f, 333t, 333–335, 334f, 335t, 413f
Botulinum toxin, 350
Botulism. *See also Clostridium botulinum.*
  infant, 175f, 175t, 175–177
Bovine spongiform encephalopathy (BSE), 353–354
Brain abscess, 239
Bronchiolitis, 2f, 57f, 57t, 57–59
Bronchitis, 2f
  chronic, acute exacerbations of
    Haemophilus influenzae, 31f, 31t, 31–33, 32t
    Streptococcus pneumoniae, 39t
Brucella, 414, 435, 439
Brucella melitensis, 367t, 367–369, 368f
Brugia malayi, 425t, 430
Bunyaviridae, 393t
Burkitt lymphoma (BL), 13

C

Calcofluor white, 442t
Caliciviridae, 393t
Campylobacter, 143t, 144, 145, 406, 434t, 439t
Campylobacter jejuni, 147f, 147t, 147–149, 413f, 455
Candida, 100t, 420t, 421, 440t, 442, 459, 460
Candida albicans, 421
  cutaneous infection by, 133t
  esophageal infection by, 133t
  genital infection by, 101t
  mucocutaneous infection by, chronic, 133t

Candida albicans (Continued)
  oropharyngeal infection by, 133t
  systemic infection by, 133t
  urinary infection by, 133t
  vulvovaginal infection by, 122f, 131f, 131t, 131–133, 133t
Capsids, 390f, 390–391
Capsofungin, 460
Capsomeres, 390
Capsules, bacterial, 414
Carbapenems, 452
Carbenicillin, 452
Carbuncles, 284, 290t
Cardiovascular system, practice questions on, 463t
Cefaclor, 453
Cefadroxil, 453
Cefazolin, 453
Cefipime, 453
Cefotaxime, 453
Cefotetan, 453
Cefoxitin, 453
Ceftazidime, 453
Ceftriaxone, 453
Cefuroxime, 453
Cell wall, bacterial, 402, 403f
Cell-mediated immunity, 399–400
Cellulitis, 284, 290t, 295t
Central nervous system (CNS)
  infections of, 236–239, 237t, 238t
  practice questions on, 463t
Cephalexin, 453
Cephalosporins, 453
Cephalothin, 453
Cerebrospinal fluid (CSF) analysis, in bacterial meningitis, 237, 238t
Cervicitis
  mucopurulent, 101, 101t, 434t
  Neisseria gonorrhoeae, 121t
Cestodes, 425t, 426
Chagas disease, 427
Chagomas, 427
Chancroid, 101, 101t
Chickenpox, 285
Chitin, fungal, 418
Chlamydia, 3, 4, 440, 445, 455, 456
Chlamydia pneumoniae, 4t, 455
Chlamydia trachomatis, 101, 101t, 413f, 434t, 455
  lymphogranuloma venereum due to, 101, 101t, 125t
  pelvic inflammatory disease due to, 123f, 123t, 123–126, 124f, 125t
Chlamydophila, 440
Chloramphenicol, 454
Chloroquine, 460
Cholera, 167f, 167t, 167–169, 168t
Cholera toxin, 416
Chromatid bars, 422
Chromomycosis, 419–420
Chronic mucocutaneous candidiasis (CMC), 133t

Cidofovir, 458
Ciliates, 423t, 424
Ciprofloxacin, 456
Clarithromycin, 455
Clindamycin, 455
Clonorchis sinensis, 425t
Clostridium, 404–405, 406, 411
Clostridium botulinum, 415t
  bioterrorism and, 350
  gastrointestinal infection by, 142, 143t, 175f, 175t, 175–177
  neurologic infection by, 239
Clostridium difficile, 142, 143t, 144, 145, 179t, 179–180, 415t, 456
Clostridium perfringens, 415t
  gastrointestinal infection by, 143t
  soft tissue infection by, 297t, 297–299, 298f
Clostridium tetani, 239, 279f, 279t, 279–281, 413f, 415t
Clotrimazole, 459
Cloxacillin, 451
CNS lymphoma, 13
Cocci, 401, 402f, 402t
Coccidioides, 5, 420t, 440t
Coccidioides immitis, 75f, 75t, 75–77, 76f, 237, 420t, 421, 442–443
Colitis
  amebic, 144
  Clostridium difficile, 144
  cytomegalovirus, 331t
  hemorrhagic, 142, 143t, 144
Colonization, bacterial, 413, 414f
Commensalism, 421
Common cold, 2, 2f
Complement fixation (CF), 449
Complementation, viral, 397
Confirmatory tests, 434
Congenital rubella syndrome (CBS), 319t
Conjugation, bacterial, 410, 412f
Conjunctivitis
  with bacterial pneumonia, *Staphylococcus aureus*, 95f, 95t, 95–97, 96f
  Neisseria gonorrhoeae, 121t
  neonatal, *Chlamydia trachomatis*, 125t
Connective tissue, practice questions on, 463t
Coronaviruses, 3t, 387f, 387t, 387–389, 389f, 394t
Corynebacterium, 405, 439t
Corynebacterium diphtheriae, 3t, 19f, 19t, 19–21, 20f, 401, 415t, 435, 451
Coxiella, 437
Coxiella burnetii, 342t
Creutzfeldt-Jakob disease (CJD), 353–354, 383f, 383t, 383–385, 384t
Cross-linking, bacterial, 403, 404f
Croup, 3
Cryptococcus, 238, 420t, 440t, 442
Cryptococcus neoformans, 420t, 460
  neurologic infection by, 237, 237t, 263t, 263–265, 264f
  respiratory infection by, 5

Cryptosporidium, 144, 145, 422, 423t, 437, 445

Cryptosporidium parvum, 142, 143t, 203f, 203t, 203–205, 204f, 424, 429, 444f

Cutaneous mycoses, 419, 420t

Cycloserine, 457

Cyclospora, 144, 437, 445

Cyclospora cayetanesis, 143t, 423t, 424

Cystitis, 100
  Escherichia coli, 103f, 103t–106t, 103–106, 104
  hemorrhagic, acute, adenoviral, 9t

Cytokines, proinflammatory, 399

Cytomegalovirus (CMV)
  gastrointestinal infection by, 143t, 145
  respiratory infection by, 4, 329f, 329t, 329–331, 331t

Cytopathic effect, 399

Cytopathogenesis, 399

Cytoplasmic inclusion bodies, 399

Cytoplasmic membrane, 401

Cytotoxic T lymphocytes (CTLs), 400

D

Dalfopristin, 455

Defense mechanisms, 142

Delavirdine, 458

Dermatitis, allergic, and *Schistosoma mansoni*, 221t

Dermatophytes, 285, 419, 420t

Dermis, 284
  skin, hair, and nail infections, 285

Diagnostic methods, 433–450
  for bacterial infections, 435–441
    antibiotic susceptibility tests and, 439–440, 440f
    cultivation of microorganisms and, 437t, 437–439, 438f, 439t
    microbial antigen detection and, 440
    microscopy as, 435f, 435–437, 436t, 437f
    serology as, 440–441, 441f
  diagnostic cycle and, 433, 433f
  for fungal infections, 441t, 441–443
    laboratory methods for, 441–443, 442t, 443f
  for parasitic infections, 443–445, 444f
    for helminth infections, 445, 446f
  performance of laboratory tests and, 433f, 433–434
  specimen collection for, 434t, 434–435
  for viral infections, 445–450
    cell culture as, 446–447, 447f, 448f
    electron microscopy as, 446
    light microscopy as, 445–446, 447f
    molecular, 449, 450f
    serology as, 447, 449

Diarrhea, 142, 143t, 144, 145
  Clostridium difficile-associated, 179t, 179–180

Dimorphism, fungal, 421

Diphtheria, 19f, 19t, 19–21, 20f

Diphtheria toxin, 416

Diphyllobothrium latum, 425t, 430

Diplococci, 401, 402f, 402t

Disinfection, 407

Disseminated gonococcal infection (DGI), 121t

DNA viruses, 392t, 395, 397
  oncogenic, 399

Doxycycline, 454

Dysentery, 142, 143t, 144
  amebic, 195f, 195t, 195–197, 196f
  bacillary, 159f, 159t, 159–161, 160f

E

Eastern equine encephalitis virus, 239

Echinocandins, 460

Echinococcus granulosus, 143t, 144, 215f, 215t, 215–217, 216f, 425t, 430, 445, 461

Echovirus type 9, 253f, 253t, 253–255, 254t

Ectoparasites, 421

Efavirenz, 458

Elephantiasis, 430

Emerging infectious diseases, 352–354, 353t

Empyema, *Streptococcus pneumoniae*, 39t

Encephalitis, 236, 239
  herpes simplex virus, 261t, 261–262
  rabies, 275f, 275t, 275–277
  St. Louis encephalitis virus, 257t, 257–259, 258f
  Toxoplasma gondii, 267f, 267t, 267–269, 268f, 269t
  West Nile virus, 239, 354, 379f, 379t, 379–381

Endemic typhus, 342t

Endocarditis
  native-valve, 325f, 325t, 325–327
  Streptococcus pneumoniae, 39t

Endolimax nana, 422, 423t, 445

Endoparasites, 421–422

Endotoxins, bacterial, 417t

Enfuvertide, 458

Entamoeba coli, 444–445

Entamoeba dispar, 445

Entamoeba gingivalis, 422, 423t

Entamoeba hartmanni, 422, 423t, 444–445

Entamoeba histolytica, 143t, 144, 195f, 195t, 195–197, 196f, 422, 423t, 426, 429f, 444f, 444–445

Entamoeba polecki, 422, 444–445

Enteric fever, 145, 155f, 155t, 155–157, 156f

Enteric infection(s), *Escherichia coli*, 106t

Enteritis
  campylobacter, 147f, 147t, 147–149
  salmonella, 151f, 151t, 151–153

Enterobacter, 452

Enterobacteriaceae, 439t, 453
  as normal flora, 411
  respiratory infection by, 4t

Enterobius, 461

Enterobius vermicularis, 425t, 445, 461

Enterococci, 411, 452
  vancomycin-resistant, 455
  viridans, 285. *See also* Streptococci, viridans.

Enterococcus, 451, 454

Enterocolitis, 144

Enteroviruses
  neurologic infection by, 239
  respiratory infection by, 3t

Enveloped viruses, 390, 390f

Enzymes, essential, 391

Eperezolid, 455

Epidemic keratoconjunctivitis, adenoviral, 9t

Epidemic typhus, 342t

Epidermis, 284

Epidermophyton, 285, 420t, 440t

Epiglottitis, *Haemophilus influenzae*, 32t

Epstein-Barr virus, 3t, 11f, 11t, 11–13, 13t

Ertapenem, 452

Erysipelas, 284, 295t

Erysipelothrix, 284

Erythromycin, 455

Escherichia coli, 406, 411, 413, 413f, 414, 415t, 422, 423t, 452
  enteroadherent (EAEC), 143t, 165t
  enterohemorrhagic (EHEC), 143t, 144, 163f, 163t, 163–165, 165t
  enteroinvasive (EIEC), 143t, 165t
  enteropathogenic (EPEC), 165t
  enterotoxigenic (ETEC), 143t, 165t, 415
  gastrointestinal infection by, 143t, 144, 163f, 163t, 163–165, 165t
  neurologic infection by, 237t, 238t
  respiratory infection by, 4t
  urogenital infection by, 100, 100t, 103f, 103t–106t, 103–106, 104

Esophagitis
  candidal, 133t
  cytomegalovirus, 331t

Essential enzymes, 391

Ethambutol, 456

Evasion mechanisms, bacterial, 414–415

Exotoxins, bacterial, 415t, 415–416, 417t

F

Facultative anaerobes, 406

Facultative rods, cutaneous infection by, 284

False negative results, 433

False positive results, 433

Famciclovir, 458

Fascia, 284

Filariasis, 430

Filoviridae, 394t
Fimbriae, bacterial, 404
Flagella
   bacterial, 404, 405f
   protozoan, 423
Flagellates, 423t, 423–424
Flaviviridae, 393t
Fluconazole, 460
Flucytosine, 460
Flukes, 425t, 426
Fluoroquinolones, 456
Folliculitis, 284, 290t
   Pseudomonas aeruginosa, 93t
Food poisoning, 143t, 171t, 171–173, 172t
Foscarnet, 458
Fournier gangrene, 284
Frameshift mutations, 408, 409f
Francisella, 439
Francisella tularensis, 371f, 371t, 371–372, 402t, 413f
   bioterrorism and, 350
Fungal infection(s)
   diagnostic methods for, 441t, 441–443
      laboratory methods for, 441–443, 442t, 443f
   respiratory, 3, 4t, 5
   of skin, hair, and nails, 285
Fungal pathogens, 418–421
   diseases and classification of, 419–420, 420t
   physiology of, 418, 419f
   reproduction by, 418, 420f
   structure of, 418, 419f
   transmission and pathogenesis of, 421
Furuncles, 284
Furunculosis, chronic, 290t
Fusion inhibitors, 458
Fusobacterium, 284, 405–406
Fusobacterium necrophorum, 285, 302t, 402t
Fusospirochetosis, 285

G

Ganciclovir, 458
Gangrene, 284
   Fournier, 284
   gas, 284, 297t, 297–299, 298f
Gardnerella, 101t, 411
Gas gangrene, 284, 297t, 297–299, 298f
Gastritis, 142, 143t
Gastroenteritis, 143t, 434t
   adenoviral, 9t
   infantile, 183f, 183t, 183–185
   nonbacterial, epidemic, 187f, 187t, 187–189
Gastrointestinal tract infection(s), 142, 143t, 144–146
   practice questions on, 463t
Gatifloxacin, 456
Gene expression, viral, 395
Genetic code, 408, 409f
Genetic exchange, bacterial, 408–411, 412f

Genetics
   bacterial, 407–411, 409f–411f
   viral, 397
Genital ulcers, 101, 101t
Genitourinary infection(s), 100t, 100–102, 101t
   practice questions on, 463t
Genotype, 408
Gentamicin, 454
German measles, 285, 317f, 317t, 317–319, 319t
Giardia, 144
Giardia lamblia, 142, 143t, 199f, 199t, 199–201, 200f, 423, 423t, 444f, 445
Glomerulonephritis, post-streptococcal, acute, 295t
Glucans, fungal, 418
Glycocalyx, 403
Glycopeptide inhibitors, 454
Gomori methenamine silver (GMS) stain, 442t
Gonorrhea, 101, 101t, 119f, 119t, 119–121, 121t, 402t, 413, 413f, 437t, 451
Gram stain, 442t
   in bacterial meningitis, 238, 238t
Gram-negative bacteria, 401, 403–404, 405f, 436t, 454
   aerobic, 284
   bacilli, 237t
   rods, 4t
Gram-positive bacteria, 401, 402–403, 436t, 454
Granulomatous inflammation, 421
Griseofulvin, 460

H

Haemophilus, 439, 439t
Haemophilus ducreyi, 101, 101t
Haemophilus influenzae, 414, 440, 452
   neurologic infection by, 236, 237t
   respiratory infection by, 3, 4, 4t, 31f, 31t, 31–33, 32t
Hairy leukoplakia, oral, 13
Hantavirus, 4–5, 353
Hantavirus pulmonary syndrome (HPS), 375f, 375t, 375–377, 376f, 376t
H-antigens, 404
Haploidy, 408
Head and neck infection(s), 285
Helicobacter, 406, 439t
Helicobacter pylori, 142, 143t, 191f, 191t, 191–193, 192f
Helminths, 144, 424–426, 425t
   cestodes as, 425t, 426
   nematodes as, 424, 425t, 426
   trematodes as, 425t, 426
Hemoflagellates, 424
β-Hemolytic group A streptococcus (BHGAS), 284–285
Hemorrhagic colitis, 142, 143t, 144
Hemorrhagic cystitis, acute, adenoviral, 9t

Hepadnaviridae, 392t
Hepatitis, 143t, 145–146
Hepatitis A virus (HAV), 142, 143t, 223f, 223t, 223–225, 224f
Hepatitis B virus (HBV), 142, 143t, 227f, 227t, 227–230, 229f, 229t
Hepatitis C virus (HCV), 142, 143t, 231t, 231–233, 232f
Hepatitis D virus (HDV), 143t
Hepatitis E virus (HEV), 143t
Hermaphrodites, 426
Herpes simplex virus (HSV)
   genital infection by, 101, 101t, 111f, 111t, 111–113, 112f
   neurologic infection by, 239, 261t, 261–262
Herpesviridae, 392t
Heterotrophic bacteria, 405
Histoplasma, 420t, 440t
Histoplasma capsulatum, 5, 69f, 69t, 69–71, 70f, 71f, 420t, 421, 442–443
Hookworm(s), 425t, 430
Human herpesvirus (HHV), 101t
Human immunodeficiency virus (HIV), 101t, 102, 135f, 135t, 135–139, 137f, 138t. *See also* Acquired immunodeficiency syndrome (AIDS).
Human papillomavirus (HPV), 101t, 115f, 115t, 115–117, 116t
Humoral immunity, 400
Hydatid cyst disease, 215f, 215t, 215–217, 216f
Hyphae, fungal, 418, 419f

I

Imidazoles, 459
Imipenem, 452
Immune defenses, innate, 399
Immunity
   cell-mediated, 399–400
   humoral, 400
   parasites and, 431, 432t
   passive, 400
Immunoblot assay, 449
Impetigo, 284, 290t, 295t
Inactivated vaccines, 400
Independent reassortment, viral, 397
India ink, 442t
Indinavir, 459
Infectious mononucleosis (IM), 11f, 11t, 11–13, 13t, 331t
Infective endocarditis, 285
Inflammation, granulomatous, 421
Influenza virus, 3, 3t, 4, 53f, 53t, 53–55, 54t
Interferon(s) (IFNs), 399
Interferon-α (IFN-α), 459
Intra-abdominal infection(s), 284
   Escherichia coli abscesses as, 106t
Iodamoeba bütschlii, 422, 423t, 445
Isoniazid, 456
Isospora, 437, 445

Isospora belli, 423t, 424
Itraconazole, 460
Ivermectin, 461

K

K antigens, 403
Karyosomes, 422
Katayama fever (KF), 221t
Keratoconjunctivitis, epidemic, adenoviral, 9t
Ketoconazole, 459
Killed vaccines, 400
Klebsiella, 100t, 411
Klebsiella pneumoniae, 414, 452
  neurologic infection by, 237t
  respiratory infection by, 4, 4t, 41f, 41t, 41–43, 42f, 43t

L

Laboratory tests, performance of, 433f, 433–434
LaCrosse virus, 239
Lactophenol cotton blue (LPCB), 442t
Lag phase, of bacterial growth, 406
Lamivudine, 458
Laryngitis, 2f, 3
Laryngotracheitis, 3
Legionella, 4, 49f, 49t, 49–51, 50f, 51f, 413f, 414, 439, 439t, 456
Legionella pneumophila, 49f, 49t, 49–51, 50f, 51t, 413, 455
Legionnaires disease, 51t
Leishmania, 423t, 424, 428, 432t
Leishmania donovani, 444f
Lemierre syndrome, 285
Leprosy, 63t
Leptospira, 145, 285–286, 439t
Leptospira interrogans, 285–286, 345f, 345t, 345–346
Levofloxacin, 456
Lincosamide, 455
Linezolid, 455
Lipopolysaccharide (LPS), 415
Listeria, 414, 439t
Listeria monocytogenes, 237t, 238t, 249f, 249t, 249–251, 250t, 414, 451, 452
Live, attenuated vaccines, 400
Liver, practice questions on, 463t
Liver flukes, 425t
Loa loa, 425t
Loeffler syndrome, 430
Logarithmic phase, of bacterial growth, 406–407
Louse-borne typhus, 342t
Ludwig angina, 285
Lung fluke, 425t
Lyme disease, 285, 333f, 333t, 333–335, 334f, 335t
Lymphedema, 430

Lymphogranuloma venereum (LGV), 101, 101t, 125t
Lymphoma
  Burkitt, 13
  CNS, 13
  non-Hodgkin, 13
Lysogeny, 411

M

Macroconidia, 418, 420f
Macrolides, 455
Macrophages, 400
Malaria, 285, 337f, 337t, 337–340, 338f. *See also Plasmodium entries.*
Malassezia, 420t, 440t
Measles, 285, 313f, 313t, 313–315
Mebendazole, 461
Mefloquine, 460
Membranes, bacterial, 401
Meningitis, 236, 434t
  aseptic, 236
  bacterial, 106t, 236–239, 237t, 238t
  cryptococcal, 263t, 263–265, 264f
  enteroviral, 253f, 253t, 253–255, 254t
  Escherichia coli, 106t
  Haemophilus influenzae, 32t
  Listeria monocytogenes, 249f, 249t, 249–251, 250t
  meningococcal, 241f, 241t, 241–243, 242f
  Streptococcus agalactiae, 245t, 245–247, 246f, 247t
  Streptococcus pneumoniae, 39t
  type III, 245t, 245–247, 246f, 247t
Meropenem, 452
Metachromatic granules, 401
Methicillin-resistant *Staphylococcus aureus* (MRSA), 290t, 455
Metronidazole, 456, 460
Mezlocillin, 452
Micafungin, 460
Miconazole, 459
Microaerophilic aerobes, 406
Microconidia, 418, 420f
Microsporida, 143t, 145, 424
Microsporum, 285, 420t, 440t
Microsporum canis, 305f, 305t, 305–308, 306f, 307t
Minocycline, 454
Missense mutations, 408, 409f
Mobiluncus, 101t, 411
Molds, 418, 419f
Monobactams, 452
Moraxella catarrhalis, 3, 4
Moxifloxacin, 456
Mucor, 420t, 440t
Multisystem infection(s), 285–286
  practice questions on, 463t
Mumps, 23t, 23–24
Murine typhus, 342t
Musculoskeletal system, practice questions on, 463t

Mutations, 408
  frameshift, 408, 409f
  missense, 408, 409f
  nonsense, 408, 409f
  point (single base pair), 408, 409f
  silent, 408
  spontaneous, 408
Mutualism, 421
Mycelium, 418
Mycetoma, 419–420
Mycobacteria, 63t, 439t, 456
Mycobacterium, 405, 413f
Mycobacterium avium complex, 143t, 145
Mycobacterium leprae, 63t
Mycobacterium tuberculosis, 402t, 414, 437f, 456, 457
  neurologic infection by, 237, 238
  respiratory infection by, 3, 61f–63f, 61t, 61–64, 63t
Mycoplasma, 3, 4, 401, 440, 455
Mycoplasma genitalium, 101t
Mycoplasma hominis, 101t
Mycoplasma pneumoniae, 3, 3t, 4t, 45f, 45t, 45–47, 47t, 413f, 455
Mycoses
  cutaneous, 419, 420t, 441t
  opportunistic, 420, 420t, 441t
  subcutaneous, 419–420, 420t, 441t
  systemic, 420, 420t, 441t
Myositis, 295t

N

Naegleria, 422, 423t, 444
Naegleria fowleri, 422, 426
Nafcillin, 451
Naked viruses, 390, 390f
Native-valve endocarditis, 325f, 325t, 325–327
Necator americanus, 430
Necrotizing fasciitis (NF), 284–285, 293f, 293t, 293–296, 295t
Neisseria, 409, 435f, 439t
Neisseria gonorrhoeae, 101, 101t, 119f, 119t, 119–121, 121t, 402t, 413, 413f, 437t, 451
Neisseria meningitidis, 236, 237t, 238, 238t, 241f, 241t, 241–243, 242f, 414, 434t, 440
Nematodes, 424, 425t, 426, 430
Neuraminidase inhibitors, 457
Neurocysticercosis, 271f, 271t, 271–273, 272f
Neurologic infection(s), 236–239, 237t, 238t
Neutrophils, fungal infections and, 421
Nevirapine, 458
Niclosamide, 461
Nifurtimox, 461
NK cells, 399
Nocardia, 4t, 405, 413f, 437
Nocardia asteroides, 5, 79f, 79t, 79–81, 80f, 402t

Nongonococcal urethritis (NGU), 125t
Non-Hodgkin lymphoma, 13
Non-nucleoside reverse transcriptase
  inhibitors, 458
Nonsense mutations, 408, 409f
Norfloxacin, 456
Norovirus (Norwalk virus), 142, 143t, 144,
  187f, 187t, 187–189
Nosocomial infection(s), 295t
  Klebsiella pneumoniae, 43
  of surgical wounds, 290t
Nuclear inclusion bodies, 399
Nucleocapsids, 390
Nucleoside reverse transcriptase inhibitors,
  458
Nystatin, 459

O

Obligate aerobes, 405
Obstetric infection(s), 295t
Onchocerca volvulus, 425t, 430, 461
Oncogenesis, 399
Onychomycosis, 307t
Ophthalmia, neonatal, 101, 101t
Opportunistic infection(s)
  in AIDS, 137, 138t
  cytomegalovirus, 331t
  mycotic, 420, 420t
Oral hairy leukoplakia (OHL), 13
Organelles, 422
Orientia, 342t
Oropharyngeal candidiasis (OPC), 133t
Orthomyxoviridae, 393t
Oseltamivir, 457
Osteomyelitis, 284
  puncture-wound, 93t
  staphylococcal, 287f, 287t, 287–290,
    288f, 290t
Otitis externa, *Pseudomonas aeruginosa*, 93t
Otitis media, 2f, 3
  acute, *Streptococcus pneumoniae*, 39t
  Haemophilus influenzae, 32t
Oxacillin, 451
Oxazolidinones, 455

P

Papovaviridae, 392t
Paracoccidioides brasiliensis, 421
Paragonimus westermani, 425t, 445
Parainfluenza virus (PIV), 3, 25t, 25–27,
  26f
Paramyxoviridae, 394t
Parasites, 421–432
  helminths as, 424–426, 425t
    cestodes as, 425t, 426
    nematodes as, 424, 425t, 426
    trematodes as, 425t, 426
  immunity and, 431, 432t

Parasites (*Continued*)
  life cycle and pathogenesis of, 426–431,
    427f–429f, 431f
  protozoa as, 422–424, 423t
    amebae as, 422, 423t
    ciliates as, 423t, 424
    flagellates as, 423t, 423–424
    sporozoa as, 423t, 424
Parasitic infection(s)
  diagnostic methods for, 443–445, 444f
    for helminth infections, 445, 446f
  intestinal, 144
Parvoviridae, 392t
Passive immunity, 400
Pasteurella multocida, 347f, 347t, 347–348
Pasteurellosis, 286
Pathogenicity islands, 408
Pelvic inflammatory disease (PID)
  Chlamydia trachomatis, 123f, 123t,
    123–126, 124f, 125t
  Neisseria gonorrhoeae, 121t
Penicillin(s)
  broad spectrum, 452
  extended spectrum, 452
  natural, 451
  penicillinase-resistant, 451
Penicillin G, 451
Penicillin V, 451
Penicillin-binding proteins (PBPs), 403
Penile ulcers, 101, 101t
Pentamidine, 461
Pentavalent antimony, 461
Peptic ulcer disease, 191f, 191t, 191–193,
  192f
Peptide cross-links, 403, 404f
Peptidoglycan, 402, 403
Peptostreptococcus, 406
Peptostreptococcus anaerobius, 302t
Peripheral nervous system (PNS)
  infection(s), 239
  practice questions on, 463t
Pertussis, 29f, 29t, 29–30
Pharyngitis, 2f, 2–3, 3t, 434t
  Neisseria gonorrhoeae, 121t
  streptococcal, 15f, 15t, 15–17, 16f, 17t
Pharyngoconjunctival fever, 7f, 7t, 7–9, 8f,
  9t
Phenotype, 408
Phenotypic mixing, viral, 397
Picornaviridae, 393t
Pili, bacterial, 404
Pinworm, 425t
Piperacillin, 452
Plague, 363f, 363t, 363–365, 364f
  bioterrorism and, 350
Plasmids, 401, 408
  conjugative, 410
Plasmodium, 422, 423t, 424, 429–430, 432t
Plasmodium falciparum, 285, 337f, 337t,
  337–340, 338f, 429, 460
Plasmodium malariae, 429
Plasmodium ovale, 429, 461

Plasmodium vivax, 429, 461
Ploidy, viral, 391
Pneumocystis jiroveci, 5, 87f–89f, 87t, 87–89,
  421, 456
Pneumonia, 2f, 3–5, 4t, 434t
  atypical, 4
  Escherichia coli, 106t
  Haemophilus influenzae, 32t
  influenza viral, 55t
  Klebsiella pneumoniae, 41f, 41t, 41–43, 42f,
    43t
  neonatal, *Chlamydia trachomatis*, 125t
  pneumococcal, 35f, 35t, 35–39, 36f, 38f,
    39t
  pneumocystis, 87f–89f, 87t, 87–89
  Pseudomonas aeruginosa, 91t, 91–94, 92f,
    93t
  secondary, *Staphylococcus aureus*, 95f, 95t,
    95–97, 96f
  typical, 4
  "walking," *Mycoplasma pneumoniae*, 45f,
    45t, 45–47, 47t
Pneumonitis, cytomegalovirus, 329f, 329t,
  329–331, 331t
Point mutations, 408, 409f
Polymerase chain reaction (PCR), 449,
  450f
Polymyxins, 457
Pontiac fever, 51t
Pore forming toxins, 416
Porphyromonas, 284, 302t
Poxviridae, 392t
Praziquantel, 461
Predictive value, of tests, 434
Prevotella, 284, 302t, 405–406, 439t
Primaquine, 461
Prion agent infection(s), 383f, 383t,
  383–385, 384t
Prions, 353–354
Proglottids, 426
Proguanil, 460
Proinflammatory cytokines, 399
Prokaryotes, 401
Propionibacterium, 405–406
Protease inhibitors, 459
Proteus, 100t, 439t, 452
Proteus mirabilis, 452
Protozoa, 422–424, 423t
  amebae as, 422, 423t
  ciliates as, 423t, 424
  flagellates as, 423t, 423–424
  gastrointestinal infections by, 144
  sporozoa as, 423t, 424
Pseudohyphae, 418, 420f
Pseudomonas, 405, 439t, 453
  respiratory infection by, 4t
  urinary infection by, 100t
Pseudomonas aeruginosa, 402t, 403f, 406,
  413, 413f, 415t, 439t, 440f, 452
  neurologic infection by, 237t, 238t
  respiratory infection by, 91t, 91–94, 92f,
    93t

Pseudopods, 422
Pulmonary aspergillosis, invasive, 65f, 65t, 65–67, 66f, 67t
Pyelonephritis, 100
Pyrantel pamoate, 461
Pyrazinamide, 456
Pyrimethamine, 460
Pyuria, 100

Q

Q fever, 342t
Quinupristin, 455

R

Rabies virus, 239, 275f, 275t, 275–277
Recombination
  bacterial, 408, 410f
  viral, 397
Reiter syndrome, 125t
Reoviridae, 393t
Replicons, 408
Respiration, fungal, 418, 419f
Respiratory syncytial virus (RSV), 4t, 57f, 57t, 57–59
Respiratory tract infection(s), 1–97, 2f, 3t, 4t
  practice questions on, 463t
Retinitis, cytomegalovirus, 331t
Retroperitoneal abscesses, 284
Retroviridae, 393t
Reye syndrome, 55t
Rhabdoviridae, 394t
Rheumatic fever, acute, 17t
Rheumatic heart disease (RHD), 17t
Rhinitis, 2f
Rhinovirus, 2
  respiratory, 3t
Rhizopus, 420t, 440t
Ribavirin, 457
Ribosomes, 401
Rickettsia, 413f, 414, 437, 440
Rickettsia akari, 342t
Rickettsia prowazekii, 342t
Rickettsia rickettsiae, 285, 341f, 341t, 341–344, 342t, 343f
Rickettsia typhi, 342t
Rickettsialpox, 342t
Rifampin, 456
Rimantadine, 457
Ringworm, 305f, 305t, 305–308, 306f, 307t, 420t
Ritonavir, 459
RNA viruses, 393t–394t, 395–396, 397
Rocky Mountain spotted fever (RMSF), 285, 341f, 341t, 341–344, 342t, 343f
Rods, 401, 402f, 402t
Rotavirus, 142, 143t, 183f, 183t, 183–185
Roundworms, 424, 425t, 426, 430

Rubella, 285, 317f, 317t, 317–319, 319t
Rubeola, 285, 313f, 313t, 313–315

S

St. Louis encephalitis virus, 239, 257t, 257–259, 258t
Salmonella, 142, 143t, 144, 145, 406, 413f, 414, 434t, 437t, 452
Salmonella typhi, 414
Salmonella typhimurium, 143t, 144, 145, 151f, 151t, 151–153, 155f, 155t, 155–157, 156f, 414
Salpingitis, *Neisseria gonorrhoeae*, 121t
Saprophytes, 421
Sarcocystis, 424
Scarlet fever, 17t
Schistosoma, 142, 432t
Schistosoma haematobium, 425t, 430, 431, 445, 446f
Schistosoma japonicum, 425t, 430–431
Schistosoma mansoni, 143t, 219f, 219t, 219–221, 220f, 221t, 425t, 430–431, 431f
Screening tests, 434
Scrub typhus, 342t
Secretory antibodies, 400
Sensitivity, of tests, 433f, 433–434
Septate hyphae, 418
Serratia, 100t, 439t, 452
Serum antibodies, 400
Severe acute respiratory syndrome (SARS), 354, 387f, 387t, 387–389, 389f
Sexually transmitted diseases (STDs), 101t, 101–102
Shiga-toxin A, 416
Shigella, 143t, 144, 145, 406, 413f, 414, 434t, 437t
Shigella dysenteriae, 415t
Shigella flexneri, 159f, 159t, 159–161, 160f
Shingles, 321f, 321t, 321–323, 322f
Siderophores, 414
Silent mutations, 408
Sin Nombre virus, 375f, 375t, 375–377, 376f, 376t
Single base pair mutations, 408, 409f
Sinusitis, 2f, 3
  Haemophilus influenzae, 32t
Skin infection(s), 284–285
  practice questions on, 463t
Skin scrapings, 442
Sleeping sickness, 427
Slime, bacterial, 403
Smallpox virus, 350, 359f, 359t, 359–362, 360f, 361t
Soft tissue infection(s), 284
Specificity, of tests, 433f, 433–434
Specimen collection, 434t, 434–435
Spontaneous mutations, 408
Sporangiospores, 418, 420f
Spores, bacterial, 404–405

Sporothrix, 420t, 440t
Sporothrix schenckii, 285, 309f, 309t, 309–311, 310f, 419–420, 420t, 421
Sporozoa, 423t, 424
Staphylococcal scalded-skin syndrome (SSSS), 290t
Staphylococcus, 284, 406, 411, 439t, 453
Staphylococcus aureus, 285, 402t, 403, 403f, 413f, 414, 415t, 439t, 451, 453, 454, 455, 456
  bone infection by, 287f, 287t, 287–290, 288f, 290t
  cutaneous infection by, 284
  gastrointestinal infection by, 143t, 171t, 171–173, 172t
  methicillin-resistant, 290t, 455
  neurologic infection by, 237t
  respiratory infection by, 4, 4t, 95f, 95t, 95–97, 96f
  urinary infection by, 100t
Staphylococcus epidermidis, 100t, 413, 439t, 454
Staphylococcus saprophyticus, 100, 100t, 439t
Stationary phase, of bacterial growth, 407
Stavudine, 458
Stenotrophomonas maltophilia, 439t
Sterilization, 407
Stevens-Johnson syndrome, 47t
Storage granules, 401
Streptococci
  group A, cutaneous, 284
  group B
    neurologic, 236, 237t
    urinary, 100t
  viridans, 285, 411, 451
    endocarditis due to, 325f, 325t, 325–327
    multisystem infections due to, 285
Streptococcus, 406, 453
Streptococcus agalactiae, 236, 238, 245t, 245–247, 246f, 247t, 414, 452
Streptococcus mutans, 413
Streptococcus paratyphi, 143t
Streptococcus pneumoniae, 402t, 403f, 409, 414, 434t, 435f, 440, 451, 452, 453, 454, 455
  neurologic infection by, 236, 237t, 238, 238t
  respiratory infection by, 3, 4, 4t, 35f, 35t, 35–39, 36f, 38f, 39t
Streptococcus pyogenes, 402t, 403f, 413f, 415t, 434t, 451, 452, 453
  respiratory infection by, 3, 3t, 15f, 15t, 15–17, 16f, 17t
  skin and soft tissue infections by, 293f, 293t, 293–296, 295t
Streptogramins, 455
Streptomycin, 454
Strongyloides stercoralis, 142, 143t, 144, 211f, 211t, 211–213, 212f, 425t, 445, 446f, 461
Subcutaneous mycoses, 419–420, 420t

Subcutis, 284
Sulfadoxine, 460
Sulfonamides, 455
Superantigens, 415–416
Suramin, 461
Symbiosis, 421
Syphilis, 101, 101t, 434t. *See also Treponema pallidum.*
  congenital, 109t
  latent, 109t
  primary, 109t
  secondary, 107f, 107t, 107–109, 108f, 109t
  tertiary, 109t
Systemic inflammatory response syndrome (SIRS), 284
Systemic mycoses, 420, 420t

T

Taenia saginata, 425t
Taenia solium, 271f, 271t, 271–273, 272f, 425t, 430, 445, 446f, 461
Tapeworms, 425t, 426
Teichoic acids, 402
Teicoplanin, 454
Telithromycin, 455
Tenofovir, 458
Terbinafine, 460
Tetanus, 279f, 279t, 279–281
Tetracyclines, 454
Thiabendazole, 461
Ticarcillin, 452
Tinea barbae, 307t
Tinea capitis, 305f, 305t, 305–308, 306f, 307t
Tinea corporis, 307t
Tinea faciei, 307t
Tinea pedis, 307t
Tinea unguium, 307t
Tinea versicolor, 420t
Tobramycin, 454
Togaviridae, 393t
Toll-like receptor signaling, 399
Toxic shock syndrome (TSS), 290t, 295t
Toxins, bacterial, 415t, 415–416, 416f, 417t
Toxocara canis, 425t
Toxoplasma gondii, 267f, 267t, 267–269, 268f, 269t, 423t, 424, 428–429, 432t
Tracheitis, 2f
Tracheobronchitis, 3
  Mycoplasma pneumoniae, 47t
Trachoma, 125t
Transduction, bacterial, 410–411, 412f
Transformation, bacterial, 409–410, 412f
Transpeptidation reaction, 403
Transposons, 408, 411f
  conjugative, 410
Trematodes, 425t, 426
"Trench mouth," 285
Treponema, 439t

Treponema pallidum, 101t, 107f, 107t, 107–109, 108f, 109t, 403f, 413f, 434t, 437, 440
Triazoles, 460
Trichinella spiralis, 425t, 430, 432t, 461
Trichomonas vaginalis, 101t, 127f, 127t, 127–128, 128f, 423t, 423–424
Trichophyton, 285, 307t, 420t, 440t
Trichuris trichiura, 461
Trimethoprim, 456
Trophozoites, 422
True negative results, 433
True positive results, 433
Trypanosoma, 424, 432t
Trypanosoma brucei, 423t, 427
Trypanosoma cruzi, 427, 461
Tuberculosis, 5
  post-primary, *Mycobacterium tuberculosis*, 61f–63f, 61t, 61–64, 63t
Tularemia, 371f, 371t, 371–372
  bioterrorism and, 350
Typhoid fever, 145, 155f, 155t, 155–157, 156f
Typhus
  endemic (murine), 342t
  epidemic (louse-borne), 342t
  scrub, 342t

U

Ulcers
  genital, 101, 101t
  penile, 101, 101t
  peptic, 191f, 191t, 191–193, 192f
Uncoating, viral, 391
Ureaplasma, 439t
Urethritis, 101, 101t
  Chlamydia trachomatis, 125t
Urogenital infection(s), 100t, 100–101, 100–102, 101t, 434t. *See also* Cystitis; Pyelonephritis.
  Candida albicans, 133t
  practice questions on, 463t
  puncture-wound, 93t

V

Vaccines, 400
Vaginitis, 101–102
Vaginosis, 101–102
Valacyclovir, 458
Valganciclovir, 458
Vancomycin, 454
Vancomycin-resistant enterococci (VRE), 455
Varicella zoster virus (VZV)
  cutaneous infection by, 285, 321f, 321t, 321–323, 322f
  respiratory infection by, 4
Variola major, 350, 359f, 359t, 359–362, 360f, 361t
Vibrio, 406, 439t

Vibrio (Continued)
  cutaneous infection by, 284
  gastrointestinal infection by, 142
Vibrio cholerae, 142, 143t, 144, 167f, 167t, 167–169, 168t, 402f, 403f, 413, 413f
Viral hemorrhagic fevers, bioterrorism and, 350
Viral infection(s)
  diagnostic methods for, 445–450
    cell culture as, 446–447, 447f, 448f
    electron microscopy as, 446
    light microscopy as, 445–446, 447f
    molecular, 449, 450f
    serology as, 447, 449
  neurologic, 239
  pathogenesis of, 397–400, 398f
    immunity and, 399–400
  respiratory, 2, 3, 4t, 4–5
  of skin, hair, and nails, 285
Viral pathogens, 390–400
  classification of, 391, 392t–394t
  enveloped, 390, 390f
  genetics of, 397
  multiplication and life cycle of, 391, 395f, 395–397, 396f, 397t
  naked, 390, 390f
  structure of, 390f, 390–391
Virions, 390f, 390–391, 396–397
Virulence, bacterial, 411–417, 413f, 414f
Virus ploidy, 391
Voriconazole, 460

W

"Walking" pneumonia, *Mycoplasma pneumoniae*, 45f, 45t, 45–47, 47t
Warts, genital, 115f, 115t, 115–117, 116t
West Nile virus, neurologic, 239, 354, 379f, 379t, 379–381
Western equine encephalitis virus, 239
White blood cell count, in bacterial meningitis, 237–238, 238t
Wound infection(s), 284–285, 290t, 434t
  nosocomial, 295t
Wuchereria bancrofti, 425t, 430, 446f

Y

Yeasts, 418, 419f
Yersinia, 144, 145
Yersinia enterocolitica, 143t, 144, 413f
Yersinia pestis, 363f, 363t, 363–365, 364f, 413f
  bioterrorism and, 350

Z

Zanamivir, 457
Zidovudine, 458
Zoster, 321f, 321t, 321–323, 322f
Zygomycosis, 420t